MULTIVARIATE STATISTICAL METHODS

McGRAW-HILL SERIES IN PROBABILITY AND STATISTICS

DAVID BLACKWELL AND HERBERT SOLOMON, *Consulting Editors*

BHARUCHA-REID: Elements of the Theory of Markov Processes and Their Applications
DE GROOT: Optimal Statistical Decisions
DRAKE: Fundamentals of Applied Probability Theory
EHRENFELD AND LITTAUER: Introduction to Statistical Methods
GIBBONS: Nonparametric Statistical Inference
GRAYBILL: Introduction to Linear Statistical Models
HODGES, KRECH, AND CRUTCHFIELD: StatLab: An Empirical Introduction to Statistics
LI: Introduction to Experimental Statistics
MOOD, GRAYBILL, AND BOES: Introduction to the Theory of Statistics
MORRISON: Multivariate Statistical Methods
RAJ: The Design of Sample Surveys
RAJ: Sampling Theory
STEEL AND TORRIE: Introduction to Statistics
THOMASIAN: The Structure of Probability Theory with Applications
WADSWORTH AND BRYAN: Applications of Probability and Random Variables
WASAN: Parametric Estimation
WOLF: Elements of Probability and Statistics

MULTIVARIATE STATISTICAL METHODS

Donald F. Morrison
Professor of Statistics
The Wharton School
University of Pennsylvania

Second Edition

McGRAW-HILL BOOK COMPANY
New York St. Louis San Francisco Auckland Düsseldorf Johannesburg
Kuala Lumpur London Mexico Montreal New Delhi Panama Paris
São Paulo Singapore Sydney Tokoyo Toronto

MULTIVARIATE STATISTICAL METHODS

7890 DODO 832

This book was set in Modern by Bi-Comp, Incorporated.
The editors were A. Anthony Arthur and Michael Gardner;
the production supervisor was Leroy A. Young.
R. R. Donnelley & Sons Company was printer and binder.

Library of Congress Cataloging in Publication Data

Morrison, Donald F
Multivariate statistical methods.

(McGraw-Hill series in probability and statistics)
Bibliography: p.
1. Multivariate analysis. I. Title.
QA278.M68 1976 519.5'3 75-14325
ISBN 0-07-043186-8

To My Mother and
the Memory of My Father

CONTENTS

PREFACE TO THE SECOND EDITION

This edition was prepared to expand the treatments of certain important methods and to maintain the currency of the citations to the multivariate statistical literature. In this revision emphasis was given to methods in the mainstream of classical multivariate analysis which were concerned with mean structures rather than models for covariance matrices.

Among the major changes in the book are some elementary results on estimates and tests with incomplete data matrices; references have also been provided for more general missing-data techniques. The multivariate analysis of covariance has been given in greater detail, and a new section on the fitting of growth curves has been included in Chapter 5. Linear discrimination has been accorded a whole chapter. Some recent results on the estimation of error rates for the two-sample case have been summarized, and classification rules for several groups have been developed by discriminant functions and minimum-distance rules. The hypothesis tests on covariance matrices have been extended to the single-sample case and to patterns useful in the analysis of repeated measurements. The Appendix charts of the greatest-characteristic-root percentage points have been augmented with tables for the parameter s through twenty.

In the revision I was encouraged by the reception of the first edition as a reference and text. It has been especially heartening that adoptions have ranged outside the intended behavioral and life sciences audience to include courses in business, economics, and the social sciences. I am indebted to many persons who have used the book for writing about its content, or for calling my attention to errors or ambiguities.

I am grateful to E. S. Pearson and K. C. S. Pillai for permitting the reproduction of Appendix Tables 6 to 14 from *Biometrika*. The inclusion of those critical values has greatly extended the usefulness of the largest-root tests and confidence statements. I am also most appreciative of the investigators and editors who have granted permission for the use of their data in examples and exercises. In the planning of the second edition I was aided by thoughtful and detailed comments from Leon Jay Gleser on the initial outline.

Finally, a special acknowledgment should be made to my wife Phyllis for her support and assistance with the manuscript, and to our son Norman for giving up time that belonged to him.

Donald F. Morrison

PREFACE TO THE FIRST EDITION

Multivariate statistical analysis is concerned with data collected on several dimensions of the same individual. Such observations are common in the social, behavioral, life, and medical sciences: the record of the prices of a commodity, the reaction times of a normal subject to several different stimulus displays, the principal bodily dimensions of an organism, or a set of blood-chemistry values from the same patient are all examples of multidimensional data. As in univariate statistics, we shall assume that a random sample of multicomponent observations has been collected from different individuals or other independent sampling units. However, the common source of each individual observation will generally lead to dependence or correlation among the dimensions, and it is this feature that distinguishes multivariate data and techniques from their univariate prototypes.

This book was written to provide investigators in the life and behavioral sciences with an elementary source for multivariate techniques which appeared to be especially useful for the design and analysis of their experimental data. The book has also been organized to serve as the text for a course in multivariate methods at the advanced undergraduate or graduate level in the sciences. The mathematical and statistical prerequisites are minimal: a semester course in elementary statistics with a survey of the fundamental sampling distributions and an exposure to the calculus for the partial differentiations and integrals required for occasional maximizations and expectations should be sufficient. The review of the essential univariate statistical concepts in the first chapter and a detailed treatment of matrix algebra in the second make the book fairly self-contained both as a reference and a text. The standard results on the multinormal distribution, the estimation of its parameters, and correlation analysis in Chapter 3 are essential background for the developments in the remaining chapters.

The selection of techniques reflects my experiential biases and preferences. Attention has been restricted to continuous observations from multivariate normal populations: no mention has been made of the newer distribution-free tests and the methods for analysis of many-way categorical data tables. It was felt that the implications of the T^2 statistic for repeated-measurements experiments justified a lengthy discussion of tests and confidence intervals for mean vectors. The multivariate general linear hypothesis and analysis of variance has been developed through the Roy union-intersection principle for the natural ease with which simultaneous confidence statements can be obtained. In my experience the Hotelling principal-component technique has proved to be exceedingly useful for data reduction, analysis of the latent structure of multivariate systems, and descriptive purposes, and its use and properties are developed at length. My approach to factor analysis has been statistical rather than psychometric, for I prefer to think of the initial steps, at least, of a factor analysis as a problem in statistical estimation.

For the preparation of this methods text I wish to acknowledge a considerable debt to those responsible for the theoretical development of multivariate analysis: the fundamental contributions of T. W. Anderson, Harold Hotelling, D. N. Lawley, and the late S. N. Roy are evident throughout. In particular the frequent references to S. N. Roy's monograph "Some Aspects of Multivariate Analysis" are indicative of his influence on the presentation. For the many derivations beyond the level of this book the reader has usually been referred to T. W. Anderson's standard theoretical source "An Introduction to Multivariate Statistical Analysis."

It is a pleasure to acknowledge those who have assisted at different stages in the preparation of this book. My thanks are due to Samuel W. Greenhouse for initially encouraging me to undertake the project. I am especially indebted to Karen D. Pettigrew and John J. Bartko for their thoughtful reading of several chapters and for offering suggestions that have improved the clarity of the presentation. George Schink carefully checked the computations of the majority of the examples. However, the ultimate responsibility for the nature and accuracy of the contents must of course rest with the author. Finally, I wish to express my gratitude to the many investigators who graciously permitted the use of their original and published data for the examples and exercises.

I am indebted to A. M. Mood and the McGraw-Hill Book Company for permission to reproduce Table 1 from the first edition of "Introduction to the Theory of Statistics." Tables 2 and 4 have been abridged from tables originally prepared by Catherine M. Thompson and Maxine Merrington, and have been reproduced with the kind permission of the editor of *Biometrika*, E. S. Pearson. I am also grateful to Professor Pearson and to H. O. Hartley for kindly permitting the reproduction of Charts 1 to 8 from *Biometrika*. I am indebted to the literary executor of the late Sir Ronald A. Fisher, F.R.S., Cambridge, to Dr. Frank Yates, F.R.S., Rothamsted, and to Messrs. Oliver & Boyd Ltd., Edinburgh, for permission to reprint Table 3 from their book "Statistical Tables for Biological, Agricultural, and Medical Research." Charts 9 to 16 have been reproduced from the *Annals of Mathematical Statistics* with the kind permission of D. L. Heck and the managing editor, P. L. Meyer.

The preparation of parts of an earlier version of the text as class notes was made possible through the enthusiastic cooperation of the Foundation for Advanced Education in the Sciences, Inc., Bethesda, Md.

Support for the use of some chapters in mimeographed form and clerical assistance was kindly provided by Dean Willis J. Winn through funds from a grant to the Wharton School of Finance and Commerce by the New York Life Insurance Co. I am also grateful for the secretarial assistance furnished by the Department of Statistics and Operations Research and the Lecture Note Fund of the University of Pennsylvania.

Donald F. Morrison

1 SOME ELEMENTARY STATISTICAL CONCEPTS

1.1 INTRODUCTION.

In this chapter we shall summarize some important parts of univariate statistical theory to which we shall frequently refer in our development of multivariate methods. Certain concepts of statistical inference will be introduced, and some essential univariate distributions will be described. We shall assume that the reader has been exposed to the elements of probability and random variables and has an acquaintance with the basic univariate techniques as applied in some substantive discipline.

1.2 RANDOM VARIABLES

Every statistical analysis must be built upon a *mathematical model* linking observable reality with the mechanism generating the observations. This model should be a parsimonious description of nature: its functional form should be simple, and the number of its parameters and components should be a minimum. The model should be *parametrized* in such a way that each parameter can be interpreted easily and identified with some aspect of reality. The functional form should be sufficiently tractable to permit the sort of mathematical manipulations required for the estimation of its parameters and other inferences about its nature.

Mathematical models may be divided into three general classes: (1) purely deterministic, (2) static, or deterministic with simple random components, and (3) stochastic. Any observation from a deterministic model is strictly a function of its parameters and such variables as time, space, or inputs of energy or a stimulus. Newtonian physics states that the distance traveled by a falling object is directly related to the squared time of fall, and if atmospheric turbulence, observer error, and other transient effects can be ignored, the displacement can be calculated exactly for a given time and gravitational constant. In the second kind of model each observation is a function of a strictly deterministic component and a random term ascribable to measurement error or sampling variation in either the observed response or the input variables. The random components are assumed to be independent of one another for different observations. The models we shall encounter in the sequel will be mainly of this class, with the further restriction that the random component will merely be added to the deterministic part. Stochastic models are constructed from fundamental random events or components to explain dynamic or evolutionary phenomena: they range in complexity from the case of a sequence of Bernoulli trials as the model for a coin-tossing experiment to the birth-and-death process describing the size of a biological population. Most stochastic models allow for a "memory" effect, so that each observed response is dependent to some degree upon its predecessors in time or neighbors in space. We shall touch only tangentially on this kind of model.

Now let us define more precisely what is meant by the notions of random variation or the random components in the second and third kinds of models. We shall begin by defining a *discrete random variable*, or one which can assume only a countable number of values. Suppose that some experiment can result in exactly one of k outcomes $E_1, \ldots, E_k$. These outcomes are mutually exclusive, in the sense that the occurrence of one event precludes that of any other. To every event we assign some number p_i between zero and one called the *probability* $P(E_i)$ of that event. p_i is the probability that in a single trial of the experiment the outcome E_i will occur. Within the framework of our experiment we assign a probability of zero to impossible events and a probability of unity to any event which must happen with certainty. Then, by the mutual exclusiveness of the events, in a single trial

$$P(E_i \cap E_j) = 0$$

$$P(E_i \cup E_j) = p_i + p_j$$

where the *intersection* symbol $\cap$ denotes the event "E_i and E_j" and the *union* symbol $\cup$ indicates the event "E_i and/or E_j." By the additive property of the probabilities of mutually exclusive outcomes the total probability of the set of events is

$$\begin{aligned} P(E_1 \cup \cdots \cup E_k) &= p_1 + \cdots + p_k \\ &= 1 \end{aligned}$$

Now assign the numerical value x_i to the ith outcome, where for convenience the outcomes have been placed in ascending order according to their x_i values. The discrete random variable X is defined as that quantity which takes on the value x_i with probability p_i at each trial of the experiment. As an example, if the experiment consists of the toss of a coin, the score of one might be assigned to the outcome heads, while zero might be the tails score. Then $x_1 = 0$, $x_2 = 1$, and $p_1 = 1 - p$, $p_2 = p$, say. This random variable would be described by its *probability function* $f(x_i)$ specifying the probabilities with which X assumes the values 0 and 1:

X	$f(x_i)$
0	$1 - p$
1	p

We note that the total probability is unity and that we have implicitly assigned a probability of zero to such irrelevant events as the coin's landing on edge or rolling out of sight. We have chosen *not* to assign a numerical value to the single parameter p; this reflects the intrinsic qualities of the coin as well as the manner in which it is tossed. It is only for convenience or for lack of knowledge of the coin's properties that p is ever taken as $\frac{1}{2}$.

The random variables we shall encounter in the sequel will take on values over some continuous region rather than a set of countable events and will be called *continuous random variables* or *continuous variates.* Both terms will be used synonymously. The continuous random variable X defined on the domain of real numbers is characterized by its *distribution function*

(1) $$F(x) = P(X \le x) \qquad -\infty < x < \infty$$

giving the probability that X is less than or equal to some value x of its domain. Since X is continuous, $P(X = x) = 0$. If $F(x)$ is an absolutely continuous function, the continuous analogue of the discrete probability function is the *density function*

(2) $$f(x) = \frac{dF(x)}{dx}$$

Conversely, by the absolute-continuity property,

(3) $$F(x) = \int_{-\infty}^{x} f(u)\, du$$

and from this integral definition follows the equivalent term *cumulative distribution function* for $F(x)$. Note that these definitions are perfectly general: if the random variable is defined only on some interval of the real line, outside that interval $f(x)$ is defined to be zero, and to the left and right of the interval $F(x)$ is zero and one, respectively. When weighted in proportion to their density function $f(x)$, the values on the interval are said to constitute the *population* or *universe* of the random variable X.

The properties of a random variable are commonly visualized in terms of its density function, and it is to that representation that such names as *rectangular* or *exponential* refer.   Figure 1.1 illustrates the densities of three familiar variates.   The distribution functions of the rectangular and exponential variates follow by straightforward integrations; the normal distribution function can be evaluated only by numerical integration.

(a) Rectangular

$$f(x) = \begin{cases} 0 & x < a \\ \dfrac{1}{b-a} & a \le x \le b \\ 0 & b < x \end{cases}$$

(b) Exponential

$$f(x) = \begin{cases} \dfrac{1}{\beta} \exp\left(\dfrac{-x}{\beta}\right) & 0 \le x \\ 0 & x < 0 \end{cases}$$

(c) Normal

$$f(x) = \frac{1}{\sqrt{2\pi}\sigma} \exp\left[-\frac{1}{2}\left(\frac{x-\mu}{\sigma}\right)^2\right]$$

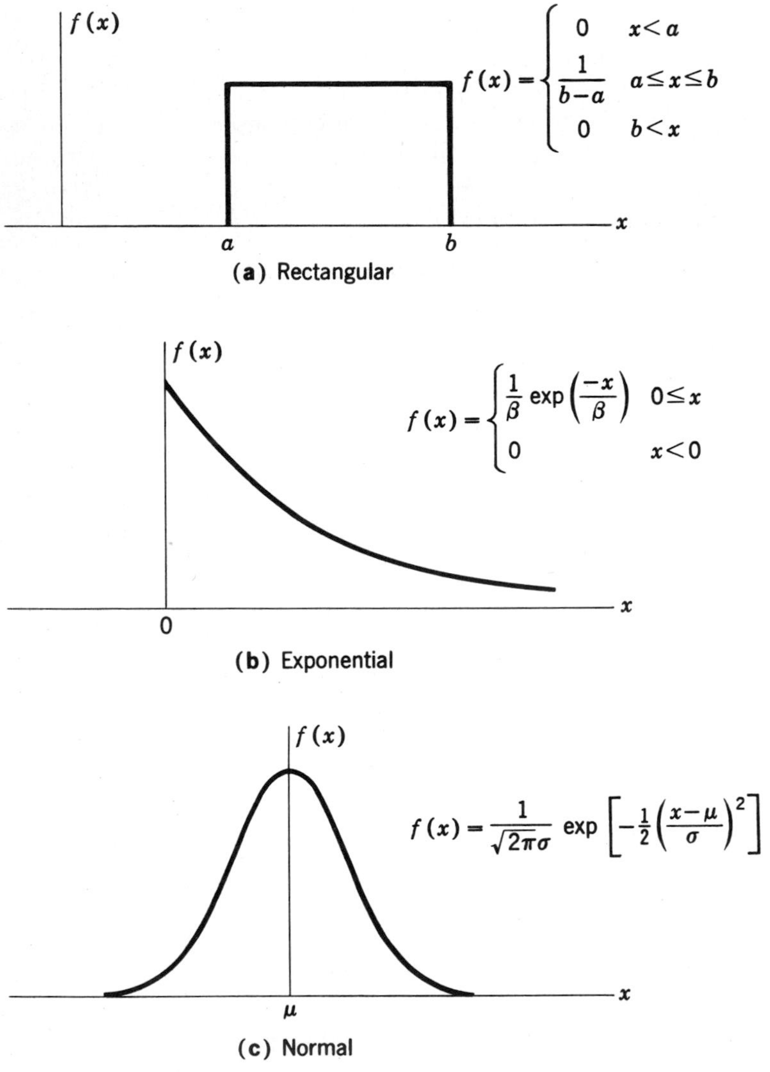

figure 1.1   Density functions.

The functions of Fig. 1.1 involve parameters that determine their positions and shapes. $\frac{1}{2}(a + b)$ of the rectangular density and μ of the normal function are *location* parameters, for their values specify the positions of the densities on the real axis. The range $b - a$ of the rectangular variate, the single exponential parameter β, and σ of the normal density are *scale* parameters, for changes in their values are equivalent to changes in the units of the variates. Larger values of these parameters imply a greater spread of the density function, and hence more variation in the random variable. In general, if the density $f(x;\alpha,\beta)$ can be written as

$$f\left(\frac{x - \alpha}{\beta}\right)$$

it follows that α is a location parameter and β is a scale parameter. We note that all density functions of that form contain the factor $1/\beta$ associated with the differential element dx.

Random variables and their densities can be characterized in another way. Let us think of the density $f(x)$ of the variate X as the function measuring the density of a continuous rod occupying the position of the x axis. The kth moment of the rod about the origin of its axis is

$$\mu'_k = \int_{-\infty}^{\infty} x^k f(x)\, dx \tag{4}$$

The first moment is the *mean, expectation,* or *expected value*

$$\mathrm{E}(X) = \int_{-\infty}^{\infty} x f(x)\, dx \tag{5}$$

of the random variable and corresponds to the physical notion of the horizontal center of gravity of the rod. The symbol E denotes the operation of computing the expected value, and is called the *expectation operator*. If c and k are nonrandom quantities, these useful properties of expectations hold:

$$\mathrm{E}(c) = c \tag{6a}$$

$$\mathrm{E}(cX) = c\mathrm{E}(X) \tag{6b}$$

$$\mathrm{E}(k + cX) = k + c\mathrm{E}(X) \tag{6c}$$

As an example, consider the rectangular density of Fig. 1.1. Then

$$\begin{aligned}\mathrm{E}(X) &= \frac{1}{b - a}\int_a^b x\, dx \\ &= \frac{1}{b - a}\frac{1}{2}x^2\Big|_a^b \\ &= \tfrac{1}{2}(b + a)\end{aligned}$$

As we might have reasoned from intuition, the mean value is the midpoint of the limits of the density, and if we wished, we might *reparametrize* the density function in terms of the mean as location parameter and the range $w = b - a$ as the scale parameter. The new density is

$$h(x) = \begin{cases} 0 & -\infty < x < \mathrm{E}(X) - \tfrac{1}{2}w \\ \dfrac{1}{w} & \mathrm{E}(X) - \tfrac{1}{2}w \le x \le \mathrm{E}(X) + \tfrac{1}{2}w \\ 0 & \mathrm{E}(X) + \tfrac{1}{2}w < x < \infty \end{cases}$$

Its functional form is unchanged, but its two parameters have different interpretations.

The *variance* of a random variable is the expected value of the squared deviations about its mean, or its *second central moment*. We shall denote the variance of X as

$$\mathrm{var}\ (X)$$

or occasionally where space is limited as

$$\sigma_x^2$$

By definition

$$\text{(7)} \qquad \begin{aligned} \mathrm{var}\ (X) &= \int_{-\infty}^{\infty} [x - \mathrm{E}(X)]^2 f(x)\, dx \\ &= \int_{-\infty}^{\infty} x^2 f(x)\, dx - [\mathrm{E}(X)]^2 \\ &= \mathrm{E}(X^2) - [\mathrm{E}(X)]^2 \end{aligned}$$

and the symbol var applied to any random variable will denote the operation of computing its variance. If c and k are constants, the variance has these properties:

$$\text{(8}a\text{)} \qquad \mathrm{var}\ (c) = 0$$

$$\text{(8}b\text{)} \qquad \mathrm{var}\ (cX) = c^2\ \mathrm{var}\ (X)$$

$$\text{(8}c\text{)} \qquad \mathrm{var}\ (X + c) = \mathrm{var}\ (X)$$

The first merely states that a nonrandom variable has zero variance. The squared-units nature of the variance is reflected in the second property, for a change of scale of X by c units changes the variance by c^2. The third result follows immediately from the definition (7) and states that the variance is unaffected by changes in the origin of the X axis. It can be shown that the variances of the rectangular, exponential, and normal densities of Fig. 1.1 are $\frac{1}{12}(b - a)^2$, β^2, and σ^2, respectively.

Frequently it is desirable to have a measure of dispersion that is in the original units of the variate. The *standard deviation* of X is the positive square root of the variance

$$\sigma_x = +\sqrt{\operatorname{var}(X)}$$

We note that the natural parameters of the normal density of Fig. 1.1*c* are the mean and standard deviation.

Independent Variates. Earlier in the chapter we referred loosely to "independent" random variables. Now we shall give a precise definition of independence. As we shall see in Chap. 3, it is possible to extend the notions of distribution and density functions to several variates, and if we write the joint distribution function of $X_1, \ldots, X_p$ as

$$(9)\quad F(x_1, \ldots, x_p) = \int_{-\infty}^{x_p} \cdots \int_{-\infty}^{x_1} f(u_1, \ldots, u_p)\, du_1 \cdots du_p$$

where $f(u_1, \ldots, u_p)$ is the joint density, the variates are said to be independent *if and only if*

$$(10)\qquad F(x_1, \ldots, x_p) = F_1(x_1) \cdots F_p(x_p)$$

where $F_i(x_i)$ is the distribution function of the single variate X_i. Alternatively, independence holds if and only if the factorization

$$(11)\qquad f(x_1, \ldots, x_p) = f_1(x_1) \cdots f_p(x_p)$$

of the joint density into the product of the individual densities $f_i(x_i)$ holds.

The product moment

$$(12)\quad \mathrm{E}(X_1^{k_1} \cdots X_p^{k_p}) = \int_{-\infty}^{\infty} \cdots \int_{-\infty}^{\infty} x_1^{k_1} \cdots x_p^{k_p} f(x_1, \ldots, x_p)\, dx_1 \cdots dx_p$$

of the variates $X_1, \ldots, X_p$ factors into the product

$$\mathrm{E}(X_1^{k_1}) \cdots \mathrm{E}(X_p^{k_p})$$

of the individual k_ith moments if the variates are independent. It follows from this result that the variance of a sum of *independent* random variables is merely the sum of the individual variances:

$$(13)\quad \operatorname{var}(X_1 + \cdots + X_p) = \operatorname{var}(X_1) + \cdots + \operatorname{var}(X_p)$$

Sources on Probability and Random Variables. Certain of the large number of basic texts on probability and random variables seem particularly relevant to the purposes of this introductory chapter. A very elementary treatment of probability as applied in biology has been written by Mosimann (1968), while another lucid introductory text is that of Goldberg (1960). Parzen's book (1960) is an excellent survey at a more intermediate level, while the two volumes of Feller (1968, 1971) are the standard source for a comprehensive study of discrete and continuous probability.

1.3 NORMAL RANDOM VARIABLES

In this section we shall describe some properties of a single normal random variable in preparation for the subsequent results and techniques that will be based upon the multivariate normal distribution. Recall that the normal density function is

$$f(x) = \frac{1}{\sqrt{2\pi}\sigma} \exp\left[-\frac{1}{2}\left(\frac{x-\mu}{\sigma}\right)^2\right] \qquad -\infty < x < \infty \tag{1}$$

and that its distribution function is given by the integral

$$\begin{aligned} P(x) &= \frac{1}{\sqrt{2\pi}\sigma} \int_{-\infty}^{x} \exp\left[-\frac{1}{2}\left(\frac{x-\mu}{\sigma}\right)^2\right] dx \\ &= \frac{1}{\sqrt{2\pi}} \int_{-\infty}^{(x-\mu)/\sigma} \exp\left(-\tfrac{1}{2}u^2\right) du \\ &= \Phi\left(\frac{x-\mu}{\sigma}\right) \end{aligned} \tag{2}$$

where $\Phi(z)$ denotes the *standard* or *unit* normal distribution function with zero mean and variance one. We shall denote the distribution of a normal variate by the Wilks symbol $N(\mu,\sigma^2)$; standardization of the variate will of course be indicated by $N(0,1)$.

Values of $\Phi(z)$ are contained in Table 1 of the Appendix. Conversely, the upper 100α percentage point of the unit normal distribution is defined as that value z_α such that

$$\begin{aligned} \alpha &= P(Z > z_\alpha) \\ &= 1 - \Phi(z_\alpha) \end{aligned} \tag{3}$$

These connections between percentage points and their probabilities are illustrated in Fig. 1.2 for the normal density and distribution functions.

Now let us offer one justification for the reliance of much of statistical methodology upon the assumption of a normal population. This is the *central-limit theorem,* which states that variates which are sums of many independent effects tend to be normally distributed as the number of effects becomes large. More formally,

If the random variables $X_1, \ldots, X_N$ *are independently distributed according to some common distribution function with mean* $\mathrm{E}(X_i) = \mu$ *and finite*

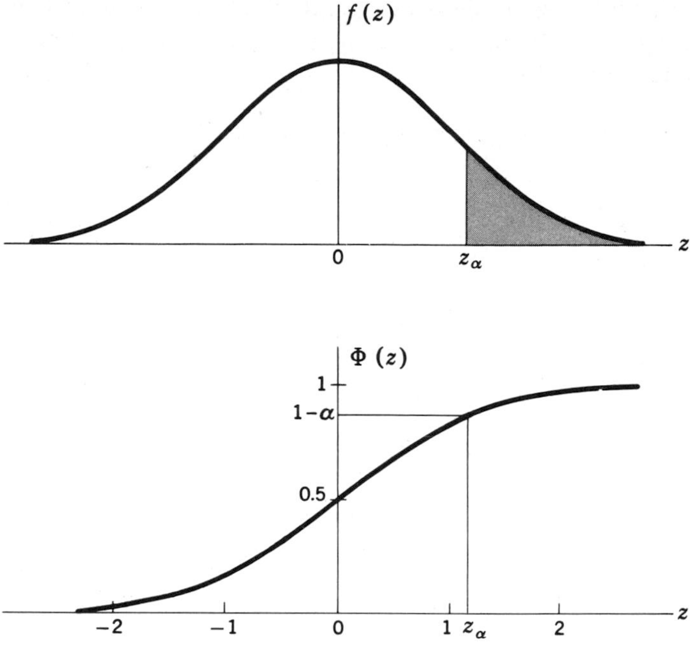

figure 1.2

variance var $(X_i) = \sigma^2$, *then as the number of variates N increases without bound, the variate*

$$Z = \frac{\sum_{i=1}^{N} X_i - N\mu}{\sqrt{N\sigma^2}}$$

converges in distribution to a normal variate with mean zero and variance one.

A proof of this version of the central-limit theorem can be found in most texts on probability and mathematical statistics. Feller (1968) has treated other more general theorems which do not assume identical distributions of the X_i. One immediate consequence of the theorem is that the function $\sqrt{N}\,(\bar{X} - \mu)$ of the sample mean of a sequence of independent random variables

$$\bar{X} = \frac{1}{N}(X_1 + \cdots + X_N)$$

whose common distribution has mean μ and variance σ^2 tends to be distributed as a $N(0,\sigma^2)$ variate as N becomes large.

We shall invoke one important property of the normal distribution continually in the sequel. If $X_1, \ldots, X_n$ are independent variates with distributions $N(\mu_1,\sigma_1^2), \ldots, N(\mu_n,\sigma_n^2)$, the linear compound

$$Y = a_1X_1 + \cdots + a_nX_n \tag{4}$$

is also normally distributed with mean $\sum_{i=1}^{n} a_i\mu_i$ and variance $\sum_{i=1}^{n} a_i^2\sigma_i^2$. We shall see in Chap. 3 that this result can be extended to a linear compound of dependent normal variates.

The Chi-squared Distribution. Many distributions can be derived from different transformations upon a set of normal variates. One of the most important is that of the *chi-squared* variate. If the variates $X_1, \ldots, X_n$ are independently and normally distributed with mean zero and unit variance, then

$$\chi^2 = X_1^2 + \cdots + X_n^2 \tag{5}$$

has the density function

$$f(\chi^2) = \frac{2^{-n/2}}{\Gamma(n/2)} (\chi^2)^{(n-2)/2} \exp\left(-\tfrac{1}{2}\chi^2\right) \qquad 0 \leq \chi^2 < \infty \tag{6}$$

and is said to be a *chi-squared variate with n degrees of freedom.* Since we started with standard normal variates, the density contains no scale or location parameters but only the single parameter n. More generally, if the X_i are independently distributed as $N(\mu_i,\sigma_i^2)$ variates, the quantity

$$\sum_{i=1}^{n} \frac{(X_i - \mu_i)^2}{\sigma_i^2}$$

has the chi-squared distribution with n degrees of freedom, for each squared term in the sum has been standardized by its mean and variance.

Table 2 of the Appendix contains percentage points of the chi-squared distribution. We shall write the 100α percentage point of the chi-squared distribution with n degrees of freedom as

$$\chi^2_{\alpha;n} \tag{7}$$

where of course $\alpha = P(\chi^2 > \chi^2_{\alpha;n})$.

The t Distribution. The t random variable with n degrees of freedom is defined as the quotient

$$t = \frac{z}{\sqrt{\chi^2/n}} \tag{8}$$

of a standard normal variate z and the square root of an independent chi-squared variate divided by its n degrees of freedom. t is a dimensionless quantity, and its density function

$$(9) \qquad f(t) = \frac{\Gamma[(n+1)/2]}{\sqrt{\pi n}\,\Gamma(n/2)} \frac{1}{(1+t^2/n)^{(n+1)/2}} \qquad -\infty < t < \infty$$

depends upon the single degrees of freedom parameter n. Percentage points of the distribution of t are given in Table 3 of the Appendix. We shall customarily write the upper 100α percentage point as $t_{\alpha;n}$ or

$$(10) \qquad \alpha = P(t > t_{\alpha;n})$$

The F Distribution. The ratio

$$(11) \qquad F = \frac{\chi_1^2/m}{\chi_2^2/n}$$

of independent chi-squared variates divided by their respective degrees of freedom has the F, or *variance-ratio*, distribution with density function

$$(12) \qquad f(F) = \frac{\Gamma[(m+n)/2]}{\Gamma(m/2)\Gamma(n/2)} \left(\frac{m}{n}\right)^{m/2} F^{(m-2)/2} \left(1 + \frac{m}{n}F\right)^{-(m+n)/2}$$

$$0 \le F < \infty$$

We shall denote the 100α upper percentage point of the F distribution with m, n degrees of freedom by $F_{\alpha;m,n}$:

$$(13) \qquad \alpha = P(F > F_{\alpha;m,n})$$

It follows from the definition (11) of the F variate that the lower percentage points can be obtained from the reciprocals of the upper values with reversed degrees of freedom:

$$(14) \qquad F_{1-\alpha;m,n} = \frac{1}{F_{\alpha;n,m}}$$

Table 4 of the Appendix gives upper percentage points of the F distribution.

Derivations and mathematical properties of these standard distributions can be found in many current sources, e.g., Hogg and Craig (1959), Kendall and Stuart (1958), or Mood *et al.* (1974). Extensive tables of the distributions and their percentage points have been compiled by Pearson and Hartley (1966), together with illustrations of their use.

1.4 RANDOM SAMPLES AND ESTIMATION

Heretofore we have discussed random variables only in terms of the abstract populations specified by their distribution or density functions. Occasionally these functions are known for some random phenomenon, and it is possible to describe the process directly from its mathematical model. More usually it is the case that neither the mathematical form of the distribution nor its parameters are known, and it is necessary to go beyond the realm of probability theory to the domain of *statistical inference* to obtain estimates of $F(x)$ or its parameters from finite *samples* of values of the random variable. In this section we shall consider one heuristic approach which leads to estimates with some desirable properties.

Let us begin by supposing that the values or "realizations" of the continuous random variable X can be observed and recorded. This is not such an obvious requirement, for many phenomena of interest in the physical or life sciences cannot be observed below threshold levels established by the organism or the measuring equipment, and above other levels the equipment may be saturated or paralyzed by the frequency or intensity of the responses. It is also clear that even if the values of X formed a continuum, the limitations of any recording or measuring device would yield discrete observations. Nevertheless, we shall treat such data as blood pressure in millimeters of mercury, percentage of a certain content of a projective test, and reaction time in milliseconds as observations from continuous populations. We shall assume that the mathematical form of the density function $f(x;\theta_1, \ldots, \theta_k)$ of X is known from substantive considerations, prior experience, or other good fortune, although the values of the parameters θ_i are unknown.

Next let us define the *parameter space* of a density. Suppose that the density depends upon the single real parameter θ, as in the case of the descending exponential or the normal distribution with mean μ and known unit variance. Then the parameter space of θ is *that portion of the real line which contains all admissible values of* θ. For the descending exponential the parameter space would be the positive half of the real line, for negative parameters would destroy the density property, and a parameter of zero would lead to the trivial "sure-number" distribution. For the normal distribution with known variance the space would be the entire real line. Similarly, the parameter space for the k-parameter density would be some region of k-dimensional euclidean space. For example, that of the normal distribution is the upper half of the (μ,σ^2) plane shown in Fig. 1.3.

Finally we must define the *sampling* or *experimental units* on which the values of X will be observed. The available units must constitute a homogeneous collection with respect to all characteristics which might affect the values of the variate. If the random variable is the blood level of free fatty acid (FFA) in normal adolescent American girls, the available

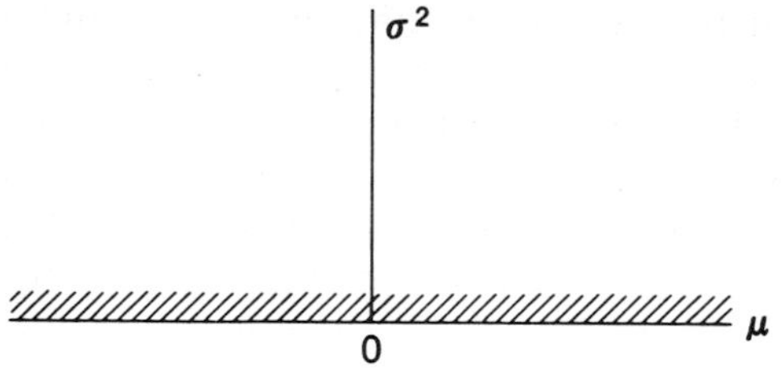

figure 1.3 Parameter space for the normal distribution.

sampling units should not include female subjects with metabolic diseases, adolescent males, or prepubertal children of either sex. The sampling units must be independent of one another and must not possess common qualities which might lead to dependent values of X. Clearly, a sequence of 10 daily FFA determinations in a single subject would not yield the same information about the biological population represented by that person as 10 single determinations of FFA obtained from as many unrelated and independent individuals. We assure such independence by drawing the sampling units *randomly* from the available collection of units. The investigator who gathers his data from the nearest convenient source of subjects, be it students in Psychology 1 or a group of paid volunteers, must risk whatever biases these nonrandom and unrepresentative samples may contain.

Now suppose that N units have been selected at random. Their observed values of X will be denoted by $x_1, \ldots, x_N$. We shall customarily distinguish such observations from the running value x of X by the presence of subscripts. Our initial problem will be to estimate the parameters θ_i by some suitable function of the observations. Such estimates will be denoted by the parameter with a hat, or $\hat{\theta}_i$. We shall call these quantities *point estimates*, for they are unequivocal single values of θ, as opposed to the *interval estimates* of the next section. But how should the function of the observations be chosen? Should we merely relate the parameters to the first few moments of the distribution and then equate the sample moments to those of the population? Would estimates using only extreme or middle values of the observations serve as well as those which depended upon all the data? Intuition might lead us to many other estimators, or classes of functions of the observations, and we should like to have some criteria for choosing among them. For example, we might ask whether the estimator possesses any of these desirable properties:

1. *Unbiasedness.* The expected value of the estimator should be its parameter, or $E(\hat{\theta}) = \theta$.

2. *Consistency.* As the sample size increases without bound, $\hat{\theta}$ should converge in probability to θ.
3. *Minimum variance.* Frequently the estimate is chosen as that one which has smallest variance among all unbiased estimators. If a minimum-variance estimator does not exist for all sample sizes, it is possible to choose among competing estimators by the ratios of their asymptotic variances. The estimator with that smallest relative variance is said to be *efficient*.
4. *Sufficiency.* An estimator $\hat{\theta}$ is said to be *sufficient* if it contains all the information in the observations for the estimation of θ. That is, knowledge of the values $x_1, \ldots, x_N$ will provide no more information than that contained in $\hat{\theta}$. Stated mathematically, the joint distribution of the x_i for a fixed value of $\hat{\theta}$ does not depend on the unknown parameter θ.

In a particular application each of these criteria must be weighed with regard to the cost of sample observations, the speed with which the data must be processed, and the consequences of small biases or larger variances on the investigator's view of nature or the policy maker's actions. As we shall see, it is sometimes possible to remove bias by a slight scaling or other redefinition of the estimate, while the common estimates of the various correlation measures to be discussed in Chap. 3 are computed in daily profusion with no concern for their biased nature. In small samples the loss of efficiency in some short-cut estimators may be offset by the ease with which they are computed.

Now let us consider a means for generating estimates known as the *method of maximum likelihood.* While that procedure is a heuristic one, or one appealing more to intuition than to the real end of producing estimates with "good" properties, it can be shown that its estimates have a number of desirable qualities. We begin by defining the *likelihood* of the random sample of observations as the joint density of the variates of the sampling units evaluated at $x_1, \ldots, x_N$:

$$L(\theta_1, \ldots, \theta_k) = \prod_{i=1}^{N} f(x_i;\theta_1, \ldots, \theta_k) \tag{1}$$

The likelihood function is a relative measure of the likelihood of the particular sample $x_1, \ldots, x_N$. The maximum-likelihood method of estimation directs that the estimates of the θ_i be chosen so as to maximize the function (1) for a given sample. If the likelihood has a relative maximum, this can be accomplished by straightforward differentiation and solution of certain equations. The vanishing of the derivatives is of course only a necessary condition for a relative maximum, and the sufficient conditions given by the second-order partial derivatives should also be verified (see, for example, Hancock, 1960, chap. 5). Absolute maxima can frequently be determined by direct inspection of the likelihood. However, since the values of the variables which maximize a function also maximize monotonic functions of it, and since the majority of likelihoods encountered in statistical inference contain exponential terms, it is usually more convenient to work with the natural logarithm of the likelihood

$$l(\theta_1, \ldots, \theta_k) = \ln L(\theta_1, \ldots, \theta_k) \tag{2}$$

Then, if the likelihood has a relative maximum, the associated estimates of the θ_i can be found by solving the system of k simultaneous equations

$$\frac{\partial l(\theta_1, \ldots, \theta_k)}{\partial \theta_j} = 0 \qquad j = 1, \ldots, k \tag{3}$$

for the estimates

$$\hat{\theta}_1, \ldots, \hat{\theta}_k \tag{4}$$

If the equations possess multiple roots, it will be necessary to choose the solution leading to the greatest likelihood.

Perhaps two simple examples will help to describe the steps in finding maximum-likelihood estimators.

Example 1.1. Let us determine the maximum-likelihood estimates of a and b in the rectangular density defined in Sec. 1.2. The likelihood function of the sample $x_1, \ldots, x_N$ is

$$L(a,b) = \frac{1}{(b-a)^N}$$

Clearly a cannot exceed the smallest observation, and b cannot be less than the largest. If we denote the ordered observations by $x_{(1)} \leq \cdots \leq x_{(N)}$,

$$a \leq x_{(1)} \leq \cdots \leq x_{(N)} \leq b$$

The likelihood will be at its greatest value when $b - a$ is as small as possible consistent with the second set of inequalities, and the estimates minimizing that range are

$$\hat{a} = x_{(1)} \qquad \hat{b} = x_{(N)}$$

Example 1.2. The likelihood of the sample $x_1, \ldots, x_N$ of N independent normal random variables is

$$L(\mu,\sigma^2) = \frac{1}{(2\pi)^{N/2}(\sigma^2)^{N/2}} \exp\left[-\frac{1}{2\sigma^2}\sum_{i=1}^{N}(x_i - \mu)^2\right]$$

and its logarithm is

$$l(\mu,\sigma^2) = -\frac{N}{2}\ln(2\pi) - \frac{N}{2}\ln\sigma^2 - \frac{1}{2\sigma^2}\sum_{i=1}^{N}(x_i - \mu)^2$$

The partial derivatives with respect to μ and σ^2 are

$$\frac{\partial l(\mu,\sigma^2)}{\partial\mu} = \frac{1}{\sigma^2}\sum_{i=1}^{N}(x_i - \mu)$$

$$\frac{\partial l(\mu,\sigma^2)}{\partial\sigma^2} = -\frac{N}{2\sigma^2} + \frac{1}{2\sigma^4}\sum_{i=1}^{N}(x_i - \mu)^2$$

If we equate these to zero and cancel any extraneous factors, the simultaneous equations (3) are

$$\sum_{i=1}^{N} x_i - N\mu = 0$$

$$\sum_{i=1}^{N}(x_i - \mu)^2 - N\sigma^2 = 0$$

Solve the first for the estimate of μ, and use that value in the second to obtain the solution for the estimate of σ^2:

$$\hat{\mu} = \frac{1}{N}\sum_{i=1}^{N} x_i$$

$$= \bar{x}$$

$$\hat{\sigma}^2 = \frac{1}{N}\sum_{i=1}^{N}(x_i - \bar{x})^2$$

$$= \frac{1}{N}\left[\sum_{i=1}^{N} x_i^2 - \frac{1}{N}(\Sigma x_i)^2\right]$$

These estimates are intuitively plausible, for the population mean is merely estimated by the sample mean $\bar{x}$, and the estimate of the variance is the average squared deviation from the sample mean.

Let us determine whether $\hat{\mu}$ and $\hat{\sigma}^2$ are unbiased estimates. Replace the observations by the random variables X_i and take expectations:

$$E(\hat{\mu}) = \frac{1}{N}\sum_{i=1}^{N} E(X_i)$$

$$= \frac{1}{N} \sum_{i=1}^{N} \mu$$

$$= \mu$$

The sample mean is an unbiased estimate of μ. The expectation of $\hat{\sigma}^2$ involves more lengthy computations:

$$\begin{aligned}
E(\hat{\sigma}^2) &= \frac{1}{N} E\left[\sum_{i=1}^{N} X_i^2 - \frac{1}{N}\left(\sum X_i\right)^2\right] \\
&= \frac{1}{N}\left[NE(X_i^2) - \frac{1}{N} E\left(\sum_{i=1}^{N}\sum_{j=1}^{N} X_i X_j\right)\right] \\
&= \frac{1}{N}\left[NE(X_i^2) - E(X_i^2) - \frac{1}{N}\sum_{i \neq j}\sum E(X_i X_j)\right] \\
&= \frac{1}{N}\left\{(N-1)E(X_i^2) - \frac{1}{N} N(N-1)[E(X_i)]^2\right\} \\
&= \frac{N-1}{N}\sigma^2
\end{aligned}$$

$\hat{\sigma}^2$ is not an unbiased estimate of the variance. However, if we replace the divisor N in the original formula by $N - 1$, the bias will be eliminated and the usual sample-variance expression

$$s^2 = \frac{1}{N-1}\left[\sum_{i=1}^{N} x_i^2 - \frac{1}{N}\left(\sum x_i\right)^2\right]$$

can be obtained.

Maximum-likelihood estimates are thus not necessarily unbiased, although it is sometimes possible to remove the bias through multiplication by an appropriate factor. Such estimates do not generally have minimum variance. However, it can be shown that all maximum-likelihood estimates are consistent and that their asymptotic distributions are normal with known parameters.

1.5 TESTS OF HYPOTHESES FOR THE PARAMETERS OF NORMAL POPULATIONS

Statistical Tests of Hypotheses. Statistical inference can be divided into two general areas. The first is concerned with the estimation of distribution functions, the parameters of such functions when their mathematical form is specified, or the parameters of models built around random variables. The second part addresses itself to the problem of testing the validity of hypotheses about distribution functions and their parameters or the parameters or components of mathematical models. In the preceding section we touched briefly on one approach to estimation through the maximum-likelihood principle. Now we shall summarize some essentials of hypothesis testing, and subsequently we shall see how these tests can be inverted to provide interval estimates for parameters.

Perhaps a simple example will help to motivate the testing problem in terms of regions in the parameter space. It is known from extensive experience that the freshman grade point indices (GPI) of students at a small liberal arts college have tended to be normally distributed with mean 2.43 and variance 0.04. However, in the selection of the present freshman class several admission standards were raised, and it is hypothesized that the mean GPI of the population out of which those students were drawn will be higher and the variance in turn will be a little smaller. We may summarize these statements about the population parameters in this fashion:

Original hypothesis:	$\mu = 2.43$	$\sigma^2 = 0.04$
Alternative hypothesis:	$\mu > 2.43$	$\sigma^2 < 0.04$

We shall designate the original description of the GPI population as the *null hypothesis;* this will be conventionally denoted by H_0. The *alternative hypothesis* will be denoted by H_1. The assumption of normality is common to both hypotheses and need not be mentioned in their statements. H_0 refers to the *single* point (2.43,0.04) in the parameter space, and therefore is called a *simple* hypothesis. The alternative designates the shaded region of Fig. 1.4, and since that set contains more than one point, H_1 is called a *composite* hypothesis. An important class of compos-

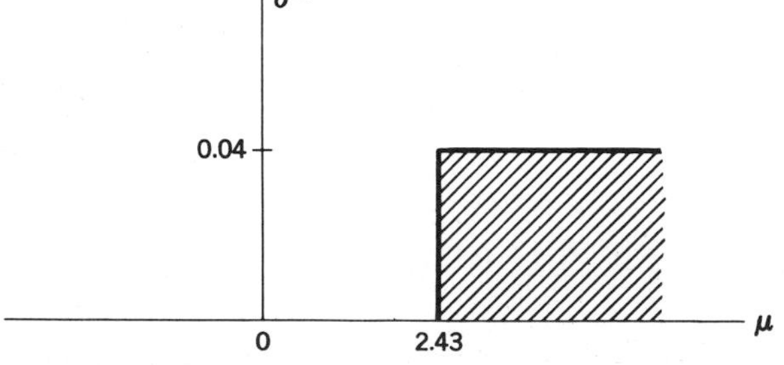

figure 1.4

ite hypotheses is formed by those statements in which the values of one or more parameters are completely unknown. For example, if the random variable X is normally distributed with *unknown* variance, the hypotheses

$$H_0: \quad \mu = \mu_0$$

and

$$H_1: \quad \mu = \mu_1 > \mu_0$$

on the mean alone specify vertical lines in the space of Fig. 1.3 terminating at $\mu = \mu_0$ and $\mu = \mu_1$ on the horizontal axis. Statistical tests of such hypotheses would have to be constructed to be unaffected by the unknown true value of σ^2.

It has been the purpose of much of the theory of statistical inference to develop tests of the validity of these hypotheses based upon sample observations. We shall prefer to consider testing only in terms of the classical, or Neyman-Pearson, approach to two-decision rules, for the subsequent multivariate tests will be of the classical sort. The reader is referred to Lindgren (1962) and Mood *et al.* (1974) for detailed discussions and examples of test construction; considerably more advanced treatments with proofs of many of the theorems can be found in Kendall and Stuart (1961), Lehmann (1959), and Wilks (1962).

For the general case let X be a continuous random variable with the real numbers as its admissible values. The density function of X depends upon the parameters $\theta_1, \ldots, \theta_k$. The *sample space* W of all possible outcomes of N observations on X is then N-dimensional euclidean space. A particular sample of observations will be written in vector form as $[x_1, \ldots, x_N]$ and will denote a point in the sample space. Let ω_1 and ω_2 be any two disjoint (that is, having no points in common) regions of the parameter space Ω. We shall set up these hypotheses about the parameters of the distribution of X:

$$(1) \qquad \begin{aligned} H_0&: \quad [\theta_1, \ldots, \theta_k] \text{ is contained in } \omega_1 \\ H_1&: \quad [\theta_1, \ldots, \theta_k] \text{ is contained in } \omega_2 \end{aligned}$$

On the basis of the sample observations we wish to decide in some "optimal" fashion which hypothesis is tenable. Since certain sets of observations would lead us to favor H_0 over H_1 and other data would support the opposite preference, our decision rule should have this form:

$$(2) \qquad \begin{aligned} &\text{Accept } H_0 \text{ if } [x_1, \ldots, x_N] \text{ falls in } W - w \\ &\text{Accept } H_1 \text{ if } [x_1, \ldots, x_N] \text{ falls in } w \end{aligned}$$

where w is a specified part of the sample space called the *critical region* or *rejection region* for H_0. If the true state of nature as described by the various parameters is specified by either H_0 or H_1, the decision maker can incur two kinds of error in the application of the decision rule. An *error of the first kind*, or a *Type I error*, consists of declaring H_1 the true state when in fact H_0 is true. An *error of the second kind*, or a *Type II error*, is made upon the acceptance of H_0 as true when H_1 describes the correct state of nature. The correctness of the actions may be summarized in this two-way table:

Action	State of nature	
	H_0 true	H_1 true
Accept truth of H_0	Correct	Type II error
Accept truth of H_1	Type I error	Correct

The probabilities of the Type I and II errors provide measures of the efficacy of the decision rule. The probability α of the Type I error is the probability that the sample observations fall in the critical region w when H_0 is true, or

$$\alpha = P([x_1, \ldots, x_N] \in w | H_0 \text{ true}) \tag{3}$$

where the symbol $\in$ denotes membership in the set w and the vertical bar indicates that the probability statement is conditional upon the truth of the null hypothesis. α is called the *size* of the test. The probability of the Type II error is

$$\beta = P([x_1, \ldots, x_N] \in W - w | H_1 \text{ true}) \tag{4}$$

The complement $1 - \beta$ of the second error probability is called the *power* of the test or decision rule. If the power is computed for a continuum of parameter values, the resulting probabilities constitute the *power function*.

If the sample size is fixed, changes in the form of the critical region that reduce α will also increase β, and conversely, minimization of β will be at the expense of larger α. The classical approach to hypothesis testing calls for a test of fixed size α whose rejection region is chosen so as to minimize β or, equivalently, to maximize the power. If both hypotheses are simple, i.e., of the form

$$H_0: \begin{matrix} \theta_1 = \theta_{10} \\ \cdot\ \cdot\ \cdot\ \cdot\ \cdot \\ \theta_k = \theta_{k0} \end{matrix} \qquad H_1: \begin{matrix} \theta_1 = \theta_{11} \\ \cdot\ \cdot\ \cdot\ \cdot\ \cdot \\ \theta_k = \theta_{k1} \end{matrix} \tag{5}$$

the *Neyman-Pearson lemma* states that the most powerful test of size α will have a critical region defined by this decision rule:

(6) $$\text{Accept } H_0 \text{ if } \lambda = \frac{f(x_1;\theta_{10}, \ldots, \theta_{k0}) \cdots f(x_N;\theta_{10}, \ldots, \theta_{k0})}{f(x_1;\theta_{11}, \ldots, \theta_{k1}) \cdots f(x_N;\theta_{11}, \ldots, \theta_{k1})} > c$$

and

$$\text{Accept } H_1 \text{ if } \lambda < c$$

where c is a constant chosen such that

$$P(\lambda < c) = \alpha$$

The lemma defines the critical region as that set of points in the sample space for which the likelihood ratio λ is less than c. The proof of the lemma is due to Neyman and Pearson (1933); numerous applications to the derivation of standard tests may be found in the text of Mood *et al.* (1974). A more general development of two-decision rules can be obtained by assigning monetary losses or other penalties to the Type I and Type II errors.

Note that the Neyman-Pearson lemma requires that both hypotheses are simple, i.e., that the regions ω_1 and ω_2 of the parameter space are points. Tests of composite hypotheses involving several parameters can be constructed by the powerful *generalized likelihood-ratio criterion*

(7) $$\lambda = \frac{L(\hat{\omega})}{L(\hat{\Omega})}$$

where

(8) $$L(\hat{\omega}) = f(x_1;\hat{\theta}_{10}, \ldots, \hat{\theta}_{k0}) \cdots f(x_N;\hat{\theta}_{10}, \ldots, \hat{\theta}_{k0})$$

is the likelihood function maximized under the assumption that

$$H_0\colon\ [\theta_1, \ldots, \theta_k] \in \omega$$

is true, and $L(\hat{\Omega})$ is the maximized likelihood for $\theta_1, \ldots, \theta_k$ permitted to take on values throughout the entire parameter space Ω. We accept H_0 if

$$\lambda > c$$

and otherwise accept the alternative

$$H_1\colon\ [\theta_1, \ldots, \theta_k] \in \Omega - \omega$$

The constant c is chosen so that $P(\lambda < c|H_0 \text{ true}) \leq \alpha$. When, as in the case of composite hypotheses, the true size of the test is actually less than or equal to α, we shall say that the test is of *level* α. It can be shown that if H_0 is true, the variate

(9) $$\chi^2 = -2 \ln \lambda$$

tends as the sample size increases to be distributed according to the chi-squared distribution with degrees of freedom equal to the difference of the dimensionalities of the parameter space Ω and the null hypothesis subspace ω or, equivalently, to the number of parameters determined by H_0. The generalized likelihood-ratio criterion is also due to Neyman and Pearson (1928); examples of its use have been given by Mood *et al.* (1974), while extensive discussions of its properties have been given by Kendall and Stuart (1961) and Wilks (1962).

Tests on the Mean of a Normal Variate with Known Variance. Let $x_1, \ldots, x_N$ be a sample of independent observations on the random variable with distribution $N(\mu,\sigma^2)$. The variance σ^2 is known, although μ is not. On the basis of the sample observations we wish to test the hypothesis

(10) $$H_0:\quad \mu = \mu_0$$

that the population mean has some specified value μ_0 against the alternative

(11) $$H_1:\quad \mu = \mu_1 > \mu_0$$

that the mean is some larger value μ_1. By the Neyman-Pearson lemma the most powerful test of size α is based upon the test statistic

(12) $$z = \frac{(\bar{x} - \mu_0)\sqrt{N}}{\sigma}$$

whose critical region is

(13) $$z > z_\alpha$$

where z_α is the upper 100α percentage point of the unit normal distribution. In terms of the mean of the original observations we should reject H_0 in favor of H_1 if

(14) $$\bar{x} > \mu_0 + \frac{\sigma}{\sqrt{N}} z_\alpha$$

and otherwise accept the null hypothesis. If the alternative hypothesis had been

(15) $$H_1':\quad \mu = \mu_1 < \mu_0$$

the same test statistic would be employed, although the critical region would be

(16) $$z < -z_\alpha$$

or, equivalently,

$$\bar{x} < \mu_0 - \frac{\sigma}{\sqrt{N}} z_\alpha \tag{17}$$

The preceding tests and alternative hypotheses H_1, H_1' are called *one-sided,* for the direction of the change from μ_0 to μ_1 is clearly indicated. When it is possible to make such predictions from subject-matter considerations or prior investigations, the power of the tests will be appreciably larger than that of the *two-sided* test with the alternative hypothesis

$$H_1'': \quad \mu = \mu_1 \neq \mu_0 \tag{18}$$

that allows for either larger or smaller alternative values of the mean. The test statistic is still (12), but the rejection region for a test of size α is

$$|z| > z_{\alpha/2} \tag{19}$$

where $z_{\alpha/2}$ is the upper 50α percentage point of the unit normal distribution. Equivalently, we reject H_0 if *either*

$$\bar{x} > \mu_0 + \frac{\sigma}{\sqrt{N}} z_{\alpha/2} \qquad \text{or} \qquad \bar{x} < \mu_0 - \frac{\sigma}{\sqrt{N}} z_{\alpha/2} \tag{20}$$

holds.

Power curves for tests against the three alternatives can be constructed from tables of the normal distribution, and since they can be found in most texts on statistical theory and methods, they will not be reproduced here. In practice one usually selects a suitable value of α and one or more tolerable β probabilities. The sample size N is then chosen to achieve the minimum β that is consistent with budgetary limitations or the size of laboratory or clinical facilities.

Finally, we note that the normal density function is a member of the exponential family cited earlier in this section, and thus tests of such composite hypotheses as

$$H_0: \quad \mu = \mu_0 \qquad H_1: \quad \mu > \mu_0$$

or

$$H_0: \quad \mu < \mu_0 \qquad H_1: \quad \mu > \mu_0$$

are uniformly most powerful.

Tests on Means When the Variance Is Unknown. In most scientific applications it is rare indeed that the population variance is known, and an important advance in statistical inference was achieved when W. S. Gosset (publishing under the pseudonym "Student") obtained the distribution of the test statistic for composite hypotheses on the mean of a normal distribution with unknown variance. The generalized likelihood-ratio criterion for testing the hypothesis (10) against the alternative (11) on the basis of N independent observations $x_1, \ldots, x_N$ with mean $\bar{x}$ and variance s^2 leads to the test statistic

$$t = \frac{(\bar{x} - \mu_0)\sqrt{N}}{s} \tag{21}$$

If H_0 is true, t has the Student-Fisher t distribution with $N - 1$ degrees of freedom, and we reject the null hypothesis for a test of size α if

$$t > t_{\alpha;N-1} \tag{22}$$

where $t_{\alpha;N-1}$ is the upper 100α percentage point of the t distribution defined in the preceding section. Similarly, if the alternative hypothesis had been (15), the rejection region would be defined by

$$t < -t_{\alpha;N-1} \tag{23}$$

and for the two-sided alternative (18) the null hypothesis would be rejected if

$$|t| > t_{\alpha/2;N-1} \tag{24}$$

The power of tests involving the t statistic can be computed from the first of the Pearson-Hartley charts of the Appendix with degrees of freedom $\nu_1 = 1$ and $\nu_2 = N - 1$ and noncentrality parameter

$$\phi = \frac{|\mu - \mu_0|}{\sigma}\left(\frac{N}{2}\right)^{1/2} \tag{25}$$

Then a sample size can be determined which will guarantee a power probability above a specified minimum for fixed α.

Now consider that two random samples have been independently drawn from normal populations with a common variance σ^2 but possibly different means μ_1 and μ_2. Let the observations of the samples be $x_1, \ldots, x_{N_1}, y_1, \ldots, y_{N_2}$. The generalized likelihood-ratio criterion for the test of the hypothesis

$$H_0\colon \quad \mu_1 = \mu_2 \tag{26}$$

of equal population means against the alternative

$$H_1\colon \quad \mu_1 > \mu_2 \tag{27}$$

leads to the test statistic

$$t = \frac{\bar{x} - \bar{y}}{\sqrt{\dfrac{\sum_{i=1}^{N_1}(x_i - \bar{x})^2 + \sum_{i=1}^{N_2}(y_i - \bar{y})^2}{N_1 + N_2 - 2}\left(\dfrac{1}{N_1} + \dfrac{1}{N_2}\right)}} \tag{28}$$

For a test of size α we reject H_0 in favor of H_1 if

$$t > t_{\alpha;\, N_1+N_2-2} \tag{29}$$

The rejection regions for the other alternative hypotheses H_1': $\mu_1 < \mu_2$ and H_1'': $\mu_1 \neq \mu_2$ are of course similar to (23) and (24) of the single-sample tests.

Confidence Intervals for Means. The investigator who has carried out a costly or intricate experiment is rarely satisfied to hear that his observations have merely rejected some hypothesis. If the findings show that the new drug or treatment has some "significant" effect beyond that of the placebo or previous standard, the experimenter and the scientific community would prefer to know not only the best estimate of the magnitude of this effect but also some range of reasonable values of the effect parameter. Such statements of possible values are called *confidence intervals* or, in contrast with the unequivocal point estimates of Sec. 1.4, *interval estimates.*

Suppose that a random sample of N observations has been drawn from some population with continuous density $f(x;\theta)$. The $100(1 - \alpha)$ percent confidence interval for θ is that set of values

$$t_1 \le \theta \le t_2 \tag{30}$$

with limits computed from the percentage points of the distribution function of the estimate $\hat{\theta}$ of θ so that

$$P(t_1 \le \theta \le t_2) = 1 - \alpha \tag{31}$$

α is usually taken as 0.05 or some smaller probability. $1 - \alpha$ is called the *confidence coefficient* of the interval. It is *essential* that the probability statement be read as "the probability that the interval with end points t_1, t_2 covers θ is $1 - \alpha$," for in our usage the parameter is hardly a random variable. We also note that an infinity of confidence intervals exists which satisfy (31); in most subsequent cases t_1 and t_2 will be chosen so that the length $t_2 - t_1$ of the interval is shortest.

Confidence intervals for the mean of a normal population or the difference of the means of two normal populations can be found from the preceding tests of hypotheses. If N independent observations with mean $\bar{x}$ have been collected on the $N(\mu,\sigma^2)$ variate with known variance, the value of μ_0 for which H_0: $\mu = \mu_0$ is just rejected in favor of

$$H_1: \quad \mu = \mu_1 > \mu_0$$

by the test of size α_1 is given by

$$\mu_0 = \bar{x} - \frac{\sigma}{\sqrt{N}} z_{\alpha_1}$$

Similarly, the largest value of μ_0 for which the size $\alpha - \alpha_1$ test of H_0 against H_1': $\mu = \mu_1 < \mu_0$ is rejected is

$$\mu_0 = \bar{x} + \frac{\sigma}{\sqrt{N}} z_{\alpha-\alpha_1}$$

z_{α_1}, $z_{\alpha-\alpha_1}$ are of course the upper $100\alpha_1$, $100(\alpha - \alpha_1)$ percentage points of the unit normal distribution. The $100(1 - \alpha)$ percent confidence interval for μ is thus

$$\bar{x} - \frac{\sigma}{\sqrt{N}} z_{\alpha_1} \leq \mu \leq \bar{x} + \frac{\sigma}{\sqrt{N}} z_{\alpha-\alpha_1}$$

and its length is $(\sigma/\sqrt{N})(z_{\alpha_1} + z_{\alpha-\alpha_1})$. But it can be shown that minimum length is achieved when $z_{\alpha_1} = z_{\alpha-\alpha_1}$ or if $\alpha_1 = \frac{1}{2}\alpha$. The shortest interval with coefficient $1 - \alpha$ has the symmetric form

$$\bar{x} - \frac{\sigma}{\sqrt{N}} z_{\alpha/2} \leq \mu \leq \bar{x} + \frac{\sigma}{\sqrt{N}} z_{\alpha/2} \tag{32}$$

Similarly, if σ^2 is unknown, the $100(1 - \alpha)$ percent confidence interval for μ is given by

$$\bar{x} - \frac{s}{\sqrt{N}} t_{\alpha/2;N-1} \leq \mu \leq \bar{x} + \frac{s}{\sqrt{N}} t_{\alpha/2;N-1} \tag{33}$$

where s is the sample standard deviation and $t_{\alpha/2;N-1}$ is the upper 50α percentage point of the t distribution with $N - 1$ degrees of freedom.

Frequently it is necessary to obtain a confidence interval for the difference of the means of two normal populations with a common, though unknown, variance. The observations from the first population might have been collected as a control for those in the second sample that had been obtained under a new treatment or experimental condition. Under the normality and common-variance assumptions the $100(1 - \alpha)$ percent confidence interval for the change attributable to the experimental condition is

$$\bar{x} - \bar{y} - s_{\bar{x}-\bar{y}} t_{\alpha/2;N_1+N_2-2} \leq \mu_1 - \mu_2 \leq \bar{x} - \bar{y} + s_{\bar{x}-\bar{y}} t_{\alpha/2;N_1+N_2-2} \tag{34}$$

where $\bar{x}$, $\bar{y}$ are the respective means of the first and second samples of N_1 and N_2 observations and

$$s_{\bar{x}-\bar{y}} = \sqrt{\frac{\Sigma(x_i - \bar{x})^2 + \Sigma(y_i - \bar{y})^2}{N_1 + N_2 - 2}\left(\frac{1}{N_1} + \frac{1}{N_2}\right)} \tag{35}$$

is the usual within-sample estimate of the standard deviation of the mean difference.

Tests and Confidence Intervals for the Variance. The multivariate tests and confidence statements of the later chapters will generally be constructed for means and other location parameters. However, for the sake of completeness we shall touch upon some hypotheses and interval estimates for the variance of a normal population. If the observations

$x_1, \ldots, x_N$ constitute a random sample from $N(\mu,\sigma^2)$, the quantity

$$\frac{(N-1)s^2}{\sigma^2} = \frac{\sum_{i=1}^{N} (x_i - \bar{x})^2}{\sigma^2} \tag{36}$$

is distributed as a chi-squared variate with $N - 1$ degrees of freedom. The generalized likelihood-ratio criterion for the test of the hypothesis

$$H_0: \quad \sigma^2 = \sigma_0^2 \tag{37}$$

against the alternative

$$H_1: \quad \sigma^2 > \sigma_0^2 \tag{38}$$

specifies that the rejection region for a test of size α is

$$\frac{(N-1)s^2}{\sigma_0^2} > \chi^2_{\alpha;N-1} \tag{39}$$

where $\chi^2_{\alpha;N-1}$ is the upper 100α percentage point of the chi-squared distribution with $N - 1$ degrees of freedom. Similarly, the rejection regions for testing the null hypothesis against the alternatives H_1': $\sigma^2 < \sigma_0^2$ and H_1'': $\sigma^2 \neq \sigma_0^2$ would be

$$\frac{(N-1)s^2}{\sigma_0^2} < \chi^2_{1-\alpha;N-1} \tag{40}$$

and

$$\frac{(N-1)s^2}{\sigma_0^2} < \chi^2_{1-\alpha_1;N-1} \qquad \text{or} \qquad \frac{(N-1)s^2}{\sigma_0^2} > \chi^2_{\alpha_2;N-1} \tag{41}$$

respectively, where in the latter case $\alpha_1 + \alpha_2 = \alpha$. In the strict sense α_1 and α_2 should be chosen so that the latter test is *unbiased*, i.e., its power function is never less than its size α, but in most applications with moderate to large sample sizes an equal split will suffice. Confidence intervals for σ^2 can be obtained directly from the rejection region (41), and will be left as an exercise for the reader.

If independent samples of N_1 and N_2 observations have been randomly drawn from the populations $N(\mu_1,\sigma_1^2)$ and $N(\mu_2,\sigma_2^2)$, the hypothesis

$$H_0: \quad \sigma_1^2 = \sigma_2^2 \tag{42}$$

can be tested against the alternative

$$H_1: \quad \sigma_1^2 > \sigma_2^2 \tag{43}$$

by the statistic

$$F = \frac{s_1^2}{s_2^2} \tag{44}$$

If the null hypothesis is true, the statistic has the F distribution with degrees of freedom $N_1 - 1$, $N_2 - 1$, and the rejection region for a test of size α is

$$F > F_{\alpha;\, N_1-1, N_2-1} \tag{45}$$

Conversely, H_0 could be tested against the other one-sided alternative H_1': $\sigma_1^2 < \sigma_2^2$ by the statistic

$$F = \frac{s_2^2}{s_1^2} \tag{46}$$

whose critical region is

$$F > F_{\alpha;\, N_2-1, N_1-1} \tag{47}$$

Finally, for the two-sided alternative H_1'': $\sigma_1^2 \neq \sigma_2^2$ the statistic (44) would be used with the critical region defined by

$$F > F_{\alpha_1;\, N_1-1, N_2-1} \quad \text{or} \quad F < \frac{1}{F_{\alpha_2;\, N_2-1, N_1-1}} \tag{48}$$

where the partitioning $\alpha_1 = \alpha_2 = \frac{1}{2}\alpha$ will generally suffice.

Some Further Reading. The source and seed of much of modern statistical methodology can be found in Fisher's classics (1969, 1972). Snedecor and Cochran's text (1967) has a tutorial quality in its style and organization, and it treats basic methods in depth from the investigator's viewpoint. Dixon and Massey (1969) cover a wide variety of techniques. Hodges and Lehmann (1964) have integrated concepts of probability and statistical theory with many practical examples of estimation and testing. Illustrations of the normal-theory methods of this section can be found in the innumerable basic texts currently available, e.g., Freund (1971).

1.6 TESTING THE EQUALITY OF SEVERAL MEANS: THE ANALYSIS OF VARIANCE

Suppose that a certain biochemical compound is known to be taken up by the brain, although some evidence is available that the amount per gram of brain tissue appears to differ among the five strains of mice commonly used in one laboratory. It will be assumed that the relative amounts assayed from the brains of sacrificed mice are normally distributed with the same variance σ^2 for each strain. Let μ_j be the population mean for the jth strain. It is possible to construct a test of

$$H_0\colon\ \mu_1 = \cdots = \mu_5 \tag{1}$$

against the alternative that some means are different by the generalized likelihood-ratio criterion, and furthermore, if H_0 is rejected, methods can be obtained for making simultaneous tests on the mean differences with a fixed Type I error probability for all comparisons.

In the general case of k strains, treatments, diagnostic categories, or experimental conditions we begin by postulating that the ith observation on the jth *treatment* (our generic term for whatever feature distinguishes the k groups) can be expressed by the mathematical model

$$x_{ij} = \mu + \tau_j + e_{ij} \qquad i = 1, \ldots, N_j, j = 1, \ldots, k \tag{2}$$

where μ = location parameter common to all observations
τ_j = effect peculiar to jth treatment
e_{ij} = normally distributed random variable with mean zero and variance σ^2

The variate terms e_{ij} are distributed independently of one another. The model is said to be linear or additive, for application of the jth treatment increases μ by the increment τ_j, and the random disturbance also merely adds or subtracts some amount from the parameters. We note that in the simpler usage of the motivating example $\mu_j \equiv \mu + \tau_j$. The hypothesis of equal treatment effects can be written as

$$H_0\colon \tau_1 = \cdots = \tau_k \tag{3}$$

and we shall take as the alternative to H_0 the general model (2) for the observations.

This is the simplest example of the *general linear model* underlying statistical experimental design. The theory of estimation and hypothesis testing in univariate linear models has been discussed in many texts; those by Scheffé (1959), Graybill (1961), and Searle (1971) are especially suitable for the theoretical background of this section and as an introduction to the treatment of multivariate models in Chap. 5.

The model is said to have *fixed* effects, for the μ_j are parameters, and any inferences from the observations can be made only with respect to the particular k treatments in the study. If the treatments constituted a random sample from a larger population (available laboratory technicians, experiments replicated at different times, or clinicians scoring projective tests), the treatment effects would be random variables, and a somewhat different approach would have to be employed in the analysis. This fundamental distinction between the fixed (model I) and random (model II) analysis-of-variance models was first made by Eisenhart (1947) and has been developed extensively by many other workers in experimental design. In the sequel we shall use multivariate techniques for the exact analysis of the *mixed model* for a fixed number of treatments applied repeatedly to each member of a random sample of experimental units.

The observations can be arranged as in the following table:

Treatment		
1	...	k
x_{11}	...	x_{1k}
.....	.....	.....
$x_{N_1 1}$	...	$x_{N_k k}$

Denote the total of the observations for the jth treatment by

$$T_j = \sum_{i=1}^{N_j} x_{ij} \tag{4}$$

and the mean of that treatment by $\bar{x}_j = T_j/N_j$. The sum of all observations will be denoted by

$$G = T_1 + \cdots + T_k \tag{5}$$

the total number of experimental units by

$$N = N_1 + \cdots + N_k \tag{6}$$

and the grand mean by $\bar{x} = G/N$. It is possible to write the total sum of squares

$$\begin{aligned} \mathrm{S} &= \sum_{j=1}^{k} \sum_{i=1}^{N_j} (x_{ij} - \bar{x})^2 \\ &= \sum_{j=1}^{k} \sum_{i=1}^{N_j} x_{ij}^2 - \frac{G^2}{N} \end{aligned} \tag{7}$$

as the sum of two independent components

$$\begin{aligned} \mathrm{SST} &= \sum_{j=1}^{k} N_j \left(\frac{T_j}{N_j} - \bar{x}\right)^2 \\ &= \sum_{j=1}^{k} \frac{T_j^2}{N_j} - \frac{G^2}{N} \end{aligned} \tag{8}$$

and

(9)
$$\begin{aligned}\text{SSE} &= \sum_{j=1}^{k} \sum_{i=1}^{N_j} (x_{ij} - \bar{x}_j)^2 \\ &= \sum_{j=1}^{k} \sum_{i=1}^{N_j} x_{ij}^2 - \sum_{j=1}^{k} \frac{T_j^2}{N_j}\end{aligned}$$

These components can be summarized in the *analysis-of-variance table* shown in Table 1.1. The statistic for the generalized likelihood-ratio test of H_0 is

(10)
$$F = \frac{N-k}{k-1} \frac{\text{SST}}{\text{SSE}}$$

and since when H_0 is true, $\text{SST}/(k-1)\sigma^2$, $\text{SSE}/(N-k)\sigma^2$ are independent chi-squared variates with $k-1$ and $N-k$ degrees of freedom, respectively, the test statistic has the F distribution with $k-1$ and

Table 1.1. **Analysis of Variance**

Source	Sum of squares	Degrees of freedom	Mean square
Treatments	SST	$k-1$	$\frac{\text{SST}}{k-1}$
Within treatments (error)	SSE	$N-k$	$\frac{\text{SSE}}{N-k}$
Total	S	$N-1$	

$N-k$ degrees of freedom. We reject H_0 with a test of level α if

(11)
$$F > F_{\alpha; k-1, N-k}$$

The power function for this analysis-of-variance test can be computed from the Pearson-Hartley noncentral F-distribution charts of the Appendix. The degrees-of-freedom parameters are $\nu_1 = k-1$ and $\nu_2 = N-k$, and the noncentrality parameter measuring the departure of the population means of the treatments from the null hypothesis (3) is

(12)
$$\phi = \frac{\sqrt{\sum_{j=1}^{k} N_j \left(\tau_j - \frac{\Sigma N_j \tau_j}{N}\right)^2}}{\sigma \sqrt{k}}$$

or, in the case of equal treatment samples usually encountered in experimental design problems,

$$\phi = \frac{\sqrt{N_j \sum_{j=1}^{k} \tau_j^2}}{\sigma \sqrt{k}} \tag{13}$$

In the latter expression the correction term has vanished from the usual restraint $\tau_1 + \cdots + \tau_k = 0$ imposed on the treatment effects. Illustrations of the use of ϕ and the power charts for selecting sample sizes for the one-way design can be found in Scheffé's text (1959, chap. 3).

Multiple Comparisons of Treatments. An analysis of variance culminating in rejection of the hypothesis of equal treatment effects still has not indicated those effects which may be equal or those which are probably different. This is the problem of *multiple comparisons,* or simultaneous inferences about the members of some family of hypotheses. The tests are constructed so that the Type I error probability for the entire family will be at most α. For an excellent treatment of the theory and methods of multiple comparisons we refer the reader to Miller's monograph (1966). Two methods will be needed frequently in the later chapters, and we shall introduce them now in the context of the one-way analysis of variance.

The first technique is due to Scheffé (1953, 1959). Define a *contrast* of the parameters τ_j of the one-way model as any linear function

$$\sum_{j=1}^{k} c_j \tau_j \tag{14}$$

whose coefficients have the property

$$\sum_{j=1}^{k} c_j = 0 \tag{15}$$

Thus $\tau_1 - \tau_2$ and $3\tau_1 - (\tau_2 + \tau_3 + \tau_4)$ are contrasts, while $\tau_2 - (\tau_3 + \tau_4)$ is not. In the particular case of the one-way analysis-of-variance model Scheffé has shown that the simultaneous confidence intervals with joint coefficient $1 - \alpha$ for all contrasts of the τ_j have the form

$$\begin{aligned} \sum c_j \bar{x}_j - s \sqrt{(k-1)F_{\alpha;k-1,N-k} \sum \frac{c_j^2}{N_j}} &\le \sum c_j \tau_j \\ &\le \sum c_j \bar{x}_j + s \sqrt{(k-1)F_{\alpha;k-1,N-k} \sum \frac{c_j^2}{N_j}} \end{aligned} \tag{16}$$

where all summations are over the k treatments. $\Sigma c_j \bar{x}_j$ is the sample estimate of the contrast $\Sigma c_j \tau_j$, and

(17)
$$s = \sqrt{\frac{\text{SSE}}{N - k}}$$

Note that $s^2\Sigma c_j^2/N_j$ is the estimate of the variance of the estimated contrast. We accept the null hypothesis

(18)
$$H_0\colon \sum_{j=1}^{k} c_j\tau_j = 0$$

at the α level if the simultaneous confidence interval for that contrast includes the value zero. If, on the other hand

$$\sum c_j\bar{x}_j > s\sqrt{(k-1)F_{\alpha;k-1,N-k}\sum \frac{c_j^2}{N_j}}$$

or

$$\sum c_j\bar{x}_j < -s\sqrt{(k-1)F_{\alpha;k-1,N-k}\sum \frac{c_j^2}{N_j}}$$

we reject H_0 in favor of the respective one-sided alternatives $H_1\colon\ \Sigma c_j\tau_j > 0$ or $H_1'\colon\ \Sigma c_j\tau_j < 0$. The joint level of *all* such tests is α.

The second multiple-comparison technique is called the Bonferroni method, for it is based on an inequality bearing that name. For this procedure we begin by restricting our attention to a family of m confidence-interval statements $H_1, \ldots, H_m$ about the parameters of the linear model. The probability that H_i is a true statement (that is, H_i covers the value of the ith parametric function) is $P(H_i)$, and the probability that all statements are simultaneously correct is $P(H_1 \cap \cdots \cap H_m)$, where the intersection notation $H_i \cap H_j$ denotes the event "both H_i and H_j are correct." This probability is the simultaneous confidence coefficient for the entire family of intervals, and we should like it to be at least equal to some specified value $1 - \alpha$, where α will be called the *error rate for the family of statements.* The calculation of the joint probability is often difficult for practical statistical problems, and its value may depend on unknown "nuisance parameters" measuring the intercorrelations of the H_i statements. Instead we usually must be content with a lower bound

(19)
$$P(H_1 \cap \cdots \cap H_m) \geq 1 - \sum_{i=1}^{m} P(\bar{H}_i)\cdot$$

where $P(\bar{H}_i) = 1 - P(H_i)$ is the probability that the ith individual statement is not true. The bound is a simple example of the Bonferroni inequalities (Miller, 1966, pp. 7–8; Feller, 1968, chap. 4; David, 1956). For a set of m simultaneous confidence intervals we usually assign each statement an error rate of α/m, so that the coefficient for the family is at least $1 - \alpha$.

The Bonferroni confidence intervals on m given contrasts

$$\Psi_i = \sum_{j=1}^{k} c_{ij}\tau_j \qquad i = 1, \ldots, m \tag{20}$$

of the one-way linear model parameters are

$$\sum_{j=1}^{k} c_{ij}\bar{x}_j - t_{\alpha/(2m);N-k}s\sqrt{\sum_{j=1}^{k}\frac{c_{ij}^2}{N_j}} \leq \Psi_i \leq \sum_{j=1}^{k} c_{ij}\bar{x}_j + t_{\alpha/(2m);N-k}s\sqrt{\sum_{j=1}^{k}\frac{c_{ij}^2}{N_j}} \tag{21}$$

The decision rule for testing H_0: $\Psi_i = 0$ is equivalent to that for the Scheffé method: if (21) does not enclose the value zero, H_0 is rejected. For all paired comparisons of the treatments, $m = \frac{1}{2}k(k - 1)$, while for successive comparisons (assuming some *a priori* ordering), $m = k - 1$. If m is small the Bonferroni intervals may be shorter on the average than those of the Scheffé technique, even though the true family confidence coefficient is greater than the nominal value $1 - \alpha$. For very large m the Bonferroni intervals will be impractically long.

A third method of multiple comparisons ascribed to Tukey (1953) is also in common use for simple experimental designs, but since it has no multivariate generalization or essential application in the later chapters we shall not discuss it here. The reader is referred to Miller (1966) or any thorough text on the analysis of variance.

Example 1.3. In a preliminary evaluation of three tranquilizing drugs time limitations and the possibility of residual effects decreed that each subject could receive only one drug. Eighteen psychiatric patients with similar diagnoses were rated with respect to anxiety on a seven-point scale. Six were randomly assigned to each of the three drugs, and after several days each patient was rated blindly on the same scale. These changes in anxiety ratings were observed:

	Drug		
	A	*B*	*C*
	4	0	1
	5	2	0
	3	1	0
	3	2	1
	4	2	2
	2	2	2
Mean	3.5	1.5	1.0

The pertinent sums are

$$T_1 = 21 \qquad T_2 = 9 \qquad T_3 = 6 \qquad G = 36 \qquad \sum_{j=1}^{3}\sum_{i=1}^{6} x_{ij}^2 = 106$$

The sum of squares due to drugs is

$$\begin{aligned} \text{SST} &= \tfrac{1}{6}(21^2 + 9^2 + 6^2) - \frac{36^2}{18} \\ &= 21 \end{aligned}$$

and the total sum of squares is S = 106 − 72 = 34. The within-drugs, or error, sum of squares follows by subtraction. These values are summarized in the analysis-of-variance table:

Source	Sum of squares	Degrees of freedom	Mean square	F
Drugs	21	2	10.5	12.1
Within drugs	13	15	0.867	
Total	34	17		

Since $F_{0.01;2,15} = 6.36$, we conclude that the hypothesis of equal drug effects is not tenable at the 1 percent level. Further consultation of the F-distribution tables reveals that the observed F also exceeds the critical value for $\alpha = 0.001$.

Now we shall use the Scheffé multiple-comparison procedure to determine which drugs are different. A simultaneous confidence coefficient of 0.99 will be chosen, and thus

$$\sqrt{(k-1)F_{0.01;\,2,15}} = 3.567$$

First compare drugs B and C by computing the confidence interval for $\tau_2 - \tau_3$. Here $c_1 = 0, c_2 = 1, c_3 = -1$, and the estimated population standard deviation of $\bar{x}_2 - \bar{x}_3$ is

$$\sqrt{0.867(\tfrac{1}{6} + \tfrac{1}{6})} = 0.5376$$

The confidence interval is

$$-1.42 \le \tau_2 - \tau_3 \le 2.42$$

and since this extends well across the zero value, we conclude that the hypothesis of equal drug B and C effects cannot be rejected at the 0.01 level.

Similarly, the 99 percent simultaneous confidence interval for the A and B effect difference is

$$0.08 \le \tau_1 - \tau_2 \le 3.92$$

and it is possible to accept the alternative hypothesis

$$H_1\colon \quad \tau_1 > \tau_2$$

at the 0.01 level. It might also be of interest to determine whether drug A is different from the average effect of drugs B and C. Here $c_1 = 1$, $c_2 = -\tfrac{1}{2}$, $c_3 = -\tfrac{1}{2}$, and the estimated standard deviation of the estimate of that contrast is

$$\sqrt{0.867(\tfrac{1}{6} + \tfrac{1}{24} + \tfrac{1}{24})} = 0.4653$$

The estimate of the contrast is of course 2.25, and the confidence interval is

$$0.59 \leq \tau_1 - \tfrac{1}{2}(\tau_2 + \tau_3) \leq 3.91$$

Drug A appears to be distinct from the essentially equivalent remaining drugs.

The Bonferroni simultaneous confidence intervals for the pair-wise contrasts with confidence coefficient at least 0.99 are

$$0.13 \leq \tau_1 - \tau_2 \leq 3.87$$

$$-1.37 \leq \tau_2 - \tau_3 \leq 2.37$$

$$0.63 \leq \tau_1 - \tau_3 \leq 4.37$$

2
MATRIX ALGEBRA

2.1 INTRODUCTION. In Chap. 3 we shall see that a multidimensional random variable is merely an ordered collection

$$[X_1, \ldots, X_p]$$

of single variates. By "ordered" we mean that the variate describing the ith facet of each sampling unit drawn from the population always appears in the ith position of the sequence. The number of variates p in the array is always specified and will remain unchanged throughout the problem or analysis at hand. For example, suppose that the components of the variate are the weights of female rats in a particular strain at birth and at 10, 20, and 30 days of age. Then the weights could be described by the random variable $[X_1,X_2,X_3,X_4]$ with some distribution in four-dimensional space. If those weights were recorded in a sample of N rats, the observations might be summarized in the array

$$\begin{bmatrix} x_{11} & \cdots & x_{14} \\ \cdot & \cdots & \cdot \\ x_{N1} & \cdots & x_{N4} \end{bmatrix}$$

whose rows correspond to different rats. Such linear and rectangular arrangements of numbers are known respectively as *vectors* and *matrices*, and rules for their manipulation constitute that part of linear systems known as *matrix algebra*. That algebra is the virtual language of multivariate analysis, particularly in its most common and important part based upon the multidimensional normal distribution. Indeed, it is almost inconceivable that the techniques could have been developed without the convenience, facility, and elegance of matrices.

In this chapter we shall summarize a number of properties, operations, and theorems of matrix algebra needed in the sequel. Further results and proofs of certain of the theorems can be found in the references cited at the end of the chapter.

2.2 SOME DEFINITIONS

Let us assume that we have at our disposal some elements which behave according to certain sets of axioms, for example, the real or complex numbers. We define a *matrix*

$$\mathbf{A} = \begin{bmatrix} a_{11} & \cdots & a_{1c} \\ \cdots & \cdots & \cdots \\ a_{r1} & \cdots & a_{rc} \end{bmatrix} \tag{1}$$

as a rectangular ordered array of the elements. The general term of the matrix will be written as a_{ij}, where the first subscript will always refer to the ith row, and the second to the jth column. The dimensions of a matrix are important, and a matrix with r rows and c columns will be referred to as $r \times c$.

In opposition to a matrix we shall call the usual numbers and variables of everyday unidimensional transactions *scalars*. A scalar is of course a 1×1 matrix. In the initial sections of this chapter we shall think of matrices as composed of real scalar elements. Later we shall treat matrices whose elements are themselves matrices of smaller dimensions.

A *vector* is a matrix with a single row or column. We shall customarily write the n-component *column vector*

$$\mathbf{x} = \begin{bmatrix} x_1 \\ \cdot \\ \cdot \\ \cdot \\ x_n \end{bmatrix} \tag{2}$$

in lowercase boldface type. Similarly, a *row vector*

$$\mathbf{x}' = [x_1, \ . \ . \ . \ , x_n] \tag{3}$$

consists of a single row of n elements. Either vector specifies the coordinates of a point in n-dimensional euclidean space, and the connection with the physical or analytical notion of a vector is immediately apparent if we think of that point as the terminus of a line segment starting from the origin of the coordinate axes.

The prime attached to a row vector means that $\mathbf{x}'$ is the *transpose* of the column vector $\mathbf{x}$. In general, if $\mathbf{A}$ is any $r \times c$ matrix, the transpose of $\mathbf{A}$ is the $c \times r$ matrix $\mathbf{A}'$ formed by interchanging the roles of rows and columns:

$$\mathbf{A}' = \begin{bmatrix} a_{11} & \cdots & a_{1c} \\ \cdot & \cdots & \cdot \\ a_{r1} & \cdots & a_{rc} \end{bmatrix}' = \begin{bmatrix} a_{11} & \cdots & a_{r1} \\ \cdot & \cdots & \cdot \\ a_{1c} & \cdots & a_{rc} \end{bmatrix} \tag{4}$$

If a matrix is square and equal to its transpose, it is said to be *symmetric*. Then $a_{ij} = a_{ji}$ for all pairs of i and j. For example,

$$\mathbf{A} = \begin{bmatrix} 3 & 0 & -1 \\ 0 & 1 & 2 \\ -1 & 2 & -4 \end{bmatrix}$$

is symmetric, while

$$\mathbf{B} = \begin{bmatrix} 2 & 3 \\ 1 & 2 \end{bmatrix}$$

is not. For brevity we shall frequently omit the duplicated lower elements. The elements a_{ii} of a square matrix occupy what are called the *main diagonal positions*. The sum of these diagonal elements is called the *trace* of $\mathbf{A}$, and will be denoted by

$$\operatorname{tr} \mathbf{A} = \sum_{i=1}^{n} a_{ii} \tag{5}$$

Certain matrices are particularly important. The *identity* matrix

$$\mathbf{I} = \begin{bmatrix} 1 & \cdots & 0 \\ \cdot & \cdots & \cdot \\ 0 & \cdots & 1 \end{bmatrix} \tag{6}$$

is a square matrix with one in each main diagonal position and zeros elsewhere. The $p \times p$ *diagonal matrix*

$$\mathbf{D}(a_i) = \begin{bmatrix} a_1 & \cdots & 0 \\ \cdot & \cdots & \cdot \\ 0 & \cdots & a_p \end{bmatrix} \tag{7}$$

has the elements $a_1, \ldots, a_p$ in its main diagonal positions and zeros in all other locations. Some of the a_i may be zero. In the sequel $\operatorname{diag}(\mathbf{A}) = \mathbf{D}(a_{ii})$ will denote the diagonal matrix formed from the square matrix $\mathbf{A}$. A $p \times p$ *triangular* matrix has the pattern

$$\begin{bmatrix} t_{11} & t_{12} & \cdots & t_{1p} \\ 0 & t_{22} & \cdots & t_{2p} \\ \cdot & \cdot & \cdots & \cdot \\ 0 & 0 & \cdots & t_{pp} \end{bmatrix} \tag{8}$$

The $r \times c$ *null matrix*

$$(9) \qquad \mathbf{0} = \begin{bmatrix} 0 & \cdots & 0 \\ \cdot & \cdots & \cdot \\ 0 & \cdots & 0 \end{bmatrix}$$

has zero in each of its positions. Occasionally we shall also need the vector

$$(10) \qquad \mathbf{j}' = [1, \ldots, 1]$$

and the matrix

$$(11) \qquad \mathbf{E} = \begin{bmatrix} 1 & \cdots & 1 \\ \cdot & \cdots & \cdot \\ 1 & \cdots & 1 \end{bmatrix}$$

with unity in every position.

2.3 ELEMENTARY OPERATIONS WITH MATRICES AND VECTORS

The operations of addition, subtraction, and multiplication of ordinary scalar arithmetic can be carried over to matrices if certain rules are followed. The matrix analogue of division is a bit more complicated and will be deferred to a later section.

Equality. Two $r \times c$ matrices $\mathbf{A}$ and $\mathbf{B}$ are said to be equal if and only if

$$(1) \qquad a_{ij} = b_{ij}$$

for all pairs of i and j.

Addition. The sum of two matrices of like dimensions is the matrix of the sums of the corresponding elements. If

$$\mathbf{A} = \begin{bmatrix} a_{11} & \cdots & a_{1c} \\ \cdot & \cdots & \cdot \\ a_{r1} & \cdots & a_{rc} \end{bmatrix} \qquad \mathbf{B} = \begin{bmatrix} b_{11} & \cdots & b_{1c} \\ \cdot & \cdots & \cdot \\ b_{r1} & \cdots & b_{rc} \end{bmatrix}$$

then

$$(2) \qquad \mathbf{A} + \mathbf{B} = \begin{bmatrix} a_{11} + b_{11} & \cdots & a_{1c} + b_{1c} \\ \cdot & \cdots & \cdot \\ a_{r1} + b_{r1} & \cdots & a_{rc} + b_{rc} \end{bmatrix}$$

One matrix is subtracted from another of like dimensions by forming the matrix of differences of the individual elements. Thus,

$$
\begin{aligned}
\mathbf{A} + \mathbf{B} &= \mathbf{B} + \mathbf{A} \\
\mathbf{A} + (\mathbf{B} + \mathbf{C}) &= (\mathbf{A} + \mathbf{B}) + \mathbf{C} \qquad (3) \\
\mathbf{A} - (\mathbf{B} - \mathbf{C}) &= \mathbf{A} - \mathbf{B} + \mathbf{C}
\end{aligned}
$$

If the dimensions of the matrices do not conform, their sums or differences are undefined.

Multiplication by a Scalar. The matrix **A** is multiplied by the scalar c by multiplying each element of **A** by c:

(4)
$$c\mathbf{A} = \begin{bmatrix} ca_{11} & \cdots & ca_{1k} \\ \cdot & \cdots\cdots & \cdot \\ ca_{r1} & \cdots & ca_{rk} \end{bmatrix}$$

Matrix Multiplication. For the matrix product **AB** to be defined it is necessary that the number of *columns* of **A** be equal to the number of *rows* of **B**. The dimensions of such matrices are said to be *conformable.* If **A** is of dimensions $p \times r$ and **B** is $r \times q$, then the ijth element of the product $\mathbf{C} = \mathbf{AB}$ is computed as

(5)
$$c_{ij} = \sum_{k=1}^{r} a_{ik}b_{kj}$$

This is the sum of the products of corresponding elements in the ith row of **A** and jth column of **B**. The dimensions of **AB** are of course $p \times q$. For example, if

$$\mathbf{A} = \begin{bmatrix} 1 & 2 & 3 \\ -1 & 0 & 1 \end{bmatrix} \qquad \mathbf{B} = \begin{bmatrix} 6 & 5 & 4 \\ -1 & 1 & -1 \\ 0 & 2 & 0 \end{bmatrix}$$

then

$$
\begin{aligned}
\mathbf{AB} &= \begin{bmatrix} 1(6) + 2(-1) + 3(0) & 1(5) + 2(1) + 3(2) & 1(4) + 2(-1) + 3(0) \\ -1(6) + 0(-1) + 1(0) & -1(5) + 0(1) + 1(2) & -1(4) + 0(-1) + 1(0) \end{bmatrix} \\
&= \begin{bmatrix} 4 & 13 & 2 \\ -6 & -3 & -4 \end{bmatrix}
\end{aligned}
$$

The product **BA** is undefined, for the two rows of **A** do not conform with the three columns of **B**.

The distributive and associative laws hold for matrix multiplication:

$$
\begin{aligned}
\mathbf{A}(\mathbf{B} + \mathbf{C}) &= \mathbf{AB} + \mathbf{AC} \qquad (6) \\
\mathbf{A}(\mathbf{BC}) &= \mathbf{AB}(\mathbf{C})
\end{aligned}
$$

However, the commutative law does *not* hold for matrix multiplication, and in general it is not true that $\mathbf{AB} = \mathbf{BA}$. For this reason the order of multiplication is crucial, and we shall speak of the product $\mathbf{AB}$ as formed from *premultiplication* of $\mathbf{B}$ by $\mathbf{A}$ or by *postmultiplication* of $\mathbf{A}$ by $\mathbf{B}$. For example, let

$$\mathbf{A} = \begin{bmatrix} 1 & 2 \\ 2 & 3 \end{bmatrix} \qquad \mathbf{B} = \begin{bmatrix} 3 & -1 \\ -1 & 1 \end{bmatrix}$$

Then

$$\mathbf{AB} = \begin{bmatrix} 1 & 1 \\ 3 & 1 \end{bmatrix} \qquad \mathbf{BA} = \begin{bmatrix} 1 & 3 \\ 1 & 1 \end{bmatrix}$$

Neither are the products equal, nor is the symmetry of the original matrices preserved in the multiplication.

Multiplication of any matrix by a conformable identity matrix leaves the matrix unchanged. Premultiplication by the diagonal matrix with elements $d_1, \ldots, d_r$ has the effect of multiplying each element in the ith row by d_i:

$$\mathbf{D}(d_i)\mathbf{A} = \begin{bmatrix} d_1a_{11} & \cdots & d_1a_{1c} \\ \cdot & \cdots & \cdot \\ d_ra_{r1} & \cdots & d_ra_{rc} \end{bmatrix} \tag{7}$$

Postmultiplication by a similar $c \times c$ diagonal matrix multiplies each element in the jth column by d_j.

A matrix can be regarded as specifying a linear transformation of the vectors in one space to those of another. If $\mathbf{x}$ has m components and $\mathbf{y}$ has n components, it is possible to express a transformation from the m-dimensional coordinate system of the elements of $\mathbf{x}$ to the n-dimensional space of those of $\mathbf{y}$ in matrix form as

$$\begin{bmatrix} y_1 \\ \cdot \\ \cdot \\ \cdot \\ y_n \end{bmatrix} = \begin{bmatrix} a_{11} & \cdots & a_{1m} \\ \cdot & \cdots & \cdot \\ a_{n1} & \cdots & a_{nm} \end{bmatrix} \begin{bmatrix} x_1 \\ \cdot \\ \cdot \\ \cdot \\ x_m \end{bmatrix} = \mathbf{Ax} \tag{8}$$

Transformation to a third set of variables specified by the p-component column vector $\mathbf{z}$ could be represented as

$$\begin{aligned} \mathbf{z} &= \mathbf{By} \\ &= \mathbf{BAx} \end{aligned} \tag{9}$$

and thus the product of two or more matrices can be thought of as the matrix of the resultant of a succession of linear transformations.

Example 2.1. It is often necessary in the social sciences to convert several disparate scores collected on individuals to a scale with a common origin and unit. If x_{ij} is the score of the ith individual on the jth measure and $\bar{x}_j$ and s_j are the sample mean and standard deviation of that measure, one common transformation is the *z score*

$$z_{ij} = \frac{x_{ij} - \bar{x}_j}{s_j}$$

The transformed observations can be computed conveniently by some simple matrix operations. Write the original scores as the $N \times p$ *data matrix*

$$\mathbf{X} = \begin{bmatrix} x_{11} & \cdots & x_{1p} \\ \cdots & \cdots & \cdots \\ x_{N1} & \cdots & x_{Np} \end{bmatrix}$$

and form the diagonal matrix

$$\mathbf{D} = \begin{bmatrix} \frac{1}{s_1} & \cdots & 0 \\ \cdots & \cdots & \cdots \\ 0 & \cdots & \frac{1}{s_p} \end{bmatrix}$$

from the sample standard deviations. If we introduce the $N \times N$ matrix

$$\mathbf{E} = \begin{bmatrix} 1 & \cdots & 1 \\ \cdots & \cdots & \cdots \\ 1 & \cdots & 1 \end{bmatrix}$$

with one in every position, the $N \times p$ matrix $\mathbf{Z}$ of standard scores can be computed as

$$\mathbf{Z} = \left(\mathbf{I} - \frac{1}{N}\mathbf{E}\right)\mathbf{XD}$$

Postmultiplication of $\mathbf{E}$ by $\mathbf{X}$ has the effect of summing each column of the data matrix.

Vector Inner Products. The inner product of two vectors with the same number of elements is defined to be the sum of the products of the corresponding elements:

$$\text{(10)} \qquad \mathbf{x}'\mathbf{y} = [x_1, \ldots, x_p]\begin{bmatrix} y_1 \\ \cdot \\ \cdot \\ \cdot \\ y_p \end{bmatrix} = \sum_{i=1}^{p} x_i y_i$$

Since the inner product is a scalar, $\mathbf{x}'\mathbf{y} = \mathbf{y}'\mathbf{x}$. The inner product of $\mathbf{x}$ with itself is the sum of squares of the elements of $\mathbf{x}$.

Inner products have important geometrical interpretations. The inner product of $\mathbf{x}$ with itself is called the *squared length* of $\mathbf{x}$, for it is the square of the distance from the origin of the p-dimensional coordinate system to the point specified by the elements of $\mathbf{x}$. More generally, the distance between the points with coordinates given by $\mathbf{x}$ and $\mathbf{y}$ is

$$d = \left[\sum_{i=1}^{p} (x_i - y_i)^2\right]^{1/2} \tag{11}$$

The cosine of the angle θ between the vectors $\mathbf{x}$ and $\mathbf{y}$ is

$$\cos\theta = \frac{\mathbf{x}'\mathbf{y}}{(\mathbf{x}'\mathbf{x})^{1/2}(\mathbf{y}'\mathbf{y})^{1/2}} \tag{12}$$

Such division of the vectors by their respective lengths is called *normalization*, and it is easy to see that as the normalized vectors become coincident, their inner product tends to unity. Similarly, vectors at right angles to each other have an inner product of zero.

Example 2.2. In three-dimensional space the vectors

$$\mathbf{x}' = [1,0,0]$$
$$\mathbf{y}' = [0,1,0]$$
$$\mathbf{z}' = [0,0,1]$$

can be construed as specifying the three coordinate axes. The vector

$$\mathbf{e}' = [1,1,1]$$

makes an angle with the same cosine $\sqrt{3}/3$ with $\mathbf{x}'$, $\mathbf{y}'$, and $\mathbf{z}'$ and characterizes the *equiangular* line in three-space. The numbers $\sqrt{3}/3$ are called the *direction cosines* of the line. Similarly, the vectors

$$\mathbf{u}' = [1,1,0]$$
$$\mathbf{v}' = [1,\ -1,\ 0]$$

make an angle of 90° in the xy plane of the space. $\mathbf{u}'$ and $\mathbf{v}'$ each have an angle of 45° with $\mathbf{x}'$, and their angles with $\mathbf{y}'$ are 45 and −45°, respectively.

Transpose of a Matrix Product. If the matrix $\mathbf{A}$ is conformable for postmultiplication by another matrix $\mathbf{B}$, it is easily verified that the transpose of their product is equal to the product of their transposes taken in the opposite order:

$$(\mathbf{AB})' = \mathbf{B}'\mathbf{A}' \tag{13}$$

More generally, if $\mathbf{A}_1, \ldots, \mathbf{A}_k$ are conformable,

$$(\mathbf{A}_1 \cdots \mathbf{A}_k)' = \mathbf{A}_k' \cdots \mathbf{A}_1' \tag{14}$$

We shall need these properties frequently in the later chapters.

2.4 THE DETERMINANT OF A SQUARE MATRIX

Associated with every square matrix is a unique scalar number called its *determinant.* The formal definition of the determinant of the $n \times n$ matrix $\mathbf{A}$ is the sum

$$\sum \cdots_{P} \sum (-1)^{\alpha} a_{1j_1} a_{2j_2} \cdots a_{nj_n} \tag{1}$$

of all products consisting of one element from each row and column and multiplied by -1 if the number of inversions of the particular permutation $j_1 j_2 \cdots j_n$ from the standard order $1, 2, \ldots, n$ is odd. The sum is taken over the set P of all $n!$ permutations of the column subscripts. The number of inversions α in a particular permutation is the total number of times in which an element is followed by numbers which would ordinarily precede it in the standard order $1, 2, \ldots, n$.

The determinant of $\mathbf{A}$ will be written as $|\mathbf{A}|$. The determinants of the three smallest square matrices follow from the formal definition as

$$\begin{aligned}
|a_{11}| &= a_{11} \\
\begin{vmatrix} a_{11} & a_{12} \\ a_{21} & a_{22} \end{vmatrix} &= a_{11}a_{22} - a_{12}a_{21} \\
\begin{vmatrix} a_{11} & a_{12} & a_{13} \\ a_{21} & a_{22} & a_{23} \\ a_{31} & a_{32} & a_{33} \end{vmatrix} &= \begin{aligned} & a_{11}a_{22}a_{33} + a_{12}a_{23}a_{31} + a_{13}a_{21}a_{32} \\ & -a_{13}a_{22}a_{31} - a_{11}a_{23}a_{32} - a_{12}a_{21}a_{33} \end{aligned}
\end{aligned} \tag{2}$$

It is more convenient to compute the determinants of larger matrices by different methods. Define the *minor* of the element a_{ij} of $\mathbf{A}$ as the determinant of the matrix formed by deleting the ith row and jth column of $\mathbf{A}$. The *cofactor* of a_{ij} is the minor multiplied by $(-1)^{i+j}$ and will be written as A_{ij}. It can be shown that the determinant of the square matrix $\mathbf{A}$ can be expressed in terms of the cofactors of the elements of any given row or column as

$$\begin{aligned}
|\mathbf{A}| &= a_{i1}A_{i1} + \cdots + a_{in}A_{in} \qquad i = 1, \ldots, n \\
&= a_{1j}A_{1j} + \cdots + a_{nj}A_{nj} \qquad j = 1, \ldots, n
\end{aligned} \tag{3}$$

For the simplest application of this rule we note that the matrix of cofactors for the general 2×2 matrix is

$$\begin{bmatrix} a_{22} & -a_{21} \\ -a_{12} & a_{11} \end{bmatrix} \tag{4}$$

and the value of the determinant follows immediately.

Example 2.3. Let us evaluate the determinant of the matrix

$$\begin{bmatrix} 3 & 0 & 0 & 0 \\ 1 & 2 & 0 & 1 \\ -1 & -3 & 2 & -1 \\ 5 & 4 & 3 & 2 \end{bmatrix}$$

Note immediately that the first row contains a single nonzero element, so that it will be necessary to compute the cofactor only of the (1,1) element. If we expand that cofactor in terms of the cofactors of its first row, the original determinant is equal to

$$3\left(2\begin{vmatrix} 2 & -1 \\ 3 & 2 \end{vmatrix} - 0\begin{vmatrix} -3 & -1 \\ 4 & 2 \end{vmatrix} + 1\begin{vmatrix} -3 & 2 \\ 4 & 3 \end{vmatrix}\right) = -9$$

The method of cofactors is efficient only for small matrices or for patterned matrices with an abundance of zero elements. We shall consider a more practical technique in Sec. 2.6.

Certain properties of determinants will prove to be useful in the later chapters:

1. The determinant of a diagonal matrix is merely the product of the diagonal elements. Similarly, the determinant of a triangular matrix is the product of its diagonal elements.
2. If the elements of a single row or column of the $n \times n$ matrix $\mathbf{A}$ are multiplied by the scalar c, the determinant of the new matrix is equal to $c|\mathbf{A}|$. If every element is multiplied by c, then $|c\mathbf{A}| = c^n|\mathbf{A}|$.
3. If two columns (or rows) of a square matrix are interchanged, the sign of the determinant is reversed.
4. It follows directly from Property 3 that if two columns or two rows of a matrix are equal, the determinant must be zero. Thus, proportional rows or columns of a matrix indicate a determinant of zero.
5. The determinant of a matrix is unchanged by adding a multiple of some column to another column. A similar result holds for rows.
6. If all elements of a row or column of a square matrix are zero, the determinant is zero.
7. If $\mathbf{A}$ and $\mathbf{B}$ are each $n \times n$ matrices, the determinant of $\mathbf{AB}$ is equal to the product of the individual determinants.
8. The sum of the products of the elements of a given row of a square matrix with the corresponding cofactors of a *different* row is equal to zero. A similar result holds for columns.

2.5 THE INVERSE MATRIX

We are now ready to define the matrix analogue of scalar division. The *inverse* of the square matrix $\mathbf{A}$ is that unique matrix $\mathbf{A}^{-1}$ with elements such that

$$\mathbf{AA}^{-1} = \mathbf{A}^{-1}\mathbf{A} = \mathbf{I} \tag{1}$$

It is possible that $\mathbf{A}^{-1}$ does not exist, just as it is not possible to perform scalar division by zero. Then $\mathbf{A}$ is said to be *singular*. Matrices whose inverses exist are called *nonsingular*.

The elements of $\mathbf{A}^{-1}$ can be computed from two results of the previous section. Form the matrix of cofactors

$$\mathbf{C} = \begin{bmatrix} A_{11} & \cdots & A_{1n} \\ \cdot & \cdots & \cdot \\ A_{n1} & \cdots & A_{nn} \end{bmatrix} \tag{2}$$

called the *adjoint* of $\mathbf{A}$. Then the inner product of the ith row of $\mathbf{C}$ and the hth row of $\mathbf{A}$ is equal to $|\mathbf{A}|$ if $i = h$ and to zero if $i \neq h$. If we take the transpose of $\mathbf{C}$ and divide each element by $|\mathbf{A}|$, we have the desired inverse

$$\mathbf{A}^{-1} = \begin{bmatrix} \frac{1}{|\mathbf{A}|}A_{11} & \cdots & \frac{1}{|\mathbf{A}|}A_{n1} \\ \cdot & \cdots & \cdot \\ \frac{1}{|\mathbf{A}|}A_{1n} & \cdots & \frac{1}{|\mathbf{A}|}A_{nn} \end{bmatrix} \tag{3}$$

It should be apparent from this definition that the inverse exists if and only if $|\mathbf{A}|$ is not zero. Computation of $\mathbf{A}^{-1}$ by cofactors is very inefficient in most practical applications, and we shall consider other methods in Sec. 2.6.

We shall frequently use these properties of inverses:

1. The inverse of a symmetric matrix is also symmetric.
2. The inverse of the transpose of $\mathbf{A}$ is the transpose of $\mathbf{A}^{-1}$.
3. The inverse of the product of several square matrices is equal to the product of the inverses in reverse order:

$$(\mathbf{ABC})^{-1} = \mathbf{C}^{-1}\mathbf{B}^{-1}\mathbf{A}^{-1} \tag{4}$$

4. If c is a nonzero scalar, $(c\mathbf{A})^{-1} = 1/c\mathbf{A}^{-1}$.
5. The inverse of a diagonal matrix is a diagonal matrix consisting of the reciprocals of the original elements.

Example 2.4. The inverse of

$$\mathbf{A} = \begin{bmatrix} 2 & 0 & 0 & 0 \\ 0 & 4 & 0 & 0 \\ 0 & 0 & 3 & 2 \\ 0 & 0 & 1 & 2 \end{bmatrix}$$

is

$$\mathbf{A}^{-1} = \begin{bmatrix} \frac{1}{2} & 0 & 0 & 0 \\ 0 & \frac{1}{4} & 0 & 0 \\ 0 & 0 & \frac{1}{2} & -\frac{1}{2} \\ 0 & 0 & -\frac{1}{4} & \frac{3}{4} \end{bmatrix}$$

2.6 THE RANK OF A MATRIX

Two p-component vectors are said to be *linearly dependent* if the elements of one vector are proportional to those of the second. Thus, the row vectors

$$\mathbf{x}' = [1, 0, -1] \qquad \mathbf{y}' = [4, 0, -4]$$

are linearly dependent, while

$$\mathbf{u}' = [2, -1, 0, 7] \qquad \mathbf{v}' = [6,2,0,0]$$

are linearly independent. A set of k vectors of equal dimensions is called linearly independent if it is impossible to write any vector of the set as some linear combination of the remaining vectors. That is, the vectors $\mathbf{x}_1, \ldots, \mathbf{x}_k$ form a linearly independent set if there do not exist scalars $c_1, \ldots, c_k$ such that for some vector $\mathbf{x}_i$ in the set

$$c_i\mathbf{x}_i = c_1\mathbf{x}_1 + \cdots + c_{i-1}\mathbf{x}_{i-1} + c_{i+1}\mathbf{x}_{i+1} + \cdots + c_k\mathbf{x}_k$$

The vectors

$$\mathbf{x}' = [1, -1, 2] \qquad \mathbf{y}' = [2, 0, -1] \qquad \mathbf{z}' = [0, -2, 5]$$

constitute a linearly dependent set, for $\mathbf{z}' = 2\mathbf{x}' - \mathbf{y}'$, while the unit vectors

$$\mathbf{t}' = [1,0,0] \qquad \mathbf{u}' = [0,1,0] \qquad \mathbf{v}' = [0,0,1]$$

are independent.

Example 2.5. Suppose that a cognitive test consists of six subscales. The first three measure certain verbal facilities, and their scores are summed to give what is called the *verbal score*. The last three reflect some motor and spatial skills, and their total is called the *performance score*. Finally, the sum of the six tests is used as a general measure of cognitive ability. The test has been administered to a large number of subjects, and it is proposed that the nine scores be used in an attempt to relate intelligence to other measures obtained on the individuals. However, if each sample of observations on a score is regarded as an N-component vector, it is immediately apparent that only six of the nine vectors are linearly independent, and no new information has been gained from the three derived scores. We shall see in the later chapters that the inclusion of these linear compounds may preclude certain kinds of statistical analyses.

Now let us formalize the degree of linear independence in a set of vectors as the *rank* of the matrix formed from the vectors. Assume for the moment that the number of vectors k does not exceed the dimensionality p of the vectors. Then the rank of

$$\mathbf{X} = \begin{bmatrix} \mathbf{x}_1' \\ \cdot \\ \cdot \\ \cdot \\ \mathbf{x}_k' \end{bmatrix} \tag{1}$$

is the number of linearly independent row vectors in the matrix. Rank may vary from zero for any null matrix to k for a matrix of *full rank*. If, on the other hand, the number of rows in $\mathbf{X}$ exceeded the number of columns, the rank of $\mathbf{X}$ would be the number of linearly independent columns. In either case, it can be shown that the rank of the matrix is a unique number, regardless of whether it is computed from rows or columns. It follows from Property 4 of Sec. 2.4 that

the matrix $\mathbf{A}$ *is of rank* r *if it contains at least one nonzero* $r \times r$ *minor, and no nonzero minor of dimensionality greater than* r.

Example 2.6. The matrix

$$\begin{bmatrix} 1 & 2 & 3 & 4 \\ 3 & 6 & 9 & 12 \\ 4 & 3 & 2 & 1 \\ -1 & 3 & 7 & 11 \\ 8 & 6 & 4 & 2 \end{bmatrix}$$

has rank two, for it can be shown that the second row is equal to three times the first row, row four is equal to the difference of rows two and three, and row five is twice the third row. The only linearly independent rows are the first and third, and these are said to form a *basis* for the five row vectors.

Rank has these important properties:

1. The rank of $\mathbf{A}'$ is equal to that of $\mathbf{A}$. This follows from the equivalence of the row and column definitions of rank.
2. The rank of $\mathbf{A}'\mathbf{A}$ is equal to that of $\mathbf{A}$. Similarly, the rank of $\mathbf{A}\mathbf{A}'$ is equivalent to the rank of $\mathbf{A}$.

3. The rank of **A** is unchanged by pre- or postmultiplication of **A** by a nonsingular matrix.

Elementary Row and Column Operations. The rank of a matrix is unchanged by these *elementary row and column operations:*

1. Interchange of any two rows (columns).
2. Multiplication of each element of a row (column) by a scalar constant.
3. Addition of a row (column) whose elements have been multiplied by a scalar constant to another row (column).

These transformations can be represented by nonsingular matrices. Row operations are performed by premultiplication of the given matrix by the elementary operation matrix, while column operations follow from postmultiplication. By Property 3 of the preceding paragraph such nonsingular transformations leave the rank invariant. By choosing the proper sequence of row or column operations we can reduce a matrix to a form consisting only of linearly independent rows or columns, and thereby determine its rank.

Example 2.7. The matrices

$$\mathbf{E}_1 = \begin{bmatrix} 0 & 1 & 0 \\ 1 & 0 & 0 \\ 0 & 0 & 1 \end{bmatrix} \qquad \mathbf{E}_2 = \begin{bmatrix} c & 0 & 0 \\ 0 & 1 & 0 \\ 0 & 0 & 1 \end{bmatrix} \qquad \mathbf{E}_3 = \begin{bmatrix} 1 & 0 & 0 \\ d & 1 & 0 \\ 0 & 0 & 1 \end{bmatrix}$$

can be used to make elementary transformations on the first two rows of

$$\mathbf{A} = \begin{bmatrix} 5 & 1 \\ -2 & 3 \\ 3 & 2 \end{bmatrix}$$

giving

$$\mathbf{E}_1\mathbf{A} = \begin{bmatrix} -2 & 3 \\ 5 & 1 \\ 3 & 2 \end{bmatrix} \qquad \mathbf{E}_2\mathbf{A} = \begin{bmatrix} 5c & c \\ -2 & 3 \\ 3 & 2 \end{bmatrix} \qquad \mathbf{E}_3\mathbf{A} = \begin{bmatrix} 5 & 1 \\ 5d - 2 & d + 3 \\ 3 & 2 \end{bmatrix}$$

Example 2.8. Let us use elementary row operations to compute the rank of the matrix

$$\mathbf{A} = \begin{bmatrix} 1 & 0 & 2 \\ -3 & 2 & 1 \\ -1 & 2 & 5 \\ 3 & 0 & 6 \end{bmatrix}$$

Note immediately that the last row is equal to the first row multiplied by 3. Row three is equal to row two plus twice row one. The appropriate row operations reduce

A successively to

$$\begin{bmatrix} 1 & 0 & 2 \\ -3 & 2 & 1 \\ -1 & 2 & 5 \\ 0 & 0 & 0 \end{bmatrix} \quad \begin{bmatrix} 1 & 0 & 2 \\ -3 & 2 & 1 \\ 0 & 0 & 0 \\ 0 & 0 & 0 \end{bmatrix}$$

Since the remaining rows are linearly independent, we conclude that **A** is of rank two.

Both row and column operations can be applied simultaneously to any $m \times n$ matrix to reduce it to its *canonical form*

(2)
$$\mathbf{F} = \begin{bmatrix} 1 & \cdots & 0 & \cdots & 0 \\ \cdot & \cdot & \cdot & \cdot & \cdot \\ 0 & \cdots & 1 & \cdots & 0 \\ \cdot & \cdot & \cdot & \cdot & \cdot \\ 0 & \cdots & 0 & \cdots & 0 \end{bmatrix}$$

of one in the first r diagonal positions and zero everywhere else. This means that for every matrix **A** there exist elementary row and column transformation matrices $\mathbf{R}_1, \ldots, \mathbf{R}_p, \mathbf{C}_1, \ldots, \mathbf{C}_q$ such that

(3)
$$\mathbf{R}_p \cdots \mathbf{R}_1\mathbf{A}\mathbf{C}_1 \cdots \mathbf{C}_q = \mathbf{PAQ}$$
$$= \mathbf{F}$$

is the canonical form of **A**. *But* if **A** is square and of full rank n, then row operations alone produce

(4)
$$\mathbf{R}_p \cdots \mathbf{R}_1\mathbf{A} = \mathbf{I}$$

and

(5)
$$\mathbf{A}^{-1} = \mathbf{R}_p \cdots \mathbf{R}_1\mathbf{I}$$

Hence, the *same* sequence of elementary row operations that reduces **A** to the identity matrix transforms the $n \times n$ identity matrix to the inverse of **A**. This property of elementary transformations provides us with a powerful tool for inverting matrices: we merely write down the $n \times 2n$ matrix

(6)
$$[\mathbf{A} \quad \mathbf{I}]$$

and apply elementary operations to each row with the aim of transforming the first n columns to the identity matrix. When this has been accomplished, $\mathbf{A}^{-1}$ will appear in the last n columns. For symmetric matrices this method has been formalized as the abbreviated Gauss-Doolittle technique (Dwyer, 1951).

Example 2.9. The successive transformations in the inversion of

$$\mathbf{A} = \begin{bmatrix} 1 & 1 & 2 \\ 2 & 3 & 3 \\ 3 & -1 & 8 \end{bmatrix}$$

are shown below:

$$\begin{bmatrix} 1 & 1 & 2 & 1 & 0 & 0 \\ 2 & 3 & 3 & 0 & 1 & 0 \\ 3 & -1 & 8 & 0 & 0 & 1 \end{bmatrix}$$

$$\begin{bmatrix} 1 & 1 & 2 & 1 & 0 & 0 \\ 0 & 1 & -1 & -2 & 1 & 0 \\ 0 & -4 & 2 & -3 & 0 & 1 \end{bmatrix}$$

$$\begin{bmatrix} 1 & 0 & 3 & 3 & -1 & 0 \\ 0 & 1 & -1 & -2 & 1 & 0 \\ 0 & 0 & -2 & -11 & 4 & 1 \end{bmatrix}$$

$$\begin{bmatrix} 1 & 0 & 0 & -13.5 & 5 & 1.5 \\ 0 & 1 & 0 & 3.5 & -1 & -0.5 \\ 0 & 0 & 1 & 5.5 & -2 & -0.5 \end{bmatrix}$$

and

$$\mathbf{A}^{-1} = \begin{bmatrix} -13.5 & 5 & 1.5 \\ 3.5 & -1 & -0.5 \\ 5.5 & -2 & -0.5 \end{bmatrix}$$

The determinant of a matrix can also be computed by elementary row or column operations, for by Property 5 of Sec. 2.4 those transformations do not alter its value. The process, called *pivotal condensation*, consists of reducing the matrix to triangular form, from which the determinant is computed as the product of the diagonal elements. If row operations are used, the matrix is written with the largest element of the first column in the (1,1) position. That element is used as the pivot for reducing the remaining elements of the column to zero. This rearrangement and reduction is continued until the triangular form is attained. Choice of the greatest element is in the interests of numerical accuracy.

Example 2.10. Let us evaluate the determinant of the matrix

$$\mathbf{A} = \begin{bmatrix} 10 & 5 & -2 & 0 \\ 6 & 3 & 2 & 1 \\ 4 & 5 & 12 & 3 \\ 5 & 1 & 3 & 8 \end{bmatrix}$$

We begin by subtracting appropriate multiples of the first row from the remaining rows:

$$\begin{bmatrix} 10 & 5 & -2 & 0 \\ 0 & 0 & 3.2 & 1 \\ 0 & 3 & 12.8 & 3 \\ 0 & -1.5 & 4 & 8 \end{bmatrix}$$

Since the (2,2) element is zero, we interchange the second and third rows of the matrix, and compensate for the resulting change of sign in the determinant by multiplying each element of the new second row by -1:

$$\begin{bmatrix} 10 & 5 & -2 & 0 \\ 0 & -3 & -12.8 & -3 \\ 0 & 0 & 3.2 & 1 \\ 0 & -1.5 & 4 & 8 \end{bmatrix}$$

A single row operation in the second column transforms this to

$$\begin{bmatrix} 10 & 5 & -2 & 0 \\ 0 & -3 & -12.8 & -3 \\ 0 & 0 & 3.2 & 1 \\ 0 & 0 & 10.4 & 9.5 \end{bmatrix}$$

Finally we have

$$\begin{bmatrix} 10 & 5 & -2 & 0 \\ 0 & -3 & -12.8 & -3 \\ 0 & 0 & 3.2 & 1 \\ 0 & 0 & 0 & 6.25 \end{bmatrix}$$

and the determinant is

$$\begin{aligned} |\mathbf{A}| &= (10)(-3)(3.2)(6.25) \\ &= -600 \end{aligned}$$

Generalized Inverse Matrices. Several generalizations of the inverse matrix have been proposed for rectangular matrices of any rank. One generalized inverse that is useful for solving systems of linear equations is defined as the matrix **G** satisfying

$$\mathbf{AGA} = \mathbf{A} \tag{7}$$

A is $p \times q$ and of rank r, while **G** is necessarily $q \times p$ and of the same rank as **A**. Other kinds of generalized inverses exist, e.g., those due to Penrose (1955) which satisfy the additional conditions

$$\mathbf{GAG} = \mathbf{G} \qquad (\mathbf{GA})' = \mathbf{GA} \qquad (\mathbf{AG})' = \mathbf{AG} \tag{8}$$

Their properties have been discussed at length by Graybill (1969), Searle (1966, 1971), and Searle and Hausman (1970); more advanced treatments have been written by Albert (1972), Boullion and Odell (1971), and Rao and Mitra (1971). However, the restricted form defined by (7) will suffice for our purposes.

Our matrix **G** can be computed from the canonical reduction (3) of **A**. We write

$$\mathbf{PAQ} = \begin{bmatrix} \mathbf{D} & \mathbf{0} \\ \mathbf{0} & \mathbf{0} \end{bmatrix} \tag{9}$$

where **D** is an $r \times r$ diagonal matrix (not necessarily the identity) and the null matrices have appropriate dimensions. **P** and **Q** are the respective products defined in (3) of the row and column elementary matrices. We introduce the $q \times p$ matrix

$$\mathbf{F}^{-} = \begin{bmatrix} \mathbf{D}^{-1} & \mathbf{0} \\ \mathbf{0} & \mathbf{0} \end{bmatrix} \tag{10}$$

Then $\mathbf{G} = \mathbf{QF^{-}P}$ is a generalized inverse of the matrix **A**, as one may verify by replacing **A** in (7) by its representation $\mathbf{P}^{-1}\mathbf{F}\mathbf{Q}^{-1}$. **G** is not a unique matrix, unless of course **A** is square and nonsingular.

Example 2.11. Let us find a generalized inverse of the matrix **A** in Example 2.8. The row operator matrices which reduce the third and fourth rows to null vectors are

$$\mathbf{R}_1 = \begin{bmatrix} 1 & 0 & 0 & 0 \\ 0 & 1 & 0 & 0 \\ 0 & 0 & 1 & 0 \\ -3 & 0 & 0 & 1 \end{bmatrix} \qquad \mathbf{R}_2 = \begin{bmatrix} 1 & 0 & 0 & 0 \\ 0 & 1 & 0 & 0 \\ -2 & -1 & 1 & 0 \\ 0 & 0 & 0 & 1 \end{bmatrix}$$

Then

$$\mathbf{P} = \mathbf{R}_2\mathbf{R}_1 = \begin{bmatrix} 1 & 0 & 0 & 0 \\ 0 & 1 & 0 & 0 \\ -2 & -1 & 1 & 0 \\ -3 & 0 & 0 & 1 \end{bmatrix}$$

The column transformations which reduce **PA** to a diagonal canonical form are

$$\mathbf{C}_1 = \begin{bmatrix} 1 & 0 & 0 \\ 3/2 & 1 & -1/2 \\ 0 & 0 & 1 \end{bmatrix} \qquad \mathbf{C}_2 = \begin{bmatrix} 1 & 0 & -2 \\ 0 & 1 & 0 \\ 0 & 0 & 1 \end{bmatrix}$$

and

$$\mathbf{Q} = \mathbf{C}_1\mathbf{C}_2 = \begin{bmatrix} 1 & 0 & -2 \\ 3/2 & 1 & -7/2 \\ 0 & 0 & 1 \end{bmatrix}$$

Then

$$\mathbf{F} = \mathbf{PAQ} = \begin{bmatrix} 1 & 0 & 0 \\ 0 & 2 & 0 \\ 0 & 0 & 0 \\ 0 & 0 & 0 \end{bmatrix} \qquad \mathbf{F}^{-} = \begin{bmatrix} 1 & 0 & 0 & 0 \\ 0 & 1/2 & 0 & 0 \\ 0 & 0 & 0 & 0 \end{bmatrix}$$

and

$$\mathbf{G} = \begin{bmatrix} 1 & 0 & 0 & 0 \\ 3/2 & 1/2 & 0 & 0 \\ 0 & 0 & 0 & 0 \end{bmatrix}$$

is one generalized inverse of **A**.

2.7 SIMULTANEOUS LINEAR EQUATIONS

The set of equations in the unknowns $x_1, \ldots, x_n$

$$
\begin{aligned}
a_{11}x_1 + \cdots + a_{1n}x_n &= c_1 \\
\cdots\cdots\cdots\cdots\cdots & \\
a_{m1}x_1 + \cdots + a_{mn}x_n &= c_m
\end{aligned} \tag{1}
$$

is called *a system of m-simultaneous linear equations in n unknowns* and can be compactly written in matrix form as

$$\mathbf{Ax} = \mathbf{c} \tag{2}$$

where $\mathbf{A}$ is the $m \times n$ matrix of the coefficients, $\mathbf{x}' = [x_1, \ldots, x_n]$, and $\mathbf{c}' = [c_1, \ldots, c_m]$. The general system $\mathbf{Ax} = \mathbf{c}$ of m equations in n unknowns possesses a solution if and only if the $m \times (n + 1)$ *augmented matrix*

$$[\mathbf{A} \quad \mathbf{c}]$$

is of the same rank r as $\mathbf{A}$. Otherwise the system is said to be inconsistent. It is essential to distinguish between homogeneous systems, for which $\mathbf{c} = \mathbf{0}$, and nonhomogeneous systems. We shall consider three types of equations that are particularly relevant for our purposes.

Nonhomogeneous System: $\mathbf{A}$ *Square and Nonsingular.* Since $\mathbf{A}^{-1}$ exists, the unique solution to the system is

$$\mathbf{x} = \mathbf{A}^{-1}\mathbf{c} \tag{3}$$

This follows from premultiplying both sides of equation (1) by $\mathbf{A}^{-1}$.

Example 2.12. The inverse matrix of the system

$$
\begin{aligned}
x_1 + x_2 - x_3 &= 1 \\
-x_1 + x_2 + x_3 &= -1 \\
x_1 - x_2 + x_3 &= 1
\end{aligned}
$$

is

$$
\begin{bmatrix}
\frac{1}{2} & 0 & \frac{1}{2} \\
\frac{1}{2} & \frac{1}{2} & 0 \\
0 & \frac{1}{2} & \frac{1}{2}
\end{bmatrix}
$$

The solution is

$$
\begin{aligned}
x_1 &= \tfrac{1}{2}(1) + 0(-1) + \tfrac{1}{2}(1) \\
&= 1 \\
x_2 &= \tfrac{1}{2}(1) + \tfrac{1}{2}(-1) + 0(1) \\
&= 0
\end{aligned}
$$

$$
\begin{aligned}
x_3 &= 0(1) + \tfrac{1}{2}(-1) + \tfrac{1}{2}(1) \\
&= 0
\end{aligned}
$$

Nonhomogeneous System: **A** *$m \times n$ and of Rank r.* If the condition on the ranks of the coefficient and augmented matrices is satisfied, a solution may be found by selecting any r linearly independent equations and solving for r of the n unknowns in terms of the constants c_i and the remaining $n - r$ variables.

Example 2.13. The equations

$$
\begin{aligned}
x_1 + x_2 &= 2 \\
x_1 + x_2 &= 1
\end{aligned}
$$

can be interpreted geometrically as defining parallel lines in the x_1x_2 plane. Since they do not intersect, the system cannot have a solution. The rank of the coefficient matrix

$$
\begin{bmatrix} 1 & 1 \\ 1 & 1 \end{bmatrix}
$$

is one, while that of the augmented matrix

$$
\begin{bmatrix} 1 & 1 & 2 \\ 1 & 1 & 1 \end{bmatrix}
$$

is two.

Example 2.14. The system

$$
\begin{aligned}
2x_1 + 3x_2 - x_3 &= 1 \\
x_2 + x_3 &= 2
\end{aligned}
$$

has coefficient and augmented matrices of rank two. Write the equations as

$$
\begin{aligned}
2x_1 + 3x_2 &= 1 + x_3 \\
x_2 &= 2 - x_3
\end{aligned}
$$

and solve for x_1 by substitution. The general solution is given by the vector

$$
[\tfrac{1}{2}(-5 + 4x_3),\ 2 - x_3,\ x_3]
$$

wherein x_3 can assume any value.

Homogeneous Systems of Equations. It is not possible for a homogeneous system of linear equations to be inconsistent, for the rank of $[\mathbf{A} \quad \mathbf{0}]$ is the same as that of **A**. Every homogeneous system is satisfied by the trivial null solution $\mathbf{x}' = [0, \ldots, 0]$, and in fact if the rank of the coefficient matrix is equal to the number of unknowns n, this is the *only* solution. Hence a homogeneous system will have a nontrivial solution if and only if the rank r of **A** is strictly less than n, and it is always possible to find $n - r$ *linearly independent* solutions of the system such that any linear combination of these solutions is itself a solution. In statistical applications we shall frequently encounter systems with as many unknowns as equations; such systems have nontrivial solutions if and only if their coefficient determinants vanish.

Example 2.15. The system

$$\begin{aligned} 3x_1 - 2x_2 &= 0 \\ 5x_1 + x_2 &= 0 \end{aligned}$$

defines two lines in the x_1x_2 plane that intersect at the origin. Hence the only solution is the trivial one [0,0].

Example 2.16. In the system

$$\begin{aligned} x_1 - x_2 + 2x_3 &= 0 \\ x_1 + 3x_2 - 2x_3 &= 0 \\ 3x_1 + x_2 + 2x_3 &= 0 \end{aligned}$$

the third equation is equal to the sum of twice the first plus the second, and the rank is thus two. The single independent solution may be determined in units of x_3 by solving the system

$$\begin{aligned} x_1 - x_2 &= -2x_3 \\ x_1 + 3x_2 &= 2x_3 \end{aligned}$$

The general solution for the original set is, in units of x_3,

$$\begin{aligned} x_1 &= -x_3 \\ x_2 &= x_3 \end{aligned}$$

Example 2.17. The system

$$\begin{aligned} x_1 + x_2 - x_3 - x_4 &= 0 \\ -2x_2 + 3x_3 + x_4 &= 0 \\ x_1 - x_2 + 2x_3 \qquad &= 0 \\ 3x_1 + x_2 \qquad - 2x_4 &= 0 \end{aligned}$$

is also of rank two. The first and third equations can be solved conveniently in terms of x_3 and x_4 to give the solution vector

$$[\tfrac{1}{2}(-x_3 + x_4), \tfrac{1}{2}(3x_3 + x_4), x_3, x_4]$$

If we set x_3 equal to zero and x_4 equal to one, and alternatively set x_3 to unity and x_4 to zero, we shall have these *linearly independent* values of the solution vector:

$$[\tfrac{1}{2}, \tfrac{1}{2}, 0, 1]$$
$$[-\tfrac{1}{2}, \tfrac{3}{2}, 1, 0]$$

Since any linear combination of these solutions is also a solution, the most general solution to the original system is

$$x_1 = \tfrac{1}{2}(a - b)$$

$$x_2 = \tfrac{1}{2}(a + 3b)$$

$$x_3 = b$$

$$x_4 = a$$

for arbitrary a and b.

Solution by Generalized Inverses. Rao (1962) has shown that the system $\mathbf{Ax} = \mathbf{c}$ of m consistent equations in n unknowns has the solution

$$\mathbf{x}^* = \mathbf{Gc} + (\mathbf{GA} - \mathbf{I})\mathbf{z} \tag{4}$$

where $\mathbf{G}$ is a generalized inverse of $\mathbf{A}$ defined by equation (7) of Sec. 2.6, $\mathbf{I}$ is the $n \times n$ identity matrix, and $\mathbf{z}$ is any $n \times 1$ vector of arbitrary constants. We note that the term containing $\mathbf{z}$ vanishes when $\mathbf{G} = \mathbf{A}^{-1}$. The algebraic properties, computing methods, and illustrations of such solutions have been treated by Searle (1971).

Example 2.18. We shall obtain solutions to the equations of Examples 2.14 and 2.16 by the method of generalized inverses. For the first system, $\mathbf{P} = \mathbf{I}$ and

$$\mathbf{Q} = \begin{bmatrix} 1 & -\tfrac{3}{2} & 2 \\ 0 & 1 & -1 \\ 0 & 0 & 1 \end{bmatrix} \qquad \mathbf{F} = \begin{bmatrix} 2 & 0 & 0 \\ 0 & 1 & 0 \end{bmatrix}$$

$$\mathbf{G} = \begin{bmatrix} \tfrac{1}{2} & -\tfrac{3}{2} \\ 0 & 1 \\ 0 & 0 \end{bmatrix} \qquad \mathbf{GA} = \begin{bmatrix} 1 & 0 & -2 \\ 0 & 1 & 1 \\ 0 & 0 & 0 \end{bmatrix}$$

The most general solution is

$$\mathbf{x}' = [-\tfrac{1}{2}(5 + z_3),\ 2 + z_3,\ -z_3]$$

For Example 2.16,

$$\mathbf{P} = \begin{bmatrix} 1 & 0 & 0 \\ -1 & 1 & 0 \\ -2 & -1 & 1 \end{bmatrix} \qquad \mathbf{Q} = \begin{bmatrix} 1 & 1 & -1 \\ 0 & 1 & 1 \\ 0 & 0 & 1 \end{bmatrix} \qquad \mathbf{F}^- = \begin{bmatrix} 1 & 0 & 0 \\ 0 & \tfrac{1}{4} & 0 \\ 0 & 0 & 0 \end{bmatrix}$$

$$\mathbf{G} = \begin{bmatrix} \tfrac{3}{4} & \tfrac{1}{4} & 0 \\ -\tfrac{1}{4} & \tfrac{1}{4} & 0 \\ 0 & 0 & 0 \end{bmatrix} \qquad \mathbf{GA} = \begin{bmatrix} 1 & 0 & 1 \\ 0 & 1 & -1 \\ 0 & 0 & 0 \end{bmatrix}$$

If we let z_3 equal $-x_3$ the general solution is $\mathbf{x}' = [-x_3, x_3, x_3]$.

Numerical Methods for Linear Equations. Perhaps the most common direct means of solving a system of rank r in r unknowns is the Gauss-Doolittle elimination method. As we have indicated before, this technique amounts to reduction of the system to triangular form by elementary row operations, followed by a backward solution of the new equations until all unknowns have been obtained. The usual form of the scheme supposes a symmetric coefficient matrix, but generalized procedures are also available. The reader is referred to Dwyer (1951), Faddeeva (1959), and Graybill (1961) for extensive discussions and worked examples of the Gauss-Doolittle method.

Another efficient direct solution is the square-root method. If the original system has been reduced to a square nonhomogeneous form $\mathbf{Ax} = \mathbf{c}$ of full rank r, it can be shown that the coefficient matrix can be factored into the product of a triangular matrix $\mathbf{T}$ and its transpose:

$$\mathbf{A} = \mathbf{TT}'$$

It is first necessary to determine the elements of $\mathbf{T}$ by a sequence of quadratic recurrence relationships. Then, letting $\mathbf{y} = \mathbf{T}'\mathbf{x}$, the triangular system

$$\mathbf{Ty} = \mathbf{c}$$

can be solved backward for the elements of $\mathbf{y}$. Finally, it is necessary to solve the other triangular system

$$\mathbf{T}'\mathbf{x} = \mathbf{y}$$

for the original unknown vector $\mathbf{x}$. The expressions for the elements of $\mathbf{T}$ and efficient arrangements of computing worksheets have been given by Dwyer (1951) and Faddeeva (1959).

Often it is necessary to find explicitly the inverse of the coefficient matrix rather than the solution to its equations. This is especially true in regression analysis, for the sampling variances and covariances of the least-squares estimates of the regression parameters are determined from the elements of the inverse. Inversion of the matrix may be formulated as the solution of r systems of equations with a common coefficient matrix $\mathbf{A}$ but with constant vectors $\mathbf{c}$, the successive columns of the $r \times r$ identity matrix. Solution of the systems can be carried out compactly by the abbreviated Gauss-Doolittle method. Details and examples of this procedure can be found in the texts of Dwyer and Graybill.

The preceding methods based upon elementary row operations are all *exact*, in the sense that a finite number of arithmetical operations will lead to a solution with an accuracy dependent only upon the precision maintained at each step in the computations and of course upon the exact or approximate nature of the coefficients in the original matrix. *Iterative* methods start with an approximation or plausible guess at the solution and repeatedly correct these trial values until they converge with a specified degree of accuracy to the actual solution. Perhaps the most familiar iterative technique for solving systems of linear equations is the Gauss-Seidel method (Faddeeva, 1959; Hotelling, 1943). Convergence of this algorithm is highly dependent upon the magnitude of the elements in the coefficient matrix. Necessary and sufficient conditions for convergence have been considered by Faddeeva, together with other iterative methods possessing accelerated or more general convergence properties.

2.8 ORTHOGONAL VECTORS AND MATRICES

In Sec. 2.3 we introduced the expression

$$\cos\theta = \frac{\mathbf{x}'\mathbf{y}}{(\mathbf{x}'\mathbf{x})^{1/2}(\mathbf{y}'\mathbf{y})^{1/2}} \tag{1}$$

for the cosine of the angle between the vectors $\mathbf{x}$ and $\mathbf{y}$ and saw that if their inner product vanished, the vectors must lie at right angles to one another. Such vectors are called *orthogonal,* or *orthonormal* if their lengths have been normalized to unity. An *orthogonal matrix* $\mathbf{T}$ is a square matrix whose rows are a set of orthonormal vectors. Hence

$$\begin{aligned} \mathbf{TT}' &= \mathbf{T}'\mathbf{T} \\ &= \mathbf{I} \end{aligned} \tag{2}$$

and the inverse of $\mathbf{T}$ is merely its transpose $\mathbf{T}'$. A simple 2×2 orthogonal matrix is

$$\begin{bmatrix} 0 & 1 \\ 1 & 0 \end{bmatrix}$$

This is the transformation matrix of a reflection of the points in the xy plane about the 45° line. A more general orthogonal matrix of that dimension is

$$\mathbf{T} = \begin{bmatrix} \cos\theta & \sin\theta \\ -\sin\theta & \cos\theta \end{bmatrix} \tag{3}$$

and is interpretable as the transformation matrix for a rotation of the xy coordinate axes through an angle θ. That is, if x and y were the coordinates of a point under the old axes, the point would have coordinates

$$\begin{aligned} u &= x\cos\theta + y\sin\theta \\ v &= -x\sin\theta + y\cos\theta \end{aligned} \tag{4}$$

after a rigid rotation of the axes through an angle of θ degrees. Larger orthogonal matrices can be constructed by the Gram-Schmidt orthogonalization process (Hohn, 1964, chap. 7) by starting with one normalized row and building up the matrix according to the requirement of mutual row orthogonality. For example, the Helmert matrix

$$\mathbf{T} = \begin{bmatrix} \frac{1}{\sqrt{3}} & \frac{1}{\sqrt{3}} & \frac{1}{\sqrt{3}} \\ \frac{1}{\sqrt{2}} & -\frac{1}{\sqrt{2}} & 0 \\ \frac{1}{\sqrt{6}} & \frac{1}{\sqrt{6}} & -\frac{2}{\sqrt{6}} \end{bmatrix}$$

is one particular 3×3 orthogonal matrix.

The $n \times n$ orthogonal matrix $\mathbf{T}$ is interpretable as the matrix of the linear transformation equivalent to a *rigid rotation* or a rotation followed by a *reflection* of the n coordinate axes about their origin. Distances in the space are unchanged by this rotation, for if we make the transformation

$$\mathbf{y} = \mathbf{Tx}$$

then

$$\begin{aligned} \mathbf{x}'\mathbf{x} &= \mathbf{y}'\mathbf{T}'\mathbf{Ty} \\ &= \mathbf{y}'\mathbf{Iy} \end{aligned}$$

and the sums of squares are invariant.

Orthogonal matrices have these useful properties:

1. The columns of an orthogonal matrix are orthogonal.
2. The determinant of an orthogonal matrix is always either 1 or -1. Since $|\mathbf{TT}'| = |\mathbf{I}| = 1$ and $|\mathbf{T}| = |\mathbf{T}'|$, $|\mathbf{T}| = \pm 1$.
3. The product of orthogonal matrices of the same dimension is itself orthogonal. That is, a succession of rigid rotations and reflections of the coordinate axes is expressible as a single rigid rotation and appropriate reflections.

2.9 QUADRATIC FORMS

A *quadratic form* in the variables $x_1, \ldots, x_n$ is an expression of the type

$$(1) \quad f(x_1, \ldots, x_n) = a_{11}x_1^2 + a_{22}x_2^2 + \cdots + a_{nn}x_n^2 + 2a_{12}x_1x_2 + \cdots + 2a_{1n}x_1x_n + \cdots + 2a_{n-1,n}x_{n-1}x_n$$

$$= \sum_{i=1}^{n} \sum_{j=1}^{n} a_{ij} x_i x_j$$

where $a_{ij} = a_{ji}$. Some of the a_{ij} may be zero. We note immediately that the quadratic form can be written in matrix notation as

$$\mathbf{x}'\mathbf{A}\mathbf{x}$$

where $\mathbf{x}' = [x_1, \ldots, x_n]$ and $\mathbf{A}$ is the $n \times n$ symmetric matrix of the coefficients. The simplest quadratic form is merely $f(x) = a_{11}x^2$, the equation of a parabola in the single variable x. Quadratic forms in which the x_i are random variables play an important role in both univariate and multivariate statistical theory. For example, the sum of squared deviations about the sample mean

$$\sum_{i=1}^{N} (x_i - \bar{x})^2 = \sum_{i=1}^{N} x_i^2 - \frac{1}{N}\left(\sum x_i\right)^2 \tag{2}$$

can be written as a quadratic form in the observations x_i with matrix

$$\mathbf{A} = \begin{bmatrix} \frac{N-1}{N} & \frac{-1}{N} & \cdots & \frac{-1}{N} \\ \frac{-1}{N} & \frac{N-1}{N} & \cdots & \frac{-1}{N} \\ \cdots & \cdots & \cdots & \cdots \\ \frac{-1}{N} & \frac{-1}{N} & \cdots & \frac{N-1}{N} \end{bmatrix} \tag{3}$$

A symmetric matrix $\mathbf{A}$ and its associated quadratic form are called *positive definite* if $\mathbf{x}'\mathbf{A}\mathbf{x} > 0$ for all nonnull $\mathbf{x}$. If $\mathbf{x}'\mathbf{A}\mathbf{x} \geq 0$, the form and its matrix are called *positive semidefinite.* Similar definitions apply for *negative definite* and *semidefinite* quadratic forms. An *indefinite* form can assume either positive, zero, or negative values. We shall need the special properties of positive definite and semidefinite matrices and quadratic forms frequently in the sequel.

Positive definite quadratic forms have matrices of *full rank*. It is possible by repeatedly completing squares to reduce such a form in n variables to the form

$$d_1 y_1^2 + \cdots + d_n y_n^2 \tag{4}$$

containing only squares of the new variables y_i and with coefficients $d_i > 0$. Similarly, a positive semidefinite quadratic form can be reduced to

(5) $$d_1y_1^2 + \cdots + d_ry_r^2$$

where all coefficients are positive, and $r \leq n$ is the rank of the form matrix. However, these properties are not convenient means of determining the nature of the form, and we shall now state a necessary and sufficient condition for positive definiteness or semidefiniteness. From the matrix form the sequence of *leading principal minor determinants*

(6) $$p_0 = 1 \qquad p_1 = a_{11} \qquad p_2 = \begin{vmatrix} a_{11} & a_{12} \\ a_{12} & a_{22} \end{vmatrix} \quad \cdots$$

$$p_i = \begin{vmatrix} a_{11} & \cdots & a_{1i} \\ \cdots & \cdots & \cdots \\ a_{1i} & \cdots & a_{ii} \end{vmatrix} \quad \cdots \quad p_n = |\mathbf{A}|$$

If $\mathbf{A}$ is of rank r it is said to be *regular* if $p_r \neq 0$ and no two consecutive p_i in the sequence are zero. It is always possible to put any symmetric matrix into regular form by interchanging rows and, simultaneously, the corresponding columns. Then, if $\mathbf{A}$ is a regularly arranged matrix,

1. A necessary and sufficient condition for positive definiteness is that $p_i > 0$ for $i = 1, \ldots, n$.
2. A necessary and sufficient condition for positive semidefiniteness is that $p_1 > 0, \ldots, p_r > 0$ and the remaining $n - r$ p_i equal zero, where r may equal n.

Example 2.19. The matrix

$$\begin{bmatrix} 4 & 6 & 0 \\ 6 & 9 & 0 \\ 0 & 0 & 2 \end{bmatrix}$$

has the leading-principal-minor-determinant sequence $p_0 = 1$, $p_1 = 4$, $p_2 = 0$, $p_3 = 0$, and so it is not regularly arranged. We can reorder the rows and columns to give the new matrix

$$\begin{bmatrix} 2 & 0 & 0 \\ 0 & 4 & 6 \\ 0 & 6 & 9 \end{bmatrix}$$

with the sequence $p_0 = 1$, $p_1 = 2$, $p_2 = 8$, $p_3 = 0$. The matrix and its quadratic form are positive semidefinite of rank two.

Example 2.20. The matrix (3) of the sample sum of squares is a particular case of the patterned matrix with a common diagonal element a and equal off-diagonal elements b. The determinant of such an $i \times i$ matrix is

$$(a - b)^{i-1}[a + (i - 1)b]$$

so that the leading principal minor determinants of (3) are

$$p_i = \frac{1}{N}[N - 1 - (i - 1)]$$

The first $N - 1$ of these are positive, while the Nth is always zero. The sum of squared deviations is then a positive semidefinite quadratic form of rank $N - 1$.

2.10 THE CHARACTERISTIC ROOTS AND VECTORS OF A MATRIX

The *characteristic roots* of the $p \times p$ matrix $\mathbf{A}$ are the solutions to the *determinantal equation*

$$|\mathbf{A} - \lambda\mathbf{I}| = 0 \tag{1}$$

The determinant is a pth-degree polynomial in λ, and thus $\mathbf{A}$ has just p characteristic roots. The Laplace expansion of the *characteristic determinant* enables us to write the *characteristic polynomial* as

$$|\mathbf{A} - \lambda\mathbf{I}| = (-\lambda)^p + S_1(-\lambda)^{p-1} + S_2(-\lambda)^{p-2} + \cdots - S_{p-1}\lambda + |\mathbf{A}| \tag{2}$$

where S_i is the sum of all $i \times i$ principal minor determinants. S_1 is merely the sum of the diagonal elements of $\mathbf{A}$, or tr $\mathbf{A}$. It follows immediately from the theory of polynomial equations that:

1. The *product* of the characteristic roots of $\mathbf{A}$ is equal to $|\mathbf{A}|$.
2. The *sum* of the characteristic roots of $\mathbf{A}$ is equal to the *trace* of $\mathbf{A}$.

Example 2.21. The matrix

$$\mathbf{A} = \begin{bmatrix} 2 & 1 & 1 \\ 1 & 2 & 1 \\ 1 & 1 & 2 \end{bmatrix}$$

has trace 6 and three 2×2 principal minor determinants each equal to

$$\begin{vmatrix} 2 & 1 \\ 1 & 2 \end{vmatrix} = 3$$

while $|\mathbf{A}| = 4$. The characteristic equation of $\mathbf{A}$ is

$$\lambda^3 - 6\lambda^2 + 9\lambda - 4 = 0$$

and its roots are 1, 1, and 4.

In the sequel we shall frequently need these properties of characteristic roots:

1. The characteristic roots of a symmetric matrix with real elements are all real.
2. The characteristic roots of a positive definite matrix are all positive.
3. If an $n \times n$ symmetric matrix is positive semidefinite of rank r, it contains exactly r positive characteristic roots and $n - r$ zero roots.
4. The nonzero characteristic roots of the product **AB** are equal to the nonzero roots of **BA**. Hence the traces of **AB** and **BA** are equal.
5. The characteristic roots of a diagonal matrix are the diagonal elements themselves.

Associated with every characteristic root λ_i of the square matrix **A** is a *characteristic vector* $\mathbf{x}_i$ whose elements satisfy the homogeneous system of equations

$$[\mathbf{A} - \lambda_i \mathbf{I}]\mathbf{x}_i = \mathbf{0} \tag{3}$$

By the definition of the characteristic root the determinant of the system vanishes, and a nontrivial solution $\mathbf{x}_i$ always exists. We note immediately that the elements of the vector are determined only up to a scale factor. Many of the characteristic vectors we shall encounter in the sequel will be computed from symmetric matrices. The characteristic roots and vectors of such matrices have these important properties:

1. If λ_i and λ_j are distinct characteristic roots of the symmetric matrix **A**, their associated vectors $\mathbf{x}_i$ and $\mathbf{x}_j$ are orthogonal. This is readily apparent if we premultiply the definitions of the vectors

$$\mathbf{A}\mathbf{x}_i = \lambda_i \mathbf{x}_i \qquad \mathbf{A}\mathbf{x}_j = \lambda_j \mathbf{x}_j$$

by $\mathbf{x}_j'$ and $\mathbf{x}_i'$, respectively. But this implies that

$$\lambda_i \mathbf{x}_j' \mathbf{x}_i = \lambda_j \mathbf{x}_i' \mathbf{x}_j$$

and since $\lambda_i \neq \lambda_j$, we conclude that the vectors are orthogonal.

2. For every real symmetric matrix **A** there exists an orthogonal matrix **P** such that

$$\mathbf{P}'\mathbf{A}\mathbf{P} = \mathbf{D} \tag{4}$$

where **D** is the diagonal matrix of the characteristic roots of **A**. The normalized characteristic vectors of **A** can be taken as the columns of **P**. Even if the characteristic roots are not distinct, it is still possible to select the elements of their vectors to give a mutually orthogonal set of characteristic vectors.

These properties of symmetric matrices have an important implication for quadratic forms. If we apply the orthogonal transformation

$$\mathbf{x} = \mathbf{P}\mathbf{y} \tag{5}$$

to the p variables in the quadratic form $\mathbf{x}'\mathbf{A}\mathbf{x}$, the form becomes

$$\begin{aligned} \mathbf{x}'\mathbf{A}\mathbf{x} &= \mathbf{y}'\mathbf{P}'\mathbf{A}\mathbf{P}\mathbf{y} \\ &= \mathbf{y}'\mathbf{D}\mathbf{y} \\ &= \lambda_1 y_1^2 + \cdots + \lambda_r y_r^2 \end{aligned} \tag{6}$$

where the λ_i are the characteristic roots of the coefficient matrix, and r is the rank of the form. Any real quadratic form can be reduced to a weighted sum of squares by computing the characteristic roots and vectors of its matrix.

Example 2.22. The characteristic vector of the largest root of Example 2.21 must satisfy the equations

$$\begin{aligned} -2x_{11} + x_{12} + x_{13} &= 0 \\ x_{11} - 2x_{12} + x_{13} &= 0 \\ x_{11} + x_{12} - 2x_{13} &= 0 \end{aligned}$$

If we arbitrarily set x_{13} equal to one and solve the first two nonhomogeneous equations, the characteristic vector is

$$\mathbf{x}' = [1,1,1]$$

and of course any nonnull vector $[a,a,a]$ is a characteristic vector for $\lambda_3 = 4$. The vector associated with the double root $\lambda_1 = \lambda_2 = 1$ must satisfy the system with the single linearly independent equation

$$x_{21} + x_{22} + x_{23} = 0$$

The most general solution to this is the vector

$$[a, b, -a - b]$$

and by properly selecting a and b we can generate two linearly independent characteristic vectors for the double root. Note that *any* choice of a and b will give vectors orthogonal to the first vector [1,1,1]. We may elect to choose a and b in the second and third vectors to continue this orthogonality. For example,

$$\begin{aligned} \mathbf{x}_1' &= [0, 1, -1] \\ \mathbf{x}_2' &= [-2, 1, 1] \end{aligned}$$

Example 2.23. Square matrices with the property

$$\mathbf{A}\mathbf{A} = \mathbf{A}$$

are called *idempotent* and play an exceedingly important role in the theory of the analysis of variance. The characteristic roots of an idempotent matrix are either zero or one, and a quadratic form with such a matrix can be reduced to a sum of r squared terms. It is easy to check that the matrix (3) of Sec. 2.9 is idempotent and that it is therefore possible to make the transformation

$$\sum_{i=1}^{N} (x_i - \bar{x})^2 = y_1^2 + \cdots + y_{N-1}^2$$

into a sum of squares of $N - 1$ new random variables whose statistical independence follows from the orthogonality of the transformation.

We shall defer our discussion of numerical methods for calculating characteristic roots and vectors until our treatment of the principal-component technique in Chap. 8. An extensive coverage of computing techniques can be found in the treatises of Bodewig (1956), Faddeeva (1959), and Householder (1953).

Bounds for characteristic roots have been obtained by a number of algebraists. Such results have been collected and summarized by Marcus and Minc (1964).

2.11 PARTITIONED MATRICES

Frequently we shall find it convenient to think of certain rows and columns of a matrix as grouped together because of common characteristics of their associated variables. Such a matrix can be written as an array

$$\mathbf{A} = \begin{bmatrix} \mathbf{A}_{11} & \cdots & \mathbf{A}_{1n} \\ \cdots & \cdots & \cdots \\ \mathbf{A}_{m1} & \cdots & \mathbf{A}_{mn} \end{bmatrix} \tag{1}$$

of submatrices $\mathbf{A}_{ij}$ containing r_i rows and c_j columns, where it is obvious that all submatrices in a given row of $\mathbf{A}$ must have the same number of rows and each column must be composed of matrices with a like number of columns. Operations with partitioned matrices are akin to those in the ungrouped case, except that the nonscalar nature of the elements must be kept in mind. The sum of the partitioned matrices $\mathbf{A}$ and $\mathbf{B}$ whose submatrices have similar dimensions is

$$\mathbf{A} + \mathbf{B} = \begin{bmatrix} \mathbf{A}_{11} + \mathbf{B}_{11} & \cdots & \mathbf{A}_{1n} + \mathbf{B}_{1n} \\ \cdots & \cdots & \cdots \\ \mathbf{A}_{m1} + \mathbf{B}_{m1} & \cdots & \mathbf{A}_{mn} + \mathbf{B}_{mn} \end{bmatrix} \tag{2}$$

The product of the partitioned matrices $\mathbf{A}$ and $\mathbf{B}$ is

(3)
$$\mathbf{AB} = \begin{bmatrix} \sum_{j=1}^{n} \mathbf{A}_{1j}\mathbf{B}_{j1} & \cdots & \sum_{j=1}^{n} \mathbf{A}_{1j}\mathbf{B}_{jp} \\ \cdots & \cdots & \cdots \\ \sum_{j=1}^{n} \mathbf{A}_{mj}\mathbf{B}_{j1} & \cdots & \sum_{j=1}^{n} \mathbf{A}_{mj}\mathbf{B}_{jp} \end{bmatrix}$$

The dimensions within each submatrix product must conform; if the submatrices of **A** have respective column numbers $c_1, \ldots, c_n$, those of **B** must have the row dimensions $c_1, \ldots, c_n$.

When the elements of partitioned matrices must be shown, as in numerical examples, it will be convenient to separate the submatrices by dashed lines or appropriate spacing.

Example 2.24. Let

$$\mathbf{A} = \left[\begin{array}{cc:c} 1 & 0 & 3 \\ -2 & 3 & 1 \\ \hdashline 1 & 1 & 2 \end{array}\right] \qquad \mathbf{B} = \left[\begin{array}{cc} 1 & 1 \\ 2 & 0 \\ \hdashline 3 & 1 \end{array}\right]$$

Then

$$\mathbf{AB} = \left[\begin{array}{cc} 10 & 4 \\ 7 & -1 \\ \hdashline 9 & 3 \end{array}\right]$$

It is also possible to express the inverse of a nonsingular partitioned matrix in terms of its submatrices. In the particularly important case of

(4)
$$\mathbf{A} = \begin{bmatrix} \mathbf{A}_{11} & \mathbf{A}_{12} \\ \mathbf{A}_{21} & \mathbf{A}_{22} \end{bmatrix}$$

where $\mathbf{A}_{11}$ and $\mathbf{A}_{22}$ are both square and by their principal-minor nature nonsingular, it can be verified that

(5) $\mathbf{A}^{-1} =$
$$\begin{bmatrix} (\mathbf{A}_{11} - \mathbf{A}_{12}\mathbf{A}_{22}^{-1}\mathbf{A}_{21})^{-1} & -(\mathbf{A}_{11} - \mathbf{A}_{12}\mathbf{A}_{22}^{-1}\mathbf{A}_{21})^{-1}\mathbf{A}_{12}\mathbf{A}_{22}^{-1} \\ -\mathbf{A}_{22}^{-1}\mathbf{A}_{21}(\mathbf{A}_{11} - \mathbf{A}_{12}\mathbf{A}_{22}^{-1}\mathbf{A}_{21})^{-1} & \mathbf{A}_{22}^{-1} + \mathbf{A}_{22}^{-1}\mathbf{A}_{21}(\mathbf{A}_{11} - \mathbf{A}_{12}\mathbf{A}_{22}^{-1}\mathbf{A}_{21})^{-1}\mathbf{A}_{12}\mathbf{A}_{22}^{-1} \end{bmatrix}$$

An alternate expression can be obtained by reversing the positions of $\mathbf{A}_{11}$ and $\mathbf{A}_{22}$ in the original matrix.

Frequently it is necessary to compute the determinant of the partitioned matrix (4). If $\mathbf{A}_{11}$ is nonsingular,

(6)
$$|\mathbf{A}| = |\mathbf{A}_{11}| \cdot |\mathbf{A}_{22} - \mathbf{A}_{21}\mathbf{A}_{11}^{-1}\mathbf{A}_{12}|$$

If $\mathbf{A}_{22}$ is nonsingular,

(7)
$$|\mathbf{A}| = |\mathbf{A}_{22}| \cdot |\mathbf{A}_{11} - \mathbf{A}_{12}\mathbf{A}_{22}^{-1}\mathbf{A}_{21}|$$

Example 2.25. The preceding expressions for the inverses and determinants are often valuable in practical computation if the submatrices are small or conveniently patterned. If

$$\mathbf{A} = \begin{bmatrix} 1 & 0 & 1 & 1 & 1 \\ 0 & 1 & 2 & 2 & 2 \\ 3 & -1 & 4 & 0 & 0 \\ 3 & -1 & 0 & 4 & 0 \\ 3 & -1 & 0 & 0 & 4 \end{bmatrix}$$

we can compute $\mathbf{A}^{-1}$ by partitioning the matrix in terms of the first and second and third to fifth rows and columns. Then

$$\mathbf{A}_{11}^{-1} = \begin{bmatrix} 1 & 0 \\ 0 & 1 \end{bmatrix} \qquad \mathbf{A}_{22}^{-1} = \begin{bmatrix} \frac{1}{4} & 0 & 0 \\ 0 & \frac{1}{4} & 0 \\ 0 & 0 & \frac{1}{4} \end{bmatrix}$$

$$\mathbf{A}_{11} - \mathbf{A}_{12}\mathbf{A}_{22}^{-1}\mathbf{A}_{21} = \begin{bmatrix} -\frac{5}{4} & \frac{3}{4} \\ -\frac{9}{2} & \frac{5}{2} \end{bmatrix}$$

$$(\mathbf{A}_{11} - \mathbf{A}_{12}\mathbf{A}_{22}^{-1}\mathbf{A}_{21})^{-1} = \begin{bmatrix} 10 & -3 \\ 18 & -5 \end{bmatrix}$$

$$\mathbf{A}_{12}\mathbf{A}_{22}^{-1} = \begin{bmatrix} \frac{1}{4} & \frac{1}{4} & \frac{1}{4} \\ \frac{1}{2} & \frac{1}{2} & \frac{1}{2} \end{bmatrix} \qquad \mathbf{A}_{22}^{-1}\mathbf{A}_{21} = \begin{bmatrix} \frac{3}{4} & -\frac{1}{4} \\ \frac{3}{4} & -\frac{1}{4} \\ \frac{3}{4} & -\frac{1}{4} \end{bmatrix}$$

Hence

$$\mathbf{A}^{-1} = \begin{bmatrix} 10 & -3 & -1 & -1 & -1 \\ 18 & -5 & -2 & -2 & -2 \\ -3 & 1 & \frac{1}{2} & \frac{1}{4} & \frac{1}{4} \\ -3 & 1 & \frac{1}{4} & \frac{1}{2} & \frac{1}{4} \\ -3 & 1 & \frac{1}{4} & \frac{1}{4} & \frac{1}{2} \end{bmatrix}$$

The determinant of $\mathbf{A}$ can be computed from (6) or (7) to be 16.

The two forms of the partitioned inverse (5) lead to a useful matrix identity. If we equate the alternative expressions for the (1,1) submatrix of $\mathbf{A}^{-1}$, we have

$$(8) \quad (\mathbf{A}_{11} - \mathbf{A}_{12}\mathbf{A}_{22}^{-1}\mathbf{A}_{21})^{-1} = \mathbf{A}_{11}^{-1} + \mathbf{A}_{11}^{-1}\mathbf{A}_{12}(\mathbf{A}_{22} - \mathbf{A}_{21}\mathbf{A}_{11}^{-1}\mathbf{A}_{12})^{-1}\mathbf{A}_{21}\mathbf{A}_{11}^{-1}$$

In particular, if $\mathbf{A}$ is $p \times p$ and nonsingular, $\mathbf{b}$ is a $p \times 1$ vector, and c is a scalar, then we have Bartlett's (1951) form of the identity:

$$(9) \quad (\mathbf{A} + c\mathbf{b}\mathbf{b}')^{-1} = \mathbf{A}^{-1} - \frac{c}{1 + c\mathbf{b}'\mathbf{A}^{-1}\mathbf{b}} \mathbf{A}^{-1}\mathbf{b}\mathbf{b}'\mathbf{A}^{-1}$$

Example 2.26. We shall use (9) to compute the inverse of the $p \times p$ matrix

$$\mathbf{K} = \begin{bmatrix} a & c & \cdots & c \\ c & a & \cdots & c \\ \cdots & \cdots & \cdots & \cdots \\ c & c & \cdots & a \end{bmatrix}$$

Let $\mathbf{A} = (a - c)\mathbf{I}$, $\mathbf{b}' = [1, \ldots, 1]$, and write $\mathbf{E} = \mathbf{bb}'$. Then

$$\mathbf{K}^{-1} = \frac{1}{a-c}\mathbf{I} - \frac{c}{(a-c)[a+c(p-1)]}\mathbf{E}$$

or a matrix with a common diagonal element

$$\frac{a + (p-2)c}{(a-c)[a+(p-1)c]}$$

and off-diagonal elements each equal to

$$-\frac{c}{(a-c)[a+(p-1)c]}$$

2.12 DIFFERENTIATION WITH VECTORS AND MATRICES

In the later chapters we shall need formulas for finding the partial derivatives of likelihoods and other scalar functions of vectors and matrices. Let $f(\mathbf{x})$ be a continuous function of the elements of the vector $\mathbf{x}' = [x_1, \ldots, x_p]$ whose first and second partial derivatives

$$\frac{\partial f(\mathbf{x})}{\partial x_i} \qquad \frac{\partial^2 f(\mathbf{x})}{\partial x_i \, \partial x_j}$$

exist for all points $\mathbf{x}$ in some region of p-dimensional euclidean space of interest to us. Define the partial derivative operator vector as

(1)
$$\frac{\partial}{\partial \mathbf{x}} = \begin{bmatrix} \dfrac{\partial}{\partial x_1} \\ \cdot \\ \cdot \\ \cdot \\ \dfrac{\partial}{\partial x_p} \end{bmatrix}$$

Application of it to $f(\mathbf{x})$ yields the vector of partial derivatives

(2)
$$\frac{\partial f(\mathbf{x})}{\partial \mathbf{x}} = \begin{bmatrix} \dfrac{\partial f(\mathbf{x})}{\partial x_1} \\ \cdot \\ \cdot \\ \cdot \\ \dfrac{\partial f(\mathbf{x})}{\partial x_p} \end{bmatrix}$$

The following functions and their derivatives are especially important:

1. If $f(\mathbf{x})$ is constant for all $\mathbf{x}$,

(3)
$$\frac{\partial f(\mathbf{x})}{\partial \mathbf{x}} = \begin{bmatrix} 0 \\ \cdot \\ \cdot \\ \cdot \\ 0 \end{bmatrix}$$

2. $f(\mathbf{x}) = \mathbf{a}'\mathbf{x}$:

(4)
$$\frac{\partial f(\mathbf{x})}{\partial \mathbf{x}} = \begin{bmatrix} a_1 \\ \cdot \\ \cdot \\ \cdot \\ a_p \end{bmatrix}$$

Note that the columnar form of the derivative vector is unchanged if we write $f(\mathbf{x}) = \mathbf{x}'\mathbf{a}$.

3. To compute the vector of derivatives of the quadratic form $\mathbf{x}'\mathbf{A}\mathbf{x}$ write

$$\mathbf{x}'\mathbf{A}\mathbf{x} = \sum_{i=1}^{p} \sum_{j=1}^{p} a_{ij}x_ix_j = a_{ii}x_i^2 + 2x_i \sum_{\substack{j=1 \\ j\neq i}} a_{ij}x_j + \sum_{\substack{h\neq i \\ j\neq i}} \sum a_{hj}x_hx_j$$

and differentiate with respect to x_i. If we let $\mathbf{a}_i'$ be the ith row of the symmetric matrix $\mathbf{A}$, the partial derivative of the quadratic form with respect to x_i is

$$2\mathbf{a}_i'\mathbf{x} = 2 \sum_{j=1}^{p} a_{ij}x_j$$

and the vector of partial derivatives is

(5)
$$\frac{\partial \mathbf{x}'\mathbf{A}\mathbf{x}}{\partial \mathbf{x}} = 2\mathbf{A}\mathbf{x}$$

In particular, the vector of derivatives of the sum of squares $\mathbf{x}'\mathbf{x}$ is merely $2\mathbf{x}$.

4. In the more general quadratic function

(6)
$$h(\mathbf{x}) = (\mathbf{a} - \mathbf{C}\mathbf{x})'\mathbf{K}(\mathbf{a} - \mathbf{C}\mathbf{x})$$

let $\mathbf{K}$ be an $N \times N$ symmetric matrix, $\mathbf{C} = [\mathbf{C}_1, \ldots, \mathbf{C}_p]$ is an $N \times p$ matrix of constants, and $\mathbf{a}$ is an $N \times 1$ vector. If we let $\mathbf{u} = \mathbf{a} - \mathbf{C}\mathbf{x}$, the derivatives of $h(\mathbf{x})$ can be computed by the chain rule:

(7)
$$\frac{\partial h(\mathbf{x})}{\partial x_i} = \sum_{j=1} \frac{\partial(\mathbf{u}'\mathbf{K}\mathbf{u})}{\partial u_j} \frac{\partial u_j}{\partial x_i}$$

$$= -2\mathbf{u}'\mathbf{K}\mathbf{C}_i \qquad i = 1, \ldots, p$$

and

$$\frac{\partial h(\mathbf{x})}{\partial \mathbf{x}} = -2\mathbf{C}'\mathbf{K}(\mathbf{a} - \mathbf{C}\mathbf{x}) \tag{8}$$

5. The matrix of second-order partial derivatives of a function of p variables is called the *hessian* matrix:

$$\mathbf{H} = \frac{\partial^2 f(\mathbf{x})}{\partial \mathbf{x}' \, \partial \mathbf{x}} = \begin{bmatrix} \dfrac{\partial^2 f(\mathbf{x})}{\partial x_1^2} & \cdots & \dfrac{\partial^2 f(\mathbf{x})}{\partial x_1 \, \partial x_p} \\ \cdots & \cdots & \cdots \\ \dfrac{\partial^2 f(\mathbf{x})}{\partial x_1 \, \partial x_p} & \cdots & \dfrac{\partial^2 f(\mathbf{x})}{\partial x_p^2} \end{bmatrix} \tag{9}$$

For example, the hessian matrix of $f(\mathbf{x}) = \mathbf{x}'\mathbf{A}\mathbf{x}$ is merely $2\mathbf{A}$. The hessian is necessarily symmetric if our original conditions of continuity and existence of all first and second derivatives are satisfied by $f(\mathbf{x})$.

Determination of Maxima and Minima. A necessary condition for a maximum or minimum of $f(\mathbf{x})$ at $\mathbf{x} = \mathbf{x}_0$ is that

$$\frac{\partial f(\mathbf{x})}{\partial \mathbf{x}} = \mathbf{0} \tag{10}$$

at that point. Such a value of the function is called a *stationary* maximum or minimum, as opposed to a *global* extremum which might exist on the boundary of the admissible region of $\mathbf{x}$, or as a cusp or other form at which the derivatives were undefined. A sufficient condition for a maximum at the point $\mathbf{x}_0$ satisfying (10) is that the hessian matrix evaluated at $\mathbf{x}_0$ is negative definite. Similarly, a positive definite hessian is a sufficient condition for a stationary minimum. If the hessian is semidefinite the test fails, and higher-order terms in the Taylor expansion of $f(\mathbf{x})$ must be examined in the vicinity of the stationary point. If the hessian is an indefinite matrix at $\mathbf{x}_0$, that point is neither a maximum nor a minimum.

As an application of these rules let us determine the stationary extremum of the function $h(\mathbf{x})$ given by (6). By equating (8) to the null vector we see that the stationary point must satisfy the system of linear equations

$$\mathbf{C}'\mathbf{K}\mathbf{C}\mathbf{x} = \mathbf{C}'\mathbf{K}\mathbf{a} \tag{11}$$

and if $\mathbf{C}'\mathbf{K}\mathbf{C}$ is of full rank p,

(12) $$\mathbf{x} = (\mathbf{C}'\mathbf{K}\mathbf{C})^{-1}\mathbf{C}'\mathbf{K}\mathbf{a}$$

If **K** is positive definite and **C** has rank p, the hessian $2\mathbf{C}'\mathbf{K}\mathbf{C}$ will be positive definite and the solution (12) will minimize $h(\mathbf{x})$.

Maximization Subject to Constraints. Frequently it will be necessary to maximize or minimize one function $f(\mathbf{x})$ subject to a constraint $g(\mathbf{x}) = c$ on the values of $\mathbf{x}$. Although one might be able to handle this problem by eliminating one variable by the constraint, a more general and efficient method is that of *Lagrange multipliers.* The mathematical basis for that technique can be found in most modern calculus texts, e.g., Courant (1966) or Hadley (1964). Essentially, we form a new function

(13) $$h(\mathbf{x},\lambda) = f(\mathbf{x}) - \lambda[g(\mathbf{x}) - c]$$

For a constrained stationary value these conditions must be met:

(14a) $$\frac{\partial h(\mathbf{x},\lambda)}{\partial \mathbf{x}} = \frac{\partial f(\mathbf{x})}{\partial \mathbf{x}} - \lambda \frac{\partial g(\mathbf{x})}{\partial \mathbf{x}} = \mathbf{0}$$

(14b) $$\frac{\partial h(\mathbf{x},\lambda)}{\partial \lambda} = -g(\mathbf{x}) + c = 0$$

The second condition is of course nothing more than the original constraint. In practice one must solve the equations (14a) for $\mathbf{x}$ after eliminating λ by algebraic manipulation or from the nature of the equations.

As an example, let us find the maximum value of the quadratic function $f(\mathbf{x}) = \mathbf{x}'\mathbf{A}\mathbf{x}$, where **A** is a positive definite matrix, subject to the constraint $\mathbf{x}'\mathbf{x} = 1$. Then

$$h(\mathbf{x},\lambda) = \mathbf{x}'\mathbf{A}\mathbf{x} - \lambda(\mathbf{x}'\mathbf{x} - 1)$$

and (14a) is the system of linear equations

(15) $$[\mathbf{A} - \lambda\mathbf{I}]\mathbf{x} = \mathbf{0}$$

defining the characteristic vectors of **A**. Premultiplication by $\mathbf{x}'$ and use of the constraint yields $\lambda = \mathbf{x}'\mathbf{A}\mathbf{x}$; clearly, if that quadratic form is to be a maximum then λ must be the *greatest* characteristic root of **A**, and $\mathbf{x}$ its associated vector. The constraint directs that the vector be normalized to length one. Similarly, the minimum value of $f(\mathbf{x})$ subject to the condition would be given by the characteristic vector corresponding to the smallest characteristic root.

Sufficient conditions for maxima and minima in the presence of one or more constraints have been given by Hancock (1960, pp. 114–116) and in certain more current texts, e.g., Goldberger (1964, pp. 45–48).

The derivative of the determinant of the $n \times n$ matrix $\mathbf{A}$ with respect to its element a_{ij} can be found from the expansion of $\mathbf{A}$ in the cofactors of the ith row or jth column. By the first expansion,

$$(16) \qquad \frac{\partial|\mathbf{A}|}{\partial a_{ij}} = \frac{\partial}{\partial a_{ij}}(a_{i1}A_{i1} + \cdots + a_{ij}A_{ij} + \cdots + a_{in}A_{in}) = A_{ij}$$

If $\mathbf{A}$ is symmetric,

$$(17) \qquad \frac{\partial|\mathbf{A}|}{\partial a_{ii}} = A_{ii}$$

and it can be shown from more general formulas for derivatives of determinants, whose elements are in turn functions of other variables, that

$$(18) \qquad \frac{\partial|\mathbf{A}|}{\partial a_{ij}} = 2A_{ij}$$

In the sequel it will be necessary to differentiate quadratic forms and other scalars with respect to elements of their constituent matrices. If $\mathbf{X}$ is an $m \times n$ matrix with general element x_{ij}, its derivative with respect to that element is

$$(19) \qquad \frac{\partial \mathbf{X}}{\partial x_{ij}} = \mathbf{J}_{ij}$$

where $\mathbf{J}_{ij}$ is the $m \times n$ matrix with a one in the ijth position and zeros everywhere else. If $\mathbf{X}$ is symmetric,

$$(20) \qquad \frac{\partial \mathbf{X}}{\partial x_{ij}} = \mathbf{J}_{ij} + \mathbf{J}_{ji} \qquad i \neq j$$

The rule for differentiating a matrix product is similar to that for scalars. Suppose that the elements of the conformable matrices $\mathbf{X}$ and $\mathbf{Y}$ are functions $x_{ij}(z)$, $y_{ij}(z)$ of some variable z. Then

$$(21) \qquad \frac{\partial \mathbf{XY}}{\partial z} = \frac{\partial \mathbf{X}}{\partial z}\mathbf{Y} + \mathbf{X}\frac{\partial \mathbf{Y}}{\partial z}$$

This formula leads to a means of differentiating the inverse of the square nonsingular matrix $\mathbf{X}$. Write

$$\mathbf{I} = \mathbf{X}\mathbf{X}^{-1}$$

Then

$$\frac{\partial \mathbf{I}}{\partial x_{ij}} = \mathbf{J}_{ij}\mathbf{X}^{-1} + \mathbf{X}\frac{\partial \mathbf{X}^{-1}}{\partial x_{ij}} = \mathbf{0}$$

and

$$\frac{\partial \mathbf{X}^{-1}}{\partial x_{ij}} = -\mathbf{X}^{-1}\mathbf{J}_{ij}\mathbf{X}^{-1} \tag{22}$$

If **X** is symmetric,

$$\frac{\partial \mathbf{X}^{-1}}{\partial x_{ij}} = \begin{cases} -\mathbf{X}^{-1}\mathbf{J}_{ii}\mathbf{X}^{-1} & i = j \\ -\mathbf{X}^{-1}(\mathbf{J}_{ij} + \mathbf{J}_{ji})\mathbf{X}^{-1} & i \neq j \end{cases} \tag{23}$$

Many other formulas for matrix and vector differentiation have been given in a series of papers by Dwyer and MacPhail (1948), Dwyer (1967), and Tracy and Dwyer (1969).

2.13 FURTHER READING

The texts of Browne (1958), Graybill (1969), Hadley (1961), Hohn (1964), Perlis (1952), Searle (1966), and Searle and Hausman (1970) serve well both as surveys of matrix algebra and as sources for the proofs of the results of this chapter. Horst (1963) has developed matrix operations at a more verbal level for workers in the social sciences. Frazer, Duncan, and Collar (1963) have discussed a wide variety of topics in the algebra and numerical-analysis methods of matrices. A more advanced treatment motivated by physical and probabilistic applications has been written by Bellman (1960). Householder (1964) has given a theoretical development of numerical techniques for the solution of systems of linear equations, matrix inversion, and extraction of characteristic roots. Roy's monograph (1957) contains in its appendices a wealth of special theorems on partitioned matrices, quadratic forms, and characteristic roots.

2.14 EXERCISES

1. Let

$$\mathbf{A} = \begin{bmatrix} 1 & 2 & -1 \\ -1 & 3 & -1 \\ 2 & 2 & 4 \end{bmatrix} \qquad \mathbf{B} = \begin{bmatrix} 3 & 2 & -1 \\ 2 & 3 & 1 \\ -1 & 1 & 3 \end{bmatrix} \qquad \mathbf{C} = \begin{bmatrix} 2 & 0 \\ -1 & 1 \\ 3 & 2 \end{bmatrix}$$

Perform those matrix operations that are defined:

a. $\mathbf{A} + \mathbf{B}$ b. $\mathbf{A} - 2\mathbf{B}$ c. $\mathbf{A}' + \mathbf{B}$ d. $\mathbf{A} + \mathbf{C}$
e. $(\mathbf{A} + \mathbf{B})'$ f. $(3\mathbf{A}' - 2\mathbf{B})'$

2. If

$$\mathbf{P} = \begin{bmatrix} 3 & 2 & 1 \\ 2 & 5 & -1 \\ 1 & -1 & 3 \end{bmatrix} \qquad \mathbf{Q} = \begin{bmatrix} 1 & 2 & 2 & 1 \\ 1 & 1 & 1 & 4 \\ 1 & 2 & 2 & 1 \end{bmatrix} \qquad \mathbf{R} = \begin{bmatrix} 1 & 2 \\ -5 & 2 \\ 3 & -1 \\ -2 & 2 \end{bmatrix}$$

$$\mathbf{x} = \begin{bmatrix} 1 \\ 0 \\ -1 \end{bmatrix} \qquad \mathbf{y} = \begin{bmatrix} 2 \\ 3 \\ 2 \end{bmatrix} \qquad \mathbf{z} = \begin{bmatrix} -1 \\ -2 \\ -3 \\ -4 \end{bmatrix}$$

compute those matrix products that are defined:

a. $\mathbf{PQ}$ b. $\mathbf{PQR}$ c. $\mathbf{QR}'$ d. $\mathbf{yx}'$
e. $\mathbf{x}'\mathbf{y}$ f. $\mathbf{x}'\mathbf{Px}$ g. $\mathbf{x}'\mathbf{Py}$ h. $\mathbf{P}(\mathbf{x} + \mathbf{y})$

3. If $\mathbf{A}$ is the general $n \times n$ matrix and $\mathbf{j}$ and $\mathbf{E}$ have the patterns defined in Sec. 2.2, interpret these expressions:

a. $\mathbf{j}'\mathbf{A}$ b. $\mathbf{Aj}$ c. diag $(\mathbf{EA})$ d. diag $(\mathbf{AE})$
e. $\mathbf{j}'\mathbf{j}$ f. $\mathbf{Ej}$ g. $\mathbf{E}^2$

4. Which of the following matrices are commutative under multiplication?

$$\mathbf{A} = \begin{bmatrix} 3 & 1 & 0 \\ 1 & 2 & 1 \\ 0 & 1 & 2 \end{bmatrix} \qquad \mathbf{B} = \begin{bmatrix} 1 & 1 & 1 \\ 1 & 1 & 1 \\ 1 & 1 & 1 \end{bmatrix} \qquad \mathbf{C} = \begin{bmatrix} 3 & 0 & 0 \\ 0 & 3 & 0 \\ 0 & 0 & 3 \end{bmatrix} \qquad \mathbf{D} = \begin{bmatrix} 2 & 1 & 0 \\ 1 & 2 & 1 \\ 0 & 1 & 2 \end{bmatrix}$$

5. Calculate the determinants of these matrices using the most convenient methods:

$$\text{a. } \begin{bmatrix} 4 & 2 & 0 \\ 5 & 3 & 0 \\ 6 & 9 & 2 \end{bmatrix} \qquad \text{b. } \begin{bmatrix} 1 & 0.8 & 0.5 \\ 0.8 & 1 & 0.6 \\ 0.5 & 0.6 & 1 \end{bmatrix} \qquad \text{c. } \begin{bmatrix} 5 & 0 & 0 \\ 0 & 3 & 0 \\ 0 & 0 & 1 \end{bmatrix}$$

$$\text{d. } \begin{bmatrix} 1 & 4 & -1 \\ 3 & 12 & -3 \\ 0 & 35 & 7 \end{bmatrix} \qquad \text{e. } \begin{bmatrix} 2 & 0 & 4 & 0 \\ 0 & 3 & 0 & 5 \\ 5 & 0 & 1 & 0 \\ 0 & 4 & 0 & 1 \end{bmatrix} \qquad \text{f. } \begin{bmatrix} 2 & 0 & 1 & 1 & 1 \\ 0 & 2 & 3 & 3 & 3 \\ 1 & 3 & 1 & 0 & 0 \\ 1 & 3 & 0 & 1 & 0 \\ 1 & 3 & 0 & 0 & 1 \end{bmatrix}$$

6. Compute the inverses of these matrices:

$$\text{a. } \begin{bmatrix} 5 & 1 & -2 \\ 2 & 6 & 3 \\ -1 & 0 & 3 \end{bmatrix} \qquad \text{b. } \begin{bmatrix} a & b & b \\ b & a & b \\ b & b & a \end{bmatrix}$$

$$\text{c. } \begin{bmatrix} 5 & 0 & 0 \\ 0 & 8 & 6 \\ 0 & 6 & 5 \end{bmatrix} \qquad \text{d. } \begin{bmatrix} 4 & 3 & 2 & 1 \\ 0 & 3 & 2 & 1 \\ 0 & 0 & 2 & 1 \\ 0 & 0 & 0 & 1 \end{bmatrix}$$

7. Use elementary row or column operations to calculate the ranks of these matrices:

$$\text{a. } \begin{bmatrix} 1 & 0 & 2 & 1 \\ 1 & 1 & 2 & 0 \\ 1 & -1 & 2 & 2 \\ 1 & 1 & 2 & 0 \end{bmatrix} \qquad \text{b. } \begin{bmatrix} 1 & 2 & 3 & 4 & 5 \\ 1 & 0 & -1 & 3 & 1 \\ 2 & 1 & 1 & 0 & 1 \\ 0 & 3 & 8 & -5 & 3 \\ -1 & 2 & 6 & -2 & 3 \\ 1 & 1 & 2 & -3 & 0 \end{bmatrix}$$

8. Find a generalized inverse of the matrix

$$\mathbf{A} = \begin{bmatrix} 3 & 4 & 5 \\ 4 & 5 & 6 \\ 5 & 6 & 7 \end{bmatrix}$$

and use it to solve the system of equations $\mathbf{x}'\mathbf{A}' = [0,1,2]$.

9. Determine the solution to the system of equations

$$\begin{aligned} 2x_1 - x_2 + x_3 &= 2 \\ x_1 + 4x_2 - 2x_3 &= -1 \\ 2x_1 + 2x_2 + x_3 &= 0 \end{aligned}$$

10. Verify that the system of homogeneous equations

$$\begin{aligned} x_1 + x_2 + x_3 &= 0 \\ -x_1 + 2x_2 - x_3 &= 0 \\ 3x_1 + 3x_3 &= 0 \end{aligned}$$

has a nontrivial solution and obtain all linearly independent solutions to the system.

11. Determine the rank of the system of equations

$$\begin{aligned} 2x_1 + x_2 + x_3 &= 0 \\ x_1 + 2x_2 - x_3 + 3x_4 &= 0 \\ x_1 - x_2 + 3x_3 - 4x_4 &= 0 \\ 3x_2 - 4x_3 + 7x_4 &= 0 \end{aligned}$$

and compute its solution by the method of generalized inverses.

12. Identify those vectors of the set

$$\begin{aligned} \mathbf{t}' &= [\tfrac{1}{2}, \tfrac{1}{2}, \tfrac{1}{2}, \tfrac{1}{2}] \\ \mathbf{u}' &= [1, 0, -1, 0] \\ \mathbf{v}' &= \left[\frac{\sqrt{2}}{2}, 0, \frac{\sqrt{2}}{2}, 0\right] \end{aligned}$$

that are orthogonal or orthonormal, and compute the angles of each pair of vectors.

13. Verify that the matrix

$$\begin{bmatrix} \frac{\sqrt{3}}{2} & \frac{1}{2} & 0 \\ -\frac{\sqrt{2}}{4} & \frac{\sqrt{6}}{4} & -\frac{\sqrt{2}}{2} \\ -\frac{\sqrt{2}}{4} & \frac{\sqrt{6}}{4} & \frac{\sqrt{2}}{2} \end{bmatrix}$$

is orthogonal.

14. Write the matrix of the quadratic form

$$2x_1^2 - 2x_1x_2 + x_2^2 + 4x_1x_3 - 3x_3^2$$

and determine whether it is positive definite.

15. Classify the following matrices as positive definite or positive semidefinite:

a. $\begin{bmatrix} 4 & 1 & 2 \\ 1 & 4 & -1 \\ 2 & -1 & 4 \end{bmatrix}$ b. $\begin{bmatrix} 1 & 0 & -1 \\ 0 & 1 & 0 \\ -1 & 0 & 1 \end{bmatrix}$ c. $\begin{bmatrix} 2 & 1 & -1 \\ 1 & 2 & 1 \\ -1 & 1 & 2 \end{bmatrix}$

16. Given the matrix

$$\mathbf{A} = \begin{bmatrix} 2 & \sqrt{2} & 0 \\ \sqrt{2} & 2 & \sqrt{2} \\ 0 & \sqrt{2} & 2 \end{bmatrix}$$

compute its characteristic roots and vectors, and determine the orthogonal matrix reducing it to diagonal form.

17. If

$$\mathbf{A} = \begin{bmatrix} 2 & -1 \\ -1 & 3 \end{bmatrix} \qquad \mathbf{B} = \begin{bmatrix} 5 & 3 \\ 3 & 6 \end{bmatrix}$$

compute sums and products of the characteristic roots of

a. $\mathbf{A}^{-1}\mathbf{B}$ b. $\mathbf{A}\mathbf{B}^{-1}$

18. From the matrix of the function

$$f(x,y,z) = 4x^2 + 4y^2 + 2z^2 + 4xy + 2xz + 2yz$$

can you determine its extremal properties at the point $x = y = z = 0$?

19. Maximize $f(\mathbf{x}) = (\mathbf{a}'\mathbf{x})^2$ subject to the constraint $\mathbf{x}'\mathbf{x} = 1$.

20. Minimize the function

$$f(x,y) = x^2 + 2axy + y^2$$

where $|a| \leq 1$, under the condition $xy = 1$. What is the geometrical interpretation of this problem?

3 SAMPLES FROM THE MULTIVARIATE NORMAL POPULATION

3.1 INTRODUCTION.

Now we are ready to extend the concept of a continuous random variable to variates defined in several dimensions. Our attention will be concentrated on the multidimensional generalization of normally distributed random variables. The older multiple and partial correlation measures for describing relations among the dimensions will be related to the parameters of the multinormal distribution. We shall consider means of estimating these parameters from random samples of observations on the variates as well as the sampling distributions of the estimates and related statistics. In the methods for hypothesis tests and confidence statements on the various parametric functions we shall emphasize procedures which control error rates for several simultaneous inferences. Finally we shall indicate what estimation techniques are available for dealing with samples containing missing observations.

3.2 MULTIDIMENSIONAL RANDOM VARIABLES

Let us define the p-dimensional random variable $\mathbf{X}$ as the vector

$$\mathbf{X}' = [X_1, \ldots, X_p] \tag{1}$$

whose elements are continuous unidimensional random variables with density functions $f_1(x_1), \ldots, f_p(x_p)$ and distribution functions $F_1(x_1), \ldots, F_p(x_p)$. In like manner $\mathbf{X}$ has the *joint distribution function*

$$F(x_1, \ldots, x_p) = P(X_1 \leq x_1, \ldots, X_p \leq x_p) \tag{2}$$

If that function is *absolutely continuous*, it is possible to write

$$F(x_1, \ldots, x_p) = \int_{-\infty}^{x_p} \cdots \int_{-\infty}^{x_1} f(u_1, \ldots, u_p) \, du_1 \cdots du_p \tag{3}$$

where $f(x_1, \ldots, x_p)$ is the *joint density function* of the elements of $\mathbf{X}$. If those quantities are *independent* random variables,

$$\begin{aligned} f(x_1, \ldots, x_p) &= f_1(x_1) \cdots f_p(x_p) \\ F(x_1, \ldots, x_p) &= F_1(x_1) \cdots F_p(x_p) \end{aligned} \tag{4}$$

Conversely, such factorizations of the density and distribution functions imply that the X_i are independent variates. Through the assumption of independence the forms of the joint distribution and density have been exceedingly simplified, and it is for this reason that classical statistical methods require random samples of observations to permit the various joint distributions to have this product form. However, the multivariate statistical models we shall encounter in the sequel will nearly always assume that the elements of the random vector are dependent, and the resulting analytical procedures and distribution theory will be constructed to be valid in the presence of this dependence. Still, as in the univariate case, we shall require that successive sample observation vectors from the multidimensional population have been drawn in such a way that they can be construed as realizations of independent random vectors.

The joint density of any subset of the elements of $\mathbf{X}$ is found by integrating the original joint density over the domain of the variates not in the subset. If the variates have been numbered conveniently, so that the subset consists of the elements $X_1, \ldots, X_p$, and the complement of that set contains $X_{p+1}, \ldots, X_{p+q}$, joint density of the first variates is

$$g(x_1, \ldots, x_p) = \int_{-\infty}^{\infty} \cdots \int_{-\infty}^{\infty} f(x_1, \ldots, x_{p+q}) \, dx_{p+1} \cdots dx_{p+q} \tag{5}$$

and the joint distribution function can be computed by setting the variates of the second subset equal to their upper limits in the original joint distribution function:

$$\begin{aligned} G(x_1, \ldots, x_p) &= P(X_1 \leq x_1, \ldots, X_p \leq x_p) \\ &= F(x_1, \ldots, x_p, \infty, \ldots, \infty) \end{aligned} \tag{6}$$

We note in particular that the marginal density of any single element of $\mathbf{X}$ is

$$f_i(x_i) = \int_{-\infty}^{\infty} \cdots \int_{-\infty}^{\infty} f(x_1, \ldots, x_p) \, dx_1 \cdots dx_{i-1} \, dx_{i+1} \cdots dx_p \tag{7}$$

In such multiple integrals as (5) and (7) it is essential to recognize that the density functions have always been properly defined so that the limits of integration can be stated formally as $-\infty$ and ∞. For example, if Y_1 and Y_2 are independent variates with the common density and distribution functions $g(y)$ and $G(y)$, respectively, the joint density of the new random variables

$$\begin{aligned} X_1 &= \text{smaller of } (Y_1,Y_2) \\ X_2 &= \text{larger of } (Y_1,Y_2) \end{aligned} \tag{8}$$

can be shown to be

$$f(x_1,x_2) = \begin{cases} 2g(x_1)g(x_2) & -\infty < x_1 \leq x_2 < \infty \\ 0 & \text{otherwise} \end{cases} \tag{9}$$

The marginal densities are

$$\begin{aligned} f_1(x_1) &= \int_{-\infty}^{\infty} f(x_1,x_2)\,dx_2 \\ &= \int_{-\infty}^{x_1} 0\,dx_2 + 2\int_{x_1}^{\infty} g(x_1)g(x_2)\,dx_2 \\ &= 2g(x_1)[1 - G(x_1)] \qquad -\infty < x_1 < \infty \end{aligned} \tag{10}$$

$$\begin{aligned} f_2(x_2) &= \int_{-\infty}^{\infty} f(x_1,x_2)\,dx_1 \\ &= 2\int_{-\infty}^{x_2} g(x_1)g(x_2)\,dx_1 + \int_{x_2}^{\infty} 0\,dx_1 \\ &= 2g(x_2)G(x_2) \qquad -\infty < x_2 < \infty \end{aligned} \tag{11}$$

We observe in passing that while $f(x_1,x_2)$ factored into the product of two densitylike functions, the variates X_1 and X_2 are *not* independent, for these factors are not the marginal densities (10) and (11).

Conditional Distributions. It is frequently necessary in multivariate analysis to know the distribution of one set of random variables, given that the variates of a second group have been set equal to specified constant values or have been constrained to lie in some subregion of their complete space. Such distribution and density functions are called *conditional.* The density function of the conditional distribution of $X_1, \ldots, X_p$ given $X_{p+1} = x_{p+1}, \ldots, X_{p+q} = x_{p+q}$ can be shown to be

$$h(x_1, \ldots, x_p|x_{p+1}, \ldots, x_{p+q}) = \frac{f(x_1, \ldots, x_{p+q})}{g(x_{p+1}, \ldots, x_{p+q})} \tag{12}$$

where $f(x_1, \ldots, x_{p+q})$ is the joint density of the complete set of $p + q$ variates and $g(x_{p+1}, \ldots, x_{p+q})$ is the joint density of the q fixed variates. If the two sets of variates are independent, factorization of the joint density implies that the conditional density of the first set is merely the joint density of those random variables. For any admissible set of values of the fixed variates the function (12) has all the properties of a density function, and if the original distribution function is absolutely continuous, the conditional distribution function can be computed in the usual manner as

$$(13)\quad F(x_1, \ldots, x_p | x_{p+1}, \ldots, x_{p+q}) = \frac{\int_{-\infty}^{x_p} \cdots \int_{-\infty}^{x_1} f(u_1, \ldots, u_p, x_{p+1}, \ldots, x_{p+q})\, du_1 \cdots du_p}{g(x_{p+1}, \ldots, x_{p+q})}$$

The fact that $F(\infty, \ldots, \infty | x_{p+1}, \ldots, x_{p+q}) = 1$ should be immediately apparent from (5) and (13).

For an example let us refer to the random variables X_1 and X_2 whose joint density is specified by expression (9). The conditional density of X_1, given that X_2 has the value x_2, is

$$(14)\qquad f(x_1|x_2) = \begin{cases} \dfrac{g(x_1)}{G(x_2)} & -\infty < x_1 \leq x_2 < \infty \\ 0 & \text{elsewhere} \end{cases}$$

The conditional distribution function is

$$(15)\qquad F(x_1|x_2) = \begin{cases} \dfrac{G(x_1)}{G(x_2)} & -\infty < x_1 \leq x_2 < \infty \\ 1 & x_2 \leq x_1 < \infty \end{cases}$$

When x_2 is small, the conditional probability that X_1 will be at most equal to x_1 will be larger than the unconditional probability $G(x_1)$ of that event. Through the conditional distribution function we may use knowledge of the magnitude of X_2 to improve our prediction of the value of the first random variable.

Moments of Multidimensional Variates. The expected value of the random vector **X** is merely the vector of the expectations of its elements:

$$(16)\qquad \mathrm{E}(\mathbf{X}') = [\mathrm{E}(X_1), \ldots, \mathrm{E}(X_p)]$$

Similarly the expectation of a random matrix is the matrix of expected values of the random elements. For the generalization of the variance to multidimensional variates let us first define the *covariance* of the elements X_i and X_j of **X** as the *product moment* of those variates about their respective means:

$$(17)\quad \operatorname{cov}(X_i, X_j) = \mathrm{E}\{[X_i - \mathrm{E}(X_i)][X_j - \mathrm{E}(X_j)]\}$$

$$= E(X_iX_j) - [E(X_i)][E(X_j)]$$

$$= \int_{-\infty}^{\infty}\int_{-\infty}^{\infty} x_ix_jf_{ij}(x_i,x_j)\,dx_i\,dx_j - [E(X_i)][E(X_j)]$$

$$= \sigma_{ij}$$

where $f_{ij}(x_i,x_j)$ is the joint density of X_i and X_j. If $i = j$, the covariance is the variance of X_i, and we shall customarily write $\sigma_{ii} = \sigma_i^2$. The extension of the variance notion to the p-component random vector $\mathbf{X}$ is the matrix of variances and covariances

$$\text{(18)} \qquad E\{[\mathbf{X} - E(\mathbf{X})][\mathbf{X} - E(\mathbf{X})]'\} = \begin{bmatrix} \sigma_{11} & \sigma_{12} & \cdots & \sigma_{1p} \\ \sigma_{12} & \sigma_{22} & \cdots & \sigma_{2p} \\ \cdots & \cdots & \cdots & \cdots \\ \sigma_{1p} & \sigma_{2p} & \cdots & \sigma_{pp} \end{bmatrix} = \mathbf{\Sigma}$$

We shall call this symmetric matrix the *covariance matrix* of $\mathbf{X}$.

It is easily verified from the definition (17) that the covariance of two random variables is unchanged by shifts in the origins of those variates. Thus,

$$\text{(19)} \qquad \operatorname{cov}(X_i + a, X_j + b) = \operatorname{cov}(X_i,X_j)$$

for all real constants a and b. Similarly, changes in scale of the variates affect the covariance by the same factors:

$$\text{(20)} \qquad \operatorname{cov}(cX_i,dX_j) = cd \operatorname{cov}(X_i,X_j)$$

This result leads to the expression for the variance of a linear compound $\mathbf{a}'\mathbf{X} = a_1X_1 + \cdots + a_pX_p$ of random variables:

$$\text{(21)} \qquad \operatorname{var}(\mathbf{a}'\mathbf{X}) = \sum_{i=1}^{p}\sum_{j=1}^{p} a_ia_j\sigma_{ij} = \mathbf{a}'\mathbf{\Sigma}\mathbf{a}$$

The covariance of the two linear compounds $\mathbf{a}'\mathbf{X}$ and $\mathbf{b}'\mathbf{X}$ in the same random variables is the bilinear form

$$\text{(22)} \qquad \operatorname{cov}(\mathbf{a}'\mathbf{X},\mathbf{b}'\mathbf{X}) = \sum_{i=1}^{p}\sum_{j=1}^{p} a_ib_j\sigma_{ij} = \mathbf{a}'\mathbf{\Sigma}\mathbf{b}$$

More generally, if $\mathbf{A}$ and $\mathbf{B}$ are of dimensions $r \times p$ and $s \times p$, respectively, and contain real elements, the covariances of the transformed variates

$$\mathbf{Y} = \mathbf{AX} \qquad \mathbf{Z} = \mathbf{BX}$$

will be given by the matrices

$$\begin{aligned} \text{cov}\ (\mathbf{Y},\mathbf{Y}) &= \mathbf{A\Sigma A'} \\ \text{cov}\ (\mathbf{Z},\mathbf{Z}) &= \mathbf{B\Sigma B'} \\ \text{cov}\ (\mathbf{Y},\mathbf{Z}) &= \mathbf{A\Sigma B'} \end{aligned} \tag{23}$$

The correlation coefficient of the variates X_i and X_j is defined as

$$\rho_{ij} = \frac{\text{cov}\ (X_i,X_j)}{\sqrt{\text{var}\ (X_i)\ \text{var}\ (X_j)}} \tag{24}$$

By the properties (19) and (20) it follows that the correlation is a pure-number invariant under changes of scale and origin of its variates. From the properties of the integrals defining the variance and covariance it can be shown that ρ cannot be less than -1 or greater than 1. If X_i and X_j are independently distributed, their covariance, and hence their correlation, is zero. However, the converse is not generally true, for it is possible to construct examples of highly dependent random variables whose correlation is zero.

In the sequel we shall need the matrix of population correlations

$$\mathbf{P} = \begin{bmatrix} 1 & \rho_{12} & \cdots & \rho_{1p} \\ \rho_{12} & 1 & \cdots & \rho_{2p} \\ \cdots & \cdots & \cdots & \cdots \\ \rho_{1p} & \rho_{2p} & \cdots & 1 \end{bmatrix} \tag{25}$$

If we denote by $\mathbf{D}(\sigma_i)$ the diagonal matrix of the standard deviations of the variates, the covariance and correlation matrices can be related as

$$\begin{aligned} \mathbf{P} &= \mathbf{D}\left(\frac{1}{\sigma_i}\right)\mathbf{\Sigma D}\left(\frac{1}{\sigma_i}\right) \\ \mathbf{\Sigma} &= \mathbf{D}(\sigma_i)\mathbf{PD}(\sigma_i) \end{aligned} \tag{26}$$

3.3 THE MULTIVARIATE NORMAL DISTRIBUTION

In the sequel only the multivariate normal distribution will be used to describe the population out of which our samples of observation vectors will be drawn. There are two compelling reasons for this restriction:

1. A random vector which arose as the sum of a large number of independently and identically distributed random vectors will be distributed according to the multivariate normal distribution as the number of these fundamental source vectors increases without

bound. That is, the usual *central-limit theorem,* which assures a normal distribution for variates which are summations of many independent random inputs, carries over directly to multidimensional inputs. This summation model appears to be a realistic one for many kinds of random phenomena encountered in the life and behavioral sciences.

2. Different models for the variate vectors might lead to rather different joint distributions of the elements whose mathematical complexity would prevent the development of the sampling distributions of the usual test statistics and estimates. Such distributions would have to be provided for each model's fundamental population. However, it seems likely that with the exception of rather pathological cases, the multivariate central-limit theorem would guarantee that the large-sample distributions of test statistics would lead us to similar conclusions about the state of nature.

Now let us develop the multivariate normal density function. Recall from Chap. 1 that the density of a normally distributed random variable X was

$$(1) \qquad \phi(x) = \frac{1}{\sqrt{2\pi}\sigma} \exp\left[-\frac{1}{2}\left(\frac{x-\mu}{\sigma}\right)^2\right] \qquad -\infty < x < \infty$$

The joint density of the independent normal variates is thus

$$(2) \quad \phi(x_1, \ldots, x_p) = \frac{1}{(2\pi)^{p/2}\sigma_1 \cdots \sigma_p} \exp\left[-\frac{1}{2}\sum_{i=1}^{p}\left(\frac{x_i-\mu_i}{\sigma_i}\right)^2\right]$$

If we write $\mathbf{x}' = [x_1, \ldots, x_p]$, $\boldsymbol{\mu}' = [\mu_1, \ldots, \mu_p]$, and

$$\boldsymbol{\Sigma} = \begin{bmatrix} \sigma_1^2 & \cdots & 0 \\ \cdot & \cdots & \cdot \\ 0 & \cdots & \sigma_p^2 \end{bmatrix}$$

the joint density can be given as

$$(3) \qquad \phi(\mathbf{x}) = \frac{1}{(2\pi)^{\frac{1}{2}p}|\boldsymbol{\Sigma}|^{\frac{1}{2}}} \exp\left[-\tfrac{1}{2}(\mathbf{x}-\boldsymbol{\mu})'\boldsymbol{\Sigma}^{-1}(\mathbf{x}-\boldsymbol{\mu})\right]$$

In this representation we see immediately that x has been replaced by a vector variate, μ is now a vector of means, and σ^2 has been generalized to a diagonal matrix. The squared term of the univariate density exponent is now a quadratic form in the deviations of the variates from their means, and the square root of the determinant of $\boldsymbol{\Sigma}$ has assumed the role of the univariate scale factor σ.

The general p-dimensional normal density function is obtained by permitting $\boldsymbol{\Sigma}$ in (3) to be *any* $p \times p$ *symmetric positive definite* matrix. Then $\phi(\mathbf{x})$ is positive for all finite $\mathbf{x}$, and

$$\int_{-\infty}^{\infty} \cdots \int_{-\infty}^{\infty} \phi(\mathbf{x})\, dx_1 \cdots dx_p = 1$$

for all $\boldsymbol{\mu}$, so that $\phi(\mathbf{x})$ is indeed a density function. The ith element of $\boldsymbol{\mu}$ is still the mean of x_i, the ith diagonal element of the more general matrix $\boldsymbol{\Sigma}$ is still the ith variance, and now the ijth element σ_{ij} of $\boldsymbol{\Sigma}$ can be shown to be the covariance of the ith and jth components of $\mathbf{x}$. We see immediately that if all $\frac{1}{2}p(p-1)$ covariances are zero, the p components of $\mathbf{x}$ are independently distributed.

The case of $p = 2$ is especially important in statistical theory. Here

$$\boldsymbol{\mu} = \begin{bmatrix} \mu_1 \\ \mu_2 \end{bmatrix} \qquad \boldsymbol{\Sigma} = \begin{bmatrix} \sigma_1^2 & \rho\sigma_1\sigma_2 \\ \rho\sigma_1\sigma_2 & \sigma_2^2 \end{bmatrix}$$

and the joint density is

$$(4) \quad \phi(x_1,x_2) = \frac{1}{2\pi\sigma_1\sigma_2\sqrt{1-\rho^2}} \exp\left\{-\frac{1}{2}\frac{1}{1-\rho^2}\left[\left(\frac{x_1-\mu_1}{\sigma_1}\right)^2 - 2\rho\frac{x_1-\mu_1}{\sigma_1}\frac{x_2-\mu_2}{\sigma_2} + \left(\frac{x_2-\mu_2}{\sigma_2}\right)^2\right]\right\}$$

The standardized bivariate normal density with means zero and unit variances is immediately apparent. If we let

$$(5) \qquad z_1 = \frac{x_1-\mu_1}{\sigma_1} \qquad z_2 = \frac{x_2-\mu_2}{\sigma_2}$$

the density becomes

$$(6) \quad \phi(z_1,z_2) = \frac{1}{2\pi\sqrt{1-\rho^2}} \exp\left[-\frac{1}{2}\frac{1}{1-\rho^2}(z_1^2 - 2\rho z_1 z_2 + z_2^2)\right]$$

The standardization has removed the scale factors σ_1 and σ_2 from the constant term, and the density contains only the correlation parameter.

Tables of the multinormal distribution function are rarely needed in applied multivariate analysis. The bivariate integral has been tabulated extensively by the National Bureau of Standards (1959), and certain symmetric higher-dimensional probability statements have been computed by Gupta (1963a) for the equicorrelation matrix. Gupta (1963b) has also compiled a bibliography for the distribution, while further references, properties, and specialized probabilities have been collected by Zelen and Severo (1965).

Principal Axes of the Multinormal Density. The exponent

$$(\mathbf{x} - \boldsymbol{\mu})'\boldsymbol{\Sigma}^{-1}(\mathbf{x} - \boldsymbol{\mu})$$

of the multivariate normal density specifies the equation of an ellipsoid in the p-dimensional variate space when it is set equal to some positive constant c. The family of ellipsoids generated by varying c has the common center point $\boldsymbol{\mu}$. The *first principal axis* of each ellipsoid is that line passing through its greatest dimension. If we represent any line through $\boldsymbol{\mu}$ to the surface of an ellipsoid by its coordinates $\mathbf{x}$ on the surface, the first principal axis will have coordinates that maximize its squared half-length

$$(\mathbf{x} - \boldsymbol{\mu})'(\mathbf{x} - \boldsymbol{\mu}) \tag{7}$$

subject to the constraint

$$(\mathbf{x} - \boldsymbol{\mu})'\boldsymbol{\Sigma}^{-1}(\mathbf{x} - \boldsymbol{\mu}) = c \tag{8}$$

that $\mathbf{x}$ be on the surface. For the length to be at its maximum value it is necessary that its derivatives with respect to the elements of $\mathbf{x}$ each equal zero. If we introduce the constraint (8) with the aid of the Lagrangian multiplier λ, the maximand is

$$f(\mathbf{x}) = (\mathbf{x} - \boldsymbol{\mu})'(\mathbf{x} - \boldsymbol{\mu}) - \lambda[(\mathbf{x} - \boldsymbol{\mu})'\boldsymbol{\Sigma}^{-1}(\mathbf{x} - \boldsymbol{\mu}) - c] \tag{9}$$

and its vector of first partial derivatives is

$$\frac{\partial f(\mathbf{x})}{\partial \mathbf{x}} = 2(\mathbf{x} - \boldsymbol{\mu}) - 2\lambda\boldsymbol{\Sigma}^{-1}(\mathbf{x} - \boldsymbol{\mu}) \tag{10}$$

The coordinates of the longest axis must satisfy the equation

$$[\mathbf{I} - \lambda\boldsymbol{\Sigma}^{-1}](\mathbf{x} - \boldsymbol{\mu}) = \mathbf{0} \tag{11}$$

or, since $\boldsymbol{\Sigma}$ is nonsingular, the equivalent condition

$$[\boldsymbol{\Sigma} - \lambda\mathbf{I}](\mathbf{x} - \boldsymbol{\mu}) = \mathbf{0} \tag{12}$$

The coordinates specifying the principal axis are proportional to the elements of a characteristic vector of $\boldsymbol{\Sigma}$. But to which of the p characteristic roots of $\boldsymbol{\Sigma}$ does this vector correspond? Premultiply equation (11) by $4(\mathbf{x} - \boldsymbol{\mu})'$:

$$\begin{aligned} 4(\mathbf{x} - \boldsymbol{\mu})'(\mathbf{x} - \boldsymbol{\mu}) &= 4\lambda(\mathbf{x} - \boldsymbol{\mu})'\boldsymbol{\Sigma}^{-1}(\mathbf{x} - \boldsymbol{\mu}) \\ &= 4\lambda c \end{aligned} \tag{13}$$

For a fixed value of c the length of the principal axis is maximized by taking λ as the *greatest* characteristic root of $\boldsymbol{\Sigma}$. Therefore,

The position of the first principal axis of the concentration ellipsoids is specified by direction cosines which are the elements of the normalized characteristic vector $\boldsymbol{\alpha}_1$ *associated with the greatest characteristic root* λ_1 of $\boldsymbol{\Sigma}$.

The length of the axis for a particular member of the family is $2\sqrt{\lambda_1 c}$.

The second longest axis of the family of ellipsoids has an orientation given by the elements of the vector of the second largest characteristic root. This process is repeated until the equations of the p new axes of the family have been determined. If the characteristic roots are *distinct*, so that

$$\lambda_1 > \lambda_2 > \cdots > \lambda_p > 0$$

the positions of the axes are uniquely specified, and since $\lambda_i \neq \lambda_j$ implies

$$\boldsymbol{\alpha}_i'\boldsymbol{\alpha}_j = 0$$

the p axes are mutually perpendicular. If two successive roots are equal, the ellipsoid is *circular* through the plane generated by the associated vectors, and although two perpendicular axes can be constructed for the common root, their position through the circle is hardly unique. In general, if λ_i is a characteristic root of multiplicity r_i, each of its axes can be chosen to be orthogonal to one another and to the remaining $p - r_i$ axes, although they can be rotated to an infinity of "principal" orientations. For such a characteristic root the ellipsoid has a hyperspherical shape in the r_i-dimensional subspace, and the associated random variates are said to have isotropic variation in that space. We shall defer examples of such principal axes to Chap. 8, where the concept is discussed in full as a statistical technique.

The new variate vector $\mathbf{Y}' = [Y_1, \ldots, Y_p]$ whose elements have values on the principal axes of the concentration ellipsoids is related to the original variates by the transformation

$$\mathbf{Y} = \mathbf{A}'(\mathbf{X} - \boldsymbol{\mu}) \tag{14}$$

where the ith column of $\mathbf{A}$ is the normalized characteristic vector $\boldsymbol{\alpha}_i$. The orthogonality of $\mathbf{A}$ implies that the transformation consists of a rigid rotation of the original axes into the principal axes of the ellipsoids, followed by a shift of the origin of the new axes to the center $\boldsymbol{\mu}$ of the ellipsoid. The covariance matrix of the elements of $\mathbf{Y}$ is

$$\mathbf{A}'\boldsymbol{\Sigma}\mathbf{A} \tag{15}$$

and the variance of the ith principal-axis variate is

$$\begin{aligned} \text{var}(Y_i) &= \boldsymbol{\alpha}_i'\boldsymbol{\Sigma}\boldsymbol{\alpha}_i \\ &= \lambda_i \end{aligned} \tag{16}$$

If the characteristic roots are distinct, or if the vectors associated with multiple roots have been constructed to be orthogonal,

$$\text{(17)} \qquad \operatorname{cov}(Y_i, Y_j) = \boldsymbol{\alpha}_i' \boldsymbol{\Sigma} \boldsymbol{\alpha}_j$$
$$= 0 \qquad i \neq j$$

The principal-axis transformation has resulted in uncorrelated variates whose variances are proportional to the axis lengths of any specific concentration ellipsoid.

Let us apply these results to the bivariate normal density. In the interests of simplicity we shall use the standardized density (6). If we define the family of ellipses by setting

$$\text{(18)} \qquad h = \phi(z_1, z_2)$$

their equation is

$$\text{(19)} \qquad (1 - \rho^2)c = z_1^2 - 2\rho z_1 z_2 + z_2^2$$

where $c = -2 \ln (2\pi h \sqrt{1 - \rho^2})$. The characteristic roots of

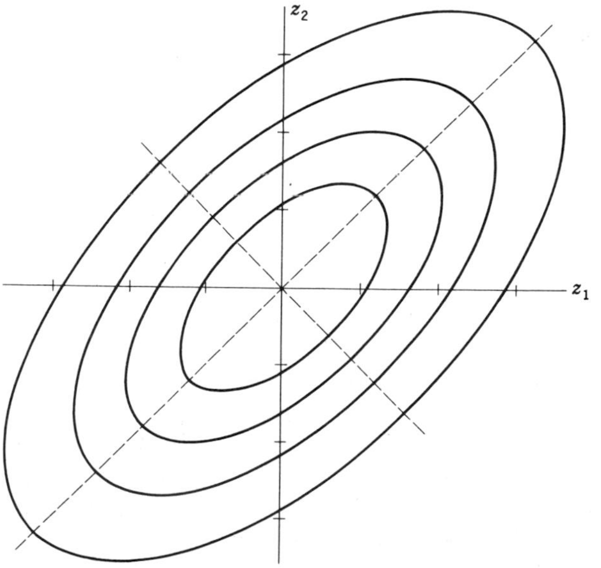

figure 3.1 Concentration ellipses of a bivariate normal density.

$$\mathbf{\Sigma} = \begin{bmatrix} 1 & \rho \\ \rho & 1 \end{bmatrix} \tag{20}$$

are $\lambda_1 = 1 + \rho$, $\lambda_2 = 1 - \rho$, and their respective normalized vectors are

$$\boldsymbol{\alpha}_1' = [\tfrac{1}{2}\sqrt{2}, \tfrac{1}{2}\sqrt{2}] \qquad \boldsymbol{\alpha}_2' = [\tfrac{1}{2}\sqrt{2}, -\tfrac{1}{2}\sqrt{2}]$$

If ρ is positive, the first principal axis is the line $z_2 = z_1$, and the minor axis runs along the line $z_2 = -z_1$. For a negative ρ the equations of major and minor axes are reversed. If $\rho = 0$, the ellipse is a circle, and an infinity of perpendicular axes (including those of the original variates) can be given as "principal." We note for this special case of equal variances that the axes have the same position for all values of ρ. These axes and concentration ellipses for values of h corresponding to 0.2, 0.4, 0.6, and 0.8 of the maximum height $\phi(0,0)$ of the standard bivariate normal density with $\rho = 0.6$ are illustrated in Fig. 3.1.

3.4 CONDITIONAL AND MARGINAL DISTRIBUTIONS OF MULTINORMAL VARIATES

In this section we shall present a number of essential results deriving from the multinormal distribution which will be needed continually in the sequel. Their proofs will usually involve more of the calculus than we have set as a prerequisite for this text, and for them the reader is referred to Anderson (1958) or Graybill (1969, chap. 10). Throughout we shall restrict our attention to the *nonsingular* multinormal distribution, or one whose covariance matrix is of full rank. We begin with this basic result.

Property 3.1. *Let the p-dimensional random vector* $\mathbf{X}$ *be distributed according to the multinormal distribution with mean vector* $\boldsymbol{\mu}$ *and covariance matrix* $\mathbf{\Sigma}$ *of rank p. If* $\mathbf{A}$ *is any* $m \times p$ *matrix of real numbers with rank* $m \le p$, *the new m-component random vector*

$$\mathbf{Y} = \mathbf{AX}$$

is a multinormal random variable with mean $\mathrm{E}(\mathbf{Y}) = \mathbf{A}\boldsymbol{\mu}$ *and covariance matrix* $\mathbf{A\Sigma A}'$.

This property has these consequences:

1. If $m = 1$, the scalar variate $\mathbf{a}'\mathbf{X}$ has the univariate normal distribution with mean $\mathbf{a}'\boldsymbol{\mu}$ and variance $\mathbf{a}'\mathbf{\Sigma}\mathbf{a}$.
2. The joint distribution of any set of elements of $\mathbf{X}$ is multinormal with mean vector and covariance matrix given by the appropriate

elements of $\boldsymbol{\mu}$ and $\boldsymbol{\Sigma}$. In particular, the marginal distribution of each element is univariate normal. However, normal marginals do *not* necessarily imply a joint multinormal distribution: examples of such distributions have been discussed extensively by Kowalski (1973).

Property 3.1 implies that the principal-axis variates defined by the transformation (14) of the preceding section have the multivariate normal distribution and, by dint of their zero correlations, are independently distributed.

The Conditional Density of the Multivariate Normal Distribution. The notions of partial and multiple correlation are doubtless familiar to the reader through their roles in data analysis. We shall now interpret these quantities as parameters of the conditional multinormal density. To obtain that density we start from the nonsingular $(p + q)$-dimensional multinormal population whose variates are written as the partitioned vector $\mathbf{X}' = [\mathbf{X}_1' \quad \mathbf{X}_2']$, where $\mathbf{X}_1'$ contains p elements and $\mathbf{X}_2'$ the remaining q. The mean vector and covariance matrix are correspondingly partitioned as

$$(1) \qquad \boldsymbol{\mu} = \begin{bmatrix} \boldsymbol{\mu}_1 \\ \boldsymbol{\mu}_2 \end{bmatrix} \qquad \boldsymbol{\Sigma} = \begin{bmatrix} \boldsymbol{\Sigma}_{11} & \boldsymbol{\Sigma}_{12} \\ \boldsymbol{\Sigma}_{12}' & \boldsymbol{\Sigma}_{22} \end{bmatrix}$$

The submatrices $\boldsymbol{\Sigma}_{11}$, $\boldsymbol{\Sigma}_{12}$, and $\boldsymbol{\Sigma}_{22}$ have dimensions $p \times p$, $p \times q$, and $q \times q$. By consequence (2) of Property 3.1, $\mathbf{X}_1$ and $\mathbf{X}_2$ are multinormal random vectors with distributions $N(\boldsymbol{\mu}_1,\boldsymbol{\Sigma}_{11})$ and $N(\boldsymbol{\mu}_2,\boldsymbol{\Sigma}_{22})$. The values of the elements of $\boldsymbol{\Sigma}_{12}$ relative to those of $\boldsymbol{\Sigma}_{11}$ and $\boldsymbol{\Sigma}_{22}$ determine the degree and pattern of dependence between the two sets of variates. In particular, we have:

Property 3.2. The random vectors $\mathbf{X}_1$ *and* $\mathbf{X}_2$ *with the multivariate normal distribution described by the parameters* (1) *are independently distributed if and only if* $\boldsymbol{\Sigma}_{12} = \mathbf{0}$.

The conditional density function of $\mathbf{X}_1$ given that the elements of $\mathbf{X}_2$ are fixed, say at $\mathbf{x}_2$, was given for general multidimensional distributions in Sec. 3.2 as

$$g(\mathbf{x}_1|\mathbf{x}_2) = \frac{f(\mathbf{x}_1,\mathbf{x}_2)}{h(\mathbf{x}_2)}$$

In the present case

$$(2) \qquad h(\mathbf{x}_2) = \frac{1}{(2\pi)^{q/2}|\boldsymbol{\Sigma}_{22}|^{1/2}} \exp\left[-\tfrac{1}{2}(\mathbf{x}_2 - \boldsymbol{\mu}_2)'\boldsymbol{\Sigma}_{22}^{-1}(\mathbf{x}_2 - \boldsymbol{\mu}_2)\right]$$

For the computation of $g(\mathbf{x}_1|\mathbf{x}_2)$ we shall need the joint density $f(\mathbf{x}_1,\mathbf{x}_2)$

expressed in terms of the submatrices of $\boldsymbol{\Sigma}$. By expression (5) of Sec. 2.11 the inverse of that matrix is

(3) $\boldsymbol{\Sigma}^{-1} =$

$$\begin{bmatrix} (\boldsymbol{\Sigma}_{11} - \boldsymbol{\Sigma}_{12}\boldsymbol{\Sigma}_{22}^{-1}\boldsymbol{\Sigma}_{12}')^{-1} & -(\boldsymbol{\Sigma}_{11} - \boldsymbol{\Sigma}_{12}\boldsymbol{\Sigma}_{22}^{-1}\boldsymbol{\Sigma}_{12}')^{-1}\boldsymbol{\Sigma}_{12}\boldsymbol{\Sigma}_{22}^{-1} \\ -\boldsymbol{\Sigma}_{22}^{-1}\boldsymbol{\Sigma}_{12}'(\boldsymbol{\Sigma}_{11} - \boldsymbol{\Sigma}_{12}\boldsymbol{\Sigma}_{22}^{-1}\boldsymbol{\Sigma}_{12}')^{-1} & \boldsymbol{\Sigma}_{22}^{-1} + \boldsymbol{\Sigma}_{22}^{-1}\boldsymbol{\Sigma}_{12}'(\boldsymbol{\Sigma}_{11} - \boldsymbol{\Sigma}_{12}\boldsymbol{\Sigma}_{22}^{-1}\boldsymbol{\Sigma}_{12}')^{-1}\boldsymbol{\Sigma}_{12}\boldsymbol{\Sigma}_{22}^{-1} \end{bmatrix}$$

and by (7) of the same section its determinant is

$$|\boldsymbol{\Sigma}| = |\boldsymbol{\Sigma}_{22}| \cdot |\boldsymbol{\Sigma}_{11} - \boldsymbol{\Sigma}_{12}\boldsymbol{\Sigma}_{22}^{-1}\boldsymbol{\Sigma}_{12}'| \tag{4}$$

The joint density can be written as

$$\begin{aligned} (5)\quad f(\mathbf{x}_1,\mathbf{x}_2) &= \frac{1}{(2\pi)^{\frac{1}{2}(p+q)}|\boldsymbol{\Sigma}_{22}|^{\frac{1}{2}} \cdot |\boldsymbol{\Sigma}_{11} - \boldsymbol{\Sigma}_{12}\boldsymbol{\Sigma}_{22}^{-1}\boldsymbol{\Sigma}_{12}'|^{\frac{1}{2}}} \\ &\cdot \exp\{-\tfrac{1}{2}[(\mathbf{x}_1 - \boldsymbol{\mu}_1)'(\boldsymbol{\Sigma}_{11} - \boldsymbol{\Sigma}_{12}\boldsymbol{\Sigma}_{22}^{-1}\boldsymbol{\Sigma}_{12}')^{-1}(\mathbf{x}_1 - \boldsymbol{\mu}_1) \\ &- (\mathbf{x}_2 - \boldsymbol{\mu}_2)'\boldsymbol{\Sigma}_{22}^{-1}\boldsymbol{\Sigma}_{12}'(\boldsymbol{\Sigma}_{11} - \boldsymbol{\Sigma}_{12}\boldsymbol{\Sigma}_{22}^{-1}\boldsymbol{\Sigma}_{12}')^{-1}(\mathbf{x}_1 - \boldsymbol{\mu}_1) \\ &- (\mathbf{x}_1 - \boldsymbol{\mu}_1)'(\boldsymbol{\Sigma}_{11} - \boldsymbol{\Sigma}_{12}\boldsymbol{\Sigma}_{22}^{-1}\boldsymbol{\Sigma}_{12}')^{-1}\boldsymbol{\Sigma}_{12}\boldsymbol{\Sigma}_{22}^{-1}(\mathbf{x}_2 - \boldsymbol{\mu}_2) \\ &+ (\mathbf{x}_2 - \boldsymbol{\mu}_2)'\boldsymbol{\Sigma}_{22}^{-1}\boldsymbol{\Sigma}_{12}'(\boldsymbol{\Sigma}_{11} - \boldsymbol{\Sigma}_{12}\boldsymbol{\Sigma}_{22}^{-1}\boldsymbol{\Sigma}_{12}')^{-1}\boldsymbol{\Sigma}_{12}\boldsymbol{\Sigma}_{22}^{-1}(\mathbf{x}_2 - \boldsymbol{\mu}_2) \\ &+ (\mathbf{x}_2 - \boldsymbol{\mu}_2)'\boldsymbol{\Sigma}_{22}^{-1}(\mathbf{x}_2 - \boldsymbol{\mu}_2)]\} \end{aligned}$$

Upon division by $h(\mathbf{x}_2)$ the term $(2\pi)^{q/2}|\boldsymbol{\Sigma}_{22}|^{\frac{1}{2}}$ in the normalizing constant and the final quadratic form of the exponent vanish. Inspection of the remaining exponential terms shows that the conditional density can be written as

$$\begin{aligned} (6)\quad g(\mathbf{x}_1|\mathbf{x}_2) &= \frac{1}{(2\pi)^{p/2}|\boldsymbol{\Sigma}_{11} - \boldsymbol{\Sigma}_{12}\boldsymbol{\Sigma}_{22}^{-1}\boldsymbol{\Sigma}_{12}'|^{\frac{1}{2}}} \\ &\cdot \exp\{-\tfrac{1}{2}[\mathbf{x}_1 - \boldsymbol{\mu}_1 - \boldsymbol{\Sigma}_{12}\boldsymbol{\Sigma}_{22}^{-1}(\mathbf{x}_2 - \boldsymbol{\mu}_2)]'(\boldsymbol{\Sigma}_{11} - \boldsymbol{\Sigma}_{12}\boldsymbol{\Sigma}_{22}^{-1}\boldsymbol{\Sigma}_{12}')^{-1} \\ &[\mathbf{x}_1 - \boldsymbol{\mu}_1 - \boldsymbol{\Sigma}_{12}\boldsymbol{\Sigma}_{22}^{-1}(\mathbf{x}_2 - \boldsymbol{\mu}_2)]\} \end{aligned}$$

This is the density of a multinormal distribution with mean vector

$$\boldsymbol{\mu}_1 + \boldsymbol{\Sigma}_{12}\boldsymbol{\Sigma}_{22}^{-1}(\mathbf{x}_2 - \boldsymbol{\mu}_2) \tag{7}$$

and covariance matrix

$$\boldsymbol{\Sigma}_{11} - \boldsymbol{\Sigma}_{12}\boldsymbol{\Sigma}_{22}^{-1}\boldsymbol{\Sigma}_{12}' \tag{8}$$

Since the elements of $\mathbf{X}_1$ and $\mathbf{X}_2$ could have been chosen in any combination from the original random vector, we may summarize the preceding result as:

Property 3.3. *The conditional distribution of any set of p variates from a multinormal population with q other variates of the population held constant is multinormal with mean vector and covariance matrix given by expressions* (7) *and* (8).

The random vector

$$\mathbf{X}_{1.2} = \mathbf{X}_1 - \boldsymbol{\mu}_1 - \boldsymbol{\Sigma}_{12}\boldsymbol{\Sigma}_{22}^{-1}(\mathbf{X}_2 - \boldsymbol{\mu}_2) \tag{9}$$

is called the set of *residual* variates, for it represents the discrepancies of the elements of $\mathbf{X}_1$ and their values as predicted from the linear relationship of the conditional distribution mean vector with the $\mathbf{X}_2$ variates. The covariances of the elements of $\mathbf{X}_1$ with those of $\mathbf{X}_{1.2}$ are

$$\begin{aligned}(10)\qquad E[(\mathbf{X}_1 - \boldsymbol{\mu}_1)\mathbf{X}'_{1.2}] &= E(\mathbf{X}_1\mathbf{X}'_{1.2}) \\ &= E\{\mathbf{X}_1[\mathbf{X}'_1 - \boldsymbol{\mu}'_1 - (\mathbf{X}_2 - \boldsymbol{\mu}_2)'\boldsymbol{\Sigma}_{22}^{-1}\boldsymbol{\Sigma}'_{12}]\} \\ &= \boldsymbol{\Sigma}_{11} - \boldsymbol{\Sigma}_{12}\boldsymbol{\Sigma}_{22}^{-1}\boldsymbol{\Sigma}'_{12}\end{aligned}$$

or merely the covariance matrix of the conditional distribution. However, the covariances of the residuals and the fixed set are

$$\begin{aligned}(11)\quad E[(\mathbf{X}_2 - \boldsymbol{\mu}_2)\mathbf{X}'_{1.2}] &= E\{(\mathbf{X}_2 - \boldsymbol{\mu}_2)[\mathbf{X}'_1 - \boldsymbol{\mu}'_1 - (\mathbf{X}_2 - \boldsymbol{\mu}_2)'\boldsymbol{\Sigma}_{22}^{-1}\boldsymbol{\Sigma}'_{12}]\} \\ &= \boldsymbol{\Sigma}'_{12} - \boldsymbol{\Sigma}_{22}\boldsymbol{\Sigma}_{22}^{-1}\boldsymbol{\Sigma}'_{12} \\ &= \mathbf{0}\end{aligned}$$

In the multinormal population the residual and fixed variates are independently distributed.

Partial Correlation. The elements of the covariance matrix (8) of the conditional distribution are called the *partial* variances and covariances, for they measure the variation and dependence of the variates in the first set conditional upon fixed values of those in the *second* set. If we denote the ijth element of the matrix by

$$\sigma_{ij.p+1,\ldots,p+q} \tag{12}$$

the *partial correlation* of the ith and jth variates of the first set with all members of the second set held constant is

$$\rho_{ij.p+1,\ldots,p+q} = \frac{\sigma_{ij.p+1,\ldots,p+q}}{\sqrt{\sigma_{ii.p+1,\ldots,p+q}\sigma_{jj.p+1,\ldots,p+q}}} \tag{13}$$

If we let $\mathbf{D} = \operatorname{diag}(\boldsymbol{\Sigma}_{11} - \boldsymbol{\Sigma}_{12}\boldsymbol{\Sigma}_{22}^{-1}\boldsymbol{\Sigma}'_{12})$, the matrix of partial correlations can be written as

$$\mathbf{D}^{-1/2}(\boldsymbol{\Sigma}_{11} - \boldsymbol{\Sigma}_{12}\boldsymbol{\Sigma}_{22}^{-1}\boldsymbol{\Sigma}'_{12})\mathbf{D}^{-1/2} \tag{14}$$

where the $-\frac{1}{2}$ power of $\mathbf{D}$ indicates that the matrix consists of the square roots of the reciprocals of the elements of $\mathbf{D}$.

If the number of fixed variates is small, or if a computer is not available for the inversion of $\mathbf{\Sigma}_{22}$, the partial correlations may be computed from a recursion relation. Let

$$\rho_{ij.\mathbf{c}} \equiv \rho_{ij.p+1,\ldots,h-1,h+1,\ldots,p+q} \tag{15}$$

represent the $(q-1)$st-order partial correlation with all variates of the second set *except* X_h held constant. Then it is possible to show that the partial of X_i and X_j with the *complete* set fixed can be computed as

$$\rho_{ij.h\mathbf{c}} = \frac{\rho_{ij.\mathbf{c}} - \rho_{ih.\mathbf{c}}\rho_{jh.\mathbf{c}}}{\sqrt{(1-\rho_{ih.\mathbf{c}}^2)(1-\rho_{jh.\mathbf{c}}^2)}} \tag{16}$$

For example, the $\frac{1}{2}p(p-1)(p-2)$ different first-order partials can be computed from

$$\rho_{ij.h} = \frac{\rho_{ij} - \rho_{ih}\rho_{jh}}{\sqrt{(1-\rho_{ih}{}^2)(1-\rho_{jh}{}^2)}} \tag{17}$$

for $h = 1, \ldots, p$; $i, j = 1, \ldots, p$, with $i < j$. In succession the second-order, . . . , $(p+q-2)$th-order partial correlations can be computed. We note that if all correlations of a certain order equal zero, all higher-order partials must also vanish. Derivations of the recursion formula can be found in Anderson (1958) for the case of a multinormal joint distribution of the variates or in Yule and Kendall (1950) or Kendall and Stuart (1961) for a general multidimensional distribution whose second-order moments are finite.

Example 3.1. Covariance matrices with the pattern

$$\mathbf{\Sigma} = \sigma^2 \begin{bmatrix} 1 & \rho & \rho^2 & \rho^3 \\ & 1 & \rho & \rho^2 \\ & & 1 & \rho \\ & & & 1 \end{bmatrix}$$

play an important role in stochastic processes and time-series analysis, and we shall see in Chap. 9 that they are also encountered as models for the dependence structure of intelligence tests. The partial correlation of the $(i-1)$st and $(i+1)$st variates with the ith held constant is

$$\begin{aligned} \rho_{i-1,i+1.i} &= \frac{\rho^2 - \rho \cdot \rho}{1-\rho^2} \\ &= 0 \qquad i = 2, 3 \end{aligned}$$

Similarly the correlation of the first and fourth variates with either the second or third fixed is zero. In the time-series context this implies that a variate X_i is dependent upon its predecessors with smaller subscripts only through its immediate neighbor X_{i-1}. Fixing the value of that random variable leaves X_i and X_{i-2} uncorrelated.

Multiple Correlation. Suppose that in the joint distribution specified by the partitioned mean vector and covariance matrix of (1) the first set contains a single variate X_1 and the second set contains q variates. It is desired to find that linear compound

$$Y = \boldsymbol{\beta}'\mathbf{X}_2 \tag{18}$$

of the second set having the greatest correlation with X_1. That correlation is

$$\frac{\boldsymbol{\beta}'\boldsymbol{\sigma}_{12}}{\sigma_1 \sqrt{\boldsymbol{\beta}'\boldsymbol{\Sigma}_{22}\boldsymbol{\beta}}} \tag{19}$$

where σ_1^2 is the single element of $\boldsymbol{\Sigma}_{11}$ and $\boldsymbol{\sigma}_{12} = \boldsymbol{\Sigma}_{12}'$ is a q-component column vector. The correlation is invariant under changes of location and scale in X_1 and Y, and so it will be necessary to impose some constraints before the linear compound can be uniquely defined. Let these constraints be

$$\sigma_1^2 = 1 \qquad \boldsymbol{\beta}'\boldsymbol{\Sigma}_{22}\boldsymbol{\beta} = 1 \tag{20}$$

The correlation can then be maximized by determining the greatest value of its covariance term subject to the second constraint. Introduction of that restriction through the Lagrangian multiplier λ yields the function

$$f(\boldsymbol{\beta}) = \boldsymbol{\beta}'\boldsymbol{\sigma}_{12} - \tfrac{1}{2}\lambda(\boldsymbol{\beta}'\boldsymbol{\Sigma}_{22}\boldsymbol{\beta} - 1) \tag{21}$$

whose vector of partial derivatives with respect to the elements of $\boldsymbol{\beta}$ is

$$\frac{\partial f(\boldsymbol{\beta})}{\partial \boldsymbol{\beta}} = \boldsymbol{\sigma}_{12} - \lambda\boldsymbol{\Sigma}_{22}\boldsymbol{\beta} \tag{22}$$

and for each derivative to vanish it is necessary that

$$\boldsymbol{\beta} = \frac{1}{\lambda}\boldsymbol{\Sigma}_{22}^{-1}\boldsymbol{\sigma}_{12} \tag{23}$$

It follows from the second constraint of (20) that $\lambda^2 = \boldsymbol{\sigma}_{12}'\boldsymbol{\Sigma}_{22}^{-1}\boldsymbol{\sigma}_{12}$. However, by the invariance property of the correlation any positive value of λ will give the same maximum correlation, and we shall take

$$\boldsymbol{\beta} = \boldsymbol{\Sigma}_{22}^{-1}\boldsymbol{\sigma}_{12} \tag{24}$$

as the vector of *regression coefficients* of the variate X_1 upon the elements of $\mathbf{X}_2$. Frequently, as in Sec. 3.7, it will be necessary to consider all regression coefficients of one set of p variates on a second set of q. We shall write that set of parameters as the $p \times q$ matrix $\mathbf{B} = \boldsymbol{\Sigma}_{12}\boldsymbol{\Sigma}_{22}^{-1}$.

The rather curious use of "regression" is due to Galton (1889), who introduced it in his studies of the correlation of heights of fathers and sons: the heights of sons of either unusually tall or short fathers tend to be closer to the sons' average height than their deviant fathers' values were to the mean for their generation. Galton referred to this phenomenon as a "regression to mediocrity" and named the parameter of the linear relationship accordingly.

The maximum correlation

$$\rho_{1.2\cdots(q+1)} = \frac{\sqrt{\boldsymbol{\sigma}_{12}'\boldsymbol{\Sigma}_{22}^{-1}\boldsymbol{\sigma}_{12}}}{\sigma_1} = \frac{\sqrt{\boldsymbol{\sigma}_{12}'\boldsymbol{\beta}}}{\sigma_1} \tag{25}$$

is called the *multiple correlation coefficient* of X_1 with the variates in $\mathbf{X}_2$. By expression (8) for the single partial variance

$$\sigma_1^2 - \boldsymbol{\sigma}_{12}'\boldsymbol{\Sigma}_{22}^{-1}\boldsymbol{\sigma}_{12} = \sigma_1^2(1 - \rho_{1.2\cdots(q+1)}^2) \tag{26}$$

and the squared multiple correlation coefficient is equal to the proportion of the variance of X_1 which can be attributed to the regression relationship with the variates in X_2.

For the special case of the bivariate normal density, $q = 1$ and the regression coefficient is

$$\beta_{12} = \rho\frac{\sigma_1}{\sigma_2} \tag{27}$$

If the free and fixed variates are interchanged, the new coefficient is

$$\beta_{21} = \rho\frac{\sigma_2}{\sigma_1} \tag{28}$$

These regression coefficients and their associated linear functions giving the means of the conditional distributions are distinguished by referring to the first as the *regression of X_1 on X_2*, and to the second as the *regression of X_2 on X_1*. In either case

$$\rho_{1.2} = \rho_{2.1} = |\rho|$$

for by definition the coefficient can have only nonnegative values.

The multiple correlation is invariant under nonsingular transformations of the original variates. Let

$$Y_1 = aX_1 + b \qquad (29)$$

$$\mathbf{Y}_2 = \mathbf{C}\mathbf{X}_2 + \mathbf{d}$$

where a and b are scalars, the $q \times q$ matrix $\mathbf{C}$ is of full rank, and $\mathbf{d}$ is a column vector of q real constants. The squared multiple correlation of Y_1 with the elements of $\mathbf{Y}_2$ is

$$\rho^2_{1.2\cdots(q+1)} = \frac{a\boldsymbol{\sigma}'_{12}\mathbf{C}'(\mathbf{C}\boldsymbol{\Sigma}_{22}\mathbf{C}')^{-1}\mathbf{C}\boldsymbol{\sigma}_{12}a}{a^2\sigma_1{}^2} \qquad (30)$$

$$= \frac{\boldsymbol{\sigma}'_{12}\boldsymbol{\Sigma}_{22}{}^{-1}\boldsymbol{\sigma}_{12}}{\sigma_1{}^2}$$

This property implies that the same correlation will be obtained from the matrix of correlations as from the covariance matrix. The regression coefficients of the transformed variates (29) are

$$\boldsymbol{\gamma} = a(\mathbf{C}\boldsymbol{\Sigma}_{22}\mathbf{C}')^{-1}\mathbf{C}\boldsymbol{\sigma}_{12} \qquad (31)$$

$$= a\mathbf{C}'^{-1}\boldsymbol{\Sigma}_{22}{}^{-1}\boldsymbol{\sigma}_{12}$$

$$= a\mathbf{C}'^{-1}\boldsymbol{\beta}$$

If $\mathbf{C}$ is a diagonal matrix of scale factors, e.g., the reciprocals of the standard deviations of the variates in $\mathbf{X}_2$, the effect of that scale change is the multiplication of each element of $\boldsymbol{\beta}$ by a/σ_i. For example, the bivariate regression coefficient (27) becomes the dimensionless quantity ρ under the scaling $a = 1/\sigma_1$, $c = 1/\sigma_2$.

If $\boldsymbol{\Sigma}$ and $\boldsymbol{\Sigma}_{22}$ both have rank q, X_1 can be expressed exactly as a linear compound of the q variates of the second set. Then the multiple correlation is exactly unity, and $\boldsymbol{\beta}$ is the coefficient vector of the linear relationship. If $\boldsymbol{\Sigma}_{22}$ has rank less than q, the usual expression for the multiple correlation is indeterminate, and the multiple correlation must be redefined as the maximum correlation of X_1 with a linear compound of a subset of the $\mathbf{X}_2$ variates whose rank is equal to that of $\boldsymbol{\Sigma}_{22}$.

3.5 SAMPLES FROM THE MULTINORMAL POPULATION

In the preceding sections we have discussed the properties of the multivariate normal distribution as though the values of its parameters were known and under our control. Occasionally this information is available, as in the case of models constructed to explain random phenomena in the physical sciences. Here the structure of the model, and usually the response characteristics of any "black boxes" through which its outputs must pass, may be known to the investigator. For such phenomena it is often possible to obtain very large samples of data in a short time and thereby estimate any other unknown parameters. Workers in the life and behavioral sciences rarely have these advantages, and in this section we shall consider methods for estimating the various parameters of the multinormal distribution from the data of small random samples and the sampling properties of these estimates.

Once again we must emphasize that the qualification "random" is essential for the validity of the estimates and their associated significance tests and confidence statements. In practice, randomness means that the sampling units of the observation vectors were drawn independently of one another from some homogeneous population. The units must not have common characteristics or traits which might induce dependence among their vectors: an investigation of mean levels of four biochemical compounds in the brains of a certain strain of mice should not be based upon assays of mice from a few large litters but from a random sample of mice in the strain. Similarly, biased and highly unreliable estimates of the general effect of a tranquilizing drug would be obtained if the sampling units were the daily mood-status responses of a small number of psychiatric patients studied over several weeks. Just as each litter of mice probably has common genetic and biological features, it is reasonable to assume that a patient's affective condition has slowly varying features which will make the ratings on successive days highly dependent. Furthermore, severe shifts in mood of one person on the ward probably induce changes in the other subjects of the study. These sources of dependence reduce the true number of independent observations, and the treatment of such data by classical procedures will lead to spuriously short confidence intervals and overly significant tests of hypotheses.

Let us suppose that these requirements of independence and homogeneity have been met by the sampling units and that N observation vectors have been recorded on the p responses of interest. These values have been summarized in the *data matrix**

$$(1) \qquad \mathbf{X} = \begin{bmatrix} x_{11} & \cdots & x_{1p} \\ \cdot & \cdots & \cdot \\ x_{N1} & \cdots & x_{Np} \end{bmatrix} = \begin{bmatrix} \mathbf{x}_1' \\ \cdot \\ \cdot \\ \cdot \\ \mathbf{x}_N' \end{bmatrix}$$

* In the sequel **X** will denote either the abstract random vector or its matrix of observed values. In each case the meaning will be made explicit from the context.

Let these data be realizations of a p-dimensional random variable distributed according to the multinormal law with mean vector $\boldsymbol{\mu}$ and nonsingular covariance matrix $\boldsymbol{\Sigma}$. The likelihood of the observations (1) is

$$(2)\qquad L(\boldsymbol{\mu},\boldsymbol{\Sigma}) = \frac{1}{(2\pi)^{\frac{1}{2}Np}|\boldsymbol{\Sigma}|^{\frac{1}{2}N}} \exp\left[-\frac{1}{2}\sum_{i=1}^{N}(\mathbf{x}_i - \boldsymbol{\mu})'\boldsymbol{\Sigma}^{-1}(\mathbf{x}_i - \boldsymbol{\mu})\right]$$

or the direct extension of the likelihood notion of Sec. 1.4 to vector-valued observations. As in the univariate case, we shall choose as estimates of the elements of $\boldsymbol{\mu}$ and $\boldsymbol{\Sigma}$ those functions of the observations which maximize (2). If we introduce the sample mean vector

$$(3)\qquad \bar{\mathbf{x}} = \frac{1}{N}\sum_{h=1}^{N}\mathbf{x}_h$$

and the matrix of sums of squares and cross products

$$(4)\qquad \begin{aligned}\mathbf{A} &= \sum_{h=1}^{N}(\mathbf{x}_h - \bar{\mathbf{x}})(\mathbf{x}_h - \bar{\mathbf{x}})' \\ &= \sum_{h=1}^{N}\mathbf{x}_h\mathbf{x}_h' - N\bar{\mathbf{x}}\bar{\mathbf{x}}'\end{aligned}$$

the natural logarithm of the likelihood can be written as

$$(5)\qquad \begin{aligned} l(\boldsymbol{\mu},\boldsymbol{\Sigma}) = -\tfrac{1}{2}Np\ln(2\pi) - \tfrac{1}{2}N\ln|\boldsymbol{\Sigma}| - \tfrac{1}{2}\operatorname{tr}\mathbf{A}\boldsymbol{\Sigma}^{-1} \\ - \tfrac{1}{2}N(\bar{\mathbf{x}} - \boldsymbol{\mu})'\boldsymbol{\Sigma}^{-1}(\bar{\mathbf{x}} - \boldsymbol{\mu})\end{aligned}$$

Since $\boldsymbol{\Sigma}^{-1}$ is a symmetric positive definite matrix, the quadratic-form term will be a minimum only when $\boldsymbol{\mu}$ is equal to $\bar{\mathbf{x}}$, so that the maximum-likelihood estimate of the mean vector is $\bar{\mathbf{x}}$.

The estimate of the covariance matrix is more easily found by computing the maximum-likelihood estimate of $\boldsymbol{\Sigma}^{-1}$ and applying this useful result:

Lemma 3.1. *Let the distribution function of a random variable (univariate or vector-valued) depend upon the parameters*

$$\begin{aligned}\theta_1 &= g_1(\lambda_1, \ldots, \lambda_k) \\ &\cdots\cdots\cdots \\ \theta_k &= g_k(\lambda_1, \ldots, \lambda_k)\end{aligned}$$

which are in turn unique functions of other parameters $\lambda_1, \ldots, \lambda_k$. *The transformation from the* θ_i *set to the* λ_i *is one to one, with inverse functions*

$$\lambda_1 = h_1(\theta_1, \ldots, \theta_k)$$
$$\cdots\cdots\cdots\cdots$$
$$\lambda_k = h_k(\theta_1, \ldots, \theta_k)$$

Then, if the maximum-likelihood estimates of the θ_i *are* $\hat{\theta}_1, \ldots, \hat{\theta}_k$*, those of the* λ_i *are given by*

$$\hat{\lambda}_i = h_i(\hat{\theta}_1, \ldots, \hat{\theta}_k) \qquad i = 1, \ldots, k$$

A proof of the lemma can be found in Anderson's text (1958, pp. 47–48). In the present context, $k = \frac{1}{2}p(p+1)$, the θ_i are the elements θ_{ij} of $\boldsymbol{\Sigma}^{-1}$, and the λ_i are those of $\boldsymbol{\Sigma}$. To find the maximum-likelihood estimate of $\boldsymbol{\theta} \equiv \boldsymbol{\Sigma}^{-1}$ we compute the derivatives

$$\begin{aligned} \frac{\partial l(\boldsymbol{\mu},\boldsymbol{\theta})}{\partial \theta_{ii}} &= \tfrac{1}{2}N\theta^{ii} - \tfrac{1}{2}a_{ii} \\ \frac{\partial l(\boldsymbol{\mu},\boldsymbol{\theta})}{\partial \theta_{ij}} &= N\theta^{ij} - a_{ij} \qquad i < j \end{aligned} \tag{6}$$

where θ^{ii}, θ^{ij} are the elements of $\boldsymbol{\theta}^{-1}$. Hence, the estimate is

$$\hat{\boldsymbol{\theta}} = \left(\frac{1}{N}\mathbf{A}\right)^{-1} \tag{7}$$

and by Lemma 3.1, the maximum-likelihood estimate of $\boldsymbol{\Sigma}$ is

$$\hat{\boldsymbol{\Sigma}} = \frac{1}{N}\mathbf{A} \tag{8}$$

It is easily shown by an application of the expectation operator that $\hat{\boldsymbol{\Sigma}}$ is a biased estimate. We shall adopt instead the unbiased modification

$$\mathbf{S} = \frac{1}{N-1}\mathbf{A} \tag{9}$$

$\mathbf{S}$ will be referred to in the sequel as the *sample covariance matrix.*

Frequently observations are collected on k independent groups of sampling units. The responses are described by a multinormal random variable with mean vector $\boldsymbol{\mu}_h$ in the hth group and a covariance matrix $\boldsymbol{\Sigma}$ *common* to all groups. Write the data matrix for group h as

$$\mathbf{X}_h = \begin{bmatrix} x_{11h} & \cdots & x_{1ph} \\ \cdots & \cdots & \cdots \\ x_{N_h1h} & \cdots & x_{N_hph} \end{bmatrix} \tag{10}$$

with ith row $\mathbf{x}'_{ih}$. The maximum-likelihood estimate of $\boldsymbol{\mu}_h$ is the sample mean vector $\bar{\mathbf{x}}_h$ of the hth group, and the unbiased estimate of $\mathbf{\Sigma}$ is

$$(11) \qquad \mathbf{S} = \frac{1}{N-k} \sum_{h=1}^{k} \sum_{i=1}^{N_h} (\mathbf{x}_{ih} - \bar{\mathbf{x}}_h)(\mathbf{x}_{ih} - \bar{\mathbf{x}}_h)'$$

$$= \frac{1}{N-k} \sum_{h=1}^{k} \mathbf{A}_h$$

where $\mathbf{A}_h$ is the matrix (4) of sums of squares and cross products within the hth group and $N = N_1 + \cdots + N_k$.

Distribution of the Sample Mean Vector. It is easily shown by Property 3.1 of the preceding section that the sample mean vector of N independent observations from the multinormal population $N(\boldsymbol{\mu},\mathbf{\Sigma})$ is also a multinormal random variable with parameters $\boldsymbol{\mu}$ and $1/N\mathbf{\Sigma}$. This is a direct generalization of the distribution of the univariate mean, and its implications for constructing significance tests and confidence regions for $\boldsymbol{\mu}$ when $\mathbf{\Sigma}$ is known will be exploited in the next chapter.

The Wishart Distribution. The matrices $\mathbf{A}$ and $\sum_{h=1}^{k} \mathbf{A}_h$ defined by (4) and (11) can be written as the sums of products of $N-1$ and $N-k$ independent p-dimensional random vectors with the common distribution $N(\mathbf{0},\mathbf{\Sigma})$. In general, any symmetric positive definite matrix $\mathbf{A}$ of quadratic and bilinear forms which can be transformed to the sum

$$(12) \qquad \sum_{i=1}^{n} \mathbf{y}_i\mathbf{y}_i'$$

whose p-component vectors $\mathbf{y}_i$ are independently distributed according to the distribution $N(\mathbf{0},\mathbf{\Sigma})$ is said to have the *Wishart* distribution (Wishart, 1928). The density function of $\mathbf{A}$ is

$$(13) \qquad w(\mathbf{A};\mathbf{\Sigma},n) = \begin{cases} \dfrac{|\mathbf{A}|^{\frac{1}{2}(n-p-1)} \exp\left(-\frac{1}{2} \operatorname{tr} \mathbf{A}\mathbf{\Sigma}^{-1}\right)}{2^{np/2}\pi^{\frac{1}{4}p(p-1)}|\mathbf{\Sigma}|^{\frac{1}{2}n} \prod_{i=1}^{p} \Gamma[\frac{1}{2}(n+1-i)]} & \mathbf{A} \text{ positive definite} \\ 0 & \text{elsewhere} \end{cases}$$

For the single-sample matrix (4) $n = N - 1$, and for the k-sample within-groups matrix in expression (11) $n = N - k$. We shall refer to n as the degrees-of-freedom parameter of the Wishart distribution; together with $\mathbf{\Sigma}$ it specifies the form of the density function.

If $p = 1$ and $\mathbf{\Sigma} = 1$, the Wishart density becomes that of the chi-squared distribution with n degrees of freedom. Wishart-distributed matrices possess many of the properties of chi-squared variates. In particular the following will be useful in the sequel:

Property 3.4. *The sample mean vector* $\bar{\mathbf{x}}$ *and the matrix of sums of squares and cross products* $\mathbf{A}$ *computed from the same sample are independently distributed.*

Property 3.5. *If* $\mathbf{A}_1, \ldots, \mathbf{A}_k$ *are independently distributed as Wishart matrices with the common parameter matrix* $\boldsymbol{\Sigma}$ *and respective degrees of freedom* $n_1, \ldots, n_k$, *their sum has the Wishart distribution with parameters* $\boldsymbol{\Sigma}$ *and* $n = n_1 + \cdots + n_k$.

Systematic treatments of these and many other results for the Wishart distribution have been given by Anderson (1958) and Press (1972).

3.6 CORRELATION AND REGRESSION

If we reparametrize the covariance matrix into a new set of parameters consisting of the p variances and the $\frac{1}{2}p(p-1)$ correlations, Lemma 3.1 states that the maximum-likelihood estimate of ρ_{ij} is given by the familiar formula

$$(1) \qquad r_{ij} = \frac{s_{ij}}{\sqrt{s_{ii}s_{jj}}}$$

$$= \frac{\sum_{h=1}^{N} x_{hi}x_{hj} - \frac{1}{N}\left(\sum_{h=1}^{N} x_{hi}\right)\left(\sum_{h=1}^{N} x_{hj}\right)}{\sqrt{\left[\sum_{h=1}^{N} x_{hi}^2 - \frac{1}{N}\left(\sum_{h=1}^{N} x_{hi}\right)^2\right]\left[\sum_{h=1}^{N} x_{hj}^2 - \frac{1}{N}\left(\sum_{h=1}^{N} x_{hj}\right)^2\right]}}$$

The matrix of sample correlations can be computed as

$$(2) \qquad \mathbf{R} = \mathbf{D}\left(\frac{1}{s_i}\right)\mathbf{S}\mathbf{D}\left(\frac{1}{s_i}\right)$$

$$= \mathbf{D}\left(\frac{1}{\sqrt{a_{ii}}}\right)\mathbf{A}\mathbf{D}\left(\frac{1}{\sqrt{a_{ii}}}\right)$$

where $\mathbf{D}(\cdot)$ denotes the diagonal matrix containing the reciprocals of either the standard deviations or the square roots of the diagonal elements of $\mathbf{A}$. As in the population matrix, $\mathbf{R}$ is always positive semidefinite or definite with unit diagonal elements. Its correlations are similarly bounded as $-1 \leq r_{ij} \leq 1$. The r_{ij} are biased estimates, although alternative minimum-variance unbiased estimates have been developed by Olkin and Pratt (1958).

In the case of a random sample of N observation pairs from the bivariate normal population with $\rho = 0$ the distribution of r has this density function:

$$(3) \qquad p(r) = \begin{cases} \dfrac{1}{\sqrt{\pi}} \dfrac{\Gamma[(N-1)/2]}{\Gamma[(N-2)/2]} (1 - r^2)^{\frac{1}{2}(N-4)} & -1 \leq r \leq 1 \\ 0 & \text{elsewhere} \end{cases}$$

The density is symmetric about $r = 0$. The variate

$$(4) \qquad t = r\sqrt{\frac{N-2}{1-r^2}}$$

has the Student-Fisher t distribution with $N - 2$ degrees of freedom. Given the sample correlation r, we can test the hypothesis

$$(5) \qquad H_0: \quad \rho = 0$$

of no correlation in the population against the alternative

$$(6) \qquad H_1: \quad \rho \neq 0$$

with the probability α of a Type I error by the rule

$$(7) \qquad \begin{aligned} &\text{Accept } H_0 \text{ if } |r|\sqrt{\frac{N-2}{1-r^2}} \leq t_{\frac{1}{2}\alpha; N-2} \\ &\text{Accept } H_1 \text{ if } |r|\sqrt{\frac{N-2}{1-r^2}} > t_{\frac{1}{2}\alpha; N-2} \end{aligned}$$

$t_{\frac{1}{2}\alpha; N-2}$ is the upper 50α percentage point of the t distribution with $N - 2$ degrees of freedom. In the choice of the alternative H_1 we made no prediction of the sign of the correlation. If, through the nature of the variates or the physical or experimental setting in which the observations were obtained, only positive values of ρ could be admitted to consideration, the alternative

$$(8) \qquad H_1': \quad \rho > 0$$

might have been specified. For a test of level α the decision rule would be

$$(9) \qquad \begin{aligned} &\text{Accept } H_0 \text{ if } r\sqrt{\frac{N-2}{1-r^2}} \leq t_{\alpha; N-2} \\ &\text{Accept } H_1' \text{ if } r\sqrt{\frac{N-2}{1-r^2}} > t_{\alpha; N-2} \end{aligned}$$

Similarly, if only negative correlations were admissible, the alternative would be

$$H_1'':\quad \rho < 0 \tag{10}$$

and the decision rule

$$\begin{aligned} &\text{Accept } H_0 \text{ if } r\sqrt{\frac{N-2}{1-r^2}} \geq -t_{\alpha;N-2} \\ &\text{Accept } H_1'' \text{ if } r\sqrt{\frac{N-2}{1-r^2}} < -t_{\alpha;N-2} \end{aligned} \tag{11}$$

would be employed. These latter tests have greater power than the first decision rule and should be used whenever restrictions can be made on the alternative value of ρ.

If $\rho \neq 0$ the distribution of r has a complicated form. The variance tends to zero as ρ becomes large, and at the same time the density becomes very skewed. Although the distribution approaches normality for large samples, that limit is of little practical use because of the skewness and variance properties. Fisher (1921) showed that the monotonic transformation

$$\begin{aligned} z &= \tfrac{1}{2} \ln \frac{1+r}{1-r} \\ &= \tanh^{-1} r \end{aligned} \tag{12}$$

produced an asymptotically normal variate with mean

$$\zeta \approx \tfrac{1}{2} \ln \frac{1+\rho}{1-\rho} \tag{13}$$

and variance

$$\operatorname{var}(z) \approx \frac{1}{N-3} \tag{14}$$

The large-sample variance of z does not involve the unknown parameter ρ, and it was this variance-stabilizing property that led Fisher to his z transformation. Values of the transformation are given in Table 5 of the Appendix.

With the aid of the z transformation it is possible to test

$$H_0:\quad \rho = \rho_0 \tag{15}$$

against the alternative

$$H_1:\quad \rho \neq \rho_0$$

If r has been computed from N independent pairs of observations and the test is to be of level α, the decision rule is

$$\text{(16)} \qquad \begin{aligned} &\text{Accept } H_0 \text{ if } |z - \zeta_0| \sqrt{N-3} \le z_{\frac{1}{2}\alpha} \\ &\text{Accept } H_1 \text{ if } |z - \zeta_0| \sqrt{N-3} > z_{\frac{1}{2}\alpha} \end{aligned}$$

ζ_0 is the z transformation evaluated for $r = \rho_0$, and $z_{\frac{1}{2}\alpha}$ is the upper 50α percentage point of the standard normal distribution function. Similar decision rules for one-sided alternatives can be obtained by analogy with the rules (9) and (11). The acceptance region of (16) leads to the $100(1 - \alpha)$ percent confidence interval for ρ:

$$\text{(17)} \qquad \tanh\left(z - \frac{z_{\frac{1}{2}\alpha}}{\sqrt{N-3}}\right) \le \rho \le \tanh\left(z + \frac{z_{\frac{1}{2}\alpha}}{\sqrt{N-3}}\right)$$

If two independent random samples of sizes N_1 and N_2 have been drawn from bivariate normal populations, the test of the hypothesis

$$\text{(18)} \qquad H_0\colon \quad \rho_1 = \rho_2$$

of the equality of their correlations can be made by applying the z transformation to each sample coefficient and computing

$$\text{(19)} \qquad d = \frac{z_1 - z_2}{\sqrt{1/(N_1 - 3) + 1/(N_2 - 3)}}$$

For a test of level α and the two-sided alternative

$$H_1\colon \quad \rho_1 \ne \rho_2$$

the decision rule would be

$$\text{(20)} \qquad \begin{aligned} &\text{Accept } H_0 \text{ if } |d| \le z_{\frac{1}{2}\alpha} \\ &\text{Accept } H_1 \text{ if } |d| > z_{\frac{1}{2}\alpha} \end{aligned}$$

The rules for one-sided alternatives should be readily apparent. Rao (1965, pp. 364–365) has offered a test of the equality of the population correlations of k independent samples through an analysis of variance of their z transformations.

The large-sample normal distribution of z can be used to obtain approximate power functions of tests of hypotheses on correlations. The power of the single-sample test of H_0: $\rho = \rho_0$ against H_1: $\rho \ne \rho_0$ is

$$(21)\quad 1 - \beta(\rho) = \Phi[-z_{\frac{1}{2}\alpha} + (\zeta_0 - \zeta)\sqrt{N-3}] + 1 - \Phi[z_{\frac{1}{2}\alpha} + (\zeta_0 - \zeta)\sqrt{N-3}]$$

$\Phi(\cdot)$ is the standard normal distribution function. To compute the power or the probability $\beta(\rho)$ of a Type II error for a particular value of ρ we first find ζ_0 and ζ from Appendix Table 5 and then evaluate (21) from Appendix Table 1. Similarly, the power functions for the α-level tests of H_0: $\rho = \rho_0$ against the one-sided alternatives

$$H_1': \rho > \rho_0 \qquad H_1'': \rho < \rho_0$$

are

$$(22)\quad \begin{aligned} 1 - \beta(\rho) &= 1 - \Phi[z_\alpha + (\zeta_0 - \zeta)\sqrt{N-3}] \\ 1 - \beta(\rho) &= \Phi[-z_\alpha + (\zeta_0 - \zeta)\sqrt{N-3}] \end{aligned}$$

respectively.

Example 3.2. It is suspected that the excretion levels of two biochemical compounds during a stress situation are correlated. The biological processes that are involved imply that this correlation could only be positive. To test the null hypothesis H_0: $\rho = 0$ at the $\alpha = 0.01$ level against the alternative H_1': $\rho > 0$ we wish to design an experiment based on a sufficient number N of independent subjects' assays so that a population correlation as low as $\rho = 0.20$ could be detected with power 0.95. The asymmetric values of the α and β probabilities reflect the relative consequences of reporting a spurious correlation or failing to detect a real, though small, association. Now $z_{0.01} = 2.33$, $\zeta_0 = 0$, and $\zeta = 0.2027$, and the power of the test must satisfy the condition

$$0.95 \leq 1 - \Phi(2.33 - 0.2027\sqrt{N-3})$$

Since $\Phi(-1.645) = 0.05$, it is necessary that the minimum sample size satisfy the equation $-1.645 = 2.33 - 0.2027\sqrt{N-3}$, and therefore at least 389 independent pairs of observations must be collected in the study. This is a formidable number, but it is the price we must pay to achieve the rather stringent α and β probability levels.

The distribution of the correlation coefficient was derived by Fisher (1915). The distribution in the noncentral case of ρ unequal to zero was tabulated by David (1938). Expressions for the distributions and moments of r and z have been intensively investigated by Hotelling (1953).

Partial Correlation. In Sec. 3.4 we defined the partial correlations of the variates $X_1, \ldots, X_p$ with $X_{p+1}, \ldots, X_{p+q}$ as the correlations computed from the conditional covariance matrix

$$\mathbf{\Sigma}_{11.2} = \mathbf{\Sigma}_{11} - \mathbf{\Sigma}_{12}\mathbf{\Sigma}_{22}^{-1}\mathbf{\Sigma}_{12}'$$

If we have a random sample of N complete observation vectors on the $p + q$ variates with correspondingly partitioned covariance and correlation matrices

(23)
$$\mathbf{S} = \begin{bmatrix} \mathbf{S}_{11} & \mathbf{S}_{12} \\ \mathbf{S}_{12}' & \mathbf{S}_{22} \end{bmatrix} \qquad \mathbf{R} = \begin{bmatrix} \mathbf{R}_{11} & \mathbf{R}_{12} \\ \mathbf{R}_{12}' & \mathbf{R}_{22} \end{bmatrix}$$

respectively, by Lemma 3.1 the maximum-likelihood estimate of the partial correlation matrix $\mathbf{P}_{11.2}$ is

(24)
$$\begin{aligned} \mathbf{R}_{11.2} &= \mathbf{D}_s^{-\frac{1}{2}}(\mathbf{S}_{11} - \mathbf{S}_{12}\mathbf{S}_{22}^{-1}\mathbf{S}_{12}')\mathbf{D}_s^{-\frac{1}{2}} \\ &= \mathbf{D}_r^{-\frac{1}{2}}(\mathbf{R}_{11} - \mathbf{R}_{12}\mathbf{R}_{22}^{-1}\mathbf{R}_{12}')\mathbf{D}_r^{-\frac{1}{2}} \end{aligned}$$

where

(25)
$$\begin{aligned} \mathbf{D}_s &= \operatorname{diag}(\mathbf{S}_{11} - \mathbf{S}_{12}\mathbf{S}_{22}^{-1}\mathbf{S}_{12}') \\ \mathbf{D}_r &= \operatorname{diag}(\mathbf{R}_{11} - \mathbf{R}_{12}\mathbf{R}_{22}^{-1}\mathbf{R}_{12}') \end{aligned}$$

In keeping with the notation of Sec. 3.4 we denote the ijth element of $\mathbf{R}_{11.2}$ as $r_{ij.p+1,\ldots,p+q}$. We note that the recursion relation (16) of that section also holds for sample correlations.

The distribution of any partial correlation is the same as that of the simple coefficient with modified degrees of freedom and population correlation parameters. The distribution follows from this useful property of sums of squares and products matrices in multinormal variates:

Property 3.6. *If the* $(p + q) \times (p + q)$ *partitioned positive definite matrix*

(26)
$$\mathbf{A} = \begin{bmatrix} \mathbf{A}_{11} & \mathbf{A}_{12} \\ \mathbf{A}_{12}' & \mathbf{A}_{22} \end{bmatrix}$$

has the Wishart distribution with a similarly partitioned covariance matrix $\mathbf{\Sigma}$ *and degrees-of-freedom parameter* n, *then*

(27)
$$\mathbf{A}_{11.2} = \mathbf{A}_{11} - \mathbf{A}_{12}\mathbf{A}_{22}^{-1}\mathbf{A}_{12}'$$

has the Wishart distribution with degrees of freedom $n - q$ *and matrix parameter* $\mathbf{\Sigma}_{11.2} = \mathbf{\Sigma}_{11} - \mathbf{\Sigma}_{12}\mathbf{\Sigma}_{22}^{-1}\mathbf{\Sigma}_{12}'$.

It follows immediately from this property that the ijth partial correlation of $\mathbf{R}_{11.2}$ has the usual density (3) with N replaced by $N - q$ when $\rho_{ij.p+1,\ldots,p+q} = 0$. The statistic (4) with a similar change from N to $N - q$ has the t distribution with $N - q - 2$ degrees of freedom under the null hypothesis. Corresponding changes in the parameters also give the noncentral distribution. In that way all the sampling properties, hypothesis tests, and confidence statements for simple correlation can be carried over to the partial case by reducing the sample size or degrees of freedom by the number of variates held constant. This result is due to Fisher (1924).

Multiple Correlation and Regression. Let us assume that our $(q+1)$-dimensional multinormal distribution has the form introduced in Sec. 3.4, in which the first variate plays a role of "dependence" upon the remaining q "independent" variates. If a random sample of N observations has been drawn, the partitioned population and sample covariance matrices can be written as

$$\text{(28)} \qquad \mathbf{\Sigma} = \begin{bmatrix} \sigma_1^2 & \boldsymbol{\sigma}_{12}' \\ \boldsymbol{\sigma}_{12} & \mathbf{\Sigma}_{22} \end{bmatrix} \qquad \mathbf{S} = \begin{bmatrix} s_1^2 & \mathbf{s}_{12}' \\ \mathbf{s}_{12} & \mathbf{S}_{22} \end{bmatrix}$$

and the estimate of the squared multiple correlation coefficient of the first variate with the other q is

$$\text{(29)} \qquad R^2 \equiv R^2_{1.2\ldots(q+1)} = \frac{\mathbf{s}_{12}'\mathbf{S}_{22}^{-1}\mathbf{s}_{12}}{s_1^2}$$

The estimate of the regression coefficient vector is

$$\text{(30)} \qquad \hat{\boldsymbol{\beta}} = \mathbf{S}_{22}^{-1}\mathbf{s}_{12}$$

When the first variate is independently distributed of the remaining q, $\boldsymbol{\sigma}_{12} = \mathbf{0}$ and the population multiple correlation is of course equal to zero. In that case

$$\text{(31)} \qquad F = \frac{N-q-1}{q}\,\frac{R^2}{1-R^2}$$

has the F distribution with degrees of freedom q and $N-q-1$. To test at the α level the equivalent hypotheses

$$\text{(32)} \qquad H_0\colon\ \boldsymbol{\sigma}_{12} = \mathbf{0} \qquad H_0\colon\ \boldsymbol{\beta} = \mathbf{0} \qquad H_0\colon\ \rho_{1.2\ldots(q+1)} = 0$$

we compute R^2 and its associated F ratio and reject H_0 if

$$F > F_{\alpha;q,N-q-1}$$

The distribution of R^2 computed from samples from multinormal populations was originally derived by Fisher (1928), and its form is a rather complicated one for $\rho > 0$. Wishart (1931) has determined its moments, and in particular if $\rho = 0$,

$$\text{(33)} \qquad E(R^2) = \frac{q}{N-1}$$

and

$$\text{var}(R^2) = \frac{2(N-q-1)q}{(N^2-1)(N-1)} \tag{34}$$

We see immediately that as the number of responses $q+1$ in the system approaches the sample size, the multiple correlation coefficient tends to unity even when the population value is zero. Without probability statements as to their significance, sample coefficients computed from a large set of responses can be misleading. Furthermore, as the determinant of $\mathbf{S}_{22}$ becomes close to zero as q approaches N, the sample regression coefficients become unstable and frequently bear no resemblance to the original dimensions and scales of the responses.

Frequently it is necessary to test hypotheses of the sort

$$H_0\colon\ \beta_j = \beta_j^0 \qquad j = 2, \ldots, q+1 \tag{35}$$

on certain predetermined elements of the regression vector $\boldsymbol{\beta} = \boldsymbol{\Sigma}_{22}^{-1}\boldsymbol{\sigma}_{12}$. It is essential that the coefficients were selected before examining their sample estimates; otherwise the multiple testing methods of Sec. 3.7 must be used for proper protection against Type I errors. For convenience let us assume that the variates have been arranged in the order X_1, X_j, $\mathbf{X}_c' = [X_2, \ldots, X_{j-1}, X_{j+1}, \ldots, X_{q+1}]$, where the symbol $\mathbf{c}$ will stand for the sequence of subscripts $2, \ldots, j-1, j+1, \ldots, q+1$. We partition the covariance matrix as

$$\mathbf{S} = \begin{bmatrix} s_1^2 & s_{1j} & \mathbf{s}_{1c}' \\ s_{1j} & s_j^2 & \mathbf{s}_{jc}' \\ \mathbf{s}_{1c} & \mathbf{s}_{jc} & \mathbf{S}_{cc} \end{bmatrix} \tag{36}$$

$$= \begin{bmatrix} s_1^2 & \mathbf{s}_{12}' \\ \mathbf{s}_{12} & \mathbf{S}_{22} \end{bmatrix}$$

and define the sample conditional variances

$$s_{j.c}^2 = s_j^2(1 - R_{j.c}^2) \tag{37}$$

$$= s_j^2 - \mathbf{s}_{jc}'\mathbf{S}_{cc}^{-1}\mathbf{s}_{jc}$$

and

$$s_{1.j,c}^2 = s_1^2(1 - R_{1.j,c}^2) \tag{38}$$

$$= s_1^2 - (s_{1j}, \mathbf{s}_{1c}') \begin{bmatrix} s_j^2 & \mathbf{s}_{jc}' \\ \mathbf{s}_{jc} & \mathbf{S}_{cc} \end{bmatrix}^{-1} \begin{bmatrix} s_{1j} \\ \mathbf{s}_{1c} \end{bmatrix}$$

$$= s_1^2 - \mathbf{s}_{12}'\mathbf{S}_{22}^{-1}\mathbf{s}_{12}$$

Bartlett (1933) has shown that

$$t = \sqrt{N - q - 1}\,\frac{s_{j.c}}{s_{1.j,c}}\,(\hat{\beta}_j - \beta_j) \tag{39}$$

has the Student-Fisher t distribution with $N - q - 1$ degrees of freedom when β_j is the true parameter. We can test the hypothesis (35) at the α level for specified j by the usual t critical values and decision rules. Finally, we note that (39) is algebraically and distributionally equivalent to the statistic for tests on a single regression coefficient in the standard linear model of full rank with *fixed* "independent" variables.

Example 3.3. In the course of an analysis of the relation of the subtest scores of the Wechsler Adult Intelligence Scale (WAIS) to $N = 933$ examinees' chronological age and years of formal education (Birren and Morrison, 1961), these correlations were obtained among the variates performance subtests total, verbal subtests total, age, and education:

$$\begin{bmatrix} 1 & 0.72 & -0.44 & 0.60 \\ & 1 & -0.13 & 0.68 \\ & & 1 & -0.29 \\ & & & 1 \end{bmatrix}$$

The sample standard deviations of the four responses were 11.228, 14.404, 10.926, and 3.095, respectively. Let us begin our study of this four-dimensional system by computing the sample partial correlations with age and education successively held constant:

$$\begin{array}{c} \textit{Age constant} \\ \begin{matrix} P \\ V \\ E \end{matrix}\begin{bmatrix} 1 & 0.74 & 0.55 \\ & 1 & 0.68 \\ & & 1 \end{bmatrix} \end{array} \qquad \begin{array}{c} \textit{Education constant} \\ \begin{matrix} P \\ V \\ A \end{matrix}\begin{bmatrix} 1 & 0.53 & -0.33 \\ & 1 & 0.10 \\ & & 1 \end{bmatrix} \end{array}$$

From these we may calculate the second-order partial correlation of the subtest totals with age and education held constant:

$$r_{12.34} = 0.62$$

For such a large sample the critical values of the t distribution are nearly equal to those of the standard normal, and the loss of degrees of freedom occasioned by holding one or two variates constant can be neglected. The critical values of the distribution of r for a two-sided test of H_0: $\rho = 0$ at the $\alpha = 0.05$ level are $r = \pm 0.064$, and we note that this hypothesis would be rejected if tested separately for all the above partial correlations.

Next we shall obtain the multiple correlation of performance subtests total with the remaining three variates:

$$\mathbf{R}_{22}^{-1} = \begin{bmatrix} 1.877 & -0.138 & -1.316 \\ & 1.102 & 0.413 \\ & & 2.015 \end{bmatrix}$$

$$R^2_{1.234} = 0.644 \qquad \hat{\boldsymbol{\beta}}' = [0.485, -0.346, 0.289]$$

The regression coefficients are in the units of the original variates. The hypothesis of zero multiple correlation in the population would be resoundingly rejected at any reasonable level.

Finally let us test the hypothesis

$$H_0: \quad \beta_{14.23} = 0$$

that the population regression coefficient of education is zero. We shall need the residual standard deviations $s_{1.234}$ and $s_{4.23}$; these are computed from

$$\begin{aligned} s_{1.234} &= 11.228(1 - R^2_{1.234})^{1/2} \\ &= 6.699 \end{aligned}$$

and

$$\begin{aligned} s_{4.23} &= 3.095 \left\{1 - (-0.29, 0.68) \begin{bmatrix} 1 & -0.13 \\ -0.13 & 1 \end{bmatrix}^{-1} \begin{bmatrix} -0.29 \\ 0.68 \end{bmatrix}\right\}^{1/2} \\ &= 2.180 \end{aligned}$$

The value of the test statistic (39) is

$$\begin{aligned} t &= \frac{2.180}{6.699} \sqrt{929}\,(0.289 - 0) \\ &= 2.87 \end{aligned}$$

If we adopt the one-sided alternative to H_0 that the population regression coefficient has some positive value, the null hypothesis of no regression would be rejected handily at the 0.05 or even the 0.005 level. We conclude that years of formal education play a significant role in the multiple correlation of the performance subtests total variate with the remaining quantities.

3.7 SIMULTANEOUS INFERENCES ABOUT REGRESSION COEFFICIENTS

In the preceding section we stated a test of the hypothesis

$$H_0: \quad \boldsymbol{\beta} = \mathbf{0}$$

or its equivalent

$$H_0: \quad \boldsymbol{\sigma}_{12} = \mathbf{0}$$

of the independence of one random variable and the remaining q other variates in a multinormal distribution. However, that test does not answer the next question, namely, *which* of the q variates have contributed to its significance and which may be considered at some joint level α not to be related to the first variate. The qualification "joint" is essential: any procedure for picking out the meaningful variates in the regression relationship should be constructed to permit successive tests on all the coefficients and perhaps even linear combinations of them with a *fixed* probability α of falsely rejecting any of the hypotheses. Such simultaneous tests are of the same sort as those encountered through multiple comparisons in the analysis of variance in Chap. 1.

One solution to this problem is provided by the set of simultaneous confidence intervals that permit the experimenter to make confidence statements on *all* linear compounds of the regression coefficients of one set of "dependent" variates upon those of an "independent" set with the assurance that the probability that all statements are true is equal to some specified probability $1 - \alpha$. Let us suppose that the first p responses of some multinormal population constitute the dependent set, while the last q play the independent role. The covariance matrix of that population has the partitioned form

$$\mathbf{\Sigma} = \begin{bmatrix} \mathbf{\Sigma}_{11} & \mathbf{\Sigma}_{12} \\ \mathbf{\Sigma}'_{12} & \mathbf{\Sigma}_{22} \end{bmatrix} \tag{1}$$

the dimensions of whose submatrices follow from the division of the responses. The vectors of the population regression coefficients are given by the rows of the $p \times q$ matrix

$$\mathbf{B} = \mathbf{\Sigma}_{12}\mathbf{\Sigma}_{22}^{-1} \tag{2}$$

That is, the parameters

$$\beta_{1,p+1.c}, \ldots, \beta_{1,p+q.c} \tag{3}$$

relating X_1 to $X_{p+1}, \ldots, X_{p+q}$ are given by the first row of $\mathbf{B}$, and so on, until the last row containing those relating X_p to $X_{p+1}, \ldots, X_{p+q}$. If $p = 1$, $\mathbf{B} = \boldsymbol{\beta}'$ of expression (24) in Sec. 3.4. From this population N observation vectors are drawn randomly, and from them the partitioned covariance matrix

$$\mathbf{S} = \begin{bmatrix} \mathbf{S}_{11} & \mathbf{S}_{12} \\ \mathbf{S}'_{12} & \mathbf{S}_{22} \end{bmatrix} \tag{4}$$

is computed. The estimates of the regression parameters are given by

$$\hat{\mathbf{B}} = \mathbf{S}_{12}\mathbf{S}_{22}^{-1} \tag{5}$$

These estimates will be used to obtain a set of simultaneous confidence intervals on all double linear compounds

$$\mathbf{c}'\mathbf{B}\mathbf{a} \tag{6}$$

of the elements of the population regression matrix.

For a *given* p-component vector $\mathbf{c}$, Bartlett (1933) has shown that the $100(1 - \alpha)$ percent joint confidence region on all elements of $\mathbf{c}'\mathbf{B}$ is given by the surface and interior of the q-dimensional ellipsoid

$$\frac{\mathbf{c}'(\hat{\mathbf{B}} - \mathbf{B})\mathbf{S}_{22}(\hat{\mathbf{B}} - \mathbf{B})'\mathbf{c}}{\mathbf{c}'\mathbf{S}_{11.2}\mathbf{c}} \leq \frac{q}{N-q-1} F_{\alpha;q,N-q-1} \tag{7}$$

where

$$\mathbf{S}_{11.2} = \mathbf{S}_{11} - \mathbf{S}_{12}\mathbf{S}_{22}^{-1}\mathbf{S}_{12}' \tag{8}$$

and in our usual notation $F_{\alpha;q,N-q-1}$ denotes the 100α percentage point of the F distribution with q and $N - q - 1$ degrees of freedom. The elements of $\mathbf{c}$ are usually suggested by the values of the sample regression coefficients rather than from *a priori* considerations, and we should prefer a confidence region that would hold with constant probability for all nonnull vectors $\mathbf{c}$. Roy (1957) has shown that such a region can be constructed by replacing the right-hand side of (7) by the similar percentage point of the distribution of the largest characteristic root of the matrix

$$\mathbf{S}_{11.2}^{-1}(\hat{\mathbf{B}} - \mathbf{B})\mathbf{S}_{22}(\hat{\mathbf{B}} - \mathbf{B})' \tag{9}$$

The percentage points of such greatest-root statistics are given by Charts 9 to 16 and Tables 6 to 14 of the Appendix, where $x_\alpha \equiv x_{\alpha;s,m,n}$ will denote the 100α percentage point with parameters

$$\begin{aligned} s &= \min(p,q) \\ m &= \frac{|p-q|-1}{2} \\ n &= \frac{N-p-q-2}{2} \end{aligned} \tag{10}$$

The percentage point for the largest characteristic root of (9) can be shown to be $x_\alpha/(1 - x_\alpha)$, and the ellipsoidal region

$$\frac{\mathbf{c}'(\hat{\mathbf{B}} - \mathbf{B})\mathbf{S}_{22}(\hat{\mathbf{B}} - \mathbf{B})'\mathbf{c}}{\mathbf{c}'\mathbf{S}_{11.2}\mathbf{c}} \leq \frac{x_\alpha}{1 - x_\alpha} \tag{11}$$

will provide confidence statements on *all* row vectors $\mathbf{c}'\mathbf{B}$ with coefficient $1 - \alpha$. For $s = 1$ the right-hand critical value is of course the same as that of expression (7).

Finally, the set of all simultaneous confidence intervals on linear compounds of the elements of $\mathbf{c}'\mathbf{B}$ can be found from the ellipsoidal region (11) by an argument due to Scheffé (1959, appendix 3). Scheffé defines a *plane of support* of the ellipsoid $(\mathbf{x} - \mathbf{x}_0)'\mathbf{M}(\mathbf{x} - \mathbf{x}_0) = 1$ as a plane entirely to one side of the figure with at least one point in common with it. Then if $\mathbf{a}$ is a nonnull vector, the point specified by $\mathbf{x}$ lies between the two parallel supporting planes orthogonal to **a** if and only if

$$|\mathbf{a}'(\mathbf{x} - \mathbf{x}_0)| \leq \sqrt{\mathbf{a}'\mathbf{M}^{-1}\mathbf{a}} \tag{12}$$

In the present case let $\mathbf{x}' = \mathbf{c}'\hat{\mathbf{B}}$, $\mathbf{x}_0' = \mathbf{c}'\mathbf{B}$, and let $\mathbf{M}$ be proportional to $\mathbf{S}_{22}$. Expansion of (12) gives the general member of the set of $100(1 - \alpha)$ percent simultaneous confidence intervals:

$$\mathbf{c}'\hat{\mathbf{B}}\mathbf{a} - \sqrt{\frac{\mathbf{c}'\mathbf{S}_{11.2}\mathbf{c}\mathbf{a}'\mathbf{S}_{22}^{-1}\mathbf{a}x_\alpha}{1 - x_\alpha}} \leq \mathbf{c}'\mathbf{B}\mathbf{a} \leq \mathbf{c}'\hat{\mathbf{B}}\mathbf{a} + \sqrt{\frac{\mathbf{c}'\mathbf{S}_{11.2}\mathbf{c}\mathbf{a}'\mathbf{S}_{22}^{-1}\mathbf{a}x_\alpha}{1 - x_\alpha}} \tag{13}$$

The reader should verify that for the bivariate case of $p = q = s = 1$ and $m = -\frac{1}{2}$ the usual single confidence interval for β_{12} follows.

In an application of the confidence intervals one begins by selecting the vectors **c** and **a** from an examination of the prominence of the pq regression coefficients. In the most common case of a single "dependent" variate, or $p = 1$, one might test all q regression coefficients with the aid of the $1 \times q$ unit vectors $[1,0, \ldots, 0], \ldots, [0, \ldots, 1]$, as well as any linear compounds of the parameters that appear to be relevant to the nature of the responses or are otherwise interesting. If the interval for $\mathbf{c}'\mathbf{B}\mathbf{a}$ contains the value zero, the hypothesis

$$H_0\colon\ \mathbf{c}'\mathbf{B}\mathbf{a} = 0 \tag{14}$$

would be acceptable at the α level in the simultaneous testing sense. Alternatively, one might test this hypothesis directly by referring the statistic

$$z = \frac{\mathbf{c}'(\hat{\mathbf{B}} - \mathbf{B})\mathbf{a}}{\sqrt{\mathbf{c}'\mathbf{S}_{11.2}\mathbf{c}\mathbf{a}'\mathbf{S}_{22}^{-1}\mathbf{a}}} \tag{15}$$

directly to the critical value. If

$$|z| < \sqrt{\frac{x_\alpha}{1 - x_\alpha}}$$

the hypothesis would be accepted, and rejected otherwise.

We may also obtain approximate simultaneous confidence intervals by the Bonferroni method introduced in Sec. 1.6. These intervals may be shorter than those of (13) when their number is small. For a set of m intervals or tests we merely use the single-regression statistic (39) of Sec. 3.6 with its critical value modified so that by Bonferroni's inequality the overall confidence coefficient will not be less than $1 - \alpha$. The general interval for the ijth element $\beta_{ij} \equiv \beta_{i,p+j.\mathbf{c}}$ of **B** is

$$\mathbf{c}'\hat{\mathbf{B}}\mathbf{a} - t_{\alpha/(2m);N-q-1}\sqrt{\frac{\mathbf{c}'\mathbf{S}_{11.2}\mathbf{c}\mathbf{a}'\mathbf{S}_{22}^{-1}\mathbf{a}}{N - q - 1}} \leq \mathbf{c}'\mathbf{B}\mathbf{a} \tag{16}$$

$$\leq \mathbf{c}'\hat{\mathbf{B}}\mathbf{a} + t_{\alpha/(2m);N-q-1}\sqrt{\frac{\mathbf{c}'\mathbf{S}_{11.2}\mathbf{c}\mathbf{a}'\mathbf{S}_{22}^{-1}\mathbf{a}}{N-q-1}}$$

where $\mathbf{c}$ and $\mathbf{a}$ are p- and q-component vectors with ones in the ith and jth positions, respectively, and zeros elsewhere. If intervals are obtained on all elements of $\mathbf{B}$, $m = pq$. We note that identical intervals could be obtained from expression (39) of the preceding section with the modified critical value.

Example 3.4. We shall now obtain simultaneous confidence intervals for the regression coefficients of the WAIS performance and verbal total scores on age and education from the data of Example 3.3. The matrix of the sample regression coefficients is

$$\hat{\mathbf{B}} = \begin{bmatrix} -54.27 & 20.96 \\ -20.46 & 30.31 \end{bmatrix} \begin{bmatrix} 0.0091462 & 0.0093658 \\ 0.0093658 & 0.113975 \end{bmatrix}$$

$$= \begin{bmatrix} -0.300 & 1.881 \\ 0.097 & 3.263 \end{bmatrix}$$

The first row of $\hat{\mathbf{B}}$ contains the coefficients relating performance to age and education while the second row contains the similar estimates for the verbal score.

Let us begin by computing simultaneous intervals for all four coefficients. We shall choose $1 - \alpha = 0.95$. $\mathbf{S}_{22}^{-1}$ is the second matrix in the first expression for $\hat{\mathbf{B}}$, and

$$\mathbf{S}_{11.2} = \begin{bmatrix} 71.72 & 53.93 \\ 53.93 & 110.56 \end{bmatrix}$$

The parameters of the greatest-characteristic-root distribution are $s = 2$, $m = -\frac{1}{2}$, and $n = 463.5$, and from Chart 9.1 of the Appendix we read that the critical value is approximately equal to 0.01. To obtain the interval for the regression coefficient of performance on age we use the vectors $\mathbf{c}' = [1,0]$ and $\mathbf{a}' = [1,0]$, with similar combinations of the unit vectors for the other coefficients. These are the simultaneous confidence intervals:

1. *Performance*
   - Age $\quad -0.381 \leq \beta_{13.4} \leq -0.219$
   - Education $\quad 1.593 \leq \beta_{14.3} \leq 2.168$
2. *Verbal*
   - Age $\quad -0.004 \leq \beta_{23.4} \leq 0.198$
   - Education $\quad 2.906 \leq \beta_{24.3} \leq 3.620$

The subscripts of the parameters refer to the original ordering in the matrix of Example 3.3. We note that the interval for the coefficient of age in the verbal regression barely includes zero, so that the significance of that parameter at the 0.05 level is in doubt.

Now we shall find the simultaneous confidence intervals for the parameters of the separate regressions. In those cases $s = 1$, $m = 0$, and $n = 464$. The critical value can be obtained from Table 4 of the Appendix by means of the relation

$$\frac{x_{0.05}}{1 - x_{0.05}} = \tfrac{1}{465} F_{0.05;\,2,930}$$
$$= 0.00645$$

The vector c is now the scalar one. S_{22} is unchanged, but $S_{11.2}$ is a scalar with value 71.72 for the performance regression and 110.56 for the verbal case. The following intervals were obtained:

1. *Performance*

Age	$-0.365 \le \beta_{13.4} \le -0.235$
Education	$1.651 \le \beta_{14.3} \le 2.110$

2. *Verbal*

Age	$0.016 \le \beta_{23.4} \le 0.178$
Education	$2.978 \le \beta_{24.3} \le 3.548$

By treating the rows of $\hat{\mathbf{B}}$ separately we have shortened each interval. As in other statistical techniques the generality of the first set was obtained at the expense of the greater uncertainty.

Shorter confidence intervals for the regression coefficients could also be constructed by the Bonferroni method. Because of the large sample size the respective critical values for the two cases can be approximated by the normal distribution percentage points:

$$\frac{z_{0.05/8}}{\sqrt{930}} = 0.0818 \qquad \frac{z_{0.05/4}}{\sqrt{930}} = 0.0735$$

The Bonferroni intervals for the first set of coefficients would be about 18.5 percent shorter than the given ones, and we would conclude that the coefficient of verbal total on age is not zero in the population. In the second case of separate performance and verbal regression confidence sets, the Bonferroni intervals would be about 8.5 percent shorter.

3.8 INFERENCES ABOUT THE CORRELATION MATRIX

In the later chapters we shall consider several methods for analyzing the covariance structures of multivariate populations. If all $\tfrac{1}{2}p(p-1)$ population correlations are equal to zero, such analyses based upon the sample covariance matrices would be in vain or even misleading, and it would be prudent to begin each study with a test of the hypothesis

$$(1) \qquad H_0\colon\ \mathbf{P} = \mathbf{I}$$

or its equivalent statement that $\boldsymbol{\Sigma}$ is a diagonal matrix. When the variates are multinormal the test is one of complete independence of the responses. The alternative hypothesis

$$H_1: \quad \mathbf{P} \neq \mathbf{I} \tag{2}$$

states that at least one correlation or covariance does not vanish; if it is tenable we would like some multiple-comparisons method for determining which correlations are probably different from zero in the population.

We shall begin by constructing a test for hypothesis (1) by the generalized likelihood-ratio criterion introduced in Sec. 1.5. Since many of the tests in the sequel can be constructed by that principle, the steps in the derivation should serve a didactic purpose. Let us assume that N observation vectors $\mathbf{x}_1, \ldots, \mathbf{x}_N$ have been drawn independently from the p-dimensional multinormal population with parameters $\boldsymbol{\mu}$ and nonsingular covariance matrix $\boldsymbol{\Sigma}$. The parameter space Ω of the distribution is that region of $\frac{1}{2}p(p+3)$-dimensional euclidean space in which the elements of $\boldsymbol{\mu}$ are finite, and those of $\boldsymbol{\Sigma}$ constitute a symmetric positive definite matrix. Then the likelihood of the random sample is given by expression (2) of Sec. 3.5. When the estimates $\hat{\boldsymbol{\mu}} = \bar{\mathbf{x}}$ and (8) of that section are inserted into the likelihood, its maximized value in Ω is

$$\begin{aligned} L(\hat{\Omega}) &= \frac{1}{(2\pi)^{\frac{1}{2}Np}|\hat{\boldsymbol{\Sigma}}|^{\frac{1}{2}N}} \exp\left[-\frac{1}{2}\sum_{i=1}^{N}(\mathbf{x}_i - \bar{\mathbf{x}})'\hat{\boldsymbol{\Sigma}}^{-1}(\mathbf{x}_i - \bar{\mathbf{x}})\right] \\ &= \frac{1}{(2\pi)^{\frac{1}{2}Np}|\hat{\boldsymbol{\Sigma}}|^{\frac{1}{2}N}} \exp\left[-\tfrac{1}{2}\operatorname{tr}\sum_{i=1}^{N}(\mathbf{x}_i - \bar{\mathbf{x}})(\mathbf{x}_i - \bar{\mathbf{x}})'\hat{\boldsymbol{\Sigma}}^{-1}\right] \\ &= \frac{1}{(2\pi)^{\frac{1}{2}Np}|\hat{\boldsymbol{\Sigma}}|^{\frac{1}{2}N}} \exp\left(-\tfrac{1}{2}N\operatorname{tr}\hat{\boldsymbol{\Sigma}}\hat{\boldsymbol{\Sigma}}^{-1}\right) \\ &= \frac{1}{(2\pi)^{\frac{1}{2}Np}|\hat{\boldsymbol{\Sigma}}|^{\frac{1}{2}N}} \exp\left(-\tfrac{1}{2}Np\right) \end{aligned} \tag{3}$$

The subspace ω corresponding to the null hypothesis (1) is the $2p$-dimensional region of Ω for which

$$\begin{aligned} -\infty < \mu_i &< \infty \\ 0 < \sigma_i^2 &< \infty \qquad \text{all } i \\ \text{and} \qquad \sigma_{ij} &= 0 \qquad i \neq j \end{aligned} \tag{4}$$

The likelihood of the sample is now the product of the likelihoods of the individual responses:

$$L(\boldsymbol{\mu},\boldsymbol{\Sigma}) = \frac{1}{(2\pi)^{\frac{1}{2}Np}\left(\prod_{i=1}^{p}\sigma_i^2\right)^{\frac{1}{2}N}} \tag{5}$$

$$\cdot \exp\left[-\frac{1}{2}\sum_{i=1}^{N}(\mathbf{x}_i-\boldsymbol{\mu})'\mathbf{D}\left(\frac{1}{\sigma_i^2}\right)(\mathbf{x}_i-\boldsymbol{\mu})\right]$$

The estimates of the means and variances follow from Example 1.2, and the maximized likelihood is

$$L(\hat{\omega}) = \frac{1}{(2\pi)^{\frac{1}{2}Np}\left(\prod_{i=1}^{p}\hat{\sigma}_i^2\right)^{\frac{1}{2}N}} \exp\left(-\tfrac{1}{2}Np\right) \tag{6}$$

The test statistic is

$$\begin{aligned} \lambda &= \frac{L(\hat{\omega})}{L(\hat{\Omega})} \\ &= \frac{|\hat{\mathbf{\Sigma}}|^{\frac{1}{2}N}}{(\hat{\sigma}_1^2 \cdots \hat{\sigma}_p^2)^{\frac{1}{2}N}} \\ &= |\mathbf{R}|^{\frac{1}{2}N} \end{aligned} \tag{7}$$

or merely the $\frac{1}{2}N$th power of the determinant of the correlation matrix. The difference of the dimensionalities of Ω and ω is $\frac{1}{2}p(p-1)$, and we can use the large-sample chi-squared distribution of $-2 \ln \lambda$ cited in expression (9) of Sec. 1.5 to obtain a decision rule for accepting or rejecting the independence hypothesis at a specified significance level. Bartlett (1954) has improved the approximation of the limiting distribution by adopting the statistic

$$\chi^2 = -\left(N - 1 - \frac{2p+5}{6}\right)\ln|\mathbf{R}| \tag{8}$$

The degrees of freedom remain unchanged, and for a test of level α we should accept the hypothesis (1) if

$$\chi^2 < \chi^2_{\alpha;\frac{1}{2}p(p-1)} \tag{9}$$

and accept the alternative (2) otherwise. $\chi^2_{\alpha;\frac{1}{2}p(p-1)}$ is clearly the upper 100α percentage point of the chi-squared distribution with $\frac{1}{2}p(p-1)$ degrees of freedom.

We note that the term $N-1$ in the statistic (8) is attributable to the N independent observation vectors in the sample. If the elements of the sample covariance matrix had been computed in some other fashion, e.g., from the within-group sums of squares and products of k independent groups, leading to a Wishart distribution with degrees-of-freedom parameter n, that value would replace $N-1$ in the statistic.

Lawley (1940) has shown through a multiple Maclaurin expansion of ln $|\mathbf{R}|$ that the test statistic can be approximated by

$$\chi^2 = \left(N - 1 - \frac{2p + 5}{6}\right) \sum_{i<j} \sum r_{ij}^2 \tag{10}$$

The summation extends only over the $\frac{1}{2}p(p - 1)$ correlations in the upper portion of the matrix. This approximation is best for small correlations, although for a large sample size it is doubtful that it would lead to a different conclusion than that obtained from the determinantal formula (8).

Example 3.5. In the course of the construction of a mathematical model for atherosclerotic changes in arterial walls Opatowski (1964) found these correlations among the severity scores of atherosclerotic changes in the abdominal aorta, thoracic aorta, and coronary arteries of $N = 54$ normotensive American men:

$$\mathbf{R} = \begin{bmatrix} 1 & 0.69 & 0.43 \\ & 1 & 0.30 \\ & & 1 \end{bmatrix}$$

$|\mathbf{R}| = 0.427$, and the exact chi-squared test statistic has the value 44.4. Even for α as small as 0.001 we would reject the hypothesis of independence of degrees of atherosclerotic changes in the three arteries of the population. The same conclusion would be reached from the approximate value of the statistic of 39.2.

Simultaneous Inferences about the ρ_{ij}. The joint density of the sample correlations for a general positive definite $\mathbf{P}$ is unavailable in a closed and practical form. While the asymptotic distribution is multinormal, its parameters (see, for example, Aitkin, 1969, 1971) depend upon the ρ_{ij} in ways too complicated to permit construction of simultaneous confidence intervals from the implicitly defined joint confidence region. However, it is possible to obtain large-sample conservative multiple tests and intervals by the Bonferroni inequality and the Fisher z transformation. For a family of m confidence statements with nominal coefficient $1 - \alpha$ the general interval is

$$\tanh\left[z_{ij} - \frac{z_{\alpha/(2m)}}{\sqrt{N - 3}}\right] \le \rho_{ij} \le \tanh\left[z_{ij} + \frac{z_{\alpha/(2m)}}{\sqrt{N - 3}}\right] \tag{11}$$

where $z_{ij} = \tanh^{-1} r_{ij}$. We accept H_{0ij}: $\rho_{ij} = 0$ if the interval contains the value zero, and we reject in favor of the usual two-sided alternative if it does not. More directly, we should reject H_{0ij} if

$$|r_{ij}| > \tanh\left[\frac{z_{\alpha/(2m)}}{\sqrt{N - 3}}\right]$$

The function tanh (·) is of course the inverse of Fisher's z transformation. m is equal to $\frac{1}{2}p(p-1)$ when bounds are obtained for all members of **P**; otherwise it is the number of correlations in some predetermined set.

Aitkin (1969) has proposed an alternative method of simultaneous confidence intervals for all elements of **P** based on S. N. Roy's results for the covariance matrix. However, these intervals are extremely wide because their family must also include statements for the variances.

Hills (1969) has used "half-normal plots" for detecting significantly large elements of the correlation matrix. In addition to the normality assumptions of the Bonferroni confidence intervals this method also requires independence of the z_{ij}. An application of the Bonferroni method to Hills's illustrative data led to slightly more conservative conclusions about significance than those of the graphical approach.

Example 3.6. In a study of the interrelations of psychometric test scores for healthy aged men, Birren *et al.* (1963) obtained a 32×32 correlation matrix for $N = 47$ subjects. The first eleven variables were the usual WAIS subtest scores, the next twelve were various measures of perception, cognition, reaction time, and motor ability, while the remaining nine were audiometric measurements. Although the WAIS intercorrelations were predictably high, the other correlations in the matrix of the first 23 psychological variables were often small. We shall use the Bonferroni method to isolate the significant correlations in that set with protection at least at the 5 percent level. Here $m = \frac{1}{2}(23)(22) - \frac{1}{2}(11)(10) = 198$, so that the critical value for the z_{ij} is $z_{0.000126}/\sqrt{44} = 0.552$, and the corresponding one for the r_{ij} follows from the inverse of Fisher's transformation to be 0.502. Only 14 of the 198 correlations actually exceeded 0.502 while 85 exceeded the individual two-sided 5 percent critical value of 0.288. The conclusions of the simultaneous tests lead to a more cautious assessment of the strengths of the relations between the WAIS and the other variables. Whether the Bonferroni approach is excessively conservative will have to await study of its true Type I error and power properties.

3.9 SAMPLES WITH INCOMPLETE OBSERVATIONS

Frequently multivariate data matrices contain missing values which preclude the use of the preceding standard techniques. In longitudinal studies participants may move away, in laboratory experiments animals may die from causes unrelated to the conditions, or in large interdisciplinary investigations new variates may be added as the study progresses. Of course, in every case it is essential that the causes of the missing data are completely independent of the nature or values of the response variates. If the number of complete vectors is small and the cost of obtaining data is appreciable, we would prefer to use methods of estimation and hypothesis testing which make maximal use of the available information.

Maximum-likelihood estimation of the multinormal mean vector and covariance matrix from a data matrix with a general random pattern of missing values leads to systems of nonlinear equations and their associated problems in numerical analysis. A number of workers have considered estimators since Wilks's initial investigation (1932b); surveys of the literature have been made by Elashoff and Afifi (1966), Timm (1970), and Hartley and Hocking (1971). In the case of known $\mathbf{\Sigma}$ Basickes (1972) has developed a notational scheme for finding an explicit estimate of the mean vector.

One special case of missing data will permit the direct calculation of the maximum-likelihood estimates of both $\boldsymbol{\mu}$ and $\mathbf{\Sigma}$. This is the data matrix with the *monotonic* or *nested* pattern

$$(1) \qquad \mathbf{X} = \begin{bmatrix} \mathbf{X}_{11} & \mathbf{X}_{12} & \cdots & \mathbf{X}_{1,k-1} & \mathbf{X}_{1k} \\ \mathbf{X}_{21} & \mathbf{X}_{22} & \cdots & \mathbf{X}_{2,k-1} & - \\ \cdots & \cdots & \cdots & \cdots & \cdots \\ \mathbf{X}_{k-2,1} & \mathbf{X}_{k-2,2} & \cdots & - & - \\ \mathbf{X}_{k-1,1} & \mathbf{X}_{k-1,2} & \cdots & - & - \\ \mathbf{X}_{k1} & - & \cdots & - & - \end{bmatrix}$$

where the dashes indicate blocks of missing observations. $\mathbf{X}_{ij}$ is an $N_i \times r_j$ submatrix, where $\sum_{i=1}^{k} N_i = N$, the total number of independent sampling units, and $\sum_{j=1}^{k} r_j = p$, the number of responses. The mean vector is correspondingly partitioned as $\boldsymbol{\mu}' = (\boldsymbol{\mu}_1', \ldots, \boldsymbol{\mu}_k')$, and the covariance matrix consists of the $r_i \times r_j$ submatrices $\mathbf{\Sigma}_{ij}$, $i, j = 1, \ldots, k$. Because of the monotonic pattern we can write the likelihood of the sample as

$$(2) \quad L(\boldsymbol{\mu},\mathbf{\Sigma}) = f(\mathbf{X})$$
$$= \left[\prod_{i=1}^{k} f(\mathbf{X}_{i1})\right]\left[\prod_{i=1}^{k-1} f(\mathbf{X}_{i2}|\mathbf{X}_{i1})\right]\left[\prod_{i=1}^{k-2} f(\mathbf{X}_{i3}|\mathbf{X}_{i1},\mathbf{X}_{i2})\right]$$
$$\cdots f(\mathbf{X}_{1k}|\mathbf{X}_{11}, \ldots, \mathbf{X}_{1,k-1})$$

where the right-hand side has been written in terms of conditional densities rather than likelihoods to avoid the introduction of notation unnecessarily complicated for our present aims. $f(\mathbf{X}_{ij}|\mathbf{X}_{i1}, \ldots, \mathbf{X}_{i,j-1})$ denotes the conditional likelihood of the ijth submatrix; it can be found in the usual manner as the product of N_i independent r_i-dimensional multinormal densities whose parameters can be obtained by an appropriate application of expressions (7) and (8) of Sec. 3.4. One proceeds by successively maximizing the factors in the square brackets: the first merely yields the usual estimates $\hat{\boldsymbol{\mu}}_1$ and $\hat{\boldsymbol{\Sigma}}_{11}$ based on all N sampling units. The second factor gives estimates of $\boldsymbol{\mu}_2 - \boldsymbol{\Sigma}_{12}'\boldsymbol{\Sigma}_{11}^{-1}\boldsymbol{\mu}_1$, the regression parameters $\Sigma_{12}'\Sigma_{11}^{-1}$, and the conditional covariance matrix $\boldsymbol{\Sigma}_{22} - \boldsymbol{\Sigma}_{12}'\boldsymbol{\Sigma}_{11}^{-1}\boldsymbol{\Sigma}_{12}$. These can be readily converted with the aid of the first estimates into the maximum-likelihood estimates of $\boldsymbol{\mu}_2$, $\boldsymbol{\Sigma}_{12}$, and $\boldsymbol{\Sigma}_{22}$. We continue in this fashion until the unconditional parameters of the kth set have been computed. This approach is due to Anderson (1957), with an extension to more than two sets of variates by Bhargava (1962).

We shall illustrate these estimates with the simplest monotonic pattern of two variate sets. For notational simplicity let $\mathbf{Y}' \equiv (\mathbf{X}_{11}',\mathbf{X}_{21}')$ and $\mathbf{X} \equiv \mathbf{X}_{12}$. Then

$$\text{(3)} \qquad \hat{\boldsymbol{\mu}}_1 = \frac{1}{N}\sum_{i=1}^{N}\mathbf{y}_i \qquad \hat{\boldsymbol{\Sigma}}_{11} = \frac{1}{N}\sum_{i=1}^{N}(\mathbf{y}_i - \bar{\mathbf{y}})(\mathbf{y}_i - \bar{\mathbf{y}})'$$

or the usual complete-data estimates. For the other estimates we define

$$\bar{\mathbf{x}} = \frac{1}{N_1}\sum_{i=1}^{N_1}\mathbf{x}_i \qquad \bar{\mathbf{y}}_1 = \frac{1}{N_1}\sum_{i=1}^{N_1}\mathbf{y}_i \qquad \bar{\mathbf{y}}_2 = \frac{1}{N_2}\sum_{i=N_1+1}^{N}\mathbf{y}_i$$

$$\text{(4)} \qquad \begin{aligned} \mathbf{A}_{11}(N_1) &= \sum_{i=1}^{N_1}(\mathbf{y}_i - \bar{\mathbf{y}}_1)(\mathbf{y}_i - \bar{\mathbf{y}}_1)' \\ \mathbf{A}_{11}(N) &= \sum_{i=1}^{N}(\mathbf{y}_i - \bar{\mathbf{y}})(\mathbf{y}_i - \bar{\mathbf{y}})' \\ \mathbf{A}_{22} &= \sum_{i=1}^{N_1}(\mathbf{x}_i - \bar{\mathbf{x}})(\mathbf{x}_i - \bar{\mathbf{x}})' \\ \mathbf{A}_{12} &= \sum_{i=1}^{N_1}(\mathbf{y}_i - \bar{\mathbf{y}}_1)(\mathbf{x}_i - \bar{\mathbf{x}})' \\ \mathbf{B} &= \mathbf{A}_{11}^{-1}(N_1)\mathbf{A}_{12} \end{aligned}$$

The estimates of the remaining parameters are

$$\text{(5)} \qquad \begin{aligned} \hat{\boldsymbol{\mu}}_2 &= \bar{\mathbf{x}} - \left(\frac{N_2}{N}\right)\mathbf{B}'(\bar{\mathbf{y}}_1 - \bar{\mathbf{y}}_2) \\ \hat{\boldsymbol{\Sigma}}_{22} &= \frac{1}{N_1}[\mathbf{A}_{22} - \mathbf{A}_{12}'\mathbf{A}_{11}^{-1}(N_1)\mathbf{A}_{12}] + \frac{1}{N}\mathbf{B}'\mathbf{A}_{11}(N)\mathbf{B} \\ \hat{\boldsymbol{\Sigma}}_{12} &= \frac{1}{N}\mathbf{A}_{11}(N)\mathbf{A}_{11}^{-1}(N_1)\mathbf{A}_{12} \end{aligned}$$

The exact expectations and some second moments of the estimates

(5) have been computed (Morrison, 1971). In particular, $\hat{\boldsymbol{\mu}}_2$ and $[N/(N-1)]\hat{\boldsymbol{\Sigma}}_{12}$ are unbiased, while it is possible to redefine the divisors of the terms in $\hat{\boldsymbol{\Sigma}}_{22}$ to obtain this unbiased estimate:

$$(6)\quad \hat{\boldsymbol{\Sigma}}_{22(U)} = \frac{1}{N_1 - r_1 - 1}\left[1 - \frac{r_1}{N_1 - r_1 - 2} + \frac{r_1(r_1+1)}{(N_1 - r_1 - 2)(N-1)}\right][\mathbf{A}_{22} - \mathbf{A}_{12}'\mathbf{A}_{11}^{-1}(N_1)\mathbf{A}_{12}] + \frac{1}{N-1}\mathbf{B}'\mathbf{A}_{11}(N)\mathbf{B}$$

Values of the variances of the estimates (5) for small N_1, N indicate that those estimates may be less efficient than the conventional ones computed from only the first N_1 complete observations when the correlations between the two sets of variates are low. Guidelines for the choice of estimate have been indicated in the previous reference (Morrison, 1971).

Example 3.7. A large university uses a multiple-regression equation to estimate the freshman grade point indices (GPI) of applicants for admission from their class rank, Scholastic Aptitude Tests (SAT), and any achievement test scores. The estimated GPI is included on the applicant's summary form for the use of the admissions committee readers. A sample of $N = 34$ applications evaluated by one faculty member contained five without estimated GPI. However, all applications had SAT mathematics and verbal test scores among the various measures of academic promise. The SAT and GPI data are an example of a monotonic missing-observation pattern, with $r_1 = 2$ complete responses, $r_2 = 1$ incomplete, $N_1 = 29$ complete sampling units, and $N_2 = 5$ incomplete. We shall assume that the verbal (Y_1), mathematics (Y_2), and GPI (X) scores are a random sample from a trivariate normal distribution whose parameters will be estimated by the method of this section.

We begin by preparing quick scatter plots of GPI against the SAT scores of the 29 complete vectors. The plots indicate a modest amount of positive correlation, and it would appear that the missing-value estimates of the mean and variance of the GPI variate might be more efficient than univariate estimates based on the 29 GPI scores. The required statistics are

$$\bar{x} = 2.43 \qquad \bar{\mathbf{y}}_1 = \begin{bmatrix} 56.38 \\ 61.31 \end{bmatrix} \qquad \bar{\mathbf{y}}_2 = \begin{bmatrix} 55.80 \\ 55.60 \end{bmatrix} \qquad \bar{\mathbf{y}} = \begin{bmatrix} 56.29 \\ 60.47 \end{bmatrix}$$

$$\mathbf{A}_{11}(N_1) = \begin{bmatrix} 1{,}292.83 & 206.59 \\ 206.59 & 1{,}328.21 \end{bmatrix} \qquad \mathbf{A}_{11}(N) = \begin{bmatrix} 1{,}409.06 & 238.29 \\ 238.29 & 1{,}668.47 \end{bmatrix}$$

$$a_{22} = 2.9455 \qquad \mathbf{A}_{11}^{-1}(N_1) = 10^{-4}\begin{bmatrix} 7.9321 & -1.2338 \\ -1.2338 & 7.7208 \end{bmatrix} \qquad \mathbf{B} = 10^{-2}\begin{bmatrix} 1.6862 \\ 2.9142 \end{bmatrix}$$

and the maximum-likelihood estimates of the GPI parameters are

$$\hat{\mu}_x = 2.40 \qquad \hat{\sigma}_x^2 = 0.1034$$

The estimate of the mean is slightly less than the usual complete-data estimate $\bar{x}$. For comparison with $\hat{\sigma}_x^2$ we list these alternative estimates of the variance:

Estimate	Value
Biased univariate MLE a_{22}/N	0.1016
Unbiased multivariate MLE (6)	0.1067
Unbiased univariate $s^2 = a_{22}/(N-1)$	0.1052

The multivariate missing-value estimates of the variance are slightly larger than their usual univariate counterparts.

3.10 EXERCISES

1. Let X have the uniform density function of Fig. 1.1, Sec. 1.2, with $a = -1$, $b = 1$. Show that X and the new variate $Y = X^2$ have zero correlation, although X and Y are not independently distributed.

2. The variates $\mathbf{X}' = [X_1,X_2,X_3]$ and $\mathbf{Y}' = [Y_1,Y_2,Y_3]$ are distributed independently according to the trivariate normal populations with respective parameters

$$\boldsymbol{\mu}_1 = \begin{bmatrix} 2 \\ 2 \\ 2 \end{bmatrix} \qquad \boldsymbol{\Sigma}_1 = \begin{bmatrix} 3 & 2 & 1 \\ 2 & 4 & 1 \\ 1 & 1 & 2 \end{bmatrix} \qquad \boldsymbol{\mu}_2 = \begin{bmatrix} 3 \\ 4 \\ 2 \end{bmatrix} \qquad \boldsymbol{\Sigma}_2 = \begin{bmatrix} 4 & 2 & 0 \\ 2 & 4 & 2 \\ 0 & 2 & 4 \end{bmatrix}$$

Determine the distributions of these new variates:

a. $[\mathbf{X} - \mathbf{Y}]$ b. $[\mathbf{X}' - \mathbf{Y}' \quad \mathbf{X}' + \mathbf{Y}']$ c. $[\mathbf{X}' \quad \mathbf{Y}']$

3. Let

$$\mathbf{A} = \begin{bmatrix} 2 & -1 & -1 \\ 0 & 1 & -1 \end{bmatrix} \qquad \mathbf{B} = \begin{bmatrix} 1 & 1 & 1 \\ 1 & -1 & 0 \\ 0 & 1 & -1 \end{bmatrix}$$

and let $\mathbf{X}$ and $\mathbf{Y}$ be the random vectors of Exercise 2. What are the distributions of the following linear transformations of those variates?

a. $\mathbf{AX}$ b. $\mathbf{BX}$ c. $[\mathbf{X}'\mathbf{A}' \quad \mathbf{Y}'\mathbf{B}']$ d. $[\mathbf{X}'\mathbf{A}' \quad \mathbf{X}'\mathbf{B}']$

4. If the random vector $\mathbf{Z}$ is distributed according to the trivariate normal law with mean vector $\boldsymbol{\mu}' = [0,0,0]$ and covariance matrix

$$\boldsymbol{\Sigma} = \begin{bmatrix} 3 & 1 & 1 \\ 1 & 3 & 1 \\ 1 & 1 & 3 \end{bmatrix}$$

answer these questions about its family of concentration ellipsoids:

a. Show that the first principal axis passes through the point [1,1,1].

b. What is the length of that axis for a particular ellipsoid?

c. Are the orientations of the remaining axes unique?

d. Determine the lengths of the second and third axes of the ellipsoid.

5. Using the formula for the inverse of a partitioned covariance matrix, show that the multiple correlation of the ith variate with the remaining $p - 1$ can be computed as

$$\rho_{i.\mathbf{c}}^2 = 1 - \frac{1}{\sigma_i^2 \sigma^{ii}}$$

where $\mathbf{c}$ denotes the set of subscripts $1, \ldots, i-1, i+1, \ldots, p$, and σ^{ii} is the ith diagonal element of $\mathbf{\Sigma}^{-1}$. Find a similar expression for the regression coefficients of X_i on the other variates in terms of the off-diagonal elements of the ith row of $\mathbf{\Sigma}^{-1}$.

6. Use the results of Exercise 5 and Sec. 2.11 to find the multiple-regression properties of a set of p variates with the equal-variance, equal-covariance matrix.

7. Use the formulas for vector and matrix differentiation given in Sec. 2.12 to obtain the maximum-likelihood estimate of the multinormal-distribution covariance matrix by differentiating the log likelihood directly with respect to the elements of $\mathbf{\Sigma}$. Assume that the likelihood has been evaluated at $\hat{\boldsymbol{\mu}} = \bar{\mathbf{x}}$.

8. Show that expression (39) of Sec. 3.6 gives the usual bivariate normal correlation test statistic

$$t = r\sqrt{\frac{N-2}{1-r^2}}$$

when $q = 1$.

9. Hearing loss in decibels at 2,000 cps in the left and right ears was measured for 37 normal elderly males and 52 men of the same age range with a certain medical diagnosis. The respective product moment correlations between the left- and right-ear losses were 0.72 and 0.86. Test at the $\alpha = 0.05$ level the hypothesis that the population correlations are equal against the alternative that the diagnostic group has a higher correlation.

10. The following correlations were observed among the responses auditory reaction time, audiometric hearing loss, WAIS comprehension, and WAIS digit symbol for a sample of $N = 47$ elderly normal males (Birren *et al.*, 1963):

$$\begin{bmatrix} 1 & 0.60 & -0.36 & -0.53 \\ & 1 & -0.46 & -0.37 \\ & & 1 & 0.37 \\ & & & 1 \end{bmatrix}$$

a. Determine the partial correlation of reaction time and hearing loss with the two WAIS subtest scores held constant. Test the hypothesis of zero partial correlation at the 5 percent level.

b. Compute the multiple correlation of reaction time with the other three variates. Test the hypothesis of independence of the first response and the last three.

11. Use both the exact determinantal and approximate test statistics of Sec. 3.8 to test at the $\alpha = 0.001$ level the hypothesis that the four variates of Exercise 10 are uncorrelated.

12. Sinha and Lee (1970) obtained the following correlation matrix of the properties and arthropod infestation counts (log-transformed) of $N = 165$ carload samples of grain:

Variate	1	2	3	4	5	6	7	8	9
1. Grade	1	0.44	0.44	0.11	0.19	0.10	0.20	0.20	−0.24
2. Moisture		1	0.34	0.25	0.32	0.40	0.49	0.16	−0.22
3. Dockage			1	0.04	0.06	0.08	0.07	0.05	−0.07
4. *Acarus*				1	0.18	0.12	0.23	0.02	−0.20
5. *Cheyletus*					1	0.22	0.48	0.14	−0.08
6. *Glycyphagus*						1	0.40	−0.11	−0.30
7. *Tarsonemus*							1	0.15	−0.13
8. *Cryptolestes*								1	−0.10
9. *Psocoptera*									1

The variate "dockage" consisted of the weight of foreign material (weed seed, etc.) in the sample. Use the Bonferroni simultaneous testing procedure with a family error rate of 0.05 to test all hypotheses $\rho_{ij} = 0$ against the usual two-sided alternatives.

13. Compute the 95 percent simultaneous confidence intervals for the regression coefficients of GPI on SAT verbal and SAT mathematics for the 29 *complete* observation vectors of Example 3.7.

14. In an investigation of the psychological effects of marihuana on human subjects Weil *et al.* (1968) recorded changes in heart rate and a digit symbol substitution test score from baseline values 15 and 90 min after the start of the experimental session.* These observed changes of the digit symbol scores were obtained for the "naive" users of marihuana:

Subject	15 min	90 min
1	5	8
2	−17	−5
3	−7	−1
4	−3	−
5	−7	−8
6	−9	−12
7	−6	−4
8	1	−3
9	−3	−10

Similarly, the "chronic" users had these changes in heart rate (beats per minute):

Subject	15 min	90 min
10	32	4
11	36	36
12	20	12
13	8	4
14	32	12
15	54	22
16	24	–
17	60	–

The dashes indicate missing observations.

Find the maximum-likelihood estimates of the parameters of the two bivariate normal distributions. In consideration of the small sample sizes and the values of the correlations between the 15- and 90-min observations, do you feel that the use of the missing-data estimates is justified?

4 TESTS OF HYPOTHESES ON MEANS

4.1 INTRODUCTION.

In this chapter we shall consider a generalization of the familiar univariate tests on means of normal populations to tests involving the mean vector of responses drawn from the multivariate normal distribution. Not only are such hypotheses more appropriate in many situations, as when one wishes to ascertain that two samples of organisms with p measurable characteristics arose from the same population, but the exact analysis of some repeated-measurements experimental designs leads directly to these hypotheses and their appropriate test statistic. By the same development of the tests it is possible to construct confidence regions for the mean vectors and to offer one solution to the problem of multiple comparisons.

Because we usually prefer tests that are invariant under coordinate changes and have null distributions free of the unknown population covariance matrix we shall devote most of our attention to applications of the Hotelling T^2 statistic. Later in Sec. 4.8 we shall consider tests based on a known covariance matrix, as well as other variations in the standard model and assumptions.

4.2 TESTS ON MEANS AND THE T^2 STATISTIC

Let us begin by recalling the simplest form of the univariate test described in Sec. 1.5: a random sample of N observations $x_1, \ldots, x_N$ has been drawn from the normal population with mean μ and unknown variance σ^2. Armed with this sample information we wish to test the hypothesis

$$H_0: \quad \mu = \mu_0$$

against the two-sided alternative

$$H_1: \quad \mu = \mu_1 \neq \mu_0$$

The test statistic was

$$t = \frac{(\bar{x} - \mu_0)\sqrt{N}}{s}$$

and the decision rule was

$$\text{Accept } H_0 \text{ if } |t| < t_{½\alpha;\,N-1}$$
$$\text{Accept } H_1 \text{ if } |t| > t_{½\alpha;\,N-1}$$

for $t_{½\alpha;\,N-1}$ the upper 50α percentage point of the t distribution with $N - 1$ degrees of freedom. Now we shall extend this test to multidimensional random variables. Let the response variate $\mathbf{X}$ consist of p components distributed according to the multinormal law with mean vector $\boldsymbol{\mu}$ and nonsingular covariance matrix $\boldsymbol{\Sigma}$. As in the univariate case, all elements of $\boldsymbol{\Sigma}$ are unknown. The null hypothesis is now

(1) $$H_0: \begin{bmatrix} \mu_1 \\ \cdot \\ \cdot \\ \cdot \\ \mu_p \end{bmatrix} = \begin{bmatrix} \mu_{01} \\ \cdot \\ \cdot \\ \cdot \\ \mu_{0p} \end{bmatrix}$$

and its alternative is

(2) $$H_1: \begin{bmatrix} \mu_1 \\ \cdot \\ \cdot \\ \cdot \\ \mu_p \end{bmatrix} \neq \begin{bmatrix} \mu_{01} \\ \cdot \\ \cdot \\ \cdot \\ \mu_{0p} \end{bmatrix}$$

N independent observation vectors have been collected on $\mathbf{X}$, and from these the usual estimates $\bar{\mathbf{x}}$ and $\mathbf{S}$ of $\boldsymbol{\mu}$ and $\boldsymbol{\Sigma}$ defined in Sec. 3.5 have been computed.

But how should a single test statistic for H_0 be constructed from the data? It will be instructive and highly useful to attack that problem through the *union-intersection* principle of test construction of S. N. Roy (1953, 1957). Form the arbitrary linear compound

(3) $$\mathbf{a}'\mathbf{X}$$

of the p responses, where $\mathbf{a}'$ is any nonnull p-component row vector of real elements. Then $\mathbf{a}'\mathbf{X}$ is univariate normal with mean $\mathbf{a}'\boldsymbol{\mu}$ and variance

$\mathbf{a}'\mathbf{\Sigma a}$, and the univariate hypothesis

$$H_0(\mathbf{a}): \quad \mathbf{a}'\boldsymbol{\mu} = \mathbf{a}'\boldsymbol{\mu}_0 \tag{4}$$

can be tested against its two-sided alternative $H_1(\mathbf{a})$: $\mathbf{a}'\boldsymbol{\mu} = \mathbf{a}'\boldsymbol{\mu}_1$ by the statistic

$$t(\mathbf{a}) = \frac{\mathbf{a}'(\bar{\mathbf{x}} - \boldsymbol{\mu}_0)\sqrt{N}}{\sqrt{\mathbf{a}'\mathbf{Sa}}} \tag{5}$$

The acceptance region is of course

$$t^2(\mathbf{a}) \leq t^2_{\beta/2;N-1} \tag{6}$$

The original multivariate hypothesis is true if and only if $\mathbf{a}'\boldsymbol{\mu} = \mathbf{a}'\boldsymbol{\mu}_0$ holds for all nonnull $\mathbf{a}$, and acceptance of that hypothesis is equivalent to accepting *all* univariate hypotheses $H_0(\mathbf{a})$ specified by varying the elements of $\mathbf{a}$. The multivariate acceptance region is the *intersection*

$$\bigcap_{\mathbf{a}} [t^2(\mathbf{a}) \leq t^2_{\beta/2;N-1}] \tag{7}$$

of all univariate acceptance regions. But stipulation that all $t^2(\mathbf{a})$ lie in this region is equivalent to the condition that

$$\max_{\mathbf{a}} t^2(\mathbf{a}) \leq t^2_{\beta/2;N-1} \tag{8}$$

and we shall adopt as test statistic for the hypothesis on the mean vectors the maximum of $t^2(\mathbf{a})$.

To determine this maximum value we first note that $t^2(\mathbf{a})$ is dimensionless and unaffected by a change of scale of the elements of $\mathbf{a}$. This indeterminacy can be removed by imposing the constraint

$$\mathbf{a}'\mathbf{Sa} = 1$$

Introduction of the constraint by the Lagrangian multiplier λ and subsequent differentiation with respect to $\mathbf{a}$ yields the system of equations

$$[(\bar{\mathbf{x}} - \boldsymbol{\mu}_0)(\bar{\mathbf{x}} - \boldsymbol{\mu}_0)'N - \lambda\mathbf{S}]\mathbf{a} = \mathbf{0} \tag{9}$$

Premultiplication by $\mathbf{a}'$ gives

$$\begin{aligned} \lambda &= \frac{\mathbf{a}'(\bar{\mathbf{x}} - \boldsymbol{\mu}_0)(\bar{\mathbf{x}} - \boldsymbol{\mu}_0)'\mathbf{a}N}{\mathbf{a}'\mathbf{Sa}} \\ &= \frac{[\mathbf{a}'(\bar{\mathbf{x}} - \boldsymbol{\mu}_0)]^2N}{\mathbf{a}'\mathbf{Sa}} \\ &= t^2(\mathbf{a}) \end{aligned} \tag{10}$$

λ is also the only nonzero root of the determinantal equation

$$|\mathbf{S}^{-1}(\bar{\mathbf{x}} - \boldsymbol{\mu}_0)(\bar{\mathbf{x}} - \boldsymbol{\mu}_0)'N - \lambda\mathbf{I}| = 0$$

and it follows that the maximum $t^2(\mathbf{a})$ statistic is

$$\begin{aligned} \lambda &= \text{tr}\, \mathbf{S}^{-1}(\bar{\mathbf{x}} - \boldsymbol{\mu}_0)(\bar{\mathbf{x}} - \boldsymbol{\mu}_0)'N \\ &= N(\bar{\mathbf{x}} - \boldsymbol{\mu}_0)'\mathbf{S}^{-1}(\bar{\mathbf{x}} - \boldsymbol{\mu}_0) \end{aligned} \tag{11}$$

This quadratic form is the single-sample *Hotelling T^2 statistic.* It is the direct analogue of the univariate t^2 with $\bar{x} - \mu_0$ replaced by $\bar{\mathbf{x}} - \boldsymbol{\mu}_0$ and s^2 replaced by $\mathbf{S}$.

When the null hypothesis (1) is true, the quantity

$$F = \frac{N - p}{p(N - 1)} T^2 \tag{12}$$

has the F distribution with degrees of freedom p and $N - p$. Departures of $\boldsymbol{\mu}$ from $\boldsymbol{\mu}_0$ can only increase the mean of T^2, and the decision rule for a test of level α is

$$\text{Accept } H_0\text{:} \quad \boldsymbol{\mu} = \boldsymbol{\mu}_0 \text{ if } T^2 \leq \frac{p(N - 1)}{N - p} F_{\alpha;\, p, N-p} \tag{13}$$

and reject H_0 otherwise.

Now let us state the distribution of the most general statistic

$$T^2 = \mathbf{Y}'\mathbf{S}^{-1}\mathbf{Y} \tag{14}$$

$\mathbf{Y}$ is a p-dimensional random variable with nonsingular distribution $N(\boldsymbol{\mu},\boldsymbol{\Sigma})$, and $n\mathbf{S}$ has the Wishart distribution with parameters n and $\boldsymbol{\Sigma}$. That is, we can write $n\mathbf{S}$ as the sum

$$n\mathbf{S} = \mathbf{z}_1\mathbf{z}_1' + \cdots + \mathbf{z}_n\mathbf{z}_n' \tag{15}$$

whose vectors $\mathbf{z}_i$ are independently distributed as $N(\mathbf{0},\boldsymbol{\Sigma})$ variates. $\mathbf{Y}$ and $\mathbf{S}$ are independently distributed of one another. *Then*

$$F = \frac{n - p + 1}{pn} T^2 \tag{16}$$

is distributed as a noncentral F variate with degrees of freedom p and $n - p + 1$, and noncentrality parameter $\boldsymbol{\mu}'\boldsymbol{\Sigma}^{-1}\boldsymbol{\mu}$. When $\boldsymbol{\mu} = \mathbf{0}$, all the tests on mean vectors of this chapter can be produced by appropriately choosing $\mathbf{Y}$, $\mathbf{S}$, and n. The power functions of the tests can be computed from standard charts or tables of the noncentral F distribution. The choice of the T^2 statistic and the derivation of its central distribution are due to Hotelling (1931); an insightful discussion of the role played by the invariance principle in the development has been given by Hotelling (1954). The noncentral distribution of T^2 was obtained by Bose and Roy (1938) and by Hsu (1938). Concise derivations of the general distribution have been given more recently by Bowker (1960) and Rao (1965).

Confidence Regions for the Mean Vector. Just as the t test can be inverted to give a confidence interval for the population mean, so can a confidence region be constructed for the elements of $\boldsymbol{\mu}$ from T^2. In the single-sample case the $100(1 - \alpha)$ percent region consists of the vectors satisfying the inequality

$$N(\bar{\mathbf{x}} - \boldsymbol{\mu})'\mathbf{S}^{-1}(\bar{\mathbf{x}} - \boldsymbol{\mu}) \leq \frac{(N-1)p}{N-p} F_{\alpha;\, p, N-p} \tag{17}$$

The boundary of the region is an ellipsoid whose center is at the point $\bar{\mathbf{x}}' = [\bar{\mathbf{x}}_1, \ldots, \bar{\mathbf{x}}_p]$. The more general definition (14) of T^2 also leads to similar ellipsoidal confidence regions, and we shall encounter some of these in the later sections.

Example 4.1. Let us fix the several single-sample techniques introduced in this section with some applications. In an investigation of the Wechsler Adult Intelligence Scale scores of older men and women Doppelt and Wallace (1955) reported that the mean verbal and performance scores for $N = 101$ subjects aged 60 to 64 were

$$\begin{bmatrix} \bar{x}_v \\ \bar{x}_p \end{bmatrix} = \begin{bmatrix} 55.24 \\ 34.97 \end{bmatrix}$$

The sample covariance matrix of the scores was

$$\mathbf{S} = \begin{bmatrix} 210.54 & 126.99 \\ 126.99 & 119.68 \end{bmatrix}$$

We wish to test at the $\alpha = 0.01$ level the null hypothesis that the observations came from a population with mean vector

$$\boldsymbol{\mu}_0 = \begin{bmatrix} 60 \\ 50 \end{bmatrix}$$

The test statistic is

$$\begin{aligned} T^2 &= 101[55.24 - 60,\ 34.97 - 50] \begin{bmatrix} 0.01319 & -0.01400 \\ -0.01400 & 0.02321 \end{bmatrix} \begin{bmatrix} 55.24 - 60 \\ 34.97 - 50 \end{bmatrix} \\ &= 357.43 \end{aligned}$$

and $F = 176.93$. The degrees of freedom of the F statistic are 2 and 99. The nearest tabulated critical value is $F_{0.01;2,60} = 4.98$. Since the observed F is far in excess of that point, we conclude that the null hypothesis is untenable and that the sample probably arose from a population with considerably lower verbal and performance means.

We can gain some knowledge of the location parameters of this bivariate

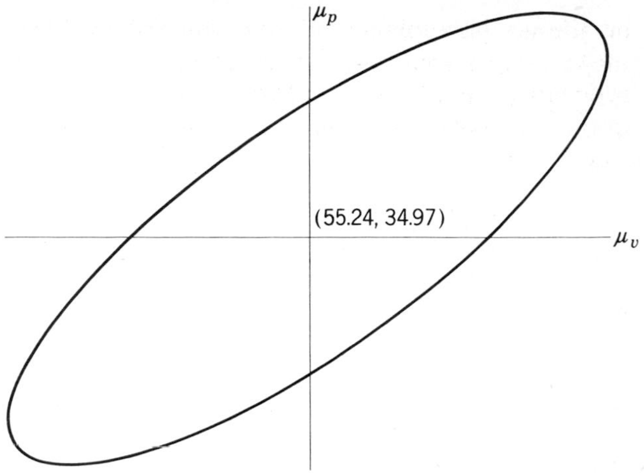

figure 4.1

population by computing the 99 percent confidence region for the means. We shall need the upper 1 percent critical value of F with 2 and 99 degrees of freedom; linear interpolation between the reciprocals of the degrees of freedom 60 and 120 gives $F_{0.01;2,99} = 4.83$. The confidence region consists of all points satisfying the inequality

$$1.33(\mu_v - 55.24)^2 - 2.83(\mu_v - 55.24)(\mu_p - 34.97) + 2.34(\mu_p - 34.97)^2 \leq 9.76$$

and is shown in Fig. 4.1.

We have introduced the Hotelling T^2 statistic for the purpose of testing that a single multinormal mean vector has specified elements. Before turning to the multiple-comparisons problem of deciding which means may have contributed to rejection of the hypothesis, and before examining more general applications of T^2, let us cite a property and a computational device which hold in general.

Invariance of T^2. Not only is T^2 unaffected by changes of units or shifts or origins of the response variates, but it is invariant under all affine transformations

$$\mathbf{y} = \mathbf{Cx} + \mathbf{d}$$

of the observations and the hypothesis, where $\mathbf{C}$ is any nonsingular $p \times p$ matrix and $\mathbf{d}$ is a $p \times 1$ vector. This is apparent when the statistic is computed from the transformed responses:

$$\begin{aligned} T^2 &= N(\mathbf{C\bar{x}} + \mathbf{d} - \mathbf{C}\boldsymbol{\mu}_0 - \mathbf{d})'(\mathbf{CSC}')^{-1}(\mathbf{C\bar{x}} + \mathbf{d} - \mathbf{C}\boldsymbol{\mu}_0 - \mathbf{d}) \\ &= N(\bar{\mathbf{x}} - \boldsymbol{\mu}_0)'\mathbf{C}'\mathbf{C}'^{-1}\mathbf{S}^{-1}\mathbf{C}^{-1}\mathbf{C}(\bar{\mathbf{x}} - \boldsymbol{\mu}_0) \\ &= N(\bar{\mathbf{x}} - \boldsymbol{\mu}_0)'\mathbf{S}^{-1}(\bar{\mathbf{x}} - \boldsymbol{\mu}_0) \end{aligned}$$

or the statistic computed from the original responses. Note that the mean vectors were also transformed to conform to the new coordinate system. It was the necessity for the invariance property that led Hotelling to his choice of the T^2 statistic as the generalization of the univariate t ratio.

Computation of T^2. It is possible to avoid inverting the covariance matrix by expressing T^2 as the quotient of two determinants. Since in practice one usually obtains the matrix **A** of sample sums of squares and cross products before **S**, we shall initially express the result in terms of that matrix. Form the $(p + 1) \times (p + 1)$ matrix

$$\mathbf{B} = \begin{bmatrix} -1 & \sqrt{N}\,(\bar{\mathbf{x}} - \boldsymbol{\mu})' \\ \sqrt{N}\,(\bar{\mathbf{x}} - \boldsymbol{\mu}) & \mathbf{A} \end{bmatrix}$$

and expand its determinant according to the results of Sec. 2.11 as

$$|\mathbf{B}| = -|\mathbf{A} + N(\bar{\mathbf{x}} - \boldsymbol{\mu})(\bar{\mathbf{x}} - \boldsymbol{\mu})'|$$

and

$$|\mathbf{B}| = -|\mathbf{A}|[1 + N(\bar{\mathbf{x}} - \boldsymbol{\mu})'\mathbf{A}^{-1}(\bar{\mathbf{x}} - \boldsymbol{\mu})]$$

We find upon equating the alternate expansions that

$$(18) \qquad \frac{T^2}{N-1} = \frac{|\mathbf{A} + N(\bar{\mathbf{x}} - \boldsymbol{\mu})(\bar{\mathbf{x}} - \boldsymbol{\mu})'|}{|\mathbf{A}|} - 1$$

Since the result merely turns on an algebraic identity, a similar expression can also be obtained in terms of **S**:

$$(19) \qquad T^2 = \frac{|\mathbf{S} + N(\bar{\mathbf{x}} - \boldsymbol{\mu})(\bar{\mathbf{x}} - \boldsymbol{\mu})'|}{|\mathbf{S}|} - 1$$

4.3 SIMULTANEOUS INFERENCES FOR MEANS

As in the univariate analysis of variance and the tests on regression and correlation coefficients in Chap. 3, the significance of T^2 still leaves unanswered the question of *which* responses have probably led to the rejection of the vector hypothesis. While we might test the individual hypotheses by referring their univariate t statistics to the Bonferroni critical value $t_{\alpha/2p;N-1}$, the union-intersection nature of the T^2 test leads directly to a way of controlling the Type I error probability for tests on all linear functions of the response means. In our usual notation suppose that a single sample of N observations has been drawn from the multinormal population with mean vector $\boldsymbol{\mu}$. Then, for any vector $\mathbf{a}' = [a_1, \ldots, a_p]$,

$$t^2(\mathbf{a}) = \frac{N[\mathbf{a}'(\bar{\mathbf{x}} - \boldsymbol{\mu})]^2}{\mathbf{a}'\mathbf{S}\mathbf{a}} \leq N(\bar{\mathbf{x}} - \boldsymbol{\mu})'\mathbf{S}^{-1}(\bar{\mathbf{x}} - \boldsymbol{\mu})$$

and from the distribution of the T^2 statistic it follows that the probability statement

(1) $$P[\text{all } t^2(\mathbf{a}) \leq T^2_{\alpha;p,N-p}] = 1 - \alpha$$

holds for all $\mathbf{a}$, where

(2) $$T^2_{\alpha;p,N-p} = \frac{(N-1)p}{N-p} F_{\alpha;p,N-p}$$

denotes the 100α percentage point of the distribution of a central T^2 variate. The inequality within the parentheses can be expanded as

(3) $$\mathbf{a}'\bar{\mathbf{x}} - \sqrt{\frac{1}{N}\mathbf{a}'\mathbf{S}\mathbf{a}}\, T_{\alpha;p,N-p} \leq \mathbf{a}'\boldsymbol{\mu} \leq \mathbf{a}'\bar{\mathbf{x}} + \sqrt{\frac{1}{N}\mathbf{a}'\mathbf{S}\mathbf{a}}\, T_{\alpha;p,N-p}$$

and of course the probability that *all* such intervals generated by different choices of the elements of $\mathbf{a}$ are simultaneously true is $1 - \alpha$. The inequalities define the family of simultaneous confidence intervals of Roy and Bose (1953) with coefficient $1 - \alpha$. For a given choice of α we may construct intervals for any and all linear compounds of the means that interest us, with the assurance that all are true with probability $1 - \alpha$. Alternatively, any hypothesis

(4) $$H_0\colon\ \mathbf{a}'\boldsymbol{\mu} = \mathbf{a}'\boldsymbol{\mu}_0$$

can be tested by referring the absolute value of the statistic

(5) $$t(\mathbf{a}) = \frac{\mathbf{a}'(\bar{\mathbf{x}} - \boldsymbol{\mu}_0)\sqrt{N}}{\sqrt{\mathbf{a}'\mathbf{S}\mathbf{a}}}$$

to the critical value $T_{\alpha;p,N-p}$. If that value is exceeded, the hypothesis is untenable at the α level in the multiple-comparison, or simultaneous-testing, sense. Any number of those tests can be made with the assurance of protection at that level.

We may also use the Bonferroni method to make simultaneous inferences about a finite number m of linear functions $\mathbf{a}'\boldsymbol{\mu}$. The family of intervals with the general member

(6) $$\mathbf{a}'\bar{\mathbf{x}} - \sqrt{\frac{1}{N}\mathbf{a}'\mathbf{S}\mathbf{a}}\, t_{\alpha/2m;N-1} \leq \mathbf{a}'\boldsymbol{\mu} \leq \mathbf{a}'\bar{\mathbf{x}} + \sqrt{\frac{1}{N}\mathbf{a}'\mathbf{S}\mathbf{a}}\, t_{\alpha/2m;N-1}$$

has a confidence coefficient at least equal to $1 - \alpha$. Similarly, we may test m individual hypotheses of the form (4) by referring the statistic (5) to the two-sided critical value $t_{\alpha/2m;N-1}$.

Table 4.1. **Ratio of Expected Lengths of 95 percent Bonferroni and Roy-Bose Simultaneous Confidence Intervals**

N	p		
	2	4	6
10	0.85	0.60	0.45
20	0.89	0.73	0.61
30	0.90	0.77	0.66
∞	0.92	0.81	0.74

The precision of the Roy-Bose and Bonferroni simultaneous confidence intervals may be compared in terms of their expected lengths. The ratio of the expected lengths is merely that of the respective critical values, or $t_{\alpha/2m;N-1}/T_{\alpha;p,N-p}$. Some values of the ratio are given in Table 4.1 for the case $m = p$. The Bonferroni intervals may be appreciably shorter when the number of responses is large, although the opposite may be true if the number of parametric functions m is greatly in excess of p.

Example 4.2. We shall use the Roy-Bose simultaneous confidence intervals to see which response in Example 4.1 appears to have contributed to rejection of the hypothesized mean vector. The difference $\bar{x}_i - \mu_i$ is much smaller for the verbal score, and furthermore its sample variance is nearly twice that of the performance mean. We shall take as coefficient vector $\mathbf{a}' = [1,0]$ and inquire whether $\mathbf{a}'\boldsymbol{\mu} = 60$ falls within the 99 percent simultaneous confidence interval set. We compute

$$\sqrt{\frac{1}{N}\mathbf{a}'\mathbf{S}\mathbf{a}} = 1.444 \qquad \sqrt{\frac{(N-1)p}{N-p}F_{0.01;2,99}} = 3.12$$

and find that the interval for the verbal mean is

$$50.73 \leq \mu_v \leq 59.75$$

Since this interval does not contain the value 60, it is reasonable to conclude at the 99 percent simultaneous confidence level that the verbal scores have contributed to our original rejection of the hypothesized mean vector. It is readily evident from inspection of the performance statistics that their confidence interval would not include a population mean of 50.

4.4 THE CASE OF TWO SAMPLES

Suppose that two independent random samples of observations on some multidimensional variate have been obtained under different experi-

Table 4.2. **Components of the Two-sample T^2 Statistic**

Sample size	Sample 1	Sample 2
	N_1	N_2
Mean vector	$\bar{\mathbf{x}}_1' = [\bar{x}_{11}, \ldots, \bar{x}_{1p}]$	$\bar{\mathbf{x}}_2' = [\bar{x}_{21}, \ldots, \bar{x}_{2p}]$
Matrix of sums of squares and products	$\mathbf{A}_1$	$\mathbf{A}_2$
Pooled covariance matrix	$\mathbf{S} = \dfrac{1}{N_1 + N_2 - 2} (\mathbf{A}_1 + \mathbf{A}_2)$	

mental or environmental conditions. We shall assume that in either condition the variates have a multivariate normal distribution with the same, though unknown, covariance matrix $\mathbf{\Sigma}$ of full rank p. However, it follows from the conditions or other evidence that the distributions may not necessarily have the same location parameters, and we now desire a test of the null hypothesis

$$H_0: \quad \boldsymbol{\mu}_1 = \boldsymbol{\mu}_2 \tag{1}$$

that the population mean vectors are identical, as opposed to the alternative hypothesis

$$H_1: \quad \boldsymbol{\mu}_1 \neq \boldsymbol{\mu}_2$$

of different means. From the data mean vectors have been computed, and the pooled estimate $\mathbf{S}$ of the common covariance matrix has been prepared according to expression (11) of Sec. 3.5. For convenience those quantities are summarized in Table 4.2. By application of the union-intersection or likelihood-ratio principles the statistic for testing the hypothesis is

$$T^2 = \frac{N_1 N_2}{N_1 + N_2} (\bar{\mathbf{x}}_1 - \bar{\mathbf{x}}_2)'\mathbf{S}^{-1}(\bar{\mathbf{x}}_1 - \bar{\mathbf{x}}_2) \tag{2}$$

The quantity

$$F = \frac{N_1 + N_2 - p - 1}{(N_1 + N_2 - 2)p} T^2 \tag{3}$$

has the variance ratio F distribution with degrees of freedom p and $N_1 + N_2 - p - 1$. In the usage of the general expression (14) of Sec. 4.2, $\mathbf{Y} = \sqrt{N_1 N_2/(N_1 + N_2)}\, (\bar{\mathbf{x}}_1 - \bar{\mathbf{x}}_2)$, $\mathbf{S}$ is the general covariance matrix, and $n = N_1 + N_2 - 2$. The decision rule for a test at the α level has this form:

Accept H_0: $\boldsymbol{\mu}_1 = \boldsymbol{\mu}_2$ if

$$(4) \qquad T^2 \leq \frac{(N_1 + N_2 - 2)p}{N_1 + N_2 - p - 1} F_{\alpha; p, N_1+N_2-p-1}$$

and reject otherwise.

We may construct an ellipsoidal confidence region for the difference $\boldsymbol{\delta} = \boldsymbol{\mu}_1 - \boldsymbol{\mu}_2$ of the population mean vectors by observing that

$$(5) \qquad T^2 = \frac{N_1 N_2}{N_1 + N_2} (\bar{\mathbf{x}}_1 - \bar{\mathbf{x}}_2 - \boldsymbol{\delta})' \mathbf{S}^{-1} (\bar{\mathbf{x}}_1 - \bar{\mathbf{x}}_2 - \boldsymbol{\delta})$$

still has the central T^2 distribution if $\boldsymbol{\delta}$ is the population mean of $\bar{\mathbf{x}}_1 - \bar{\mathbf{x}}_2$. The $100(1 - \alpha)$ percent joint confidence region is specified by those vectors $\boldsymbol{\delta}$ satisfying the inequality

$$(6) \qquad (\bar{\mathbf{x}}_1 - \bar{\mathbf{x}}_2 - \boldsymbol{\delta})' \mathbf{S}^{-1} (\bar{\mathbf{x}}_1 - \bar{\mathbf{x}}_2 - \boldsymbol{\delta}) \leq \frac{N_1 + N_2}{N_1 N_2} T^2_{\alpha; p, N_1+N_2-p-1}$$

where

$$(7) \qquad T^2_{\alpha; p, N_1+N_2-p-1} = \frac{(N_1 + N_2 - 2)p}{N_1 + N_2 - p - 1} F_{\alpha; p, N_1+N_2-p-1}$$

Similarly, the $100(1 - \alpha)$ percent *simultaneous* confidence intervals for all linear compounds $\mathbf{a}'\boldsymbol{\delta}$ of the mean differences are defined by

$$(8) \qquad \mathbf{a}'(\bar{\mathbf{x}}_1 - \bar{\mathbf{x}}_2) - \sqrt{\mathbf{a}'\mathbf{S}\mathbf{a} \frac{N_1 + N_2}{N_1 N_2}} T_{\alpha; p, N_1+N_2-p-1} \leq \mathbf{a}'\boldsymbol{\delta}$$
$$\leq \mathbf{a}'(\bar{\mathbf{x}}_1 - \bar{\mathbf{x}}_2) + \sqrt{\mathbf{a}'\mathbf{S}\mathbf{a} \frac{N_1 + N_2}{N_1 N_2}} T_{\alpha; p, N_1+N_2-p-1}$$

As in the single-sample case of Sec. 4.3, the Bonferroni intervals for m linear compounds can be obtained by replacing the T critical value of (8) by $t_{\alpha/2m; N_1+N_2-2}$.

Example 4.3. Forty-nine elderly men participating in an interdisciplinary study of human aging were classified into the diagnostic categories "senile factor present" and "no senile factor" on the basis of an intensive psychiatric examination.* The Wechsler Adult Intelligence Scale had been administered to all subjects by an independent investigator, and certain subtests showed large differences between the two groups. It was proposed that a test be made of the hypothesis that the groups arose from populations with a common mean vector, and if this hypothesis was rejected, the simultaneous confidence interval set would be used to determine which individual subtest means were significantly different.

* These diagnoses and data were collected by Seymour Perlin and Robert N. Butler, and are used here with their kind permission.

Table 4.3. **WAIS Subtest Means***

Subtest	Group	
	No senile factor, $N_1 = 37$	Senile factor, $N_2 = 12$
Information	12.57	8.75
Similarities	9.57	5.33
Arithmetic	11.49	8.50
Picture completion	7.97	4.75

* The WAIS was administered by James E. Birren and Jack Botwinick, and the scores are presented here with their kind permission.

The sample means are shown in Table 4.3. The within-group covariance matrix and its inverse were

$$\mathbf{S} = \begin{bmatrix} 11.2553 & 9.4042 & 7.1489 & 3.3830 \\ & 13.5318 & 7.3830 & 2.5532 \\ & & 11.5744 & 2.6170 \\ & & & 5.8085 \end{bmatrix}$$

and

$$\mathbf{S}^{-1} = \begin{bmatrix} 0.259064 & -0.135783 & -0.058797 & -0.064719 \\ & 0.186449 & -0.038305 & 0.014382 \\ & & 0.150964 & -0.016920 \\ & & & 0.211171 \end{bmatrix}$$

The two-sample T^2 statistic had the value 22.05; the associated F was 5.16, with degrees of freedom 4 and 44. Under the hypothesis of equal mean vectors the probability of exceeding such an F value would be less than 0.005, and we should reject that null hypothesis at the conventional 5 or 1 percent level.

Although we have rejected the null hypothesis, we still do not know which of the four mean differences may have contributed to the significant T^2 or for which it might be reasonable to conclude that their population means are equal. It would not be proper to test the four individual mean differences by univariate t statistics, for we must have protection against the effects of positive correlations among the subtests as well as the tendency for individual differences to be significant merely by chance as more responses are included in the variate vectors. The Roy-Bose simultaneous confidence intervals will be used to test the individual differences. We begin by noting that the arithmetic subtest mean difference is the smallest of the four, while the pooled variance of the score is rather large. The critical constant for this and subsequent confidence intervals with joint confidence coefficient 0.95 is

$$\sqrt{\frac{N_1 + N_2}{N_1 N_2}}\, T_{\alpha;\, p, N_1+N_2-p-1} = \sqrt{\frac{(49)(47)(4)}{(12)(37)(44)}\, 2.59}$$
$$= 1.11$$

where the 5 percent critical value of F was computed by interpolating linearly in the reciprocals of the degrees of freedom. The resulting confidence interval for the difference in the arithmetic means of the senile-factor and no-senile-factor populations is

$$-0.79 \leq \mu_{31} - \mu_{32} \leq 6.77$$

Since zero is included in the interval, we conclude at the 5 percent joint significance level that the arithmetic means are not different. The other 95 percent simultaneous confidence intervals indicate different population means, although the picture-completion subtest stands out as the most significantly different:

Information:	$0.10 \leq \mu_{11} - \mu_{12} \leq 7.54$
Similarities:	$0.16 \leq \mu_{21} - \mu_{22} \leq 8.32$
Picture completion:	$0.54 \leq \mu_{41} - \mu_{42} \leq 5.90$

Example 4.4. Mueller (1963) compared the concentration changes in plasma free fatty acid (FFA) and blood glucose (G) in 12 schizophrenic patients and 13 normal volunteers following intramuscular injections of insulin. The changes in those constituents are summarized below:

	Mean change	
	Schizophrenics	Normals
G, mg, % FFA, mequiv/liter	−25.6 −0.06	−31.1 −0.15
Sums of squares and products matrices	$\begin{bmatrix} 3,445 & 9.9492 \\ 9.9492 & 0.1105 \end{bmatrix}$	$\begin{bmatrix} 3,509 & -3.2408 \\ -3.2408 & 0.0865 \end{bmatrix}$
Pooled covariance matrix	$\mathbf{S} = \begin{bmatrix} 302 & 0.292 \\ 0.292 & 0.00856 \end{bmatrix}$	

Under the assumption that the FFA and G values have a bivariate normal distribution with a covariance matrix unaffected by psychiatric diagnoses, we wish to test the hypothesis that the normal and schizophrenic mean vectors are equivalent. We note that the assumption of a common covariance matrix appears valid with respect to the diagonal elements of the sample matrices, although less so for the covariance term. Since the null hypothesis is

$$H_0: \quad \boldsymbol{\mu}_1 - \boldsymbol{\mu}_2 = \mathbf{0}$$

the test statistic is

$$T^2 = \frac{|\mathbf{S} + N_1N_2(N_1 + N_2)^{-1}(\bar{\bar{\mathbf{x}}}_1 - \bar{\bar{\mathbf{x}}}_2)(\bar{\bar{\mathbf{x}}}_1 - \bar{\bar{\mathbf{x}}}_2)'|}{|\mathbf{S}|} - 1$$

The numerator determinant is 17.55, $|\mathbf{S}| = 2.500$, and $T^2 = 6.03$. The associated F statistic is equal to 2.88; with its 2 and 22 degrees of freedom we note that it is greater than the 10 percent critical value of 2.56 and less than the 5 percent critical value of 3.44. We conclude that at the $\alpha = 0.05$ level we should *not* reject the null hypothesis of a common mean vector for normal and schizophrenic determinations.

The Assumption of Equal Covariance Matrices. In this section we have assumed that the two samples were drawn from multinormal populations with a common covariance matrix. In practice this is a rather dubious requirement, for many experimental conditions which lead to higher mean values may also produce responses with larger variances. Ito and Schull (1964) have investigated the large-sample properties of the T^2 statistic in the presence of unequal covariance matrices $\mathbf{\Sigma}_1$, $\mathbf{\Sigma}_2$. The inequality is reflected in the computation of the true level α and the power probabilities in terms of the characteristic roots θ_i of the matrix $\mathbf{\Sigma}_1\mathbf{\Sigma}_2^{-1}$. If the elements of $\mathbf{\Sigma}_1$ are large relative to those of $\mathbf{\Sigma}_2$, at least some of the θ_i will be greater than unity, and if the opposite is true, the θ_i will tend to be less than one. Ito and Schull demonstrate that if the sample sizes N_1, N_2 are equal, the unequal matrices have no effect upon the size of the Type I error probability or the power function if the N_i are large. If $N_1 > N_2$ and the roots of $\mathbf{\Sigma}_1\mathbf{\Sigma}_2^{-1}$ are all equal to some common value $\theta > 1$, the true α probability is increasingly larger than its nominal value as N_1/N_2 increases. If $\theta > 1$, the opposite is true, and the true size of the test may be very small for N_1 several times greater than N_2. The effect upon the true power function is the same.

4.5 THE ANALYSIS OF REPEATED MEASUREMENTS

Frequently experimental observations are collected at different times on the same sampling unit. The simplest example of such a repeated-measurements design consists of the administration of a control treatment (a placebo, standard drug, or other null treatment) on one occasion and the treatment of interest on another. Here each subject serves as its own control. If it is reasonable to assume that the treatment has an *additive* effect, we can write the responses of the ith subject to the control and experimental conditions as

$$\begin{aligned} x_{i1} &= \mu + e_{i1} \\ x_{i2} &= \mu + \tau + e_{i2} \end{aligned} \tag{1}$$

respectively, where μ is a general-level effect common to all subjects and treatments, τ is the effect of the experimental condition, and e_{i1} and e_{i2} are random disturbance terms. The essential feature of this model is that the random disturbances e_{i1} and e_{i2} reflect fairly constant characteristics of the ith subject and thus must be treated as correlated. If we let the pairs $[e_{i1}, e_{i2}]$ be independently distributed according to the bivariate normal distribution with parameters

$$\begin{aligned} E(e_{i1}) &= E(e_{i2}) = 0 \\ \text{var}(e_{i1}) &= \sigma_1^2 \qquad \text{var}(e_{i2}) = \sigma_2^2 \\ \text{cov}(e_{i1}, e_{i2}) &= \rho\sigma_1\sigma_2 \end{aligned} \tag{2}$$

for all i, the sample information for testing the hypothesis

$$H_0: \quad \tau = 0 \tag{3}$$

will be contained in the treatment-control differences

$$d_i = x_{i2} - x_{i1} \qquad i = 1, \ldots, N \tag{4}$$

If we denote by $\bar{d}$ and s_d the mean and standard deviation of these differences, the test statistic for the hypothesis is

$$t = \frac{\bar{d}\sqrt{N}}{s_d} \tag{5}$$

If the null hypothesis H_0 is true, t has the Student-Fisher t distribution with $N - 1$ degrees of freedom.

Before the test of H_0 is carried out, it is also necessary to specify the alternative hypothesized values of τ if the null hypothesis is not supported by the data. Common alternatives are

$$\begin{aligned} H_{11}&: \quad \tau < 0 \\ H_{12}&: \quad \tau > 0 \\ H_{13}&: \quad \tau \neq 0 \end{aligned} \tag{6}$$

The first two imply a prediction of the direction of the treatment effect, and their forms rule out the possibility of unexpected changes in the opposite direction. As we remarked in Chap. 1, one-sided tests are more powerful in the usual sense that departures from the null hypothesis in the direction of the alternative hypothesis have a greater probability of detection. These alternatives should be used whenever it is possible to specify the direction of the treatment effect from previous investigations or other substantive considerations. The third alternative provides for changes in either direction; it and its associated test are called *two-sided.* We are permitted this additional vagueness about the possible treatment effect at the expense of a loss of power. If we denote by $t_{\alpha;N-1}$ the 100α percent critical value of the t distribution with $N - 1$ degrees of freedom, we may summarize the decision rules for the three alternatives at the α level in this fashion:

Alternative	*Reject* H_0: $\tau = 0$ *if*
H_{11}: $\tau < 0$	$t < -t_{\alpha;N-1}$
H_{12}: $\tau > 0$	$t > t_{\alpha;N-1}$
H_{13}: $\tau \neq 0$	$\|t\| > t_{\frac{1}{2}\alpha;N-1}$

Note that the probability α of falsely rejecting H_0 is equally divided between both tails of the t distribution in the third alternative.

Example 4.5. In a preliminary clinical trial of a tranquilizing drug six schizophrenic patients were randomly drawn from a group with similar characteristics. The first three patients were administered the drug and the remaining three a placebo, and subsequently each patient's anxiety level was independently scored by four members of the nursing staff on a standard scale. Two days later the drug and placebo recipients were reversed, and anxiety ratings were again made by the same personnel. These total scores were observed:

Patient	Placebo	Drug	Difference
1	27	19	−8
2	21	16	−5
3	25	14	−11
4	13	15	2
5	17	5	−12
6	11	3	−8

From pharmacological considerations it is reasonable to assume that the drug could only reduce anxiety in a patient, and the alternative hypothesis will be one-sided. The average change in the score after administration of the drug was −7 points. The standard deviation of the differences was equal to 5.060 points, and the resulting test statistic for the hypothesis of no drug-placebo difference was

$$t = \frac{-7\sqrt{6}}{5.060}$$

$$= -3.39$$

If we elect to work at the 0.05 significance level the critical value of t is

$$t_{0.05;5} = -2.015$$

and since the observed value of the statistic is less than that point, we should reject the hypothesis of no drug effect and conclude instead that at least at the 0.05 level the new drug has reduced anxiety as reflected by the rating instrument.

We remark in passing that in the interest of simplicity we have ignored tests for the *order* effect of the drug and control.

Now let us extend the paired-observation case to the general setup of p responses collected on the same experimental unit at successive times or under a variety of experimental conditions. We shall assume that the

p responses constitute a *fixed* set, in the sense that inferences about their parameters will apply *only* to the particular responses investigated and not to some larger population out of which the p levels might have been drawn. Let the observations on N independent units be arranged as in Table 4.4. The mathematical model for the jth response in the ith sampling unit is

$$x_{ij} = \mu + \mu_j + e_{ij} \qquad i = 1, \ldots, N, j = 1, \ldots, p \tag{7}$$

where μ = general-level parameter common to all observations
μ_j = measure of effect specific to jth condition
e_{ij} = random disturbance

The variate e_{ij} reflects both the interaction of the ith unit with the jth response and experimental error in that conjunction. To obtain tests of hypotheses on the μ_j it will be necessary to assume that the vector of variates $[e_{i1}, \ldots, e_{ip}]$ in the ith unit has the multivariate normal distribution with mean vector

$$\mathrm{E}[e_{i1}, \ldots, e_{ip}] = [0, \ldots, 0] \tag{8}$$

and covariance matrix

$$\mathbf{\Sigma} = \mathrm{E}\left(\begin{bmatrix} e_{i1} \\ \cdot \\ \cdot \\ \cdot \\ e_{ip} \end{bmatrix} [e_{i1}, \ldots, e_{ip}]\right) \tag{9}$$

Residual vectors of different experimental units are distributed independently of one another. Hence each row of Table 4.4 is independently distributed according to the multinormal law with mean vector

$$\boldsymbol{\mu}' = [\mu + \mu_1, \ldots, \mu + \mu_p] \tag{10}$$

and covariance matrix $\mathbf{\Sigma}$.

Table 4.4. **Observations from a Repeated-measurements Experiment**

Sampling unit	Condition		
	1	. . .	p
1	x_{11}	. . .	x_{1p}
. . .	. . .	. . .	. . .
N	x_{N1}	. . .	x_{Np}

The null hypothesis of equal response effects is

$$H_0: \quad \mu_1 = \cdots = \mu_p$$

The alternative hypothesis is

$$H_1: \quad \mu_i \neq \mu_j$$

for at least one pair of distinct treatments i and j. The vector statements of these hypotheses are

$$(11) \qquad H_0: \begin{bmatrix} \mu_1 - \mu_2 \\ \cdots \\ \mu_{p-1} - \mu_p \end{bmatrix} = \begin{bmatrix} 0 \\ \cdot \\ \cdot \\ \cdot \\ 0 \end{bmatrix} \qquad H_1: \begin{bmatrix} \mu_1 - \mu_2 \\ \cdots \\ \mu_{p-1} - \mu_p \end{bmatrix} \neq \begin{bmatrix} 0 \\ \cdot \\ \cdot \\ \cdot \\ 0 \end{bmatrix}$$

The test of H_0 can be carried out by the Hotelling T^2 statistic computed from the mean vector and sample covariance matrix of the differences

$$(12) \qquad y_{ij} = x_{ij} - x_{i,j+1}$$

of the observations on adjacent responses. If we denote by $\bar{\mathbf{y}}'$ the $(p - 1)$-component vector of differences of successive response means

$$(13) \qquad [\bar{y}_1, \ldots, \bar{y}_{p-1}] = [\bar{x}_1 - \bar{x}_2, \ldots, \bar{x}_{p-1} - \bar{x}_p]$$

and by $\mathbf{S}_y$ the $(p - 1) \times (p - 1)$ covariance matrix of the y_{ij} differences, the statistic for testing the null hypothesis of equal response effects is

$$(14) \qquad T^2 = N\bar{\mathbf{y}}'\mathbf{S}_y^{-1}\bar{\mathbf{y}}$$

and in turn if the null hypothesis is true,

$$(15) \qquad F = \frac{N - p + 1}{(N - 1)(p - 1)} T^2$$

has the F distribution with $p - 1$ and $N - p + 1$ degrees of freedom. If the level of the test is α, the null hypothesis is rejected if

$$F_{\alpha; p-1, N-p+1}$$

and accepted otherwise.

It is not essential that the null hypothesis of equal response effects be stated in terms of the differences of adjacent responses. Identical tests could be based upon the deviations of the first $p - 1$ effects from their mean, or upon the differences of the first or last effect and the remaining effects. This uniqueness of the test follows from the invariance property of the T^2 statistic discussed in Sec. 4.2. Write the differences of successive response observations in vector form as

$$\mathbf{y}_i = \mathbf{C}\mathbf{x}_i \tag{16}$$

for the ith experimental unit, where $\mathbf{y}_i$ and $\mathbf{x}_i$ are the transformed and original observation vectors and $\mathbf{C}$ is the $(p - 1) \times p$ patterned matrix

$$\begin{bmatrix} 1 & -1 & 0 & \cdots & 0 & 0 \\ 0 & 1 & -1 & \cdots & 0 & 0 \\ \cdot & \cdot & \cdot & \cdots & \cdot & \cdot \\ 0 & 0 & 0 & \cdots & 1 & -1 \end{bmatrix} \tag{17}$$

Then, in terms of the original mean vector and covariance matrix, the test statistic is

$$T^2 = N\bar{\mathbf{x}}'\mathbf{C}'(\mathbf{CSC}')^{-1}\mathbf{C}\bar{\mathbf{x}} \tag{18}$$

If $\mathbf{A}$ is any nonsingular $(p - 1) \times (p - 1)$ matrix, any transformation of the original observations of the form

$$\mathbf{y}_i = \mathbf{AC}\mathbf{x}_i$$

can be used to test the null hypothesis. We prefer the original $\mathbf{C}$ for its computational convenience, although, as we have indicated, these transformations might be used if the first or last responses play some special role:

$$\mathbf{C}_1 = \begin{bmatrix} 1 & -1 & \cdots & 0 \\ 1 & 0 & \cdots & 0 \\ \cdot & \cdot & \cdots & \cdot \\ 1 & 0 & \cdots & -1 \end{bmatrix} \qquad \mathbf{C}_2 = \begin{bmatrix} -1 & \cdots & 0 & 1 \\ 0 & \cdots & 0 & 1 \\ \cdot & \cdots & \cdot & \cdot \\ 0 & \cdots & -1 & 1 \end{bmatrix} \tag{19}$$

Example 4.6. The levels of free fatty acid (FFA) in blood were measured in $N = 15$ hypnotized normal volunteers who had been asked to experience fear, depression, and anger affects while in the hypnotic state.* During each of three sessions changes in FFA following dissimulation of the effect were recorded. Under the assumption that each effect produced the same degree of change, the investigators wished to ascertain that the first stress session did not produce different FFA changes as compared to the second and third. The mean FFA changes were

$$\bar{x}_1 = 2.699 \qquad \bar{x}_2 = 2.178 \qquad \bar{x}_3 = 2.558$$

The covariance matrix of the stress differences $y_{i1} = x_{i1} - x_{i2}$, $y_{i2} = x_{i1} - x_{i3}$ was

$$\mathbf{S} = \begin{bmatrix} 1.7343 & 1.1666 \\ 1.1666 & 2.7733 \end{bmatrix} \qquad \mathbf{S}^{-1} = \begin{bmatrix} 0.8041 & -0.3382 \\ -0.3382 & 0.5029 \end{bmatrix}$$

$T^2 = 2.68$, and $F = 1.24$ with degrees of freedom 2 and 13. The hypothesis of equal stress FFA changes, or of no order effect, is not rejected at the 0.05 level.

* These data are used with the kind permission of J. R. Fishman and P. S. Mueller.

Multiple Comparisons. If the hypothesis of equal treatment effects is rejected, the simultaneous confidence intervals developed in Sec. 4.3 can be used to determine which of the treatments differ. Let us begin by restating the definition given in Sec. 1.6 of a contrast as any linear function

$$\sum_{j=1}^{p} b_j \mu_j \tag{20}$$

of the treatment effects whose coefficients b_j sum to zero. Clearly, the elements of each row of **C** obey that property and thereby generate contrasts of adjacent treatments. Now let $\mathbf{a}'$ be any $(p - 1)$-component row vector. By the union-intersection principle it follows that all inequalities

$$[\mathbf{a}'\mathbf{C}(\bar{\mathbf{x}} - \boldsymbol{\mu})]^2 \leq \frac{1}{N}\, \mathbf{a}'\mathbf{CSC}'\mathbf{a}T^2_{\alpha;p-1,N-p+1} \tag{21}$$

formed from different choices of the elements of **a** hold simultaneously with probability $1 - \alpha$. But since the row sums of **C** each equal zero, so must the sum of the elements of $\mathbf{a}'\mathbf{C}$ be zero. $\mathbf{b}'\boldsymbol{\mu} = \mathbf{a}'\mathbf{C}\boldsymbol{\mu}$ is therefore a contrast of the treatment effects, for by the contrast property

$$\begin{aligned} \mathbf{b}'\boldsymbol{\mu} &= \sum_{j=1}^{p} b_j(\mu + \mu_j) \\ &= \sum_{j=1}^{p} b_j \mu_j \end{aligned} \tag{22}$$

Expansion of (21) leads to the set of $100(1 - \alpha)$ percent simultaneous confidence intervals whose general member is

$$\begin{aligned} \mathbf{b}'\bar{\mathbf{x}} - \sqrt{\frac{1}{N}\, \mathbf{b}'\mathbf{Sb}}\; T_{\alpha;p-1,N-p+1} &\leq \mathbf{b}'\boldsymbol{\mu} \\ &\leq \mathbf{b}'\bar{\mathbf{x}} + \sqrt{\frac{1}{N}\, \mathbf{b}'\mathbf{Sb}}\; T_{\alpha;p-1,N-p+1} \end{aligned} \tag{23}$$

If the interval includes zero, the hypothesis

$$H_0\colon \sum_{j=1}^{p} b_j \mu_j = 0 \tag{24}$$

is acceptable in the multiple comparison sense at the α level. Alternatively, that hypothesis can be tested directly by referring

$$t(\mathbf{b}) = \frac{\mathbf{b}'\bar{\mathbf{x}}\sqrt{N}}{\sqrt{\mathbf{b}'\mathbf{S}\mathbf{b}}}$$

to the critical value. If

$$|t(\mathbf{b})| \le T_{\alpha;p-1,N-p+1}$$

the hypothesis is accepted. We note that the original T^2 test rejects the hypothesis of equal treatment effects if and only if *some* contrast exists whose hypothesis (24) would be rejected by the multiple-comparison procedure.

In an actual application it is best to begin by ordering the treatments by their sample means. Starting perhaps with the smallest difference, simultaneous confidence intervals or $t(\mathbf{b})$ statistics are computed, and treatments that are not significantly different are so designated by a common underscore line. Since the variances $(1/N)\mathbf{b}'\mathbf{S}\mathbf{b}$ may vary widely, it will generally not be possible to assess the significance of all mean differences and other important contrasts without a fair amount of arithmetic.

The preceding multiple-comparison method is the extension of the Scheffé technique to repeated-measurement designs and has been developed independently by Scheffé in that connection (1953, 1959). We may also use the Bonferroni intervals for a particular set of contrasts as in Sec. 4.3.

Example 4.7. Reaction times to visual stimuli were obtained from $N = 20$ young normal men under three conditions A, B, and C of stimulus display. The mean reaction times in hundredths of seconds were 21.05, 21.65, and 28.95, respectively. The sample covariance matrix was

$$\mathbf{S} = \begin{bmatrix} 2.2605 & 2.1763 & 1.6342 \\ & 2.6605 & 1.8237 \\ & & 2.4710 \end{bmatrix}$$

We shall base our test of the hypothesis of equal stimulus condition effects upon the differences $B - A$ and $C - B$. The transformation matrix is then

$$\mathbf{C} = \begin{bmatrix} -1 & 1 & 0 \\ 0 & -1 & 1 \end{bmatrix}$$

Thus

$$\mathbf{CSC}' = \begin{bmatrix} 0.5684 & -0.2947 \\ -0.2947 & 1.4842 \end{bmatrix} \qquad (\mathbf{CSC}')^{-1} = \begin{bmatrix} 1.9611 & 0.3894 \\ 0.3894 & 0.7510 \end{bmatrix}$$

and

$$T^2 = 20[0.60, 7.30] \begin{bmatrix} 1.9611 & 0.3894 \\ 0.3894 & 0.7510 \end{bmatrix} \begin{bmatrix} 0.60 \\ 7.30 \end{bmatrix}$$
$$= 882.8$$

The associated F statistic is equal to 418.2. Since the critical value of F with 2 and 18 degrees of freedom is 10.39 for $\alpha = 0.001$, we conclude that the null hypothesis is overwhelmingly rejected even at that severe level. Now let us determine which condition effects are significantly different in the sense afforded by the simultaneous confidence bounds on all contrasts derivable from the three conditions. Because of the exceptionally high significance of T^2, let us take as the level of the simultaneous tests $\alpha = 0.001$, so that the associated confidence intervals will have coefficient 0.999. Then

$$T_{0.001;2,18} = 4.68$$

Let us begin with an examination of the contrast

$$\mu_B - \mu_A$$

This is estimated by the difference of the B and A means, 0.60. The sample variance of those response differences is the (1,1) element 0.5684 of the matrix **CSC′**, and the standard deviation of $\bar{x}_B - \bar{x}_A$ is

$$s_{\bar{x}_B - \bar{x}_A} = \sqrt{\frac{0.5684}{20}} = 0.1686$$

The limits of the 99.9 percent simultaneous confidence interval for the contrast will be determined by

$$s_{\bar{x}_B - \bar{x}_A} T_{0.001;2,18} = 0.789$$

Since the sample estimate 0.60 of the contrast does not exceed that value, the interval will contain zero, and the hypothesis of equal A and B condition effects is acceptable at the 0.001 level. We note for completeness that the actual confidence interval is

$$-0.19 \leq \mu_B - \mu_A \leq 1.39$$

Next let us consider the contrast

$$\mu_C - \mu_B$$

The sample estimate is $\bar{x}_C - \bar{x}_B = 7.30$, and the standard deviation of that estimate is

$$s_{\bar{x}_C - \bar{x}_B} = \sqrt{\frac{1.4842}{20}} = 0.2724$$

The 99.9 percent confidence interval is

$$6.02 \leq \mu_C - \mu_B \leq 8.58$$

and since the quotient of the estimate and its estimated standard deviation is larger than the critical value $T_{0.001;2,18}$, we conclude that the C condition effect is larger than that of condition B. We might continue in this vein and compute confidence intervals for the contrasts $\mu_C - \mu_A$, $\frac{1}{2}(\mu_A + \mu_B) - \mu_C$, although it would appear from our previous intervals that these would still support the significant difference of condition C from conditions A and B.

The Mixed-model Analysis of Variance. Some older tests for the equal-means hypothesis in repeated-measurements designs have used an analysis of variance based upon the linear model (7). Because the conditions were considered a fixed set and the subjects a random sample, the model was called *mixed* in reference to models I and II of Eisenhart (1947). Different forms of the two-way mixed model with n observations in each cell have been advocated by various workers: the variations in sampling assumptions and the covariance structures of the random components lead to different tests. Scheffé (1959) has discussed the justification and properties of his model at length; Hocking (1973) has described the competing models in relation to that of Scheffé. In the present brief treatment we shall merely consider the case of model (7) and Table 4.4 with a single observation for each condition–subject combination and the resulting analysis-of-variance test for the equal-means hypothesis.

For the validity of the analysis-of-variance F distribution we shall assume that the covariance matrix of the e_{ij} has the equal-variance, equal-correlation pattern

(25)
$$\mathbf{\Sigma} = \sigma^2 \begin{bmatrix} 1 & \cdots & \rho \\ \cdot & \cdots & \cdot \\ \rho & \cdots & 1 \end{bmatrix}$$

for each i. The analysis of variance is shown in Table 4.5, where

(26)
$$\begin{aligned} S_1 &= \frac{1}{N}\sum_{j=1}^{p} T_j^2 - \frac{G^2}{Np} \\ S_2 &= \frac{1}{p}\sum_{i=1}^{N} R_i^2 - \frac{G^2}{Np} \\ S_3 &= S_4 - S_1 - S_2 \\ S_4 &= \sum_{i=1}^{N}\sum_{j=1}^{p} x_{ij}^2 - \frac{G^2}{Np} \end{aligned}$$

$$T_j = \sum_{i=1}^{N} x_{ij} \qquad R_i = \sum_{j=1}^{p} x_{ij} \qquad G = \sum_{i=1}^{N}\sum_{j=1}^{p} x_{ij}$$

Under the null hypothesis of equal means,

(27)
$$F = (N-1)\frac{S_1}{S_3}$$

Table 4.5. **The Mixed-model Analysis of Variance**

Source	Sum of squares	Degrees of freedom	Mean square
Conditions	S_1	$p - 1$	$\dfrac{S_1}{p-1}$
Subjects	S_2	$N - 1$	$\dfrac{S_2}{N-1}$
$C \times S$ interaction	S_3	$(p-1)(N-1)$	$\dfrac{S_3}{(p-1)(N-1)}$
Total	S_4	$Np - 1$	-

has the F distribution with degrees of freedom $p - 1$ and $(p - 1)(N - 1)$, and we would reject that hypothesis at the α level if $F > F_{\alpha;p-1,(p-1)(N-1)}$. Under the alternative of a general mean vector the statistic has the corresponding noncentral F distribution with a noncentrality parameter

$$(28) \qquad \delta^2 = N[\sigma^2(1-\rho)]^{-1} \sum_{j=1}^{p} (\mu_j - \bar{\mu}_.)^2$$

where $\bar{\mu}_. = (1/p) \sum_{j=1}^{p} \mu_j$. Because the second degrees-of-freedom parameter is greater than that of the Hotelling T^2 F statistic, the power of the analysis-of-variance test will always be greater for the same alternative when the covariance model (25) holds.

Scheffé (1959) has given the exact $100(1 - \alpha)$ percent simultaneous confidence interval set on all contrasts $\boldsymbol{\Psi} = \mathbf{b}'\boldsymbol{\mu}$ defined by expression (22). The general interval is

$$(29) \qquad \mathbf{b}'\bar{\mathbf{x}} - \sqrt{(p-1)F_{\alpha;p-1,(p-1)(N-1)}}\, \hat{\sigma}(\hat{\psi}) \le \mathbf{b}'\boldsymbol{\mu} \le \mathbf{b}'\bar{\mathbf{x}} + \sqrt{(p-1)F_{\alpha;p-1,(p-1)(N-1)}}\, \hat{\sigma}(\hat{\psi})$$

where

$$(30) \qquad \hat{\sigma}^2(\hat{\psi}) = \frac{S_3}{N(N-1)(p-1)} \mathbf{b}'\mathbf{b}$$

is the estimated variance of the sample contrast. The expected lengths of these intervals will be shorter than those of (23) because of the more specialized covariance assumptions. Some comparisons of the lengths have been given by Morrison (1972).

The test for equal means under the covariance pattern (25) was obtained by Wilks (1946) from the generalized likelihood-ratio principle. While that pattern is a sufficient condition for the test statistic (27) to have an F distribution, it is not a necessary one. Necessary and sufficient conditions have been determined by Huynh and Feldt (1970); they may be expressed in these alternative forms:

1. $\mathbf{\Sigma}$ has the pattern defined by

$$\sigma_{ij} = \begin{cases} \alpha_i + \alpha_j + \lambda & i = j \\ \alpha_i + \alpha_j & i \neq j \end{cases}$$

2. All possible differences $X_i - X_j$ of the response variates have the same variance. This condition also implies the equality of the covariances of all pairs of differences.
3. The function ϵ of the elements of $\mathbf{\Sigma}$ defined by expression (20) of Sec. 5.6 is equal to one.

The Wilks test and the test for the necessary and sufficient pattern will be given in Sec. 7.3.

Example 4.8. Let us illustrate the mixed-model analysis of variance with the reaction-time statistics of Example 4.7. In doing so we proceed as if the equal-variance, equal-covariance assumption is tenable, although this should be supported by a formal test. The sums of squares can be computed from the means $\bar{x}_j$ and the elements s_{ij} of $\mathbf{S}$ as

$$S_1 = N \sum_{j=1}^{p} \bar{x}_j^2 - \frac{G^2}{Np}$$

$$S_2 = \frac{N-1}{p} \sum_{i=1}^{p} \sum_{j=1}^{p} s_{ij}$$

$$S_4 = (N-1) \operatorname{tr} \mathbf{S} + N \sum_{j=1}^{p} \bar{x}_j^2$$

The analysis of variance follows:

Source	Sum of squares	Degrees of freedom	Mean square
Conditions	773.734	2	386.87
Subjects	118.183	19	–
Interaction	22.264	38	0.5859
Total	914.181	59	

The F ratio of 660.3 would of course imply rejection of the equal-means hypothesis at any reasonable significance level. The 99.9 percent Scheffé simultaneous confidence interval for the difference of the means of the A and B conditions is $-0.39 \leq \mu_B - \mu_A \leq 1.59$; this is slightly wider than that obtained from the T^2 test.

Some Further References. Morrison (1972b) has given a survey of repeated-measurements methods, with particular emphasis on tests with special covariance patterns. The development of some of those tests was due to Geisser (1963). Cole (1969) has considered tests for another class of patterned matrices. Tests for the equality of bivariate normal means with missing data have been proposed and studied by Mehta and Gurland (1969, 1973), Morrison (1972, 1973), and Lin and Stivers (1974).

4.6 PROFILE ANALYSIS FOR TWO INDEPENDENT GROUPS

Let us suppose that two independent random samples of individuals have been administered the same battery of psychological tests. The scores or responses of the tests can be assumed to be observations on continuous random variables with some multivariate distribution. However, in departure from our earlier treatment of the problem of two samples we shall require that the responses be *commensurable,* or expressed in comparable units. It is convenient for counseling and interpretive purposes to record an individual's scores graphically as the *profile* of test performance. In this section we shall consider a mathematical model and certain tests of hypotheses for the analysis of group, or average, profiles.

We shall always assume that the groups have been formed from criteria other than the profiles themselves. Division of a single sample into two or more subgroups purely by an inspection of the profiles is an entirely different problem of far greater complexity. A number of methods have been proposed for resolving a sample of profiles into an unspecified number of groups, but their misclassification probabilities and other operating characteristics do not appear to have been worked out.

To motivate our treatment assume that the mean profiles of the two samples had the appearance of Fig. 4.2. These profiles suggest three questions we might ask concerning the population profiles of the samples:

1. Are the population mean profiles *similar,* in the sense that the line segments of adjacent tests are *parallel?*
2. If the two population profiles are indeed parallel, are they also at the same *level?*

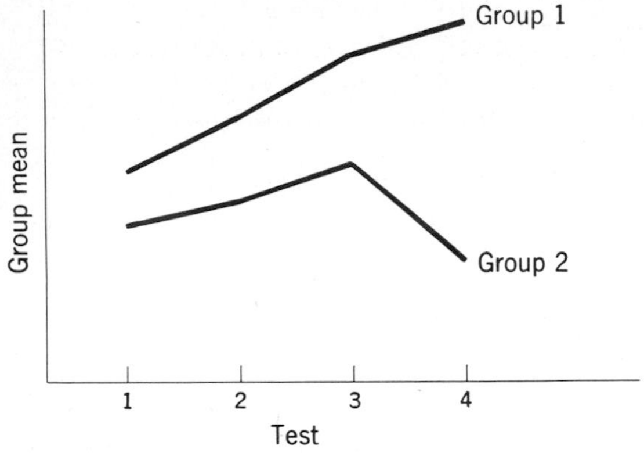

figure 4.2

3. Again assuming parallelism, are the population means of the tests different?

In the usage of experimental design question 1 refers to the hypothesis of no *response by group interaction*, while question 2 addresses itself to the hypothesis of equal group effects. Because the tests for equal levels and response effects have no meaning if a group-response interaction is present, question 1 has been accorded priority among the tests. Question 3 is treated by the repeated-measurements techniques of the previous section, with a slight modification for the presence of two samples.

Now let us consider a model for profile observations that will permit us to test the three hypotheses. We begin with the usual assumption that the responses are described by a p-dimensional multinormal random variable $\mathbf{X}$. Under the two experimental treatments, diagnostic classifications, or other conditions the respective mean vectors of $\mathbf{X}$ will be $\boldsymbol{\mu}_1' = [\mu_{11}, \ldots, \mu_{1p}]$ and $\boldsymbol{\mu}_2' = [\mu_{21}, \ldots, \mu_{2p}]$. Both populations will have the common, though unknown, covariance matrix $\boldsymbol{\Sigma}$. Through careful selection of the responses and elimination of obvious means, totals, and differences $\boldsymbol{\Sigma}$ can be assumed nonsingular. In our development the variances of the response variates need not be equal, although *whatever inferences are made about the population profile structure must be accepted in the context of the scales chosen for the analysis.* For that reason we shall prefer to give the three tests in terms of the elements of the mean vectors $\boldsymbol{\mu}_1$ and $\boldsymbol{\mu}_2$ rather than through the additive analysis-of-variance model.

Under each condition random samples of N_1 and N_2 observation vectors are now obtained independently on $N_1 + N_2$ separate sampling

Table 4.6. **Observations for the Profile Analysis of Two Independent Groups**

	Group 1 responses			Group 2 responses		
	1	. . .	p	1	. . .	p
	x_{111}	. . .	x_{11p}	x_{211}	. . .	x_{21p}
	. . .	. . .	. . .	. . .	. . .	. . .
	x_{1N_11}	. . .	x_{1N_1p}	x_{2N_21}	. . .	x_{2N_2p}
Response means	$\bar{x}_{11}$	. . .	$\bar{x}_{1p}$	$\bar{x}_{21}$	. . .	$\bar{x}_{2p}$

units. It will be convenient to think of the data as arranged in Table 4.6, with each row corresponding to the vector from a different unit. We shall denote the mean vectors by $\bar{\mathbf{x}}_1' = [\bar{x}_{11}, \ldots, \bar{x}_{1p}]$ and $\bar{\mathbf{x}}_2' = [\bar{x}_{21}, \ldots, \bar{x}_{2p}]$, and as before $\mathbf{S}$ will stand for the unbiased estimate of $\mathbf{\Sigma}$ obtained from the sums of squares and products computed within the sample of each condition. Those vectors and that covariance matrix will supply the information needed for the three tests.

Let us begin with the parallelism, or no groups-by-tests interaction, hypothesis. This states that the slopes of the population profile segments are the same under each condition:

(1) $$H_0: \begin{bmatrix} \mu_{11} - \mu_{12} \\ \cdots \\ \mu_{1,p-1} - \mu_{1p} \end{bmatrix} = \begin{bmatrix} \mu_{21} - \mu_{22} \\ \cdots \\ \mu_{2,p-1} - \mu_{2p} \end{bmatrix}$$

This can be written in matrix form as

(2) $$H_0: \quad \mathbf{C}\boldsymbol{\mu}_1 = \mathbf{C}\boldsymbol{\mu}_2$$

where $\mathbf{C}$ is the $(p - 1) \times p$ transformation matrix (17) of Sec. 4.5. The estimates of those segments for the ith sample are

(3) $$\bar{\mathbf{x}}_i'\mathbf{C}' = [\bar{\mathbf{x}}_{i1} - \bar{\mathbf{x}}_{i2}, \ldots, \bar{\mathbf{x}}_{i,p-1} - \bar{\mathbf{x}}_{ip}]$$

and their sample covariance matrix is $\mathbf{CSC}'/N_i$. The statistic for testing the parallelism hypothesis is the two-sample T^2 computed from the $p - 1$ differences of the successive responses:

(4) $$T^2 = \frac{N_1N_2}{N_1 + N_2}(\bar{\mathbf{x}}_1 - \bar{\mathbf{x}}_2)'\mathbf{C}'(\mathbf{CSC}')^{-1}\mathbf{C}(\bar{\mathbf{x}}_1 - \bar{\mathbf{x}}_2)$$

We refer

(5) $$F = \frac{N_1 + N_2 - p}{(N_1 + N_2 - 2)(p - 1)} T^2$$

to a table of the F distribution with degrees of freedom $p - 1$ and $N_1 + N_2 - p$ and reject the hypothesis (1) at the α level if the observed F exceeds the critical value $F_{\alpha; p-1, N_1+N_2-p}$.

If the hypothesis of parallel profiles is tenable, we may test the hypothesis (question 2) of equal condition levels. In our development of profile analysis this is

(6) $$H_0: \quad \mathbf{j}'\boldsymbol{\mu}_1 = \mathbf{j}'\boldsymbol{\mu}_2$$

where $\mathbf{j}' = [1, \ldots, 1]$ is the p-component vector with unity in each position. This hypothesis is tested by computing the usual two-sample t statistic from the sums of the observations on all responses in each sampling unit. If we denote those row sums of Table 4.6 by

(7) $$y_{1j} = \sum_{h=1}^{p} x_{1jh} \qquad y_{2j} = \sum_{h=1}^{p} x_{2jh}$$

and their means by

(8) $$\bar{y}_1 = \frac{1}{N_1} \sum_{j=1}^{N_1} \sum_{h=1}^{p} x_{1jh} \qquad \bar{y}_2 = \frac{1}{N_2} \sum_{j=1}^{N_2} \sum_{h=1}^{p} x_{2jh}$$

the statistic is

(9) $$t = \frac{\bar{y}_1 - \bar{y}_2}{\sqrt{\dfrac{\Sigma(y_{1j} - \bar{y}_1)^2 + \Sigma(y_{2j} - \bar{y}_2)^2}{N_1 + N_2 - 2}\left(\dfrac{1}{N_1} + \dfrac{1}{N_2}\right)}}$$

or, in matrix notation,

(10) $$t = \frac{\mathbf{j}'(\bar{\mathbf{x}}_1 - \bar{\mathbf{x}}_2)}{\sqrt{\mathbf{j}'\mathbf{S}\mathbf{j}(1/N_1 + 1/N_2)}}$$

If the hypothesis (6) is true, the statistic has the t distribution with $N_1 + N_2 - 2$ degrees of freedom. The usual rules for acceptance or rejection of the null hypothesis apply: at the α level with no prejudice that one group should necessarily be higher than the other we accept if

$$|t| < t_{\frac{1}{2}\alpha; N_1+N_2-2}$$

and otherwise reject. If prior information or experimental considerations cause us to believe that the profile of one treatment should be higher than that of the other, the one-sided t rules described in Secs. 1.5 and 4.5 should be used.

Finally, let us consider the hypothesis (question 3) of equal response means under the two conditions. As in the preceding test, we must require that the parallelism hypothesis be acceptable. If it is not, the test must be carried out separately for the two samples. Under the assumption of parallel mean profiles the hypothesis is

$$H_0: \quad \mathbf{C}(\boldsymbol{\mu}_1 + \boldsymbol{\mu}_2) = \mathbf{0} \tag{11}$$

It is easily verified that this hypothesis is unaffected by a difference in the profile levels. For its test we compute the grand mean vector

$$\bar{\mathbf{x}} = \frac{N_1}{N_1 + N_2}\bar{\mathbf{x}}_1 + \frac{N_2}{N_1 + N_2}\bar{\mathbf{x}}_2 \tag{12}$$

and from it the single-sample T^2 statistic

$$T^2 = (N_1 + N_2)\bar{\mathbf{x}}'\mathbf{C}'(\mathbf{CSC}')^{-1}\mathbf{C}\bar{\mathbf{x}} \tag{13}$$

$\mathbf{S}$ is of course the original $p \times p$ within-samples covariance matrix. When the hypothesis (11) is true, the quantity

$$F = \frac{N_1 + N_2 - p}{(N_1 + N_2 - 2)(p - 1)} T^2 \tag{14}$$

has the F distribution with degrees of freedom $p - 1$ and $N_1 + N_2 - p$, and we reject at the α level the notion of equal response means if the observed F exceeds the upper critical value $F_{\alpha; p-1, N_1+N_2-p}$.

Example 4.9. Let us reexamine through the profile model the WAIS data originally considered in Example 4.3. The profiles of the means of Table 4.3 are given in Fig. 4.3 and appear to be exceptionally parallel. The covariance matrix of the differences of adjacent subtests can be computed to be

$$\mathbf{CSC}' = \begin{bmatrix} 5.99 & -3.90 & -1.06 \\ & 10.34 & -4.12 \\ & & 12.14 \end{bmatrix}$$

and its inverse is

$$(\mathbf{CSC}')^{-1} = \begin{bmatrix} 0.2585 & 0.1232 & 0.0644 \\ & 0.1705 & 0.0686 \\ & & 0.1113 \end{bmatrix}$$

The vectors of mean differences are

$$\bar{\mathbf{x}}_1'\mathbf{C}' = [3.00, -1.92, 3.52]$$

$$\bar{\mathbf{x}}_2'\mathbf{C}' = [3.42, -3.17, 3.75]$$

and the test statistic for the parallelism hypothesis is

$$T^2 = \frac{(12)(37)}{49}[-0.42, 1.25, -0.23]\begin{bmatrix} 0.2585 & 0.1232 & 0.0644 \\ 0.1232 & 0.1705 & 0.0686 \\ 0.0644 & 0.0686 & 0.1113 \end{bmatrix}\begin{bmatrix} -0.42 \\ 1.25 \\ -0.23 \end{bmatrix}$$

($\bar{x}_1 - \bar{x}_2$)

$$= 1.46$$

$$F = \frac{(1.464)(45)}{(47)(3)}$$

$$= 0.47$$

As we might have surmised from the profiles, this F statistic is hardly significant, and we conclude that the hypothesis of parallel mean profiles in the population is indeed most tenable.

Next we shall test the hypothesis of equal profile levels in the two diagnostic populations. Since it is rather unlikely that the presence of any condition relating to senility could change cognitive processes in the direction of *higher* WAIS scores, we shall test the null hypothesis against the alternative

$$H_a: \quad \mathbf{j}'\boldsymbol{\mu}_1 > \mathbf{j}'\boldsymbol{\mu}_2$$

of a higher no-senile-factor group profile. The means of the total scores of the four subtests are

$$\bar{y}_1 = 41.60 \qquad \bar{y}_2 = 27.33$$

and

$$\mathbf{j}'\mathbf{S}\mathbf{j} = 107.13$$

The test statistic is

$$t = \frac{41.60 - 27.33}{\sqrt{107.13(\frac{1}{12} + \frac{1}{37})}}$$

$$= 4.15$$

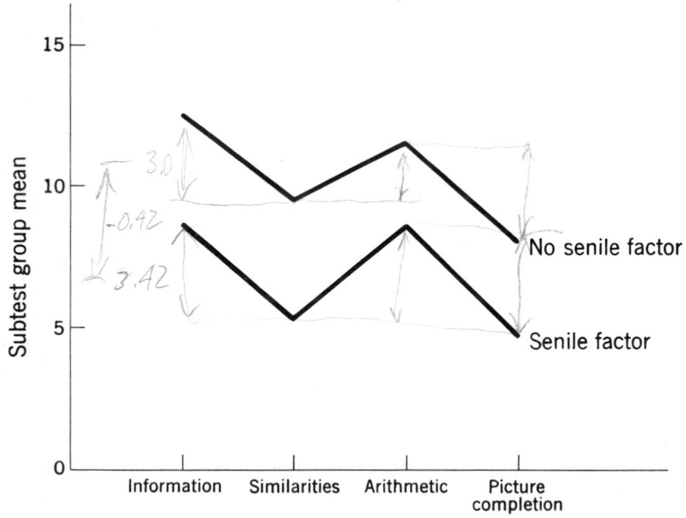

figure 4.3 WAIS subtest profiles.

Since the probability of a t variate with 47 degrees of freedom exceeding 4.0 is nearly 10^{-4}, we can accept the alternative hypothesis. The one-sided 99 percent confidence interval for the level difference can be computed to be

$$5.98 < \mathbf{j}'(\boldsymbol{\mu}_1 - \boldsymbol{\mu}_2) < \infty$$

Finally we shall test the hypothesis of equivalent parameters in the four subtests. The grand mean vector can be computed from the original observations to be

$$\bar{\mathbf{x}}' = [11.63, 8.53, 10.76, 7.18]$$

The test statistic (13) is

$$T^2 = 49[3.10, \ -2.23, \ 3.58](\mathbf{CSC}')^{-1}\begin{bmatrix} 3.10 \\ -2.23 \\ 3.58 \end{bmatrix}$$
$$= 166.07$$

and $F = 53.01$. Since $F_{0.001;3,45} \approx 6.50$, we handily reject the hypothesis of equal subtest effects. The four subtests can be compared by the simultaneous confidence intervals for their mean differences. For the confidence coefficient 0.999, $T_{0.001;3,45} = 4.51$. The sample variance of the difference of the grand means of tests i and j can be found from the elements of $\mathbf{S}$ by the familiar relation

$$\frac{s_{ii} + s_{jj} - 2s_{ij}}{N_1 + N_2}$$

The confidence intervals on the differences of the elements of the population grand mean vector $\boldsymbol{\mu} = \boldsymbol{\mu}_1 + \boldsymbol{\mu}_2$ are

Information-similarities:	$1.52 \leq \mu_1 - \mu_2 \leq 4.68$
Information-arithmetic:	$-1.01 \leq \mu_1 - \mu_3 \leq 2.75$
Information–picture completion:	$2.38 \leq \mu_1 - \mu_4 \leq 6.52$
Similarities-arithmetic:	$-4.30 \leq \mu_2 - \mu_3 \leq -0.16$
Similarities–picture completion:	$-1.08 \leq \mu_2 - \mu_4 \leq 3.78$
Arithmetic–picture completion:	$1.34 \leq \mu_3 - \mu_4 \leq 5.82$

We note that only the information-arithmetic and similarities–picture completion differences are not significant at the 0.001 level.

Finally, how should we reconcile the profile analysis with the original T^2 test of Example 4.3? If we are willing to accept the absence of a diagnosis-by-WAIS scale interaction based on the units of the actual scores, we can decompose the gross conclusion of unequal mean vectors into differences attributable to shape or level of the profiles. The units of the four subtests are commensurable, and the standard deviations are nearly equal, and it would seem more meaningful to report the analysis in terms of the profile model.

A general discussion of the problems of profile analysis and some useful approximate methods have been given by Greenhouse and Geisser (1959). Several applications of their technique to experimental psychology have been described by Winer (1971). We shall treat the approximate tests in Chap. 5 for the general case of several groups.

4.7 THE POWER OF TESTS ON MEAN VECTORS

Heretofore in our discussion of multivariate hypothesis testing we have only been concerned with the control of the probability α of erroneously rejecting the null hypothesis. In a properly designed experiment or sampling inquiry the probability of the Type II error should also be controlled. That probability β can be obtained from the distribution of T^2 when the null hypothesis is not true. Recall from Sec. 4.2 that if

$$T^2 = \mathbf{y}'\mathbf{S}^{-1}\mathbf{y} \tag{1}$$

is computed from the p-dimensional vector $\mathbf{y}$ with distribution $N(\boldsymbol{\mu},\boldsymbol{\Sigma})$ and the independent estimate $\mathbf{S}$ of $\boldsymbol{\Sigma}$ based upon n degrees of freedom, the quantity

$$F = T^2 \frac{n - p + 1}{np} \tag{2}$$

has the *noncentral F* distribution with degrees of freedom p and $n - p + 1$, and *noncentrality parameter*

$$\delta^2 = \boldsymbol{\mu}'\boldsymbol{\Sigma}^{-1}\boldsymbol{\mu} \tag{3}$$

In particular, if the hypothesis H_0: $\boldsymbol{\mu} = \boldsymbol{\mu}_0$ is being tested against the alternative H_1: $\boldsymbol{\mu} \neq \boldsymbol{\mu}_0$ by the single-sample T^2 statistic, then $\mathbf{y}$ is the sample mean vector of N independent observations from $N(\boldsymbol{\mu},\boldsymbol{\Sigma})$, $n = N - 1$, and the noncentrality parameter is

$$\delta^2 = N(\boldsymbol{\mu} - \boldsymbol{\mu}_0)'\boldsymbol{\Sigma}^{-1}(\boldsymbol{\mu} - \boldsymbol{\mu}_0) \tag{4}$$

For the two-sample test of the hypothesis H_0: $\boldsymbol{\mu}_1 = \boldsymbol{\mu}_2$ against the alternative H_1: $\boldsymbol{\mu}_1 \neq \boldsymbol{\mu}_2$ the statistic (3) of Sec. 4.4 has the noncentral F distribution with parameter

$$\frac{N_1N_2}{N_1 + N_2}(\boldsymbol{\mu}_1 - \boldsymbol{\mu}_2)'\boldsymbol{\Sigma}^{-1}(\boldsymbol{\mu}_1 - \boldsymbol{\mu}_2) \tag{5}$$

and degrees of freedom p and $N_1 + N_2 - p - 1$. In either case the power function of the test is

$$1 - \beta(\delta^2) = P(F' > F_{\alpha;p,n-p+1}) \tag{6}$$

where F' has the noncentral F distribution with the prescribed parameters. Pearson and Hartley (1951) have prepared charts of the power function for tests of size 0.05 and 0.01 and a range of degrees of freedom. These charts are reproduced in the Appendix, and are entered with p, α, $n - p + 1$, and the argument

$$\phi = \frac{\delta}{\sqrt{p+1}} \tag{7}$$

In the design of an experiment one first specifies the magnitude of the smallest changes in the mean vector components that are desired to be detected with probability greater than $1 - \beta$ by a test of size α. $\mathbf{\Sigma}$ must be approximated from earlier data or perhaps some reasonable model for the structure of the responses. From these the argument ϕ is computed, and upon entering the appropriate chart and family of curves the curve with fewest degrees of freedom is found whose power at ϕ is greater than $1 - \beta$.

Example 4.10. Let us illustrate these power calculations by a physiological experiment. Electroencephalograms and continuous records of systolic blood pressure, heart rate, and respiration are to be recorded on sleeping normal volunteers. The EEG patterns will be scored as stage I, II, or III, where stage I is relatively light sleep with rapid eye movements indicative of dreaming, and stages II and III suggest successively deeper sleep. Among other comparisons it is desired to investigate differences in blood pressure among the three stages. Although all stages occur in a subject's EEG many times each night, the physiological responses are undoubtedly correlated over time, and for a very conservative study of the stage differences we shall elect to work with the mean blood pressures in each stage for N subjects, where (1) the stages have been drawn from the same time of night for biological reasons and (2) each subject has contributed the same number of observations to a particular stage mean. The observations will have the form of a simple repeated-measurements design for testing

$$H_0\colon \quad \mu_{\mathrm{I}} = \mu_{\mathrm{II}} = \mu_{\mathrm{III}}$$

against the alternative of different stage means, and we wish to determine the number N of volunteers. Let us set $\alpha = 0.05$ and require that β be less than 0.10 that the minimum stage differences

$$\mu_{\mathrm{I}} - \mu_{\mathrm{II}} = 4 \qquad \mu_{\mathrm{II}} - \mu_{\mathrm{III}} = 2$$

will not be detected. The covariance matrix will be taken as

$$\mathbf{\Sigma} = \begin{bmatrix} 10 & 4 & 4 \\ & 10 & 4 \\ & & 10 \end{bmatrix}$$

and the covariance matrix of the response differences of adjacent stages will then be

$$\mathbf{C\Sigma C}' = \begin{bmatrix} 12 & -6 \\ -6 & 12 \end{bmatrix}$$

with inverse

$$(\mathbf{C\Sigma C'})^{-1} = \begin{bmatrix} 1/9 & 1/18 \\ 1/18 & 1/9 \end{bmatrix}$$

The noncentrality parameter is

$$\begin{aligned} \delta^2 &= N\boldsymbol{\mu}'\mathbf{C}'(\mathbf{C\Sigma C'})^{-1}\mathbf{C}\boldsymbol{\mu} \\ &= \frac{N}{18}[4,2]\begin{bmatrix} 2 & 1 \\ 1 & 2 \end{bmatrix}\begin{bmatrix} 4 \\ 2 \end{bmatrix} \\ &= \frac{28N}{9} \end{aligned}$$

and $\phi = \sqrt{28N/27}$. The second chart will be used, with $N - 2$ second degrees of freedom. Let us begin with a sample size of 8. Then $\phi = 2.87$, and since the second degrees-of-freedom parameter is 6, the power can be read as approximately 0.93. We conclude that eight or more subjects should be observed in the experiment.

The power of the T^2 test is a monotonically increasing function of the noncentrality parameter. Let us consider two properties of the power function in the bivariate case of the single-sample test where the noncentrality parameter is

$$\frac{N}{1 - \rho^2}[(\mu_1 - \mu_{10})^2 - 2\rho(\mu_1 - \mu_{10})(\mu_2 - \mu_{20}) + (\mu_2 - \mu_{20})^2] \tag{8}$$

We note that:

1. Negative values of ρ lead to higher power probabilities for the same $\mu_i - \mu_{i0}$ of like signs.
2. For constant $\mu_1 - \mu_{10}$ the minimum power occurs at

$$\mu_2 - \mu_{20} = \rho(\mu_1 - \mu_{10})$$

with a similar minimum for $\mu_2 - \mu_{20}$ held constant. For example, if $\rho = 1/2$, the powers of the tests of H_0: $[\mu_1,\mu_2] = [0,0]$ against the alternatives

$$H_1: \begin{bmatrix} \mu_1 \\ \mu_2 \end{bmatrix} = \begin{bmatrix} \mu_0 \\ 0 \end{bmatrix} \qquad H_1': \begin{bmatrix} \mu_1 \\ \mu_2 \end{bmatrix} = \begin{bmatrix} \mu_0 \\ \mu_0 \end{bmatrix} \qquad H_1'': \begin{bmatrix} \mu_1 \\ \mu_2 \end{bmatrix} = \begin{bmatrix} 0 \\ \mu_0 \end{bmatrix}$$

are all equal.

In line with result 2 let us compare the powers of similar hypotheses tested on univariate and bivariate random variables to determine the effect of including in the responses a component which is unaffected by the experimental conditions. Perhaps the following psychological example will serve to motivate this situation. A cognitive test is made up of verbal and spatial subtests, and an investigator wishes to test the hypothesis that in the presence of a certain debility the scores of one particular spatial test have a lower mean than the norm for healthy subjects. This hypothesis is in fact true. However, the investigator has noted that a certain verbal test tends to be correlated with the spatial test, although its mean for the debilitated sample is only slightly less than the mean of a healthy population. Actually, the population difference is zero. In weighing the choice of a univariate test on the spatial scale against a bivariate test of both means the investigator wishes to know the powers of each test against various alternatives.

In the interests of simplicity let us assume that the population variances and correlation coefficient of the two tests are known and furthermore that the test scores have been scaled so that the healthy population means are zero and the variances are unity in every case. The univariate hypothesis and its alternative are

$$H_0: \quad \mu_s = 0 \qquad H_1: \quad \mu_s = \mu_{s1} < 0$$

and those of the multivariate extension are

$$H_0: \quad \begin{bmatrix} \mu_s \\ \mu_v \end{bmatrix} = \begin{bmatrix} 0 \\ 0 \end{bmatrix} \qquad H_1: \quad \begin{bmatrix} \mu_s \\ \mu_v \end{bmatrix} = \begin{bmatrix} \mu_{s1} \\ 0 \end{bmatrix}$$

Both tests will be conducted at the 0.05 level. The power function of the univariate test based on a sample of size N is

$$1 - \beta(\sqrt{N}\,\mu_{s1}) = \Phi(-\sqrt{N}\,\mu_{s1} - 1.645)$$

where $\Phi(\cdot)$ is the unit normal distribution function. Some power probabilities for the two tests are summarized in Table 4.7. For zero cor-

Table 4.7. **Comparison of Univariate and Multivariate Powers**

$\sqrt{N}\,\mu_{s1}$	Univariate	Multivariate		
		$\lvert\rho\rvert = 0$	$\lvert\rho\rvert = 0.5$	$\lvert\rho\rvert = 0.8$
0.0	0.05	0.05	0.05	0.05
−2.0	0.64	0.42	0.53	0.86
−2.5	0.80	0.62	0.73	0.97
−3.0	0.91	0.79	0.89	1.00
−4.0	0.99	0.96	0.99	1.00

relation between the verbal and spatial tests the T^2 test has appreciably less power when the verbal-test means are not different in the healthy and ill populations. As the absolute value of ρ increases, the multivariate power eventually surpasses that of the univariate test. It should be noted that the T^2 test is inherently two-sided, but even in comparison with the two-sided univariate test its power is less for small $|\rho|$.

The loss in power of the multivariate test in comparison with those for the separate responses is sometimes referred to as Rao's paradox, after its discussion by C. R. Rao (1966) in the context of the linear discriminant function. Healy (1969) has given a useful description of the basis of the "paradox."

4.8 SOME TESTS WITH KNOWN COVARIANCE MATRICES

In the preceding sections our general methods for hypothesis tests and confidence statements were based on the Hotelling T^2 because of its properties of invariance and independence of the unknown covariance matrix. Now we shall consider some tests on multinormal mean vectors when $\mathbf{\Sigma}$ is either known or the sample size is sufficiently large (perhaps in excess of one hundred observations for tests at the 5 percent level) to treat it as such. The first two tests will be the usual one- and two-sample analogues of T^2, while the remaining two will be extensions of them to the case of incomplete observations in the monotonic pattern. In each case the test can be constructed from the generalized likelihood-ratio principle, while the distribution of the statistic follows from the fact that the quadratic form of the multinormal density has a chi-squared distribution with degrees of freedom equal to its dimensionality (Anderson, 1958, p. 54).

For the single-sample test we assume that N independent observations with mean vector $\bar{\mathbf{x}}$ have been drawn from the population $N(\boldsymbol{\mu},\mathbf{\Sigma})$. The hypothesis H_0: $\boldsymbol{\mu} = \boldsymbol{\mu}_0$ can be tested against a general alternative by the statistic

$$\chi^2 = N(\bar{\mathbf{x}} - \boldsymbol{\mu}_0)'\mathbf{\Sigma}^{-1}(\bar{\mathbf{x}} - \boldsymbol{\mu}_0) \tag{1}$$

We accept H_0 at the α level if $\chi^2 < \chi^2_{\alpha;p}$ and reject the hypothesis if the statistic exceeds that critical value. Similarly, if two samples of independent observations with mean vectors $\bar{\mathbf{x}}_1$ and $\bar{\mathbf{x}}_2$ have been drawn from multinormal populations with a known common covariance matrix $\mathbf{\Sigma}$, we can test the hypothesis

$$H_0: \quad \boldsymbol{\mu}_1 = \boldsymbol{\mu}_2$$

that the mean vectors $\boldsymbol{\mu}_1$ and $\boldsymbol{\mu}_2$ are equal by the statistic

(2) $$\chi^2 = \frac{N_1N_2}{N_1 + N_2}(\bar{\mathbf{x}}_1 - \bar{\mathbf{x}}_2)'\mathbf{\Sigma}^{-1}(\bar{\mathbf{x}}_1 - \bar{\mathbf{x}}_2)$$

If H_0 is true, this quantity has the chi-squared distribution with p degrees of freedom, and the usual acceptance and rejection rules hold. It is possible in both the single- and the two-sample situations to construct confidence ellipsoids and simultaneous confidence intervals in direct analogy with the regions and intervals of Secs. 4.3 and 4.4.

Incomplete Data. These tests can be extended to the case of missing observations in the data matrix. For simplicity we shall consider only the case of the monotonic pattern defined by expression (1) of Sec. 3.9 with two groups of responses. The tests follow from the likelihood factorization (2) of that section and the application of the likelihood-ratio principle to it when the covariance matrix is known. Although exact tests with a general missing-data pattern have been developed for known $\mathbf{\Sigma}$ (Basickes, 1972), and those for the general monotonic case with unknown $\mathbf{\Sigma}$ have been constructed by Bhargava (1962), their distributions and percentage points calculated by Bhoj (1971, 1973a, 1973b), and certain of their power probabilities computed by Morrison and Bhoj (1973), the scope we have set for this text will only permit the description of the two-group monotonic pattern with known $\mathbf{\Sigma}$.

For the single-sample test of H_0: $\boldsymbol{\mu} = \boldsymbol{\mu}_0$ the first r_1 responses have been observed on all $N = N_1 + N_2$ sampling units, while the remaining $r_2 = p - r_1$ have observations on only the first N_1 units. The data matrix has the form

$$\begin{bmatrix} \mathbf{Y}_1 & \mathbf{X} \\ \mathbf{Y}_2 & - \end{bmatrix}$$

The mean vector and covariance matrix have been correspondingly partitioned as $\boldsymbol{\mu}' = [\boldsymbol{\mu}_1',\boldsymbol{\mu}_2']$ and

$$\mathbf{\Sigma} = \begin{bmatrix} \mathbf{\Sigma}_{11} & \mathbf{\Sigma}_{12} \\ \mathbf{\Sigma}_{12}' & \mathbf{\Sigma}_{22} \end{bmatrix}$$

Similarly, the hypothesis can be written as H_0: $\boldsymbol{\mu}' = [\boldsymbol{\mu}_{10}',\boldsymbol{\mu}_{20}']$. The maximum-likelihood estimates of the mean vector components under the general alternative are

(3) $$\hat{\boldsymbol{\mu}}_1 = \frac{1}{N}\sum_{i=1}^{N} \mathbf{y}_i \qquad \hat{\boldsymbol{\mu}}_2 = \bar{\mathbf{x}} - \frac{N_2}{N}\mathbf{\Sigma}_{12}'\mathbf{\Sigma}_{11}^{-1}(\bar{\mathbf{y}}_1 - \bar{\mathbf{y}}_2)$$

where $\mathbf{y}_i$ is the ith row of $\mathbf{Y}' = [\mathbf{Y}_1',\mathbf{Y}_2']$ and $\bar{\mathbf{x}}$, $\bar{\mathbf{y}}_1$, and $\bar{\mathbf{y}}_2$ are the mean vectors defined by (4) of Sec. 3.9. The test statistic is

$$(4)\quad \chi^2 = N_1[\bar{\mathbf{x}} - \boldsymbol{\mu}_{20} - \boldsymbol{\Sigma}'_{12}\boldsymbol{\Sigma}_{11}^{-1}(\bar{\mathbf{y}}_1 - \boldsymbol{\mu}_{10})]'(\boldsymbol{\Sigma}_{22} - \boldsymbol{\Sigma}'_{12}\boldsymbol{\Sigma}_{11}^{-1}\boldsymbol{\Sigma}_{12})^{-1} \cdot [\bar{\mathbf{x}} - \boldsymbol{\mu}_{20} - \boldsymbol{\Sigma}'_{12}\boldsymbol{\Sigma}_{11}^{-1}(\bar{\mathbf{y}}_1 - \boldsymbol{\mu}_{10})] + N(\bar{\mathbf{y}} - \boldsymbol{\mu}_{10})'\boldsymbol{\Sigma}_{11}^{-1}(\bar{\mathbf{y}} - \boldsymbol{\mu}_{10})$$

When the null hypothesis is true this statistic has the chi-squared distribution with p degrees of freedom, and we would reject H_0 at the α level if $\chi^2 > \chi^2_{\alpha;p}$. Under an alternative hypothesis of a general mean vector the distribution is noncentral chi-squared with a noncentrality parameter computed by replacing the sample mean vectors in (4) by their expected values.

The two-sample test of the equality of mean vectors also leads to a chi-squared statistic when the common covariance matrix $\boldsymbol{\Sigma}$ is known. Let the data matrices have the monotonic patterns

$$(5)\qquad \begin{bmatrix} \mathbf{Y}_{11} & \mathbf{X}_1 \\ \mathbf{Y}_{21} & - \end{bmatrix} \qquad \begin{bmatrix} \mathbf{Y}_{12} & \mathbf{X}_2 \\ \mathbf{Y}_{22} & - \end{bmatrix}$$

where $\mathbf{Y}_{ik}$ is $N_{ik} \times r_1$, $\mathbf{X}_k$ is $N_{1k} \times r_2$, and $p = r_1 + r_2$. Let $N_{i.} = N_{i1} + N_{i2}$, $i = 1, 2$, and $N_{.k} = N_{1k} + N_{2k}$, $k = 1, 2$. The maximum-likelihood estimates of the mean vector components for the kth sample are

$$(6)\qquad \hat{\boldsymbol{\mu}}_{1k} = \frac{1}{N_{.k}} \sum_{i=1}^{N_k} \mathbf{y}_{ik} \qquad \hat{\boldsymbol{\mu}}_{2k} = \bar{\mathbf{x}}_k - \frac{N_{2k}}{N_{.k}} \boldsymbol{\Sigma}'_{12}\boldsymbol{\Sigma}_{11}^{-1}(\bar{\mathbf{y}}_{1k} - \bar{\mathbf{y}}_{2k})$$

and the statistic for testing

$$H_0\colon \begin{bmatrix} \boldsymbol{\mu}_{11} \\ \boldsymbol{\mu}_{21} \end{bmatrix} = \begin{bmatrix} \boldsymbol{\mu}_{12} \\ \boldsymbol{\mu}_{22} \end{bmatrix}$$

is

$$(7)\quad \chi^2 = \frac{N_{11}N_{12}}{N_{1.}} [\hat{\boldsymbol{\mu}}_{21} - \hat{\boldsymbol{\mu}}_{22} - \boldsymbol{\Sigma}'_{12}\boldsymbol{\Sigma}_{11}^{-1}(\hat{\boldsymbol{\mu}}_{11} - \hat{\boldsymbol{\mu}}_{12})]'(\boldsymbol{\Sigma}_{22} - \boldsymbol{\Sigma}'_{12}\boldsymbol{\Sigma}_{11}^{-1}\boldsymbol{\Sigma}_{12})^{-1} \cdot [\hat{\boldsymbol{\mu}}_{21} - \hat{\boldsymbol{\mu}}_{22} - \boldsymbol{\Sigma}'_{12}\boldsymbol{\Sigma}_{11}^{-1}(\hat{\boldsymbol{\mu}}_{11} - \hat{\boldsymbol{\mu}}_{12})] + \frac{N_{.1}N_{.2}}{N_{.1} + N_{.2}} (\hat{\boldsymbol{\mu}}_{11} - \hat{\boldsymbol{\mu}}_{12})'\boldsymbol{\Sigma}_{11}^{-1}(\hat{\boldsymbol{\mu}}_{11} - \hat{\boldsymbol{\mu}}_{12})$$

We would reject H_0 at the α level if $\chi^2 > \chi^2_{\alpha;p}$.

These tests can be extended to a monotonic pattern with any number of response groups by means of the likelihood factorization (2) of Sec. 3.9.

4.9 EXERCISES

1. Show that the various formulas and confidence statements of Secs. 4.2 to 4.4 involving T^2 reduce to those for univariate t when $p = 1$.
2. With the aid of expression (9) of Sec. 2.11 verify the identity

$$\mathbf{C}'(\mathbf{CSC}')^{-1}\mathbf{C} = \mathbf{S}^{-1} - (\mathbf{j}'\mathbf{S}^{-1}\mathbf{j})^{-1}\mathbf{S}^{-1}\mathbf{j}\mathbf{j}'\mathbf{S}^{-1}$$

for $\mathbf{j}' = [1, \ldots, 1]$ and $\mathbf{C}$ any $(p-1) \times p$ variant of the transformation matrix (17) of Sec. 4.5 (Williams, 1970).

3. Show that the noncentrality parameter (28) of Sec. 4.5 follows from the evaluation of $\delta^2 = N\boldsymbol{\mu}'\boldsymbol{\Sigma}^{-1}\boldsymbol{\mu}$, where $\boldsymbol{\mu}$ is a general $p \times 1$ vector and $\boldsymbol{\Sigma}$ has the equal-variance, equal-correlation pattern.

4. The blackbody CIE (Commission Internationale d'Éclairage) chromaticity specification for a color temperature of 4000°K is $x = 0.3804$ and $y = 0.3768$. In 10 color-matching trials (Jackson, 1959, 1962) one subject had mean chromaticity values $\bar{x} = 0.3745$ and $\bar{y} = 0.3719$, and sample covariance matrix

$$\mathbf{S} = 10^{-5}\begin{bmatrix} 1.843 & 1.799 \\ 1.799 & 1.836 \end{bmatrix}$$

Test the null hypothesis that the observations came from the bivariate normal population with mean vector the 4000°K standards.

5. In the experiment of Exercise 4 two independent observers of another color density had these mean vectors and covariance matrices for $N_1 = N_2 = 10$ observations:

$$\bar{\mathbf{x}}_1' = [0.3781, 0.3755] \qquad \bar{\mathbf{x}}_2' = [0.3772, 0.3750]$$

$$\mathbf{S}_1 = 10^{-6}\begin{bmatrix} 2.64 & 3.18 \\ 3.18 & 11.42 \end{bmatrix} \qquad \mathbf{S}_2 = 10^{-6}\begin{bmatrix} 6.48 & 0.45 \\ 0.45 & 8.26 \end{bmatrix}$$

Test the hypothesis that the population mean vectors of the observers are the same.

6. It is felt that a certain drug may lead to changes in the level of three biochemical compounds found in the brain. Twenty-four mice of the same strain were randomly divided into equal groups, with the second receiving periodic administrations of the drug. Both samples received the same care and diet, although two of the control group mice died of natural causes during the experiment. Assays of the brains of the sacrificed mice revealed these amounts of the compounds in micrograms per gram of brain tissue:

Control			Drug		
A	*B*	*C*	*A*	*B*	*C*
1.21	0.61	0.70	1.40	0.50	0.71
0.92	0.43	0.71	1.17	0.39	0.69
0.80	0.35	0.71	1.23	0.44	0.70
0.85	0.48	0.68	1.19	0.37	0.72
0.98	0.42	0.71	1.38	0.42	0.71
1.15	0.52	0.72	1.17	0.45	0.70
1.10	0.50	0.75	1.31	0.41	0.70
1.02	0.53	0.70	1.30	0.47	0.67
1.18	0.45	0.70	1.22	0.29	0.68
1.09	0.40	0.69	1.00	0.30	0.70
			1.12	0.27	0.72
			1.09	0.35	0.73

Test the hypothesis of no difference in the control and drug mean vectors under the assumption of multinormal distributions with a common unknown covariance matrix. If that hypothesis is rejected at the 0.05 level, use the appropriate multiple-comparison procedure to determine which compound means are different.

7. Three cottage parents at an institution for retarded boys were asked to rate their charges on a number of behavioral and affective dimensions. One seven-point affect scale had these means and covariance matrix for the 64 boys seen by the three parents:

	Cottage parent		
	A	B	C
Mean rating	3.05	3.31	2.92
Covariance matrix	$\begin{bmatrix} 1.28 & 1.05 & 0.75 \\ & 1.35 & 0.93 \\ & & 1.12 \end{bmatrix}$		

Test the hypothesis of no difference in cottage-parent mean ratings at the 0.05 level, and if the hypothesis is rejected, specify which parents are different by computing the 95 percent simultaneous confidence intervals for the relevant parent contrasts. All inferences will be made *only* to the three parents A, B, and C.

8. In a psychiatric study of the families of schizophrenic children two Rorschach scores of the mothers and fathers were of particular import. The observed values of these scores for the parents of ten psychotic adolescents and six normal control children of similar age and socioeconomic status are given below:

Schizophrenic families				Control families			
M_1	M_2	F_1	F_2	M_1	M_2	F_1	F_2
30	35	25	33	20	15	26	25
21	41	15	21	11	3	18	19
27	32	25	34	7	2	11	8
35	34	31	36	21	15	24	22
20	37	14	21	15	11	17	13
23	38	19	25	13	12	20	15
28	27	26	30				
32	42	29	37				
26	36	27	33				
29	35	24	32				

Carry out a complete analysis of these profiles, testing in turn the hypotheses of similar shapes, equal levels, and equal response effects. Use appropriate multiple-comparison techniques to determine which scores or linear compounds of scores have contributed to differences in the mean profiles.

9. Box (1950) gave the following initial weights x_0 in grams and weekly gains in weights d_i for rats in a control sample and in a sample whose drinking water contained thiouracil:

Control					Thiouracil				
x_0	d_1	d_2	d_3	d_4	x_0	d_1	d_2	d_3	d_4
57	29	28	25	33	61	25	23	11	9
60	33	30	23	31	59	21	21	10	11
52	25	34	33	41	53	26	21	6	27
49	18	33	29	35	59	29	12	11	11
56	25	23	17	30	51	24	26	22	17
46	24	32	29	22	51	24	17	8	19
51	20	23	16	31	56	22	17	8	5
63	28	21	18	24	58	11	24	21	24
49	18	23	22	28	46	15	17	12	17
57	25	28	29	30	53	19	17	15	18

Determine whether the hypotheses of equal shapes and levels can be accepted for these growth profiles, and if not, which increments are different. Make the usual assumptions of multinormality and a common covariance matrix, and conduct all tests with $\alpha = 0.05$.

10. A psychologist wishes to rate children of normal and neurotic parents on three behavioral scales. The ratings on such scales can be assumed to be multinormally distributed in both populations with the common covariance matrix

$$\mathbf{\Sigma} = \begin{bmatrix} 10 & 5 & 1 \\ & 8 & 2 \\ & & 10 \end{bmatrix}$$

It is desired to detect normal-neurotic mean differences as small as 5, 2, and 3 units on the respective scales, with a probability of rejection of the null hypothesis at least equal to 0.95 for a test of size 0.01.

a. If equal numbers of normal and neurotic families are to be randomly selected, determine the sample size that will satisfy the above power requirement.

b. The covariance matrix was in fact arrived at by guess and the results of a similar study. If the matrix

$$\mathbf{\Sigma} = \begin{bmatrix} 10 & 6 & 0 \\ & 12 & 6 \\ & & 10 \end{bmatrix}$$

were employed what would the sample size be?

5 THE MULTIVARIATE ANALYSIS OF VARIANCE

5.1 INTRODUCTION. In Chap. 4 the one- and two-sample tests on the means of normal variates were extended to multinormal mean vectors. Now we shall consider the multivariate generalization of the analysis of variance for testing the equality of mean vectors of several populations. This technique is but a special case of the *multivariate general linear hypothesis*, and we shall see that very general methods are available for carrying out tests on the parameters in that model. These tests include an extension of the earlier profile analysis to several groups of subjects.

Unlike the T^2 statistic, different methods of test construction for the multivariate analysis of variance do not lead to the same test statistic. We shall prefer to follow the union-intersection approach of S. N. Roy introduced in Sec. 4.2, for the percentage points of the distribution of that statistic are available and it is possible to construct simultaneous confidence intervals for the determination of the significance of group or response differences. We shall give some other statistics in Sec. 5.8 and compare their power properties with those of Roy's test.

5.2 THE MULTIVARIATE GENERAL LINEAR MODEL

In Sec. 1.6 we discussed the essentials of the one-way fixed-effects analysis of variance: independent observations x_{ij} were drawn from each of k normal populations with means $\mu_1, \ldots,$

μ_k and a common unknown variance σ^2. The ith observation from the jth population was generated by the linear model

$$x_{ij} = \mu + \tau_j + e_{ij} \tag{1}$$

in which μ was a general-level parameter common to all observations, τ_j was an effect due to a treatment or condition associated with the jth population, and e_{ij} was a normal random variable with zero mean and variance σ^2 for all pairs of i and j. The residuals e_{ij} were also required to be independently distributed. This is the simplest example of a linear model, and the hypothesis

$$H_0\colon \quad \tau_1 = \cdots = \tau_k \tag{2}$$

of equal population means is an elementary case of the *general linear hypothesis*. Alternatively, we might have written the model and hypothesis as

$$x_{ij} = \mu_j + e_{ij}$$

and

$$H_0\colon \quad \mu_1 = \cdots = \mu_k$$

through the *reparametrization* $\mu_j = \mu + \tau_j$ of the model.

Now let us introduce the notion of a *design matrix* as a means of writing the linear models of all observations in compact matrix form. Let the number of sampling units in the jth group be N_j, and for convenience let $N = N_1 + \cdots + N_k$. Write the observations and residuals as the N-component vectors

$$\begin{aligned} \mathbf{x}' &= [x_{11}, \ldots, x_{N_1 1}, \ldots, x_{1k}, \ldots, x_{N_k k}] \\ \boldsymbol{\varepsilon}' &= [e_{11}, \ldots, e_{N_1 1}, \ldots, e_{1k}, \ldots, e_{N_k k}] \end{aligned} \tag{3}$$

and the parameters as

$$\boldsymbol{\mu}' = [\tau_1, \ldots, \tau_k, \mu] \tag{4}$$

Then the design matrix for the one-way analysis-of-variance model is

$$\mathbf{A} = \begin{bmatrix} 1 & 0 & \cdots & 0 & 1 \\ \cdot & \cdot & \cdots & \cdot & \cdot \\ 1 & 0 & \cdots & 0 & 1 \\ 0 & 1 & \cdots & 0 & 1 \\ \cdot & \cdot & \cdots & \cdot & \cdot \\ 0 & 1 & \cdots & 0 & 1 \\ \cdot & \cdot & \cdots & \cdot & \cdot \\ 0 & 0 & \cdots & 1 & 1 \\ \cdot & \cdot & \cdots & \cdot & \cdot \\ 0 & 0 & \cdots & 1 & 1 \end{bmatrix} \tag{5}$$

A has been partitioned into k $N_j \times (k+1)$ submatrices. In the jth of these the jth and last columns contain a one in each position, while the remaining columns contain only zeros. The pattern of ones and zeros in a design matrix follows from the experimental plan and its linear model. For the analysis-of-covariance model the design matrix also contains the values of the concomitant variables in the appropriate regression equation. In the present case the model (1) can be written as

$$\mathbf{x} = \mathbf{A}\boldsymbol{\mu} + \boldsymbol{\varepsilon} \tag{6}$$

since postmultiplication of the design matrix by the parameter vector assures that the ijth observation will involve only the constant $\mu + \tau_j$.

The matrix (5) has rank k, for its last column is merely the sum of the first k columns, and since the latter are linearly independent, they form a basis for the design matrix. In the sequel we shall write the $N \times q$ design matrix **A** of rank r as the partitioned matrix

$$\mathbf{A} = [\mathbf{A}_1 \quad \mathbf{A}_2] \tag{7}$$

where $\mathbf{A}_1$ has r columns and is a basis for **A**. The parameter vector will also be partitioned into r and $q - r$ components as

$$\boldsymbol{\mu}' = [\boldsymbol{\mu}_1' \quad \boldsymbol{\mu}_2'] \tag{8}$$

to conform with the dimensions of $\mathbf{A}_1$ and $\mathbf{A}_2$. Alternatively, the partitioning can be avoided by reparametrizing the model so that its design matrix is $\mathbf{A}_1$. For example, if we had adopted the parametrization $\mu_j = \mu + \tau_j$ for the one-way model (1), the matrix form (6) would become

$$\mathbf{x} = \mathbf{A}_1\boldsymbol{\beta} + \boldsymbol{\varepsilon} \tag{9}$$

for $\boldsymbol{\beta}' = [\mu_1, \ldots, \mu_k]$. At the expense of less parametric detail we have obtained a model whose design matrix has no superfluous linearly dependent columns.

Next we shall consider the estimation of the parameters of the general linear model

$$\mathbf{x} = \mathbf{A}\boldsymbol{\mu} + \boldsymbol{\varepsilon} \tag{10}$$

The linear compound

$$\mathbf{a}'\boldsymbol{\mu} \tag{11}$$

of the parameters is said to be *estimable* (Bose, 1944) if a function $\mathbf{b}'\mathbf{x}$ of the observations exists with the property that

(12) $$E(\mathbf{b}'\mathbf{x}) = \mathbf{a}'\boldsymbol{\mu}$$

Various conditions and tests for the estimability of a parametric function have been developed (see, for example, Graybill, 1961, chap. 11), but we shall confine our attention to one due to S. N. Roy (1957). Partition the compounding vector as $\mathbf{a}' = [\mathbf{a}_1' \quad \mathbf{a}_2']$ and the parameter vector as $\boldsymbol{\mu}' = [\boldsymbol{\mu}_1' \quad \boldsymbol{\mu}_2']$ in accordance with the partitioning (7) of the design matrix. Then $\mathbf{a}'\boldsymbol{\mu} = \mathbf{a}_1'\boldsymbol{\mu}_1 + \mathbf{a}_2'\boldsymbol{\mu}_2$ is estimable only if the condition

(13) $$\mathbf{a}_2' = \mathbf{a}_1'(\mathbf{A}_1'\mathbf{A}_1)^{-1}\mathbf{A}_1'\mathbf{A}_2$$

is satisfied by the elements of $\mathbf{a}$. The actual minimum-variance estimate of $\mathbf{a}'\boldsymbol{\mu}$ can be shown to be

(14) $$\mathbf{a}'\hat{\boldsymbol{\mu}} = \mathbf{a}_1'(\mathbf{A}_1'\mathbf{A}_1)^{-1}\mathbf{A}_1'\mathbf{x}$$

As an application of these general results let us obtain estimates of some parametric functions in the one-way classification model. Here

$$(\mathbf{A}_1'\mathbf{A}_1)^{-1} = \begin{bmatrix} \frac{1}{N_1} & \cdots & 0 \\ \cdot & \cdot\;\cdot\;\cdot & \cdot \\ 0 & \cdots & \frac{1}{N_k} \end{bmatrix}$$

$$(\mathbf{A}_1'\mathbf{A}_1)^{-1}\mathbf{A}_1'\mathbf{A}_2 = \begin{bmatrix} 1 \\ \cdot \\ \cdot \\ \cdot \\ 1 \end{bmatrix}$$

and

$$(\mathbf{A}_1'\mathbf{A}_1)^{-1}\mathbf{A}_1'\mathbf{x} = \begin{bmatrix} \bar{x}_1 \\ \cdot \\ \cdot \\ \cdot \\ \bar{x}_k \end{bmatrix}$$

or the vector of the group sample means. If $\mathbf{a}'\boldsymbol{\mu}$ is any estimable function, its minimum variance unbiased estimate is the linear compound $\mathbf{a}_1'\bar{\mathbf{x}}$ of the sample means. Thus, if $\mathbf{a}_1' = [1,0, \ldots, 0]$ and $a_2 = 1$, the function $\mu + \tau_1$ is estimable, and its estimate is $\bar{x}_1$. Similarly, any contrast

$$\varphi = \sum_{i=1}^{k} c_i\tau_i$$

of the treatment effects is estimable, for by definition of a contrast $\sum_{i=1}^{k} c_i = 0$, and thereby the condition

$$a_2 = [c_1, \ . \ . \ . \ , c_k] \begin{bmatrix} 1 \\ \cdot \\ \cdot \\ \cdot \\ 1 \end{bmatrix} = 0$$

for estimability is met. Finally, let us see whether it is possible to estimate merely μ or a single τ_j. In the first case $\mathbf{a}_1' = [0, \ . \ . \ . \ , 0]$, $a_2 = 1$, and the estimability condition (13) is not satisfied. We must conclude that no nonnull linear function of the observations exists whose expected value is μ and therefore that μ is inestimable. A similar appeal to (13) shows that any single-group effect is inestimable; in general the only estimable functions of the τ_j are contrasts.

Now let us use the notion of estimability for tests of the general linear hypothesis

$$H_0\colon \ \mathbf{C}\boldsymbol{\mu} = \mathbf{0} \tag{15}$$

$\mathbf{C}$ is a $g \times q$ matrix of rank g describing the hypothesis on the q components of the parameter vector in the general linear model (10). The alternative hypothesis is merely

$$H_1\colon \ \mathbf{C}\boldsymbol{\mu} \neq \mathbf{0} \tag{16}$$

If the hypothesis matrix is partitioned in conformance with $\boldsymbol{\mu}_1$, $\boldsymbol{\mu}_2$ as

$$\mathbf{C} = [\mathbf{C}_1 \quad \mathbf{C}_2] \tag{17}$$

where $\mathbf{C}_1$ and $\mathbf{C}_2$ have respective dimensions $g \times r$ and $g \times q - r$, the null hypothesis can be written as

$$H_0\colon \ \mathbf{C}_1\boldsymbol{\mu}_1 + \mathbf{C}_2\boldsymbol{\mu}_2 = \mathbf{0} \tag{18}$$

This hypothesis is said to be *testable* in the sense of S. N. Roy (1953) if the rows of the matrix $\mathbf{C}_2$ satisfy the estimability condition (13) in terms of their counterparts in $\mathbf{C}_1$. If this requirement is met, the statistic for testing (18) has been shown by Roy to be

$$F = \frac{(N - r)\mathbf{x}'\mathbf{A}_1(\mathbf{A}_1'\mathbf{A}_1)^{-1}\mathbf{C}_1'[\mathbf{C}_1(\mathbf{A}_1'\mathbf{A}_1)^{-1}\mathbf{C}_1']^{-1}\mathbf{C}_1(\mathbf{A}_1'\mathbf{A}_1)^{-1}\mathbf{A}_1'\mathbf{x}}{g\mathbf{x}'[\mathbf{I} - \mathbf{A}_1(\mathbf{A}_1'\mathbf{A}_1)^{-1}\mathbf{A}_1']\mathbf{x}} \tag{19}$$

If the hypothesis is true, this ratio has the F variance-ratio distribution with degrees of freedom g and $N - r$, and for a test of level α we accept H_0 if

$$F \leq F_{\alpha; g, N-r} \tag{20}$$

and reject otherwise.

The null hypothesis (2) of the one-way analysis of variance can be expressed in matrix form in an infinity of ways. One convenient form consists of

$$\mathbf{C}_1 = \begin{bmatrix} 1 & 0 & 0 & \cdots & 0 & -1 \\ 0 & 1 & 0 & \cdots & 0 & -1 \\ \cdot & \cdot & \cdot & \cdots & \cdot & \cdot \\ 0 & 0 & 0 & \cdots & 1 & -1 \end{bmatrix} \tag{21}$$

with dimensions $k - 1$ and k, and $\mathbf{C}_2$ the column vector of $k - 1$ zeros. Then $g = k - 1$ and $r = k$. The actual evaluation of the F ratio (19) is rather lengthy but can be accomplished with the assistance of some results by Roy and Sarhan (1956) on the inverses of patterned matrices. The ratio is of course equal to the statistic (10) of our original description of the one-way analysis of variance in Sec. 1.6.

Now we are ready to extend the linear model and hypothesis to several variates. Let the responses of interest be described by the p-dimensional multinormal random variable $\mathbf{x}$. N observation vectors $\mathbf{x}_1, \ldots, \mathbf{x}_N$ have been collected on $\mathbf{x}$ under some experimental design or sample survey plan, and from that situation we hypothesize that the ith observation on the hth response was generated by the linear model

$$x_{ih} = a_{i1}\xi_{1h} + \cdots + a_{im}\xi_{mh} + e_{ih} \tag{22}$$

Each response in the ith vector has the same coefficients a_{ij}, so that the same design matrix $\mathbf{A}$ holds for *all* dimensions. The residual variates $e_{i1}, \ldots, e_{ip}$ of the ith observation are distributed according to the multinormal law with null mean vector and covariance matrix $\mathbf{\Sigma}$ of full rank p. The model for all observations is

$$\mathbf{X} = \mathbf{A}\boldsymbol{\xi} + \boldsymbol{\varepsilon} \tag{23}$$

where the $N \times p$ matrix $\mathbf{X}$ has the N observation vectors for rows, $\mathbf{A}$ is the appropriate design matrix,

$$\boldsymbol{\xi} = \begin{bmatrix} \xi_{11} & \cdots & \xi_{1p} \\ \cdot\ \cdot & \cdots & \cdot\ \cdot \\ \xi_{q1} & \cdots & \xi_{qp} \end{bmatrix} \tag{24}$$

is the matrix of unknown parameters, and the $N \times p$ matrix $\boldsymbol{\varepsilon}$ contains the residual variates e_{ih}. The estimability condition and the minimum-variance estimates for linear compounds of the elements of a particular column of $\boldsymbol{\xi}$ follow as in the univariate model.

The multivariate extension of the general linear hypothesis is

$$H_0\colon \quad \mathbf{C}\boldsymbol{\xi}\mathbf{M} = \mathbf{0} \tag{25}$$

and its alternative is

$$H_1\colon \quad \mathbf{C}\boldsymbol{\xi}\mathbf{M} \neq \mathbf{0} \tag{26}$$

As in the univariate case the hypothesis matrix $\mathbf{C}$ has dimensions $g \times q$ and rank $g \leq r$. It is partitioned as $\mathbf{C} = [\mathbf{C}_1 \quad \mathbf{C}_2]$, where its submatrices have r and $q - r$ columns, respectively, to conform with the partitioned parameter matrix $[\boldsymbol{\xi}_1' \quad \boldsymbol{\xi}_2']$. $\mathbf{M}$ has dimensions $p \times u$ and rank $u \leq p$. Under these conditions the hypothesis can be stated equivalently as

$$H_0\colon \quad \mathbf{C}_1\boldsymbol{\xi}_1\mathbf{M} + \mathbf{C}_2\boldsymbol{\xi}_2\mathbf{M} = \mathbf{0} \tag{27}$$

As in the unidimensional hypothesis, the matrix $\mathbf{C}$ refers to hypotheses on the elements within given columns of the parameter matrix, while the postfactor $\mathbf{M}$ permits the generation of hypotheses among the different response parameters. In our initial treatment of the multivariate linear model we shall be concerned only with the first kind of hypothesis, and $\mathbf{M}$ will be taken as the appropriate identity matrix. Other kinds of $\mathbf{C}$ and $\mathbf{M}$ matrices will come into play in the extended discussion of profile analysis in Sec. 5.6.

Now let us sketch the derivation of the union-intersection test of the hypothesis (25). That multivariate hypothesis is true if and only if the univariate hypotheses

$$H_0\colon \quad \mathbf{C}\boldsymbol{\xi}\mathbf{M}\mathbf{a} = \mathbf{0} \tag{28}$$

hold for *all* nonnull u-component vectors $\mathbf{a}$. The test statistic for any one of these univariate hypotheses is given by expression (19):

$$F(\mathbf{a}) = \frac{(N - r)\mathbf{a}'\mathbf{M}'\mathbf{X}'\mathbf{A}_1(\mathbf{A}_1'\mathbf{A}_1)^{-1}\mathbf{C}_1'[\mathbf{C}_1(\mathbf{A}_1'\mathbf{A}_1)^{-1}\mathbf{C}_1']^{-1}\mathbf{C}_1(\mathbf{A}_1'\mathbf{A}_1)^{-1}\mathbf{A}_1'\mathbf{X}\mathbf{M}\mathbf{a}}{g\mathbf{a}'\mathbf{M}'\mathbf{X}'[\mathbf{I} - \mathbf{A}_1(\mathbf{A}_1'\mathbf{A}_1)^{-1}\mathbf{A}_1']\mathbf{X}\mathbf{M}\mathbf{a}} \tag{29}$$

For a univariate test of level β we would accept (28) if

$$F(\mathbf{a}) \leq F_{\beta;g,N-r} \tag{30}$$

and accept the original multivariate hypothesis (25) at some other level α, say, if

$$\bigcap_{\mathbf{a}} [F(\mathbf{a}) \leq F_{\beta;g,N-r}] \tag{31}$$

for all nonnull $\mathbf{a}$. As in the case of the T^2 statistic, this acceptance region is equivalent to that defined by

$$\max_{\mathbf{a}} F(\mathbf{a}) \leq F_{\beta;\,g,N-r} \tag{32}$$

for if the greatest F ratio falls in the acceptance region, so must those of all other compounding vectors. If the constraint of a unit denominator is introduced by a Lagrangian multiplier, the maximum value of $F(\mathbf{a})$ can be shown to be proportional to the greatest root of the determinantal equation

$$|\mathbf{H} - \lambda\mathbf{E}| = 0 \tag{33}$$

where

$$\begin{aligned} \mathbf{H} &= \mathbf{M}'\mathbf{X}'\mathbf{A}_1(\mathbf{A}_1'\mathbf{A}_1)^{-1}\mathbf{C}_1'[\mathbf{C}_1(\mathbf{A}_1'\mathbf{A}_1)^{-1}\mathbf{C}_1']^{-1}\mathbf{C}_1(\mathbf{A}_1'\mathbf{A}_1)^{-1}\mathbf{A}_1'\mathbf{X}\mathbf{M} \\ \mathbf{E} &= \mathbf{M}'\mathbf{X}'[\mathbf{I} - \mathbf{A}_1(\mathbf{A}_1'\mathbf{A}_1)^{-1}\mathbf{A}_1']\mathbf{X}\mathbf{M} \end{aligned} \tag{34}$$

The diagonal terms of these respective symmetric matrices are the numerator and denominator quadratic forms of (19) evaluated for each of the u variates, and the off-diagonal terms are bilinear forms in the observations on all pairs of variates. Within $\mathbf{H}$ and $\mathbf{E}$ both kinds of forms have the same coefficient matrix, and we see immediately that *any univariate hypothesis with statistic* (19) *can be generalized to the multivariate situation merely by replacing the F-ratio sums of squares by their matrix extensions* $\mathbf{H}$ *and* $\mathbf{E}$.

Let us denote the greatest root of (33) by c_s, where

$$s = \min\,(g,u) \tag{35}$$

or the smaller of the parameters g and u. We accept the multivariate hypothesis (25) at the α level if

$$c_s \leq c(\alpha) \tag{36}$$

and reject otherwise. $c(\alpha)$ is the 100α percentage point of the greatest-root distribution when the hypothesis is true.

Upper percentage points of the greatest-root distribution have been computed by Heck (1960), Pillai and Bantegui (1959), and Pillai (1964, 1965, 1967), and are given in part by Charts 9 to 16 and Tables 6 to 14 of the Appendix. The distributions are those of the largest root of $|\mathbf{H} - \lambda(\mathbf{H} + \mathbf{E})| = 0$, and for that reason the charts and tables must be entered with the test statistic

$$\theta_s = \frac{c_s}{1 + c_s} \tag{37}$$

The parameters of the distribution of θ_s under the null hypothesis are

$$\begin{aligned} s &= \min(g,u) \\ m &= \frac{|g - u| - 1}{2} \\ n &= \frac{N - r - u - 1}{2} \end{aligned} \tag{38}$$

and for an α-level test of H_0: $\mathbf{C\xi M} = 0$ the acceptance region is

$$\theta_s \leq x_{\alpha;s,m,n} \tag{39}$$

where $x_{\alpha;s,m,n} \equiv x_\alpha$ is the upper 100α percentage point obtained from the appropriate chart or table. The hypothesis is rejected if θ_s exceeds that value.

The nonzero roots of $|\mathbf{H} - \lambda\mathbf{E}| = 0$ are equal to the nonzero characteristic roots of $\mathbf{HE}^{-1}$, and in practice it is usually more efficient to extract the roots from that matrix with a standard computer program. It is important to note that $\mathbf{HE}^{-1}$ is not necessarily symmetric, although its roots can be shown to be real and nonnegative.

Frequently it follows from the experimental design that $s = 1$, and then the single positive root θ of $\mathbf{H}(\mathbf{H} + \mathbf{E})^{-1}$ has the beta distribution with density function

$$f(\theta) = \frac{1}{B(m + 1, n + 1)} \theta^m(1 - \theta)^n \qquad 0 \leq \theta \leq 1 \tag{40}$$

For that case the critical values of θ must be determined from tables of the incomplete beta function (K. Pearson, 1968) or, alternatively, through the transformation

$$F = \frac{n + 1}{m + 1} \frac{\theta}{1 - \theta} \tag{41}$$

to an F variate with degrees of freedom $2m + 2$ and $2n + 2$. The statistic can also be calculated directly from the single nonzero root $c = \operatorname{tr} \mathbf{HE}^{-1}$ as $F = [(n + 1)/(m + 1)]c$.

Characteristic Root Distributions. The joint distribution of the characteristic roots of matrices of the type $\mathbf{HE}^{-1}$ was found simultaneously by Fisher (1939), Girshick (1939), Hsu (1939), Roy (1939), and Mood (1951) for the "central" case of equal population roots. Roy (1945) found expressions for the marginal distributions of individual roots; Nanda (1948, 1951) obtained recursion formulas and tables of certain percentage points. Pillai (1955, 1956a, 1956b, 1965) also found reduction formulas and approximations to the cumulative distribution function of the greatest root near its upper percentage points; these results were essential for the computation of the critical values by Heck and Pillai. The noncentral distribution of the characteristic roots was first derived by Roy (1942); its expression in a computationally tractable form was accomplished by Constantine (1963) and James (1964). The power of the greatest root and alternative tests obtained from the noncentral distribution will be discussed in Sec. 5.8.

5.3 THE MULTIVARIATE ANALYSIS OF VARIANCE

The general results of the last section will now be used to extend the analysis of variance for some common linear models to the case of multiple responses.

The One-way Classification. Here a fixed set of k treatments, experimental conditions, or diagnostic classifications are of interest to the investigator, and in the presence of each some p response variates will be recorded on the sampling units. Those measurements or ratings are assumed to be independent observations on the p-dimensional multinormal variates with mean vectors $\boldsymbol{\mu}_1, \ldots, \boldsymbol{\mu}_k$ under the different treatments and a common unknown covariance matrix $\boldsymbol{\Sigma}$ for all k conditions. The linear model for this layout has the same design matrix $\mathbf{A}$ as that given by (5) of Sec. 5.2 for its unidimensional prototype, but its parameter matrix is now

$$(1) \qquad \boldsymbol{\xi} = \begin{bmatrix} \tau_{11} & \cdots & \tau_{1p} \\ \cdot\;\cdot & \cdots & \cdot\;\cdot \\ \tau_{k1} & \cdots & \tau_{kp} \\ \mu_1 & \cdots & \mu_p \end{bmatrix}$$

The hypothesis to be tested is that of equal-treatment-effect vectors:

$$(2) \qquad H_0: \begin{bmatrix} \tau_{11} \\ \cdot \\ \cdot \\ \cdot \\ \tau_{1p} \end{bmatrix} = \cdots = \begin{bmatrix} \tau_{k1} \\ \cdot \\ \cdot \\ \cdot \\ \tau_{kp} \end{bmatrix}$$

Hence the postfactor matrix $\mathbf{M}$ is the $p \times p$ identity matrix, and the matrix representation H_0: $\mathbf{C}\boldsymbol{\xi} = \mathbf{0}$ of the hypothesis has the same hypothesis matrix $\mathbf{C} = [\mathbf{C}_1 \quad \mathbf{C}_2]$ as in the univariate motivation. If those values of $\mathbf{A}_1$ and $\mathbf{C}_1$ are used to evaluate the general $\mathbf{H}$ and $\mathbf{E}$ matrices defined by (34) of the preceding section, their rsth elements will be found to be

$$\text{(3)}\qquad \begin{aligned} h_{rs} &= \sum_{j=1}^{k} \frac{1}{N_j} T_{jr}T_{js} - \frac{1}{N} G_r G_s \\ e_{rs} &= \sum_{j=1}^{k} \sum_{i=1}^{N_j} x_{ijr}x_{ijs} - \sum_{j=1}^{k} \frac{1}{N_j} T_{jr}T_{js} \end{aligned}$$

where $x_{ijr} = i$th observation on response r under treatment j

$T_{jr} = \sum_{i=1}^{N_j} x_{ijr} =$ sum of all observations on rth response in presence of treatment j

$G_r = \sum_{j=1}^{k} T_{jr} =$ grand total of all observations on rth response

$N = N_1 + \cdots + N_k$

The analogy with the treatment and error sums of squares in the univariate analysis should be evident. To test the hypothesis of equal treatment effects we extract the greatest characteristic root c_s from $\mathbf{HE}^{-1}$ and refer the quantity $\theta_s = c_s/(1 + c_s)$ to the appropriate percentage-point chart or table for the specified level α and parameters

$$\text{(4)}\qquad s = \min(k-1, p) \qquad m = \frac{|k-p-1|-1}{2} \qquad n = \frac{N-k-p-1}{2}$$

The reader should verify that the case of $k = 2$ leads to the two-sample T^2 whose F statistic follows from expression (41) of Sec. 5.2.

Example 5.1. In his initiation of the linear-discriminant-function technique Fisher (1936) considered as an application the case of random samples of flowers from the iris species *virginica, versicolor,* and *setosa.* The responses were the four measurements sepal length, sepal width, petal length, and petal width. Within each species observations were recorded on those dimensions for $N_j = 50$ plants. Through these data we shall test the hypothesis of a common mean vector for the four dimensions in the three species populations. We shall of course assume tacitly that these populations are quadrivariate normal with a common covariance matrix. The sample means for the dimensions of the three types of flower had these values:

Table 5.1. **Sample Means**

Species	Dimension			
	SL	*SW*	*PL*	*PW*
Virginica	6.588	2.974	5.552	2.026
Versicolor	5.936	2.770	4.260	1.326
Setosa	5.006	3.428	1.462	0.246

The totals T_{jr} for each species-dimension combination can be computed handily by multiplying each mean by 50 and will be omitted. The matrix of sums of squares and products among species is

$$\mathbf{H} = \begin{bmatrix} 63.21 & -19.95 & 165.25 & 71.28 \\ & 11.35 & -57.24 & -22.93 \\ & & 437.11 & 186.78 \\ & & & 80.41 \end{bmatrix}$$

The error matrix is

$$\mathbf{E} = \begin{bmatrix} 38.96 & 13.63 & 24.62 & 5.64 \\ & 16.96 & 8.12 & 4.81 \\ & & 27.22 & 6.27 \\ & & & 6.16 \end{bmatrix}$$

and its inverse is

$$\mathbf{E}^{-1} = 10^{-2}\begin{bmatrix} 7.3732 & -3.6590 & -6.1141 & 2.3295 \\ & 9.6865 & 1.8163 & -6.0623 \\ & & 10.0573 & -6.0572 \\ & & & 25.0001 \end{bmatrix}$$

Finally,

$$\mathbf{HE}^{-1} = \begin{bmatrix} -3.052 & -5.565 & 8.075 & 10.493 \\ 1.079 & 2.180 & -2.942 & -3.418 \\ -8.096 & -14.975 & 21.505 & 27.538 \\ -3.452 & -6.312 & 9.140 & 11.840 \end{bmatrix}$$

The rank of this matrix is two, and its nonzero characteristic roots can be found conveniently by the expansion (2) of Sec. 2.10 for the characteristic polynomial. The necessary coefficients of that function are

$$S_1 = \operatorname{tr} \mathbf{HE}^{-1} = 32.473$$

and

$$S_2 = \text{sum of all } 2 \times 2 \text{ principal minor determinants} = 9.131$$

and the equation

$$\lambda^2 - 32.473\lambda + 9.131 = 0$$

yields the greatest root $c_2 = 32.188$ of $\mathbf{HE}^{-1}$. The statistic required for the Heck charts is $\theta_2 = 0.970$; the parameters for its distribution are $s = 2$, $m = \frac{1}{2}$, and $n = 71$. If we choose to test at the $\alpha = 0.01$ level, the critical value for θ_2 can be read by interpolation in the second chart as approximately $x_{0.01} = 0.11$. Since the test statistic greatly exceeds that value, we must reject the hypothesis of a common mean vector for the three species of iris. The question of *which* species and dimensions have contributed to this rejection will be deferred to the treatment of simultaneous confidence intervals in Sec. 5.5.

Randomized Blocks. In a laboratory four standard strains of mice are regularly used for experiments measuring the effects of drugs on the levels of biochemical constituents of certain organs. In one trial of six variants of a drug the brains of the sacrificed mice are assayed for two products whose levels are suspected to be influenced in unknown degrees by the drugs. It is conceivable that the levels are correlated, and that their values, or perhaps logarithmic transformations of them, are distributed according to the bivariate normal distribution. Strain differences in the assays are of no interest, but it is desired to plan the experiment to remove that source of variation from the estimates of experimental error. To test the equality of the six treatment-effect vectors six mice will be randomly selected from each strain, and within these four blocks the drugs will be assigned at random.

Let us develop this experimental plan for the general case of b blocks containing each of the k treatments or conditions. Denote the observation on the rth response under the ith treatment in block j by x_{ijr}. In a randomized block design the mathematical model for that observation is

$$x_{ijr} = \mu_r + \tau_{ir} + \beta_{jr} + e_{ijr} \tag{5}$$

where μ_r = usual general-level effect for response r
τ_{ir} = effect of ith treatment on rth response
β_{jr} = effect of jth block on rth response
e_{ijr} = random effect specific to ijrth combination of treatment, block, and response

The quantities μ_r, τ_{ir}, and β_{jr} are *parameters*, for in this model inferences from the observations will be restricted to the k treatments as applied to just those b blocks employed in the experiment. The randomized design is thus a special case of the two-way fixed-effects analysis of variance with a single observation on each treatment-block combination and no provision for a blocks-by-treatment interaction parameter in its model. If the blocks had been selected randomly from a larger available population, e.g., the universe of all litters of mice from one strain, the β_{jr} would be random variables, and a different approach would be required (see, for example, Scheffé, 1959, chap. 9). The random terms can be written as the vectors $\mathbf{e}'_{ij} = [e_{ij1}, \ldots, e_{ijp}]$; they are assumed to have the p-dimensional multinormal distribution with null mean vector and common-covariance matrix $\mathbf{\Sigma}$ for all combinations of i and j. The $\mathbf{e}_{ij}$ corresponding to different sampling units in any block are independently distributed. If the x_{ijr} are written in row-vector form $\mathbf{x}'_{ij}$ for the responses at the ijth treatment-block combination and then grouped by blocks and treatments within blocks into the $bk \times p$ matrix $\mathbf{X}$, the general linear model (23) of Sec. 5.2 has design matrix

$$\text{(6)} \qquad \mathbf{A} = \begin{bmatrix} 1 & \cdots & 0 & 1 & \cdots & 0 & 1 \\ \cdot & \cdots & \cdot & \cdot & \cdots & \cdot & \cdot \\ 1 & \cdots & 0 & 0 & \cdots & 1 & 1 \\ \cdot & \cdots & \cdot & \cdot & \cdots & \cdot & \cdot \\ 0 & \cdots & 1 & 1 & \cdots & 0 & 1 \\ \cdot & \cdots & \cdot & \cdot & \cdots & \cdot & \cdot \\ 0 & \cdots & 1 & 0 & \cdots & 1 & 1 \end{bmatrix}$$

The first b columns contain a succession of ones in the first, . . . , last k positions and zeros elsewhere. The next k columns form $k \times k$ identity matrices corresponding to each of the blocks, while the last column consists of unity in each position. The dimensions of $\mathbf{A}$ are $bk \times (b + k + 1)$, and its rank is $b + k - 1$. The parameter matrix is

$$\text{(7)} \qquad \boldsymbol{\xi} = \begin{bmatrix} \beta_{11} & \cdots & \beta_{1p} \\ \cdot & \cdots & \cdot \\ \beta_{b1} & \cdots & \beta_{bp} \\ \tau_{11} & \cdots & \tau_{1p} \\ \cdot & \cdots & \cdot \\ \tau_{k1} & \cdots & \tau_{kp} \\ \mu_1 & \cdots & \mu_p \end{bmatrix}$$

Finally the residual matrix $\boldsymbol{\varepsilon}$ has as rows the bk vectors $\mathbf{e}'_{ij}$. Since we are interested only in the hypothesis

$$\text{(8)} \qquad H_0\colon \begin{bmatrix} \tau_{11} \\ \cdot \\ \cdot \\ \cdot \\ \tau_{1p} \end{bmatrix} = \cdots = \begin{bmatrix} \tau_{k1} \\ \cdot \\ \cdot \\ \cdot \\ \tau_{kp} \end{bmatrix}$$

of equal treatment effects, its matrix form (25) of Sec. 5.2 has hypothesis matrix

$$\mathbf{C} = \begin{bmatrix} 0 & \cdots & 0 & 1 & -1 & 0 & \cdots & 0 & 0 & 0 \\ \cdot & \cdots & \cdot & \cdots & \cdots & \cdots & \cdots & \cdots & \cdots & \cdot \\ 0 & \cdots & 0 & 0 & 0 & 0 & \cdots & 1 & -1 & 0 \end{bmatrix} \tag{9}$$

of rank $k - 1$. As in the one-way analysis-of-variance hypothesis, the postfactor matrix $\mathbf{M}$ is merely the $p \times p$ identity matrix. $\mathbf{C}$ is of course partitioned according to our choice of the basis matrix $\mathbf{A}_1$ for $\mathbf{A}$; for example, $\mathbf{A}_1$ could consist of any $b - 1$ of the first b columns of $\mathbf{A}$ and the next k columns forming the $k \times k$ identity matrices.

The hypothesis and error matrices can now be obtained as an exercise in the evaluation of the general expressions of Sec. 5.2 or by straightforward extension of the univariate analysis of variance sums of squares. Introduce the notation

$$B_{jr} = \sum_{i=1}^{k} x_{ijr} \tag{10}$$

for the total of the observations on the rth response in block j. As in the one-way analysis of variance, let

$$T_{ir} = \sum_{j=1}^{b} x_{ijr}$$

be the total of the rth observations under treatment i and let

$$\begin{aligned} G_r &= B_{1r} + \cdots + B_{br} \\ &= T_{1r} + \cdots + T_{kr} \end{aligned}$$

be the grand total of the values of that response in all sampling units. Then the rsth elements of $\mathbf{H}$ and $\mathbf{E}$ can be written as

$$\begin{aligned} h_{rs} &= \frac{1}{b} \sum_{i=1}^{k} T_{ir}T_{is} - \frac{1}{bk} G_rG_s \\ e_{rs} &= \sum_{i=1}^{k} \sum_{j=1}^{b} x_{ijr}x_{ijs} - \frac{1}{b} \sum_{i=1}^{k} T_{ir}T_{is} - \frac{1}{k} \sum_{j=1}^{b} B_{jr}B_{js} + \frac{1}{bk} G_rG_s \\ &= \sum_{i=1}^{k} \sum_{j=1}^{b} x_{ijr}x_{ijs} - \frac{1}{bk} G_rG_s - h_{rs} - \left(\frac{1}{k} \sum_{j=1}^{b} B_{jr}B_{js} - \frac{1}{bk} G_rG_s\right) \end{aligned} \tag{11}$$

The final parenthetical term of e_{rs} is the sum of squares or cross products attributable to block differences, and the $\mathbf{H}$ matrix for testing the equality of the block parameter vectors would be constructed from those quantities. Since the blocks were chosen for the homogeneity and similarity of their sampling units with no regard for the severity of block differences, this hypothesis is rarely of interest.

When the hypothesis of equal treatment effects is true, the statistic θ_s computed from the largest root of $\mathbf{HE}^{-1}$ has the greatest-characteristic-root distribution with parameters

$$(12) \quad s = \min(k-1, p) \quad m = \frac{|k-1-p|-1}{2} \quad n = \frac{k(b-1)-b-p}{2}$$

and the hypothesis is rejected if θ_s exceeds $x_{\alpha;s,m,n}$.

The Two-way Analysis of Variance. Let us motivate this experimental design through a rather simplified example of a drug trial. Three drugs A, B, C are known to have toxic effects that are apparent in several measurable features of an organism ingesting them. In the present case let these characteristics be the weight losses in the first and second weeks of the trial. It is also thought that the effects of the drugs might be different in male animals from those in females, and this sex difference will be measured by the second way of classification in the experimental plan. For the trial $3n$ male rats of essentially equal ages and weights are drawn from a common laboratory strain and divided randomly and equally among the three drugs. $3n$ female rats of comparable ages and weights are obtained from litters of the original strain and similarly assigned drugs. Under the assumption that the weight decrements have a bivariate normal distribution with the same covariance matrix for all sex and drug combinations, we wish to answer the following questions:

1. Are the drug mean vectors equal?
2. Are the sex mean vectors equal?
3. Do some drugs interact with sex to produce inordinately high or low weight decrements?

In the common usage of the analysis of variance questions 1 and 2 refer to the drug and sex *main effects*, while question 3 is concerned with a possible *interaction* of the drug and sex factors. For example, drug A might induce a severe loss of weight in the male rats but only a negligible change in the females, while drugs B and C had similar effects in both sexes. As in the two-sample case of Chap. 4, it is also possible to have interactions of the ways of classification with the responses, and we shall treat that extension in Sec. 5.6.

The tests of the hypotheses suggested by questions 1 to 3 can be produced from the general results of the preceding section, and we shall now present them for the two-way analysis of variance on p responses with n independent observation vectors in each combination. The restriction to equal, or at least proportional, cell numbers is essential: otherwise, as in the univariate analysis, the various $\mathbf{H}$ and $\mathbf{E}$ matrices

Table 5.2. **Observations from the Two-way Layout**

Row treatments	Column treatments 1	. . .	Column treatments c
1	$\mathbf{x}'_{111}$ . . . $\mathbf{x}'_{11n}$	. . .	$\mathbf{x}'_{1c1}$ . . . $\mathbf{x}'_{1cn}$
. . .	. . .	. . .	. . .
r	$\mathbf{x}'_{r11}$ . . . $\mathbf{x}'_{r1n}$	. . .	$\mathbf{x}'_{rc1}$ . . . $\mathbf{x}'_{rcn}$

will not sum to the total sums of squares and products matrix. The case of unequal numbers of observations can still be handled for the individual hypotheses by the general expressions of Sec. 5.2. Let x_{ijkr} denote the kth observation on the rth response obtained under the ith treatment of the first (or row) way of classification and the jth treatment of the second (or column) set. If we let

$$\mathbf{x}'_{ijk} = [x_{ijk1}, \ . \ . \ . \ , x_{ijkp}] \tag{13}$$

the data obtained from a two-way design with r rows and c columns can be arranged according to Table 5.2.

The linear model for the general observation is

$$x_{ijkh} = \mu_h + \alpha_{ih} + \tau_{jh} + \eta_{ijh} + e_{ijkh} \tag{14}$$

where μ_h = general-level parameter of hth response
α_{ih} = effect of ith row treatment on hth response
τ_{jh} = effect of jth column treatment on hth response
η_{ijh} = effect of interaction of ith and jth treatments on hth response
e_{ijkh} = usual multinormal random variable term

The transpose of the parameter matrix can be written as

(15) $\boldsymbol{\xi}' =$

$$\begin{bmatrix} \alpha_{11} & \cdots & \alpha_{r1} & \tau_{11} & \cdots & \tau_{c1} & \eta_{111} & \cdots & \eta_{1c1} & \cdots & \eta_{r11} & \cdots & \eta_{rc1} & \mu_1 \\ \cdots & \cdots & \cdots & \cdots & \cdots & \cdots & \cdots & \cdots & \cdots & \cdots & \cdots & \cdots & \cdots & \cdots \\ \alpha_{1p} & \cdots & \alpha_{rp} & \tau_{1p} & \cdots & \tau_{cp} & \eta_{11p} & \cdots & \eta_{1cp} & \cdots & \eta_{r1p} & \cdots & \eta_{rcp} & \mu_p \end{bmatrix}$$

and the design matrix can be expressed in partitioned form as

$$(16)\quad \mathbf{A} = \begin{bmatrix} \mathbf{j} & \cdots & \mathbf{0} & \mathbf{j} & \cdots & \mathbf{0} & \mathbf{j} & \cdots & \mathbf{0} & \cdots & \mathbf{0} & \cdots & \mathbf{0} & \mathbf{j} \\ \cdot & \cdot & \cdot & \cdot & \cdot & \cdot & \cdot & \cdot & \cdot & \cdot & \cdot & \cdot & \cdot & \cdot \\ \mathbf{j} & \cdots & \mathbf{0} & \mathbf{0} & \cdots & \mathbf{j} & \mathbf{0} & \cdots & \mathbf{j} & \cdots & \mathbf{0} & \cdots & \mathbf{0} & \mathbf{j} \\ \cdot & \cdot & \cdot & \cdot & \cdot & \cdot & \cdot & \cdot & \cdot & \cdot & \cdot & \cdot & \cdot & \cdot \\ \mathbf{0} & \cdots & \mathbf{j} & \mathbf{j} & \cdots & \mathbf{0} & \mathbf{0} & \cdots & \mathbf{0} & \cdots & \mathbf{j} & \cdots & \mathbf{0} & \mathbf{j} \\ \cdot & \cdot & \cdot & \cdot & \cdot & \cdot & \cdot & \cdot & \cdot & \cdot & \cdot & \cdot & \cdot & \cdot \\ \mathbf{0} & \cdots & \mathbf{j} & \mathbf{0} & \cdots & \mathbf{j} & \mathbf{0} & \cdots & \mathbf{0} & \cdots & \mathbf{0} & \cdots & \mathbf{j} & \mathbf{j} \end{bmatrix}$$

$\mathbf{j}$ is the column vector of n ones, while $\mathbf{0}$ is the column vector containing n zeros. The first horizontal block of such vectors in $\mathbf{A}$ corresponds to the n observation vectors in each of the c column treatments appearing with the first row treatment. As we may see from the form of the parameter matrix, the first column of that block introduces the parameters of the first row α_{1h}, while the next c columns successively bring in the appropriate column parameters $\tau_{1h}, \ldots, \tau_{ch}$. This pattern is repeated for the r blocks of column vectors. The columns $r+c+1, \ldots, r+c+rc$ of the matrix introduce the interaction-effect parameters η_{ijh}, while the final column of ones accounts for the general-level parameters μ_h.

The hypothesis matrices for the tests on row and column effects are

$$(17)\qquad \mathbf{C}^{(1)} = \begin{bmatrix} 1 & 0 & \cdots & 0 & -1 & \cdots & 0 \\ 0 & 1 & \cdots & 0 & -1 & \cdots & 0 \\ \cdot & \cdot & \cdot & \cdot & \cdot & \cdot & \cdot \\ 0 & 0 & \cdots & 1 & -1 & \cdots & 0 \end{bmatrix}$$

and

$$(18)\quad \mathbf{C}^{(2)} = \begin{bmatrix} 0 & \cdots & 0 & 1 & 0 & \cdots & 0 & -1 & \cdots & 0 \\ 0 & \cdots & 0 & 0 & 1 & \cdots & 0 & -1 & \cdots & 0 \\ \cdot & \cdot & \cdot & \cdot & \cdot & \cdot & \cdot & \cdot & \cdot & \cdot \\ 0 & \cdots & 0 & 0 & 0 & \cdots & 1 & -1 & \cdots & 0 \end{bmatrix}$$

$\mathbf{C}^{(1)}$ has dimensions $r-1$ and $(r+1)(c+1)$. Its first r columns have the usual hypothesis-matrix form introduced in Sec. 5.2, while its remaining elements are zeros. $\mathbf{C}^{(2)}$ is a $(c-1) \times (r+1)(c+1)$ matrix with zeros in its first r columns, the usual contrast matrix in its next c columns, and zeros in the last $rc+1$ columns. For the interaction-hypothesis matrix we first note that the η_{ijh} have been grouped first by rows and then by columns within each row in the columns of the parameter matrix:

$$(19)\qquad \eta_{11h}, \ldots, \eta_{1ch}, \ldots, \eta_{r1h}, \ldots, \eta_{rch}$$

Then, if **C** is the $(c-1) \times c$ form of the hypothesis matrix (21) of Sec. 5.2, the interaction-hypothesis matrix can be written as

$$\mathbf{C}^{(3)} = \begin{bmatrix} \mathbf{0} & \mathbf{C} & \mathbf{0} & \cdots & \mathbf{0} & -\mathbf{C} & \mathbf{0} \\ \mathbf{0} & \mathbf{0} & \mathbf{C} & \cdots & \mathbf{0} & -\mathbf{C} & \mathbf{0} \\ \cdot & \cdot & \cdot & \cdots & \cdot & \cdot & \cdot \\ \mathbf{0} & \mathbf{0} & \mathbf{0} & \cdots & \mathbf{C} & -\mathbf{C} & \mathbf{0} \end{bmatrix} \tag{20}$$

$\mathbf{C}^{(3)}$ has dimensions $(r-1)(c-1) \times (r+1)(c+1)$. The null submatrices of the first column have dimensions $(c-1) \times (r+c)$, while those of the next r columns must be $(c-1) \times c$ to conform with **C**. Finally, the submatrices of the last column are $(c-1)$-component null column vectors. The rank of $\mathbf{C}^{(3)}$ is $(r-1)(c-1)$.

For the row and column hypotheses to be testable it is necessary that $\mathbf{C}^{(3)}\boldsymbol{\xi} = \mathbf{0}$, or that the hypothesis of no interaction effect is true. For that reason it is usually best in practice to test the interaction hypothesis first, since its rejection implies that the row and column treatment tests must be made *separately* for each of the c column and r row categories.

From the preceding matrices and the general expressions of the last section the hypotheses and error matrices could be calculated for the two-way multivariate analysis of variance. However, it is far easier to obtain those quantities as the matrix generalizations of the sums of squares of the univariate analysis. Let us begin by defining the following totals of the observations on each response:

1. *Cells:*

$$C_{ijh} = \sum_{k=1}^{n} x_{ijkh}$$

2. *Column treatments:*

$$T_{jh} = \sum_{i=1}^{r} \sum_{k=1}^{n} x_{ijkh}$$

3. *Row treatments:*

$$R_{ih} = \sum_{j=1}^{c} \sum_{k=1}^{n} x_{ijkh}$$

4. *Grand total:*

$$G_h = \sum_{i=1}^{r} \sum_{j=1}^{c} \sum_{k=1}^{n} x_{ijkh}$$

These totals are computed for all combinations of the treatment and response subscripts. Then the matrices of error and hypotheses sums of squares and products are computed as in Table 5.3. $u, v = 1, \ldots, p$

Table 5.3. **Matrix Elements for the Two-way Analysis of Variance**

Source	Sums of squares and products	
	Matrix	General element
Row treatments	$\mathbf{H}_1$	$h_{1uv} = \frac{1}{cn} \sum_{i=1}^{r} R_{iu}R_{iv} - \frac{G_uG_v}{rcn}$
Column treatments	$\mathbf{H}_2$	$h_{2uv} = \frac{1}{rn} \sum_{j=1}^{c} T_{ju}T_{jv} - \frac{G_uG_v}{rcn}$
Interaction	$\mathbf{H}_3$	$h_{3uv} = t_{uv} - h_{1uv} - h_{2uv} - e_{uv}$
Error	$\mathbf{E}$	$e_{uv} = \sum_{i=1}^{r} \sum_{j=1}^{c} \sum_{k=1}^{n} x_{ijku}x_{ijkv} - \frac{1}{n} \sum_{i=1}^{r} \sum_{j=1}^{c} C_{iju}C_{ijv}$
Total	$\mathbf{T}$	$t_{uv} = \sum_{i=1}^{r} \sum_{j=1}^{c} \sum_{k=1}^{n} x_{ijku}x_{ijkv} - \frac{G_uG_v}{rcn}$

for each of the five matrices. For the tests of the three hypotheses it is first necessary to invert $\mathbf{E}$ and then to form the products

$$\mathbf{H}_1\mathbf{E}^{-1} \qquad \mathbf{H}_2\mathbf{E}^{-1} \qquad \mathbf{H}_3\mathbf{E}^{-1}$$

From these the greatest characteristic roots c_{1s}, c_{2s}, c_{3s} are extracted and their statistics $c_{is}/(1 + c_{is})$ referred to the Heck charts. The parameters of the relevant greatest-characteristic-root distributions are summarized in Table 5.4.

Table 5.4. **Greatest-root Distribution Parameters**

Source	Statistic	Parameters		
		s	m	n
Rows	$\frac{c_{1s}}{1 + c_{1s}}$	min $(r - 1, p)$	$\frac{\lvert r - 1 - p\rvert - 1}{2}$	$\frac{rc(n - 1) - p - 1}{2}$
Columns	$\frac{c_{2s}}{1 + c_{2s}}$	min $(c - 1, p)$	$\frac{\lvert c - 1 - p\rvert - 1}{2}$	$\frac{rc(n - 1) - p - 1}{2}$
Interaction	$\frac{c_{3s}}{1 + c_{3s}}$	min $[(r - 1)(c - 1), p]$	$\frac{\lvert (r - 1)(c - 1) - p\rvert - 1}{2}$	$\frac{rc(n - 1) - p - 1}{2}$

Example 5.2. Let us return to the original situation used as a motivating example for the two-way multivariate analysis of variance. There the first way of classification consisted of $r = 2$ sexes and the second involved $c = 3$ drugs. Under the original homogeneity and randomization specifications suppose that $n = 4$ rats of each sex were assigned to each drug. From this hypothetical experiment pairs of weight losses in grams were observed for the first and second weeks (see Table 5.5). The total and error sums of squares and products matrices are

$$\mathbf{T} = \begin{bmatrix} 410.5 & 196.0 \\ 196.0 & 183.333 \end{bmatrix} \qquad \mathbf{E} = \begin{bmatrix} 94.5 & 76.5 \\ 76.5 & 114.0 \end{bmatrix}$$

The hypothesis matrices for testing the sex, drug, and interaction effects are

$$\mathbf{H}_1 = \begin{bmatrix} 0.667 & 0.667 \\ 0.667 & 0.667 \end{bmatrix} \qquad \mathbf{H}_2 = \begin{bmatrix} 301.0 & 97.5 \\ 97.5 & 36.333 \end{bmatrix}$$

and

$$\mathbf{H}_3 = \begin{bmatrix} 14.333 & 21.333 \\ 21.333 & 32.333 \end{bmatrix}$$

The inverse of the error matrix is

$$\mathbf{E}^{-1} = \begin{bmatrix} 0.02317 & -0.01555 \\ -0.01555 & 0.01920 \end{bmatrix}$$

Table 5.5. **Weight Losses**

	Drug			R_{ih}
	A	*B*	*C*	
Male	[5,6] [5,4] [9,9] [7,6]	[7,6] [7,7] [9,12] [6,8]	[21,15] [14,11] [17,12] [12,10]	
C_{1jh}	[26,25]	[29,33]	[64,48]	[119,106]
Female	[7,10] [6,6] [9,7] [8,10]	[10,13] [8,7] [7,6] [6,9]	[16,12] [14,9] [14,8] [10,5]	
C_{2jh}	[30,33]	[31,35]	[54,34]	[115,102]
T_{jh}	[56,58]	[60,68]	[118,82]	

$$[G_1,G_2] = [234,208]$$

$$\Sigma\Sigma\Sigma x_{ijk1}^2 = 2{,}692 \qquad \Sigma\Sigma\Sigma x_{ijk1}x_{ijk2} = 2{,}224 \qquad \Sigma\Sigma\Sigma x_{ijk2}^2 = 1{,}986$$

The matrix product needed for the interaction test is

$$\mathbf{H}_3\mathbf{E}^{-1} = \begin{bmatrix} 0.00041 & 0.18686 \\ -0.00843 & 0.28928 \end{bmatrix}$$

and its characteristic equation is $\lambda^2 - 0.28969\lambda + 0.001693 = 0$. The greatest root is $c_{32} = 0.284$, and the test statistic in the form required by the percentage-point charts is $\theta_2 = 0.221$. The distribution parameters are $s = 2$, $m = -\frac{1}{2}$, and $n = 7.5$, and if we choose to test at the 0.05 level, the critical value for θ_2 can be found from the first chart to be $x_{0.05} = 0.39$. Since the observed statistic does not exceed that number, we do not reject the hypothesis of no sex and drug interaction. The additive model for sex and drug effects can be assumed to hold for all combinations of treatments.

With the possibility of an interaction dismissed, we are now ready to test the main effects. Comparison of the diagonal elements of $\mathbf{H}_1$ with those of $\mathbf{E}$ shows that a significant sex effect could not be obtained, and that hypothesis will not be tested formally. To test the equality of drug effects we compute

$$\mathbf{H}_2\mathbf{E}^{-1} = \begin{bmatrix} 5.45753 & -2.80700 \\ 1.69395 & -0.81800 \end{bmatrix}$$

and extract its greatest characteristic root $c_{22} = 4.576$. The associated test statistic is $\theta_2 = 0.8207$. Since for parameters $s = 2$, $m = -\frac{1}{2}$, and $n = 7.5$, the $\alpha = 0.01$ critical value is $x_{0.01} = 0.50$, we reject the hypothesis of equal drug effects. The multiple-comparisons problem of determining *which* drugs are different will be postponed until Sec. 5.5.

Multivariate Regression Analysis. The univariate linear regression model

$$(21) \quad x_i = \beta_0 + \beta_1(u_{i1} - \bar{u}_1) + \cdots + \beta_t(u_{it} - \bar{u}_t) + e_i \qquad i = 1, \ldots, N$$

expresses the ith observation x_i on a "dependent" variate as a linear function of the values of t nonrandom "independent" variables and a random disturbance term e_i. We prefer to express the independent variables as deviations about their respective averages $\bar{u}_j$, for that form of the model leads to an estimate of β_0 equal to the mean of the x_i and to the inversion of a $t \times t$ matrix for the other estimates. As in the univariate analysis-of-variance model, the e_i are independently and normally distributed with mean zero and a common variance σ^2. The multivariate general linear model can be used to extend (21) to a vector of p variates simultaneously related to the same set of "independent" observations. In matrix form the multivariate regression model is

4th Fl 10:3

$$\mathbf{X} = [\mathbf{j} \quad \mathbf{U}] \begin{bmatrix} \boldsymbol{\beta}_0' \\ \mathbf{B}_1 \end{bmatrix} + \boldsymbol{\varepsilon} \tag{22}$$

where $\mathbf{X}$ is the usual $N \times p$ data matrix, $\mathbf{j}$ is an $N \times 1$ vector of ones,

$$\mathbf{U} = \begin{bmatrix} u_{11} - \bar{u}_1 & \cdots & u_{1t} - \bar{u}_t \\ \cdot & & \cdot \\ \cdot & & \cdot \\ \cdot & & \cdot \\ u_{N1} - \bar{u}_1 & \cdots & u_{Nt} - \bar{u}_t \end{bmatrix} \tag{23}$$

is the matrix of values of the independent variables expressed in deviation form, $\boldsymbol{\beta}_0' = [\beta_{01}, \ldots, \beta_{0p}]$ contains the intercept parameters for the p models, and the hjth element of

$$\mathbf{B}_1 = \begin{bmatrix} \beta_{11} & \cdots & \beta_{1p} \\ \cdot & & \cdot \\ \cdot & & \cdot \\ \cdot & & \cdot \\ \beta_{t1} & \cdots & \beta_{tp} \end{bmatrix} \tag{24}$$

is the regression coefficient of the jth response variate on the hth independent variable. The rows of $\boldsymbol{\varepsilon}$ are independently distributed $N(\mathbf{0},\boldsymbol{\Sigma})$ vectors, where $\boldsymbol{\Sigma}$ is the usual unknown positive definite covariance matrix.

The minimum-variance unbiased estimates of the regression parameters are

$$\hat{\boldsymbol{\beta}}_0' = [\bar{x}_1, \ldots, \bar{x}_p] \qquad \hat{\mathbf{B}}_1 = (\mathbf{U}'\mathbf{U})^{-1}\mathbf{U}'\mathbf{X} \tag{25}$$

These are merely vector and matrix generalizations of the univariate estimates and could have been obtained by maximizing the likelihood of $\mathbf{X}$ without recourse to the development of Sec. 5.2. We also note that the covariance matrix does not enter into the estimates. $\hat{\mathbf{B}}_1$ is identical with the regression matrix (5) of Sec. 3.7 obtained by maximum likelihood under the assumption that the dependent and independent variables came from a $(p + t)$-dimensional multinormal distribution.

The hypothesis H_0: $\mathbf{B}_1 = \mathbf{0}$ of no regression relations can be tested by the multivariate analysis of variance. Since $\mathbf{A} \equiv \mathbf{A}_1 = [\mathbf{j} \quad \mathbf{U}]$ and $\mathbf{C} \equiv \mathbf{C}_1 = [\mathbf{0} \quad \mathbf{I}]$, where $\mathbf{I}$ is the $t \times t$ identity matrix, the required matrices are

$$\begin{aligned} \mathbf{H} &= \mathbf{X}'\mathbf{U}(\mathbf{U}'\mathbf{U})^{-1}\mathbf{U}'\mathbf{X} \\ \mathbf{E} &= \mathbf{X}'\left[\mathbf{I} - \frac{1}{N}\mathbf{j}\mathbf{j}' - \mathbf{U}(\mathbf{U}'\mathbf{U})^{-1}\mathbf{U}'\right]\mathbf{X} \end{aligned} \tag{26}$$

The parameters of the greatest-root distribution are

$$(27) \quad s = \min(t,p) \qquad m = \frac{|t-p|-1}{2} \qquad n = \frac{N-t-p-2}{2}$$

The statistic and its distribution are identical with those of Sec. 3.7 for the conditional multinormal regression model.

5.4 THE MULTIVARIATE ANALYSIS OF COVARIANCE

In the example of the two-way experimental design in the preceding section the rats were required to be of nearly equal weights and ages to assure that the losses in weight during the trial would not be affected by extreme initial weights or concomitants of the growth process. Such a requirement might be difficult to satisfy, for the available pool of rats or other sampling units might be small and heterogeneous. However, if certain assumptions can be made on the effects of the concomitant variables, it is possible to incorporate that relationship as part of the general linear model, and by so doing to adjust the treatment mean vectors for different values of the secondary variables. This technique is called the *analysis of covariance.*

Let us suppose that t characteristics have been selected as concomitant variables and that each of the p response variates can be represented as a linear function of those covariables and the various treatment effects of the experimental plan. Denote by $u_{i1}, \ldots, u_{it}$ the values of the concomitant variables associated with the observation vector obtained under the ith treatment combination of the design. Then the linear model for the hth response observation in that vector is

$$(1) \quad x_{ih} = a_{i1}\xi_{1h} + \cdots + a_{iq}\xi_{qh} + \beta_{1h}u_{i1} + \cdots + \beta_{th}u_{it} + e_{ih}$$

In our model the regression parameters $\beta_{1h}, \ldots, \beta_{th}$ are the same for all values of i: the relationship of responses and covariates is unaffected by the particular treatment combinations. Unlike the model (21) of the last section we shall not express the concomitant observations as deviations about their means. The matrix form of the linear model is now

$$(2) \quad \begin{aligned} \mathbf{X} &= \mathbf{A}\boldsymbol{\xi} + \mathbf{UB} + \boldsymbol{\varepsilon} \\ &= [\mathbf{A} \quad \mathbf{U}]\begin{bmatrix} \boldsymbol{\xi} \\ \mathbf{B} \end{bmatrix} + \boldsymbol{\varepsilon} \end{aligned}$$

where $\mathbf{A}$ is the $N \times q$ design matrix of rank r of the particular layout, $\mathbf{U}$ is the $N \times t$ matrix of concomitant observations, and $\mathbf{B}$ is the $t \times p$ matrix of regression parameters. $\mathbf{A}$ contains the $N \times r$ basis $\mathbf{A}_1$ and is linearly independent of $\mathbf{U}$.

Although hypothesis tests on the elements of $\boldsymbol{\xi}$ and $\mathbf{B}$ could be obtained from the general expressions (34) of Sec. 5.2 or by matrix generalizations of the univariate analysis-of-covariance formulas, we shall prefer to express the **H** and **E** matrices in terms of those for the original design without the concomitant variables. For the test of

$$H_0\text{:}\quad \mathbf{C}\boldsymbol{\xi} = \mathbf{0} \tag{3}$$

where **C** is a $g \times q$ hypothesis matrix of rank g, the sums of squares and products matrices can be written as

$$\begin{aligned} \mathbf{H}_1 &= \mathbf{H}_{xx} - (\mathbf{H}_{xu} + \mathbf{E}_{xu})(\mathbf{H}_{uu} + \mathbf{E}_{uu})^{-1}(\mathbf{H}_{ux} + \mathbf{E}_{ux}) + \mathbf{E}_{xu}\mathbf{E}_{uu}^{-1}\mathbf{E}_{ux} \\ \mathbf{E} &= \mathbf{E}_{xx} - \mathbf{E}_{xu}\mathbf{E}_{uu}^{-1}\mathbf{E}_{ux} \end{aligned} \tag{4}$$

Here

$$\begin{aligned} \mathbf{H}_{xx} &= \mathbf{X}'\mathbf{K}\mathbf{X} \qquad \mathbf{H}_{uu} = \mathbf{U}'\mathbf{K}\mathbf{U} \qquad \mathbf{H}_{xu} = \mathbf{H}'_{ux} = \mathbf{X}'\mathbf{K}\mathbf{U} \\ \mathbf{E}_{xx} &= \mathbf{X}'\mathbf{L}\mathbf{X} \qquad \mathbf{E}_{uu} = \mathbf{U}'\mathbf{L}\mathbf{U} \qquad \mathbf{E}_{xu} = \mathbf{E}'_{ux} = \mathbf{X}'\mathbf{L}\mathbf{U} \end{aligned} \tag{5}$$

for

$$\begin{aligned} \mathbf{K} &= \mathbf{A}_1(\mathbf{A}_1'\mathbf{A}_1)^{-1}\mathbf{C}_1'[\mathbf{C}_1(\mathbf{A}_1'\mathbf{A}_1)^{-1}\mathbf{C}_1']^{-1}\mathbf{C}_1(\mathbf{A}_1'\mathbf{A}_1)^{-1}\mathbf{A}_1' \\ \mathbf{L} &= \mathbf{I} - \mathbf{A}_1(\mathbf{A}_1'\mathbf{A}_1)^{-1}\mathbf{A}_1' \end{aligned} \tag{6}$$

$\mathbf{H}_{xx}$ and $\mathbf{E}_{xx}$ are the usual sums of squares and products matrices for testing (3) in the linear model without the concomitant variables, while $\mathbf{H}_{uu}$ and $\mathbf{E}_{uu}$ are matrices of the same quadratic and bilinear forms computed from the covariable observations. $\mathbf{H}_{xu}$ and $\mathbf{E}_{xu}$ are corresponding matrices of sums of products of the response and concomitant observations. Under the null hypothesis (3) the greatest characteristic root of $\mathbf{H}_1\mathbf{E}^{-1}$ would have the usual distribution with parameters

$$s = \min(g,p) \qquad m = \frac{|g-p|-1}{2} \qquad n = \frac{N-r-t-p-1}{2} \tag{7}$$

If $\mathbf{a}'\boldsymbol{\xi}$ is an estimable function of the design parameters the unbiased minimum-variance estimates of its elements are given by

$$\widehat{\mathbf{a}'\boldsymbol{\xi}} = \mathbf{a}'(\mathbf{A}_1'\mathbf{A}_1)^{-1}\mathbf{A}_1'(\mathbf{X} - \mathbf{U}\hat{\mathbf{B}}) \tag{8}$$

where

$$\hat{\mathbf{B}} = \mathbf{E}_{uu}^{-1}\mathbf{E}_{ux} \tag{9}$$

is the estimate of the regression coefficient matrix. Because the regression model is the same for each sampling unit in the design, $\hat{\mathbf{B}}$ is computed from the within-treatments or "error" sums of squares and products matrices rather than the total matrices of the conventional estimates (25) of Sec. 5.3.

The hypothesis

$$H_0\colon\ \mathbf{B} = \mathbf{0} \tag{10}$$

of no linear relation of the response and concomitant variables can be tested by computing the hypothesis sums of squares and products matrix

$$\mathbf{H}_2 = \mathbf{E}_{xu}\mathbf{E}_{uu}^{-1}\mathbf{E}_{ux} \tag{11}$$

and extracting the greatest characteristic root of $\mathbf{H}_2\mathbf{E}^{-1}$, where $\mathbf{E}$ is the same matrix defined by (4). The parameters for entering the charts or tables of the percentage points of the statistic are

$$s = \min\,(t,p) \qquad m = \frac{|t-p|-1}{2} \qquad n = \frac{N-r-t-p-1}{2} \tag{12}$$

The One-way Design with a Single Covariate. The linear model for the ith observation on the rth response in the jth treatment group is

$$x_{ijr} = \mu_r + \tau_{jr} + \beta_r u_{ij} + e_{ijr} \tag{13}$$

where, as in Sec. 5.3, $i = 1, \ldots, N_j$, $j = 1, \ldots, k$, and $r = 1, \ldots, p$. The matrices $\mathbf{H}_1$ and $\mathbf{E}$ of (4) have rsth elements

$$\begin{aligned} h_{rs}^{(1)} &= h_{rs} - \frac{(h_{ru}+e_{ru})(h_{su}+e_{su})}{h_{uu}+e_{uu}} + \frac{e_{ru}e_{su}}{e_{uu}} \\ e_{rs}^{(1)} &= e_{rs} - \frac{e_{ru}e_{su}}{e_{uu}} \end{aligned} \tag{14}$$

h_{rs} and e_{rs} are the usual one-way sums of squares and products of the responses defined by (3) of Sec. 5.3. If we write the sums of the concomitant observations under treatment j and for the entire experiment as

$$U_j = \sum_{i=1}^{N_j} u_{ij} \qquad G_u = \sum_{j=1}^{k} U_j \tag{15}$$

then

$$\begin{aligned} h_{ru} &= \sum_{j=1}^{k} \frac{1}{N_j} T_{jr}U_j - \frac{1}{N} G_r G_u \\ h_{uu} &= \sum_{j=1}^{k} \frac{1}{N_j} U_j^2 - \frac{1}{N} G_u^2 \\ e_{ru} &= \sum_{j=1}^{k}\sum_{i=1}^{N_j} x_{ijr}u_{ij} - \sum_{j=1}^{k} \frac{1}{N_j} T_{jr}U_j \end{aligned} \tag{16}$$

$$e_{uu} = \sum_{j=1}^{k} \sum_{i=1}^{N_j} u_{ij}^2 - \sum_{j=1}^{k} \frac{1}{N_j} U_j^2$$

The parameters for the greatest-root percentage points follow from (7) with $g = k - 1$.

The estimate of the mean $\mu_r + \tau_{jr}$ of the rth response under the jth treatment is the adjusted average

$$\bar{x}_{jr} - \hat{\beta}_r \bar{u}_j \tag{17}$$

where

$$\bar{x}_{jr} = \frac{1}{N_j} T_{jr} \qquad \bar{u}_j = \frac{1}{N_j} U_j \qquad \hat{\beta}_r = \frac{e_{ru}}{e_{uu}} \tag{18}$$

In the present case of a single covariate the test statistic for the null hypothesis (10) of zero regression coefficients can be expressed as

$$F = \frac{n+1}{m+1} \operatorname{tr} \mathbf{H}_2 \mathbf{E}^{-1}$$

$$= \frac{N-k-p}{p} \frac{\mathbf{E}'_{xu} \mathbf{E}_{xx}^{-1} \mathbf{E}_{xu}}{e_{uu} - \mathbf{E}'_{xu} \mathbf{E}_{xx}^{-1} \mathbf{E}_{xu}} \tag{19}$$

When (10) is true the statistic has the F distribution with degrees of freedom p and $N - k - p$. The expression (19) holds for any single-covariate experimental design with a nonsingular matrix $\mathbf{E}_{xx}$ and of course linear independence of the covariate and response observations. Its derivation follows from an application of the identity (9) of Sec. 2.11 and will be left as an exercise.

Example 5.3. We shall illustrate the structure and results of a one-way multivariate analysis of covariance with a biochemical application due to Smith *et al.* (1962). The data consisted of 13 biochemicals or characteristics of urine specimens of young men classified into four groups according to their degree of obesity or underweight. Most subjects contributed two or more specimens, and the total of $N = 45$ observation vectors was considered a random sample from a multinormal population with no provision for the effect of possible correlations within a subject's successive determinations. Two measures were selected as concomitant variables:

$$u_1 = \text{volume, ml}$$
$$u_2 = (\text{specific gravity} - 1) \times 10^3$$

The 11 response variates were pH, a modified creatinine coefficient, pigment creatinine, and the concentrations of phosphate, calcium, phosphorous, creatinine, chloride, boron, choline, and copper. The linear model for the ith observation on the response h in the jth weight group was

$$x_{ijh} = \mu_h + \tau_{jh} + \beta_{1h}u_{ij1} + \beta_{2h}u_{ij2} + e_{ijh}$$

The hypotheses

$$H_{01}: \begin{bmatrix} \tau_{11} \\ \cdot \\ \cdot \\ \cdot \\ \tau_{1,11} \end{bmatrix} = \cdots = \begin{bmatrix} \tau_{41} \\ \cdot \\ \cdot \\ \cdot \\ \tau_{4,11} \end{bmatrix}$$

$$H_{02}: \quad \mu_h = 0 \qquad \text{all } h$$

$$H_{03}: \begin{bmatrix} \beta_{11} \\ \cdot \\ \cdot \\ \cdot \\ \beta_{1,11} \end{bmatrix} = \begin{bmatrix} 0 \\ \cdot \\ \cdot \\ \cdot \\ 0 \end{bmatrix} \qquad H_{04}: \begin{bmatrix} \beta_{21} \\ \cdot \\ \cdot \\ \cdot \\ \beta_{2,11} \end{bmatrix} = \begin{bmatrix} 0 \\ \cdot \\ \cdot \\ \cdot \\ 0 \end{bmatrix}$$

were tested by Smith and his co-workers through a general computer program for the necessary **H** and **E** matrices and the greatest-characteristic-root statistic. The tests are summarized in Table 5.6. For each hypothesis $s = g$, the rank of its **C** matrix. The inclusion of the concomitant variables is vindicated in part by the rejection of the hypotheses H_{03}, H_{04} of no linear regression.

Table 5.6. Analysis of Covariance

Hypothesis	Parameters			$x_{0.05;\,s,m,n}$	θ_s	Action
	s	m	n			
H_{01}	3	3.5	13.5	0.56	0.85	Reject
H_{02}	1	4.5	13.5	0.45	0.92	Reject
H_{03}	1	4.5	13.5	0.45	0.53	Reject
H_{04}	1	4.5	13.5	0.45	0.73	Reject

5.5 MULTIPLE COMPARISONS IN THE MULTIVARIATE ANALYSIS OF VARIANCE

As in the univariate analysis of variance, rejection of some hypothesis on the parameters of the multivariate linear model does not indicate which treatments or treatment combinations are different and which could be considered as coming from common populations. In the multivariate model we would also like to determine which responses or response-treatment combinations may have led to the rejection: our multiple comparisons must be made among the columns of the parameter matrix as well as among the rows. For such purposes the Roy union-intersection approach to testing is distinctly superior, for the nature of its construction leads directly to simultaneous confidence bounds on all double linear compounds of the elements of the matrix $\mathbf{C\xi M}$.

It will be helpful in the presentation of those confidence intervals to recall the forms and dimensions of the matrices in the linear model and hypotheses of Sec. 5.2. The $N \times q$ design matrix was partitioned as $\mathbf{A} = [\mathbf{A}_1 \quad \mathbf{A}_2]$, where the $N \times r$ matrix $\mathbf{A}_1$ constituted a basis for $\mathbf{A}$. In conformance with the column dimensions of $\mathbf{A}_1$ and $\mathbf{A}_2$ the parameter matrix was partitioned as

$$\xi = \begin{bmatrix} \xi_1 \\ \xi_2 \end{bmatrix} \tag{1}$$

where the dimensions of ξ_1 and ξ_2 are of course $r \times p$ and $(q - r) \times p$. Similarly, the $g \times q$ hypothesis matrix $\mathbf{C}$ was partitioned into $g \times r$ and $g \times (q - r)$ submatrices as $[\mathbf{C}_1 \quad \mathbf{C}_2]$, so that the null and alternative hypotheses can be written as

$$H_0\colon \quad [\mathbf{C}_1 \quad \mathbf{C}_2] \begin{bmatrix} \xi_1 \\ \xi_2 \end{bmatrix} \mathbf{M} = \mathbf{0} \qquad H_1\colon \quad [\mathbf{C}_1 \quad \mathbf{C}_2] \begin{bmatrix} \xi_1 \\ \xi_2 \end{bmatrix} \mathbf{M} \neq \mathbf{0} \tag{2}$$

$\mathbf{M}$ is some $p \times u$ matrix of rank $u \leq p$, and the null matrices have dimensions $g \times u$. Now let $\mathbf{a}$ and $\mathbf{b}$ be any nonnull vectors of respective dimensions $u \times 1$ and $g \times 1$. Roy and Bose (1953) and Roy (1957) have shown that the $100(1 - \alpha)$ percent simultaneous confidence bounds on all functions $\mathbf{b'C\xi Ma} = \mathbf{b'}(\mathbf{C}_1\xi_1 + \mathbf{C}_2\xi_2)\mathbf{Ma}$ of the parameters of the multivariate general linear model are given by

$$\begin{aligned} \mathbf{b'C}_1(\mathbf{A}_1'\mathbf{A}_1)^{-1}\mathbf{A}_1'\mathbf{XMa} - \sqrt{\frac{x_\alpha}{1 - x_\alpha}\,\mathbf{a'M'EMab'C}_1(\mathbf{A}_1'\mathbf{A}_1)^{-1}\mathbf{C}_1'\mathbf{b}} \\ \leq \mathbf{b'}(\mathbf{C}_1\xi_1 + \mathbf{C}_2\xi_2)\mathbf{Ma} \\ \leq \mathbf{b'C}_1(\mathbf{A}_1'\mathbf{A}_1)^{-1}\mathbf{A}_1'\mathbf{XMa} + \sqrt{\frac{x_\alpha}{1 - x_\alpha}\,\mathbf{a'M'EMab'C}_1(\mathbf{A}_1'\mathbf{A}_1)^{-1}\mathbf{C}_1\mathbf{b}} \end{aligned} \tag{3}$$

where $x_\alpha \equiv x_{\alpha;s,m,n}$ is the 100α percentage point from the Heck charts or Pillai tables. If $s = 1$, $x_\alpha/(1 - x_\alpha) = [(m + 1)/(n + 1)]F_{\alpha;2m+2,2n+2}$. The term $\mathbf{b'C}_1(\mathbf{A}_1'\mathbf{A}_1)^{-1}\mathbf{A}_1'\mathbf{XMa}$ is the unbiased minimum-variance estimate of the parametric function enclosed by the bounds.

These general bounds will now be evaluated for a few common experimental plans.

The One-way Analysis of Variance. For the usual system of p responses, k treatments, and N_j observations in the jth treatment group, assume that the design matrix has the form (5) of Sec. 5.2, where $\mathbf{A}_1$ consists of the first k columns. $\mathbf{M}$ will merely be the $p \times p$ identity matrix. Recall that $(\mathbf{A}_1'\mathbf{A}_1)^{-1}$ is the diagonal matrix with jth element $1/N_j$. Let

$$\text{(4)} \qquad \mathbf{C}_1 = \begin{bmatrix} 1 & -1 & 0 & \cdots & 0 & 0 \\ 0 & 1 & -1 & \cdots & 0 & 0 \\ \cdot & \cdot & \cdot & \cdots & \cdot & \cdot \\ 0 & 0 & 0 & \cdots & 1 & -1 \end{bmatrix}$$

be the $(k - 1) \times k$ successive difference matrix, so that

$$\text{(5)} \qquad \mathbf{c}' = [c_1, \ldots, c_k] = \mathbf{b}'\mathbf{C}_1$$

has the coefficients of a treatment-effects contrast. If $\bar{x}_{jh}$ denotes the mean of the hth response under treatment j, the $100(1 - \alpha)$ percent simultaneous confidence intervals on all linear functions of the treatment contrasts are

$$\text{(6)} \qquad \sum_{h=1}^{p} \sum_{j=1}^{k} a_h c_j \bar{x}_{jh} - \sqrt{\frac{x_\alpha}{1 - x_\alpha} \mathbf{a}'\mathbf{E}\mathbf{a} \left(\sum_{j=1}^{k} \frac{c_j^2}{N_j} \right)} \le \sum_{h=1}^{p} \sum_{j=1}^{k} a_h c_j \tau_{jh} \le \sum_{h=1}^{p} \sum_{j=1}^{k} a_h c_j \bar{x}_{jh} + \sqrt{\frac{x_\alpha}{1 - x_\alpha} \mathbf{a}'\mathbf{E}\mathbf{a} \left(\sum_{j=1}^{k} \frac{c_j^2}{N_j} \right)}$$

The parameters of the critical value x_α are $s = \min(k - 1, p)$, $m = (|k - p - 1| - 1)/2$, and $n = (N - k - p - 1)/2$. In particular, if $\mathbf{b}$ is the vector with one in the jth position and zeros elsewhere, we would have the interval for the linear compound

$$\text{(7)} \qquad \sum_{h=1}^{p} a_h(\tau_{jh} - \tau_{j+1,h})$$

of the differences of the effects of treatments j and $j + 1$. If the general confidence interval contains zero, we would conclude that

$$H_0: \sum_{h=1}^{p} \sum_{j=1}^{k} a_h c_j \tau_{jh} = 0$$

should be accepted at the α level in the simultaneous testing sense.

Example 5.4. The individual species differences of Example 5.1 can now be tested through their 99 percent simultaneous confidence intervals. The $\alpha = 0.01$ critical value is approximately

$$x_{0.01;\,2,\frac{1}{2},71} = 0.11$$

and

$$\sqrt{\frac{x_{0.01}}{1 - x_{0.01}} \left(\frac{1}{N_i} + \frac{1}{N_j} \right)} = 0.070$$

The common components of the intervals can be summarized in this form:

Dimension	$\sqrt{e_{ii}}$	$0.070\sqrt{e_{ii}}$
Sepal length	6.241	0.437
Sepal width	4.118	0.288
Petal length	5.217	0.365
Petal width	2.482	0.174

These intervals were obtained for the sepal width species differences:

Virginica and *versicolor* $\quad -0.08 \leq \tau_{12} - \tau_{22} \leq 0.49$

Setosa and *versicolor* $\quad 0.37 \leq \tau_{32} - \tau_{22} \leq 0.95$

Setosa and *virginica* $\quad 0.17 \leq \tau_{32} - \tau_{12} \leq 0.74$

The hypothesis of equal sepal widths can be accepted at the 0.01 level for the *virginica* and *versicolor* species. It is easily verified by inspection that the other confidence intervals on the dimension differences do not contain zero. Apart from the single sepal-width difference, it would appear that the three species are different on all dimensions.

The Two-way Analysis of Variance. We must begin by requiring that the interaction parameters η_{ij} in the two-way linear model vanish, for otherwise comparisons among the row and column treatments are meaningless. The design matrix then consists of the first $r + c$ columns and the last column of the original matrix (16) of Sec. 5.3. The new matrix has rank $r + c - 1$, and its basis $\mathbf{A}_1$ can be formed from the first $r + c - 1$ columns, if a test on row effects is contemplated, or from the last $r + c - 1$ columns for the column-treatments hypothesis. In either case the matrix $\mathbf{A}_1'\mathbf{A}_1$ is obtained, although other intermediate matrix products have more convenient forms if the different bases are employed. Unlike the one-way analysis, $\mathbf{A}_1'\mathbf{A}_1$ is not a diagonal matrix, although it can easily be inverted by the results of Roy and Sarhan (1956) for patterned matrices. In that manner $\mathbf{H}$, $\mathbf{E}$, and $\mathbf{C}_1(\mathbf{A}_1'\mathbf{A}_1)^{-1}\mathbf{A}_1'\mathbf{X}$ can be computed, and sets of simultaneous confidence intervals constructed for contrasts of the row or column effects. The $100(1 - \alpha)$ percent simultaneous intervals for the linear compound of the general row-effects contrast are given by

$$(8) \quad \sum_{h=1}^{p} \sum_{i=1}^{r} a_h c_i \bar{x}_{i.h} - \sqrt{\frac{x_\alpha}{cn(1 - x_\alpha)} \left(\sum_{i=1}^{r} c_i^2\right) \mathbf{a}'\mathbf{E}\mathbf{a}} \leq \sum_{h=1}^{p} \sum_{i=1}^{r} a_h c_i \alpha_{ih}$$

$$\leq \sum_{h=1}^{p} \sum_{i=1}^{r} a_h c_i \bar{x}_{i.h} + \sqrt{\frac{x_\alpha}{cn(1 - x_\alpha)} \left(\sum_{i=1}^{r} c_i^2 \right) \mathbf{a'Ea}}$$

where $\mathbf{a}' = [a_1, \ldots, a_p]$ is the usual nonnull vector, $\mathbf{c}' = [c_1, \ldots, c_r]$ is the vector of contrast coefficients, $\bar{x}_{i.h}$ is the hth response mean from row treatment i, and $\mathbf{E}$ is the error matrix whose elements were described in Table 5.3. The parameters of the critical value x_α are the same as those given in Table 5.4, for we have elected not to pool the interaction-hypothesis matrix with $\mathbf{E}$. Similarly, the simultaneous confidence intervals for compounds of the column-effects contrasts are

$$(9) \quad \sum_{h=1}^{p} \sum_{j=1}^{c} a_h c_j \bar{x}_{.jh} - \sqrt{\frac{x_\alpha}{rn(1 - x_\alpha)} \left(\sum_{j=1}^{c} c_j^2 \right) \mathbf{a'Ea}}$$

$$\leq \sum_{h=1}^{p} \sum_{j=1}^{c} a_h c_j \tau_{jh}$$

$$\leq \sum_{h=1}^{p} \sum_{j=1}^{c} a_h c_j \bar{x}_{.jh} + \sqrt{\frac{x_\alpha}{rn(1 - x_\alpha)} \left(\sum_{j=1}^{c} c_j^2 \right) \mathbf{a'Ea}}$$

where $\bar{x}_{.jh}$ is the mean of the hth response for column treatment j. The parameters of x_α are given in Table 5.4.

Example 5.5. The error matrix for the drug trial data of Example 5.2 was

$$\mathbf{E} = \begin{bmatrix} 94.5 & 76.5 \\ 76.5 & 114.0 \end{bmatrix}$$

The 1 percent critical value is $x_{0.01;\,2,-\frac{1}{2},7.5} = 0.50$. We set $\mathbf{a}' = [1,0]$ to construct intervals for the first week's drug differences. The 99 percent simultaneous confidence intervals for the $B - A$, $C - B$, and $C - A$ differences are

$$-4.36 \leq \tau_{B1} - \tau_{A1} \leq 5.36$$

$$2.39 \leq \tau_{C1} - \tau_{B1} \leq 12.11$$

$$2.89 \leq \tau_{C1} - \tau_{A1} \leq 12.61$$

The conventional column subscripts have been replaced by the drug labels. The first interval contains the zero difference, and we can conclude at the 0.01 level that the drugs A and B are not different in the multiple-comparison sense with respect to their effects on weight during the first week of the trial.

It is easily verified that the 99 percent simultaneous confidence intervals on the drug differences of the second response all contain zero. It would appear that the significance of the original analysis of variance was largely due to the effect of drug C in the first week.

Multivariate Analysis of Covariance. Confidence intervals for linear compounds of treatment contrasts in the analysis of covariance can be obtained by evaluating expression (3) for the various partitioned models and matrices of Sec. 5.4. The general $100(1 - \alpha)$ percent simultaneous confidence interval on the estimable function $\mathbf{b}'\mathbf{C}\boldsymbol{\xi}\mathbf{M}\mathbf{a} = \mathbf{b}'\mathbf{C}_1\boldsymbol{\xi}_1\mathbf{M}\mathbf{a} + \mathbf{b}'\mathbf{C}_2\boldsymbol{\xi}_2\mathbf{M}\mathbf{a}$ of the treatment parameters can be reduced to

$$(10)\quad \mathbf{b}'\mathbf{C}_1(\mathbf{A}_1'\mathbf{A}_1)^{-1}\mathbf{A}_1'(\mathbf{X} - \mathbf{U}\hat{\mathbf{B}})\mathbf{M}\mathbf{a} - \sqrt{\frac{x_\alpha}{1 - x_\alpha}(\mathbf{a}'\mathbf{M}'\mathbf{E}\mathbf{M}\mathbf{a})[\mathbf{b}'\mathbf{C}_1(\mathbf{A}_1'\mathbf{A}_1)^{-1}\mathbf{C}_1'\mathbf{b} + \mathbf{b}'\mathbf{C}_1(\mathbf{A}_1'\mathbf{A}_1)^{-1}\mathbf{A}_1'\mathbf{U}\mathbf{E}_{uu}^{-1}\mathbf{U}'\mathbf{A}_1(\mathbf{A}_1'\mathbf{A}_1)^{-1}\mathbf{C}_1\mathbf{b}]}$$

$$\leq \mathbf{b}'\mathbf{C}\boldsymbol{\xi}\mathbf{M}\mathbf{a}$$

$$\leq \mathbf{b}'\mathbf{C}_1(\mathbf{A}_1'\mathbf{A}_1)^{-1}\mathbf{A}_1'(\mathbf{X} - \mathbf{U}\hat{\mathbf{B}})\mathbf{M}\mathbf{a} + \sqrt{\frac{x_\alpha}{1 - x_\alpha}(\mathbf{a}'\mathbf{M}'\mathbf{E}\mathbf{M}\mathbf{a})[\mathbf{b}'\mathbf{C}_1(\mathbf{A}_1'\mathbf{A}_1)^{-1}\mathbf{C}_1'\mathbf{b} + \mathbf{b}'\mathbf{C}_1(\mathbf{A}_1'\mathbf{A}_1)^{-1}\mathbf{A}_1'\mathbf{U}\mathbf{E}_{uu}^{-1}\mathbf{U}'\mathbf{A}_1(\mathbf{A}_1'\mathbf{A}_1)^{-1}\mathbf{C}_1\mathbf{b}]}$$

$\mathbf{b}$ and $\mathbf{a}$ are general nonnull $g \times 1$ and $u \times 1$ vectors, while $\mathbf{M}$ is the usual $p \times u$ compounding matrix. $\mathbf{C} = [\mathbf{C}_1 \quad \mathbf{C}_2]$ is a $g \times q$ hypothesis matrix whose submatrix $\mathbf{C}_1$ has dimensions $g \times r$, $\mathbf{A}_1$ is the $N \times r$ basis of the design matrix $\mathbf{A}$ of (2), Sec. 5.4, and $\boldsymbol{\xi} = [\boldsymbol{\xi}_1 \quad \boldsymbol{\xi}_2]$ is the $q \times p$ parameter matrix of design effects with $q \times r$ submatrix $\boldsymbol{\xi}_1$. $\mathbf{U}$ is the $N \times t$ concomitant observation matrix. The matrices $\mathbf{E}$, $\mathbf{E}_{uu}$, and $\hat{\mathbf{B}}$ are defined by expressions (4) to (6) and (9) of Sec. 5.4. x_α is the greatest-characteristic-root distribution critical value with parameters s, m, n given by (7) of Sec. 5.4.

For the special case of the one-way classification let $\boldsymbol{\tau}' = [\tau_1, \ldots, \tau_k]$ be the vector of the k treatment effects and $\mathbf{c}' = \mathbf{b}'\mathbf{C}_1 = [c_1, \ldots, c_k]$ be any contrast vector. $\bar{\mathbf{x}}_j$ is the $p \times 1$ vector of response means, $\bar{\mathbf{u}}_j$ is the $t \times 1$ concomitant mean vector for the jth treatment, and $\mathbf{u} = [\bar{\mathbf{u}}_1, \ldots, \bar{\mathbf{u}}_k]$. The general interval is given by

$$(11)\quad \sum_{j=1}^{k} c_j(\bar{\mathbf{x}}_j' - \bar{\mathbf{u}}_j'\hat{\mathbf{B}})\mathbf{M}\mathbf{a} - \sqrt{\frac{x_\alpha}{1 - x_\alpha}(\mathbf{a}'\mathbf{M}'\mathbf{E}\mathbf{M}\mathbf{a})\left[\sum_{j=1}^{k}\frac{c_j^2}{N_j} + \mathbf{c}'\bar{\mathbf{u}}'\mathbf{E}_{uu}^{-1}\bar{\mathbf{u}}\mathbf{c}\right]}$$

$$\leq \mathbf{c}'\boldsymbol{\tau}\mathbf{M}\mathbf{a}$$

$$\leq \sum_{j=1}^{k} c_j(\bar{\mathbf{x}}_j' - \bar{\mathbf{u}}_j'\hat{\mathbf{B}})\mathbf{M}\mathbf{a} + \sqrt{\frac{x_\alpha}{1 - x_\alpha}(\mathbf{a}'\mathbf{M}\mathbf{E}\mathbf{M}\mathbf{a})\left[\sum_{j=1}^{k}\frac{c_j^2}{N_j} + \mathbf{c}'\bar{\mathbf{u}}'\mathbf{E}_{uu}^{-1}\bar{\mathbf{u}}\mathbf{c}\right]}$$

where the parameters of x_α are $s = \min(k-1, u), m = (|k-1-u|-1)/2$, and $n = (N-k-t-u-1)/2$.

Example 5.6. We shall illustrate these simultaneous confidence intervals with an application to the biochemical data described in Example 5.3. For simplicity we shall use only the responses pigment creatinine (X_1) and chloride (X_2), and the single covariate volume (V). The mean values of the four weight groups are as follows:

Group	X_1	X_2	V
1. ($N_1 = 12$)	15.89	6.01	216.25
2. ($N_2 = 14$)	17.82	5.21	285.71
3. ($N_3 = 11$)	16.35	5.37	333.18
4. ($N_4 = 8$)	11.91	3.98	198.38

The hypotheses and error matrices are

$$\mathbf{H}_{xx} = \begin{bmatrix} 181.07 & 40.03 \\ 40.03 & 20.03 \end{bmatrix} \qquad \mathbf{E}_{xx} = \begin{bmatrix} 496.77 & 4.51 \\ 4.51 & 152.85 \end{bmatrix}$$

$$\mathbf{H}_{xu} = \begin{bmatrix} 9{,}625.70 \\ -1{,}342.71 \end{bmatrix} \qquad \mathbf{E}_{xu} = \begin{bmatrix} 6{,}604.48 \\ -1{,}654.32 \end{bmatrix}$$

$$h_{uu} = 121{,}038.4 \qquad e_{uu} = 395{,}050.6$$

The adjusted matrices are

$$\mathbf{H} = \begin{bmatrix} 137.66 & 37.41 \\ 37.41 & 23.47 \end{bmatrix} \qquad \mathbf{E} = \begin{bmatrix} 360.65 & 32.17 \\ 32.17 & 145.92 \end{bmatrix}$$

$$\mathbf{E}^{-1}\mathbf{H} = \begin{bmatrix} 0.36603 & 0.09118 \\ 0.17568 & 0.14074 \end{bmatrix}$$

The greatest characteristic root is $c_2 = 0.423$, and the statistic $\theta_2 = 0.297$ exceeds the 0.05 critical value $x_{0.05;2,0,18.5} = 0.285$. We can proceed to calculate 95 percent confidence intervals with assurance that *some* bilinear contrast function will have an interval which does not cover zero.

The estimated within-groups regression coefficients of creatinine and chloride on volume are $\hat{\beta}_{X_1,V} = 0.016718$ and $\hat{\beta}_{X_2,V} = -0.004188$, and the adjusted response means can be calculated from them to be the following:

Group	X_1	X_2
1	12.28	6.92
2	13.04	6.41
3	10.77	6.77
4	8.60	4.81

The largest group difference of the adjusted creatinine means occurs for groups 1 and 4: its confidence interval is

$$-1.17 \leq \tau_{11} - \tau_{41} \leq 8.53$$

and we may conclude that those groups do not differ significantly in creatinine values at the 0.05 level. Similarly, the other five paired comparisons of the creatinine parameters have intervals containing zero, and we must conclude that the hypothesis of equal creatinine levels cannot be rejected in the multiple testing sense. The largest chloride adjusted mean difference also occurs between groups 1 and 4; its 95 percent interval

$$-0.98 \leq \tau_{21} - \tau_{24} \leq 5.19$$

does not support a difference in chloride level for those groups. The other paired contrasts also do not support significant differences.

Although the individual chemicals are not different for pairs of groups, we know from the multivariate analysis of covariance that there is some linear compound of the response variables for which a group contrast will be significantly different from zero. The compound coefficient vector **a** is the characteristic vector corresponding to the greatest root of $\mathbf{E}^{-1}\mathbf{H}$, and the new variable is

$$Y = X_1 + 0.6228X_2$$

M is of course the identity matrix, and $\mathbf{a}'\mathbf{M}'\mathbf{E}\mathbf{M}\mathbf{a} = 457.13$. The adjusted means of the compound have these values:

Group	Adjusted mean
1	16.58
2	17.04
3	14.99
4	11.59

Again the six paired-group intervals each contain zero, so that we must look still further for a "significant" contrast. One is given by the comparison of group 4 with groups 1 and 2: the interval is

$$0.29 \leq \tfrac{1}{2}(\tau_{11} + \tau_{21}) - \tau_{41} + 0.6228(\tfrac{1}{2}\tau_{12} + \tfrac{1}{2}\tau_{22} - \tau_{42}) \leq 10.15$$

The same group contrast for the individual creatinine and chloride variables does not imply significant differences. Unless some biochemical interpretation can be ascribed to the derived Y variable one might be inclined to speculate whether the original conclusion of the analysis of covariance is an artefact of the data. Larger samples in the extreme weight groups would probably be required to resolve this question.

5.6 PROFILE ANALYSIS

Through the general results for the multivariate linear hypothesis we can now extend the profile analysis technique of Sec. 4.6 to several groups. Suppose that p commensurable responses have been collected from independent sampling units grouped according to k treatments or experimental conditions. The observations are arranged as in Table 5.7. Let the model for the ith observation on the hth response under treatment j be

$$x_{ijh} = \xi_{jh} + e_{ijh} \tag{1}$$

We note from the table that the subscripts take on the values

$$i = 1, \ldots, N_j \qquad j = 1, \ldots, k \qquad h = 1, \ldots, p$$

Table 5.7. **Profile Observations**

Treatment	Response			Unit total
	1	$\cdots$	p	
1	x_{111}	$\cdots$	x_{11p}	R_{11}
	$\cdots$	$\cdots$	$\cdots$	$\cdots$
	$x_{N_1 11}$	$\cdots$	$x_{N_1 1p}$	$R_{N_1 1}$
Total	T_{11}	$\cdots$	T_{1p}	C_1
$\cdots$	$\cdots$	$\cdots$	$\cdots$	$\cdots$
k	x_{1k1}	$\cdots$	x_{1kp}	R_{1k}
	$\cdots$	$\cdots$	$\cdots$	$\cdots$
	$x_{N_k k1}$	$\cdots$	$x_{N_k kp}$	$R_{N_k k}$
Total	T_{k1}	$\cdots$	T_{kp}	C_k
Grand total	G_1	$\cdots$	G_p	G

i = subjects (observations)
h = trials (responses)
j = groups (treatments)

For convenience we shall write $N = N_1 + \cdots + N_k$. The vector of residuals $\mathbf{e}'_{ij} = [e_{ij1}, \ldots, e_{ijp}]$ of the ijth sampling unit has the multinormal distribution with null mean vector and some unknown nonsingular covariance matrix $\mathbf{\Sigma}$. The residual variates of different units are independently distributed. In this parametrization the design matrix

$$\mathbf{A}_1 = \begin{bmatrix} 1 & 0 & \cdots & 0 \\ \cdot & \cdot & \cdots & \cdot \\ 1 & 0 & \cdots & 0 \\ 0 & 1 & \cdots & 0 \\ \cdot & \cdot & \cdots & \cdot \\ 0 & 1 & \cdots & 0 \\ \cdot & \cdot & \cdots & \cdot \\ 0 & 0 & \cdots & 1 \\ \cdot & \cdot & \cdots & \cdot \\ 0 & 0 & \cdots & 1 \end{bmatrix} \tag{2}$$

merely consists of ones in the ith column of the successive blocks of $N_1, \ldots, N_k$ rows and zeros elsewhere. The parameter matrix is

$$\boldsymbol{\xi} = \begin{bmatrix} \xi_{11} & \cdots & \xi_{1p} \\ \cdot\ \cdot & \cdots & \cdot\ \cdot \\ \xi_{k1} & \cdots & \xi_{kp} \end{bmatrix} \tag{3}$$

Under those assumptions the general results of Sec. 5.2 will provide tests of the three profile hypotheses

$$H_{0a}: \quad \sum_{h=1}^{p} \xi_{1h} = \cdots = \sum_{h=1}^{p} \xi_{kh} \tag{4}$$

$$H_{0b}: \quad \sum_{j=1}^{k} \xi_{j1} = \cdots = \sum_{j=1}^{k} \xi_{jp} \tag{5}$$

and

$$H_{0c}: \quad \begin{bmatrix} \xi_{11} - \xi_{12} \\ \cdots \\ \xi_{1,p-1} - \xi_{1p} \end{bmatrix} = \cdots = \begin{bmatrix} \xi_{k1} - \xi_{k2} \\ \cdots \\ \xi_{k,p-1} - \xi_{kp} \end{bmatrix} \tag{6}$$

of equal treatment levels, equal response means, and parallelism of the treatment mean profiles. Let us begin with the parallelism hypothesis, for its truth is essential for the meaningfulness of the other two hypotheses. The matrix statement of (6) is H_{0c}: $\mathbf{C}_1\boldsymbol{\xi}\mathbf{M}_1 = \mathbf{0}$, where

$$\mathbf{C}_1 = \begin{bmatrix} 1 & -1 & 0 & \cdots & 0 \\ 0 & 1 & -1 & \cdots & 0 \\ \cdot & \cdot & \cdot & \cdots & \cdot \\ 0 & 0 & 0 & \cdots & -1 \end{bmatrix} \qquad \mathbf{M}_1 = \begin{bmatrix} 1 & 0 & \cdots & 0 \\ -1 & 1 & \cdots & 0 \\ 0 & -1 & \cdots & 0 \\ \cdot & \cdot & \cdots & \cdot \\ 0 & 0 & \cdots & -1 \end{bmatrix} \tag{7}$$

are the $(k-1) \times k$ and $p \times (p-1)$ cases of the transformation matrix and its transpose originally defined in Sec. 4.5. The test of H_{0c} amounts to a one-way multivariate analysis of variance on the $p-1$ differences of the observations of the adjacent responses from each sampling unit. It can be carried out in that fashion by postmultiplying the data matrix by $\mathbf{M}_1$ and computing the matrices $\mathbf{H}$ and $\mathbf{E}$ defined by (3) of Sec. 5.3. If a general multivariate linear hypothesis program is available, the basic expressions of (34), Sec. 5.2, can be used. In either case the parameters of the distribution of the greatest characteristic root of $\mathbf{HE}^{-1}$ when H_{0c} is true are

$$(8) \qquad s = \min(k-1, p-1) \qquad m = \frac{|k-p|-1}{2} \qquad n = \frac{N-k-p}{2}$$

and by the usual decision rule the parallelism hypothesis would be rejected for large values of the test statistic.

Next let us assume that the hypothesis of no treatment by response interaction was tenable. The hypothesis (4) of equal treatment effects can be expressed as H_{0a}: $\mathbf{C}_1\boldsymbol{\xi}\mathbf{m}_2 = 0$, where $\mathbf{C}_1$ is the first matrix of (7) and

$$(9) \qquad \mathbf{m}_2' = [1, \ldots, 1]$$

contains p ones. Thus $s = 1$, and the test of equal treatment effects, or identical profile heights, is carried out by a one-way univariate analysis of variance on the sums of the responses of each sampling unit across the k treatment groups. Although the matrices $\mathbf{H}$ and $\mathbf{E}$ could be computed for $\mathbf{m}_2$ as part of a general computer program and the single nonzero root referred to tables of the appropriate beta distribution, it is usually more efficient to make the transformation

$$(10) \qquad \mathbf{R} = \mathbf{X}\mathbf{m}_2$$

to the row sums of Table 5.7 at the outset and carry out the test as an analysis of variance on the response totals. Recall that

$$(11) \qquad R_{ij} = \sum_{h=1}^{p} x_{ijh}$$

is the total for the ith unit in the jth treatment group. Then, as in Sec. 1.6, the test statistic is

$$(12) \qquad F = \frac{N-k}{k-1} \frac{\text{SST}}{\text{SSE}}$$

where

$$\text{(13)} \qquad \begin{aligned} \text{SST} &= \sum_{j=1}^{k} \frac{1}{N_j} C_j^2 - \frac{1}{N}\left(\sum_{j=1}^{k} C_j\right)^2 \\ \text{SSE} &= \sum_{j=1}^{k} \sum_{i=1}^{N_j} R_{ij}^2 - \sum_{j=1}^{k} \frac{1}{N_j} C_j^2 \end{aligned}$$

As in Table 5.7, $C_j = \sum_{i=1}^{N_j} R_{ij}$. H_{0a} is accepted at the α level if $F \leq F_{\alpha;k-1,N-k}$.

Finally, we shall present the test for the hypothesis H_{0b} of equal response effects. As in the preceding test it will be assumed that the parallelism, or interaction, hypothesis is tenable. Then the hypothesis can be stated in matrix form as H_{0b}: $\mathbf{c}_2'\boldsymbol{\xi}\mathbf{M}_1 = \mathbf{0}$, where

$$\text{(14)} \qquad \mathbf{c}_2' = [1, \ . \ . \ . \ , 1]$$

contains k ones, and $\mathbf{M}_1$ is the $p \times (p - 1)$ matrix defined in (7). Again $s = 1$, and the test statistic can be found from the single nonzero root of $\mathbf{HE}^{-1}$ to be

$$\text{(15)} \qquad T^2 = N\bar{\mathbf{x}}'\mathbf{M}_1(\mathbf{M}_1'\mathbf{S}\mathbf{M}_1)^{-1}\mathbf{M}_1'\bar{\mathbf{x}}$$

$\bar{\mathbf{x}}$ is the grand mean vector with hth element $\bar{x}_h = G_h/N$, and

$$\mathbf{S} = \frac{1}{N - k}\,\mathbf{E}$$

is the usual $p \times p$ within-treatments sample covariance matrix. When H_{0b} is true,

$$\text{(16)} \qquad F = \frac{N - k - p + 2}{(N - k)(p - 1)}\,T^2$$

has the F distribution with degrees of freedom $p - 1$ and $N - k - p + 2$, and we would accept the hypothesis of equal response effects at the α level if $F \leq F_{\alpha;p-1,N-k-p+2}$.

If the hypothesis of parallel-treatment population profiles cannot be accepted, it will be necessary to test the equality of the treatment effects separately for each response by p univariate analyses of variance. Similarly, the hypothesis of equal response effects would be tested within each treatment group through the single-sample repeated-measurements T^2 statistic described in Sec. 4.5.

The reader should note that the mathematical model for the preceding tests required only of the covariance matrix $\boldsymbol{\Sigma}$ that it be of full rank p: no stipulation of equal residual variances was necessary. It is implicit in most profile studies, however, that the units of the responses are at least commensurable, if not, as in the case of psychological tests, actually scaled to equal population variances.

The simultaneous confidence intervals of Sec. 5.5 can be used to determine which treatments may have contributed to rejection of the parallelism hypothesis. Similarly, the Scheffé intervals of Sec. 1.6 can be used to sort out differences among the treatment levels. The single-sample T^2 intervals of Sec. 4.3 can be modified for the degrees of freedom for a covariance matrix computed from k samples to permit multiple comparisons among the response means.

Example 5.7. We shall now illustrate the preceding methods with some data contrived to fit a familiar situation in the behavioral sciences. Suppose that three scales A, B, C of an inventory measuring certain maternal attitudes were administered to mothers participating in a study of child development. As part of the investigation each mother had been assigned to one of four socioeconomic status classes. The profiles for the $N = 21$ subjects had the values in Table 5.8. The class mean profiles are shown in Fig. 5.1.

The analysis is begun with the parallelism test, and for that step the observations are transformed to the differences of scales A and B and B and C. The total, hypothesis, and error sums of squares and products matrices are

$$\mathbf{T} = \begin{bmatrix} 75.81 & -54.48 \\ -54.48 & 85.81 \end{bmatrix} \qquad \mathbf{H} = \begin{bmatrix} 24.61 & -20.48 \\ -20.48 & 24.31 \end{bmatrix} \qquad \mathbf{E} = \begin{bmatrix} 51.20 & -34.00 \\ -34.00 & 61.50 \end{bmatrix}$$

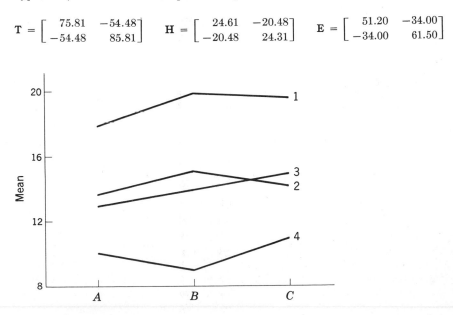

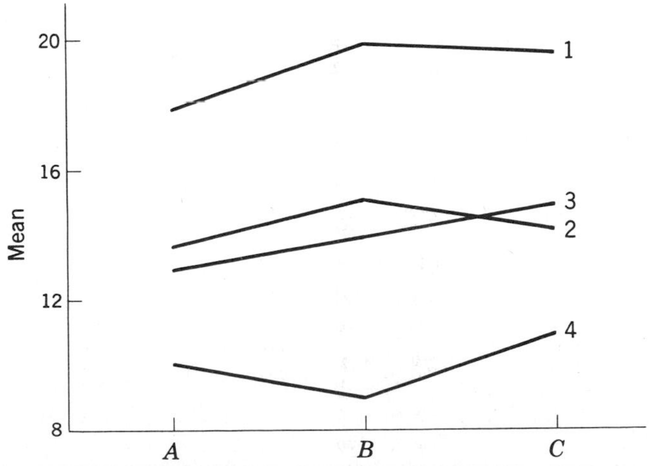

figure 5.1 Maternal attitude profiles.

Table 5.8.   **Profile Observations** $\sum\sum\sum \alpha_{ijh}^2 = 15709$

Socioeconomic class	Scale			Subject total
	A	B	C	
1	19	20	18	57
	20	21	19	60
	19	22	22	63
	18	19	21	58
	16	18	20	54
	17	22	19	58
	20	19	20	59
	15	19	19	53
T_{1h}	144	160	158	462
2	12	14	12	38
	15	15	17	47
	15	17	15	47
	13	14	14	41
	14	16	13	43
T_{2h}	69	76	71	216
3	15	14	17	46
	13	14	15	42
	12	15	15	42
	12	13	13	38
T_{3h}	52	56	60	168
4	8	9	10	27
	10	10	12	32
	11	10	10	31
	11	7	12	30
T_{4h}	40	36	44	120
G_h	305	328	333	966

and

$$\mathbf{E}^{-1} = \begin{bmatrix} 0.03086 & 0.01706 \\ 0.01706 & 0.02570 \end{bmatrix} \qquad \mathbf{HE}^{-1} = \begin{bmatrix} 0.4101 & -0.1064 \\ -0.2172 & 0.2754 \end{bmatrix}$$

The characteristic equation of $\mathbf{HE}^{-1}$ is

$$\lambda^2 - 0.6855\lambda + 0.0899 = 0$$

and its larger root can be found from the quadratic formula to be $c_2 = 0.509$. The

$$\theta_s = F \cdot \frac{C_s}{1+C_s} = \frac{.509}{1.509} = .337$$

p178

statistic for the percentage-point chart is $\theta_2 = 0.337$. The requisite parameters are $s = 2$, $m = 0$, and $n = 7$. If the $\alpha = 0.05$ level is chosen, the critical value for θ_2 can be found from the second lowest curve of the first chart to be approximately $x_{0.05;2,0,7} = 0.47$. Since the sample value does not exceed that number, we *accept* at the 5 percent level the hypothesis of parallel scale profiles in the four socioeconomic populations.

The test for equal class effects led to this analysis of variance on the subject totals:

Source	Sum of squares	Degrees of freedom	Mean square	F
Classes	2,231.7	3	743.90	70.9
Within classes	178.3	17	10.49	
Total	2,410.0	20		

Since the 5 percent critical value of the F ratio with 3 and 17 degrees of freedom is 3.20, the hypothesis could not be accepted at that level or even at any of the customary smaller values of α.

The statistic (15) for testing the equality of the scale means can be computed from the grand mean vector $\bar{\mathbf{x}}'\mathbf{M}_1 = [-1.095, -0.238]$ and the inverse of the matrix $\mathbf{E}$ used in the parallelism test as

$$T^2 = (17)(21)[-1.095, -0.238] \begin{bmatrix} 0.03086 & 0.01706 \\ 0.01706 & 0.02570 \end{bmatrix} \begin{bmatrix} -1.095 \\ -0.238 \end{bmatrix}$$
$$= 16.91$$

The associated F statistic is equal to 7.96, and since $F_{0.05;2,16} = 3.63$, the hypothesis of equal scale effects would be rejected.

Finally, we shall use the appropriate simultaneous confidence intervals to make some of the more important multiple comparisons among classes and scales. In the first case it will be a bit more convenient to work with the class means of the totals for the three scales:

	Class			
	1	2	3	4
N_j	8	5	4	4
Mean	57.75	43.20	42.00	30.00

The sample variance of the observation totals is the error mean square of the preceding analysis-of-variance table; its square root is 3.237. The critical value first introduced in the discussion of the Scheffé multiple-comparison procedure in Sec. 1.6 is

$$\sqrt{(k-1)F_{0.05;k-1,N-k}} = 3.10$$

The test statistics for the comparisons of the first and second, second and third, and third and fourth class means are

$$t = \frac{14.55}{3.237\sqrt{1/8 + 1/5}} = 7.9$$

$$t = \frac{1.2}{3.237\sqrt{1/5 + 1/4}} = 0.6$$

$$t = \frac{12}{3.237\sqrt{1/2}} = 5.2$$

We may conclude that at the 5 percent level the first socioeconomic class is different from the remaining three, the second and third classes do not appear to be different, and the fourth class is different from the others.

The 5 percent critical value for simultaneous tests on the scale effects is

$$\sqrt{\frac{(p-1)(N-k)}{N-k-p+2} F_{0.05;\,p-1,N-k-p+2}} = 2.78$$

and the differences among the three scale grand means divided by their respective standard errors are

$$t_{A-B} = -2.89 \qquad t_{B-C} = -0.57 \qquad t_{A-C} = -3.77$$

We conclude that scale A is significantly different in the multiple-comparison sense from B and C and B and C cannot be declared significantly different at any conventional α level.

The Analysis of Variance for Profile Data. In earlier treatments of profile analysis by some workers the three hypotheses associated with groups of repeated measurements were tested by a two-way mixed-model analysis of variance. As in the case of a single group, this approach is strictly valid only if the residual terms within a sampling unit have the same variance and a common correlation for all pairs. Since the residuals must be multinormally distributed for testing purposes, this requirement includes as a special case the conventional model with independent residuals. While such symmetry would rarely be encountered in the real world, nevertheless the analysis-of-variance approach has a strong appeal: its structure and interpretation are familiar to anyone aware of experimental design, its computations do not require matrix inversion and the extraction of a characteristic root, and, unlike the multivariate method, its test statistics are still defined if the number of responses exceeds the within-groups degrees of freedom. For these reasons Geisser and Greenhouse (1958, 1959) investigated the profile analysis of variance under the assumption of multinormally distributed residuals with a general covariance matrix and proposed (1) an approximate analysis of variance and (2) "conservative" tests for response differences and the parallelism hypothesis whose true level α cannot exceed some specified value.

Table 5.9. **Profile Analysis of Variance**

Source	Sum of squares	Degrees of freedom	Mean square	F	Conservative degrees of freedom
Responses Trials	S_1	$p - 1$	$\dfrac{S_1}{p-1}$	$(N-k)\dfrac{S_1}{S_5}$	$1, N - k$
Treatments GROUPS	S_2	$k - 1$	$\dfrac{S_2}{k-1}$	$\dfrac{(N-k)S_2}{(k-1)S_3}$	
Subjects (within treatments)	S_3	$N - k$	$\dfrac{S_3}{N-k}$		
Responses × treatments	S_4	$(p-1)(k-1)$	$\dfrac{S_4}{(p-1)(k-1)}$	$\dfrac{(N-k)S_4}{(k-1)S_5}$	$k - 1, N - k$
Subjects × responses (within treatments)	S_5	$(p-1)(N-k)$	$\dfrac{S_5}{(p-1)(N-k)}$		
Total	S_6	$Np - 1$			

The tests are based upon certain approximate distributions of the sums of squares under an arbitrary multinormal model. We shall begin our discussion of the method by setting out the analysis of variance in Table 5.9 for the data of Table 5.7. The six sums of squares are computed from these formulas:

p = # of trials (responses)

(17)

$$S_1 = \frac{1}{N}\sum_{h=1}^{p} G_h{}^2 - \frac{G^2}{Np}$$

$$S_2 = \frac{1}{p}\sum_{j=1}^{k} \frac{1}{N_j} C_j{}^2 - \frac{G^2}{Np}$$

$$S_3 = \frac{1}{p}\sum_{j=1}^{k}\sum_{i=1}^{N_j} R_{ij}{}^2 - \frac{1}{p}\sum_{j=1}^{k}\frac{1}{N_j} C_j{}^2$$

$$S_4 = \sum_{j=1}^{k}\sum_{h=1}^{p}\frac{1}{N_j} T_{jh}{}^2 - \frac{G^2}{Np} - S_1 - S_2$$

$$S_5 = S_6 - S_1 - S_2 - S_3 - S_4$$

$$S_6 = \sum_{h=1}^{p}\sum_{j=1}^{k}\sum_{i=1}^{N_j} x_{ijh}^2 - \frac{G^2}{Np}$$

Greenhouse and Geisser have shown that when the hypothesis of equal response effects is true, the test statistic

$$F = (N - k)\frac{S_1}{S_5} \tag{18}$$

is approximately distributed as an F variate with degrees of freedom $(p - 1)\epsilon$ and $(p - 1)(N - k)\epsilon$, and when the parallelism hypothesis is true,

$$F = \frac{(N - k)S_4}{(k - 1)S_5} \tag{19}$$

is approximately distributed as an F variate with degrees of freedom $(p - 1)(k - 1)\epsilon$ and $(p - 1)(N - k)\epsilon$. The scale factor

$$\epsilon = \frac{p^2(\bar{\sigma}_d - \bar{\sigma}_{..})^2}{(p - 1)(\Sigma\Sigma\sigma_{rs}{}^2 - 2p\Sigma\bar{\sigma}_{r.}{}^2 + p^2\bar{\sigma}_{..}{}^2)} \tag{20}$$

is computed from the elements σ_{rs} of the population covariance matrix, where $\bar{\sigma}_d$ is the mean of the p diagonal terms, $\bar{\sigma}_{..}$ is the grand mean of all variances and covariances, and $\bar{\sigma}_{r.}$ is the mean of the elements in the rth row.

The effect of permitting a general covariance structure is the reduction of the degrees of freedom for the first and third F tests by the factor ϵ. However, the elements of $\mathbf{\Sigma}$ needed for the calculation of ϵ are never known in practice, and the estimation of ϵ from the sample covariance matrix would introduce still another uncertainty into the approximate analysis of variance. Greenhouse and Geisser have shown that ϵ must satisfy

$$\epsilon > \frac{1}{p - 1} \tag{21}$$

for any covariance matrix, and therefore the degrees of freedom for the approximate tests cannot be less than 1 and $N - k$, and $k - 1$ and $N - k$, respectively. Such tests are called *conservative* because they are based upon the maximum reductions in degrees of freedom. Extensive simulation studies of the true Type I error probabilities of the conservative and ϵ-adjusted tests have been carried out by Collier *et al.* (1967).

Huynh and Feldt (1970) have found a necessary and sufficient condition on the covariance structure of the responses for the statistics (18) and (19) to have exact F distributions. For that condition we begin by permitting the k multinormal populations to have positive definite covariance matrices $\mathbf{\Sigma}_1, \ldots, \mathbf{\Sigma}_k$ rather than a single common matrix. The parallelism and equal response means tests are exact if and only if the elements of each $\mathbf{\Sigma}_h$ satisfy

$$(22) \qquad \sigma_{ijh} = \begin{cases} \bar{\sigma}_{i.h} + \bar{\sigma}_{j.h} - \bar{\sigma}_{..h} + \lambda(1 - 1/k) & i = j \\ \bar{\sigma}_{i.h} + \bar{\sigma}_{j.h} - \bar{\sigma}_{..h} - \lambda/k & i \neq j \end{cases}$$

where λ must be the same for all h. This covariance structure is exceedingly more general than that of a common equal-variance, equal-covariance matrix for all treatments. Examples of such matrices and a hypothesis test for the validity of their pattern have been given by Huynh and Feldt.

Finally, we note that the treatments F ratio is identical with expression (12) of the original multivariate analysis; its validity would of course require that the variances $\mathbf{j}'\mathbf{\Sigma}_1\mathbf{j}, \ldots, \mathbf{j}'\mathbf{\Sigma}_k\mathbf{j}$ be equal, where $\mathbf{j}$ is the usual column vector of ones.

Example 5.8. The analysis-of-variance approach will now be applied to the profiles of Example 5.7. The total sum of squares for all responses is 15,708, while the correction term is $G^2/Np = 14{,}812$. The analysis of variance is given by Table 5.10.

The hypothesis of equal scale effects would be rejected at the $\alpha = 0.01$ level under the conventional analysis of variance and at the $\alpha = 0.025$ level by the conservative-test criterion. The hypothesis of no scales-by-classes interaction would be rejected under the conventional analysis if α had been set at the rather large value of 0.10, while the conservative test would not reject the hypothesis at that level.

Table 5.10. **Analysis of Variance of Profile Data**

Source	Sum of squares	Degrees of freedom	Mean square	F	Conservative degrees of freedom
Scales	21.24	2	10.62	6.88	1, 17
Classes	743.90	3	247.97	70.93	
Subjects (within classes)	59.43	17	3.50		
Scales × classes	18.96	6	3.16	2.05	3, 17
Subjects × scales (within classes)	52.47	34	1.54		
Total	896.00	62			

Further examples of the Greenhouse-Geisser procedure may be found in the text of Winer (1971). The technique has also been extended to the case of profiles obtained at several times on the same subjects by McHugh, Sivanich, and Geisser (1961). Cole and Grizzle (1966) have illustrated and compared the multivariate and univariate profile methods with analyses of blood histamine values at different times and drug combinations; in their treatment the parallelism hypothesis was broken down into single-degrees-of-freedom contrasts measuring certain interactions of treatments, times, and profile shape. Bock (1963) has pointed out that the treatments and responses-by-treatments interaction sums of squares are not independently distributed for a general covariance structure, nor are the subjects (within treatments) and subjects by responses (within treatments) sums of squares. He has suggested an alternative multivariate analysis which leads to independent inferences.

5.7 CURVE FITTING FOR REPEATED MEASUREMENTS

When the data of a repeated-measurements experiment have been collected at known times, a natural sequel to hypothesis testing might be the fitting of a simple polynomial function to the sample means by some form of least squares. In this way substantive models for the responses might be tested, or the means summarized more concisely by a few coefficients as opposed to the entire vector. When the number of responses is very large, such a reduction to a linear or quadratic function can be useful for later multivariate analyses, e.g., the comparison of the coefficients of k independent samples as opposed to complete profile analyses.

These methods fall under the general heading of the *analysis of growth curves,* for their development has been motivated by the problem of fitting polynomials or other functions linear in their parameters to time series of the sizes or weights of an organism. Unlike the classical least-squares model, the observations are usually highly correlated from one time to another, and this correlation must be taken into account in the estimates and tests for the parameters. The problem of the correlation structure has led to several methods of increasing generality and complexity for fitting growth curves. We shall restrict our attention to one due to Khatri (1966) which is a special case of a more general family of procedures suggested by Rao (1966) and developed by Grizzle and Allen (1969). The Khatri technique is also a specialization of a model proposed by Potthoff and Roy (1964). The requirements of space and the level of algebraic complexity we have chosen for this text will prevent us from considering these alternative approaches.

For the description of Khatri's techniques it will be helpful to begin with a treatment of least-squares estimation with correlated error variates. Let

$$\mathbf{x} = \mathbf{B}\boldsymbol{\xi} + \mathbf{e} \tag{1}$$

be the full-rank univariate linear model for the $p \times 1$ random vector $\mathbf{x}$ as a function of the $q \times 1$ unknown parameter vector $\boldsymbol{\xi}$ and the error vector $\mathbf{e}$. $\mathbf{B}$ is a known $p \times q$ nonrandom matrix of rank $q \leq p$. Unlike the classical regression model we shall assume that $\mathbf{e}$ has the multinormal distribution with null mean vector and positive definite covariance matrix $E(\mathbf{ee}') = \boldsymbol{\Sigma}$. Then $\mathbf{x}$ is multinormal with the same covariance matrix, but mean vector $\mathbf{B}\boldsymbol{\xi}$. If we consider $\mathbf{x}$ a single observation from that distribution, the method of maximum likelihood described in Chaps. 1 and 3 gives the Aitken (1934) or generalized least-squares estimator

$$\hat{\boldsymbol{\xi}} = (\mathbf{B}'\boldsymbol{\Sigma}^{-1}\mathbf{B})^{-1}\mathbf{B}'\boldsymbol{\Sigma}^{-1}\mathbf{x} \tag{2}$$

of the regression parameters. As in the case of ordinary least-squares estimators, $\hat{\boldsymbol{\xi}}$ is unbiased and its linear compounds $\mathbf{a}'\hat{\boldsymbol{\xi}}$ have smallest variances among all linear unbiased functions of $\mathbf{x}$.

In most applications $\boldsymbol{\Sigma}$ is unknown, and the exact Aitken estimator cannot be found. However, unlike the case of a single time series, let us suppose that N independent observations have been taken on $\mathbf{x}$ under a generalization of model (1) which allows for the imposition of an experimental design on the sampling units. Then we may represent the $N \times p$ data matrix $\mathbf{X}$ as

$$\begin{aligned} \mathbf{X}' &= [\mathbf{x}_1, \ldots, \mathbf{x}_N] \\ &= \mathbf{B}\boldsymbol{\xi}\mathbf{A}_1' + \mathbf{e}' \end{aligned} \tag{3}$$

where $\mathbf{B}$ is a $p \times q$ known nonrandom matrix of full rank q, $\boldsymbol{\xi}$ is a $q \times r$ matrix of unknown parameters, the $N \times r$ matrix $\mathbf{A}_1$ corresponds to the basis of some design matrix of rank r, and the N columns of $\mathbf{e}'$ are distributed independently as $N(\mathbf{0},\boldsymbol{\Sigma})$ vectors. Khatri has shown that the maximum-likelihood estimator of $\boldsymbol{\xi}$ is

$$\hat{\boldsymbol{\xi}} = (\mathbf{B}'\mathbf{D}^{-1}\mathbf{B})^{-1}\mathbf{B}'\mathbf{D}^{-1}\mathbf{X}'\mathbf{A}_1(\mathbf{A}_1'\mathbf{A}_1)^{-1} \tag{4}$$

where

$$\mathbf{D} = \mathbf{X}'[\mathbf{I} - \mathbf{A}_1(\mathbf{A}_1'\mathbf{A}_1)^{-1}\mathbf{A}_1']\mathbf{X} \tag{5}$$

is the usual error sums of squares and products matrix defined by (34), Sec. 5.2. We may think of $\mathbf{D}$ as proportional to an estimator of $\boldsymbol{\Sigma}$. By casting the model (3) into one involving concomitant variables Khatri has constructed tests of the general hypothesis

$$H_0\colon\quad \mathbf{F}\boldsymbol{\xi}\mathbf{C}' = \mathbf{0} \tag{6}$$

against the alternative H_1: $\mathbf{F}\boldsymbol{\xi}\mathbf{C}' \neq \mathbf{0}$. The $c \times q$ matrix $\mathbf{F}$ has rank $c \leq q$, while $\mathbf{C}'$ has dimensions $r \times g$ and rank g. The test consists of a multivariate analysis of variance based on the error and hypothesis matrices

$$\text{(7)} \qquad \begin{aligned} \mathbf{H} &= \mathbf{F}\hat{\boldsymbol{\xi}}\mathbf{C}'(\mathbf{CRC}')^{-1}(\mathbf{F}\hat{\boldsymbol{\xi}}\mathbf{C}')' \\ \mathbf{E} &= \mathbf{F}(\mathbf{B}'\mathbf{D}^{-1}\mathbf{B})^{-1}\mathbf{F}' \end{aligned}$$

in which

$$\text{(8)} \quad \begin{aligned} \mathbf{R} &= (\mathbf{A}_1'\mathbf{A}_1)^{-1} \\ &\quad + (\mathbf{A}_1'\mathbf{A}_1)^{-1}\mathbf{A}_1'\mathbf{X}[\mathbf{D}^{-1} - \mathbf{D}^{-1}\mathbf{B}(\mathbf{B}'\mathbf{D}^{-1}\mathbf{B})^{-1}\mathbf{B}'\mathbf{D}^{-1}]\mathbf{X}'\mathbf{A}_1(\mathbf{A}_1'\mathbf{A}_1)^{-1} \\ &= (\mathbf{A}_1'\mathbf{A}_1)^{-1} + (\mathbf{A}_1'\mathbf{A}_1)^{-1}\mathbf{A}_1'\mathbf{X}\mathbf{D}^{-1}\mathbf{X}'\mathbf{A}_1(\mathbf{A}_1'\mathbf{A}_1)^{-1} - \hat{\boldsymbol{\xi}}'(\mathbf{B}'\mathbf{D}^{-1}\mathbf{B})\hat{\boldsymbol{\xi}} \end{aligned}$$

To test H_0 we calculate the greatest characteristic root c_s of $\mathbf{HE}^{-1}$ and refer $\theta_s = c_s/(1 + c_s)$ to the appropriate Heck chart or Pillai table with parameters

$$\text{(9)} \quad s = \min(c,g) \qquad m = \frac{|c - g| - 1}{2} \qquad n = \frac{N - r - p + q - c - 1}{2}$$

The $100(1 - \alpha)$ percent simultaneous confidence intervals on all bilinear compounds $\mathbf{a}'\mathbf{F}\boldsymbol{\xi}\mathbf{C}'\mathbf{b}$ are given by

$$\text{(10)} \quad \mathbf{a}'\mathbf{F}\hat{\boldsymbol{\xi}}\mathbf{C}'\mathbf{b} - \left[\frac{x_\alpha}{1 - x_\alpha}(\mathbf{a}'\mathbf{E}\mathbf{a})(\mathbf{b}'\mathbf{CRC}'\mathbf{b})\right]^{1/2} \leq \mathbf{a}'\mathbf{F}\boldsymbol{\xi}\mathbf{C}'\mathbf{b} \leq \mathbf{a}'\mathbf{F}\hat{\boldsymbol{\xi}}\mathbf{C}'\mathbf{b} + \left[\frac{x_\alpha}{1 - x_\alpha}(\mathbf{a}'\mathbf{E}\mathbf{a})(\mathbf{b}'\mathbf{CRC}'\mathbf{b})\right]^{1/2}$$

where $x_\alpha \equiv x_{\alpha;s,m,n}$ is the 100α percent Heck or Pillai critical value. If $s = 1$, $x_\alpha/(1 - x_\alpha)$ should be replaced by the critical value $[(m + 1)/(n + 1)]F_{\alpha;2m+2,2n+2}$.

Example 5.9. Let us illustrate these methods by fitting polynomials to some anatomical measurements given by Potthoff and Roy (1964). The data consist of the distances in millimeters from the center of the pituitary to the pterygomaxillary fissure for $N_1 = 11$ girls and $N_2 = 16$ boys at ages eight, ten, twelve, and fourteen. The mean vectors for the girls and boys were

$$\begin{aligned} \bar{\mathbf{x}}_1' &= [21.18, 22.23, 23.09, 24.09] \\ \bar{\mathbf{x}}_2' &= [22.88, 23.81, 25.72, 27.47] \end{aligned}$$

respectively, and the within-sex matrix of sums of squares and products was

$$\mathbf{D} = \begin{bmatrix} 135.3864 & 67.9205 & 97.7557 & 67.7557 \\ & 104.6193 & 73.1789 & 82.4289 \\ & & 161.3935 & 103.2685 \\ & & & 124.6435 \end{bmatrix}$$

Because the change of the distance during growth is of importance in orthodontal

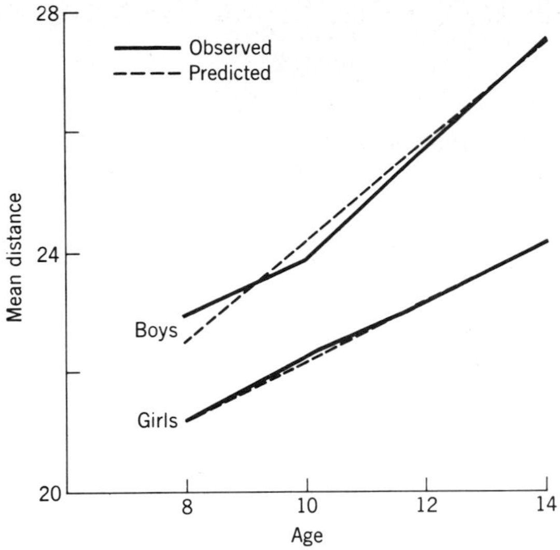

figure 5.2 Observed and predicted mean pituitary-pterygomaxillary distances (mm).

therapy we would like to describe it by a simple function and compare the functions for girls and boys.

From the plots of the mean distances in Fig. 5.2 a linear function would appear to be satisfactory for girls and probably adequate for boys. The model for the tth mean of the jth sex will be

$$\bar{x}_{tj} = \xi_{1j} + 2\xi_{2j}(t - \bar{t}) + e_{tj} \qquad t = 1, \ldots, 4 \qquad j = 1, 2$$

where $\bar{t} = 5/2$ and the factor two before ξ_{2j} has been inserted to avoid fractions. Thus,

$$\mathbf{B} = \begin{bmatrix} 1 & -3 \\ 1 & -1 \\ 1 & 1 \\ 1 & 3 \end{bmatrix} \qquad \mathbf{A}_1' = \begin{bmatrix} 1 & \cdots & 1 & 0 & \cdots & 0 \\ 0 & \cdots & 0 & 1 & \cdots & 1 \end{bmatrix}$$

and

$$(\mathbf{A}_1'\mathbf{A}_1)^{-1} = \begin{bmatrix} 1/11 & 0 \\ 0 & 1/16 \end{bmatrix}$$

Since

$$\mathbf{D}^{-1} = 10^{-2}\begin{bmatrix} 1.48523 & -0.60942 & -0.77586 & 0.23847 \\ & 2.25838 & 0.18897 & -1.31879 \\ & & 1.73229 & -1.13844 \\ & & & 2.48800 \end{bmatrix}$$

$$\mathbf{B}'\mathbf{D}^{-1}\mathbf{B} = 10^{-2}\begin{bmatrix} 1.1339 & -0.7195 \\ -0.7195 & 37.1599 \end{bmatrix} \qquad (\mathbf{B}'\mathbf{D}^{-1}\mathbf{B})^{-1} = \begin{bmatrix} 89.2883 & 1.7288 \\ 1.7288 & 2.7246 \end{bmatrix}$$

and

$$\mathbf{B}'\mathbf{D}^{-1}[\bar{\mathbf{x}}_1 \quad \bar{\mathbf{x}}_2] = \begin{bmatrix} 0.25358 & 0.27682 \\ 0.01387 & 0.12732 \end{bmatrix}$$

we have

$$\begin{bmatrix} \hat{\xi}_{11} & \hat{\xi}_{12} \\ \hat{\xi}_{21} & \hat{\xi}_{22} \end{bmatrix} = \begin{bmatrix} 22.6659 & 24.9373 \\ 0.4762 & 0.8255 \end{bmatrix}$$

The means predicted by the linear model are shown below:

Age	Girls	Boys
8	21.24	22.46
10	22.19	24.11
12	23.14	25.76
14	24.09	27.41

Their plots are indicated by the dashed lines in Fig. 5.2.

Now we shall test the hypothesis

$$H_0: \quad [\xi_{21}, \xi_{22}] = [0,0]$$

of zero slope coefficients for both sexes. Here

$$\mathbf{F} = [0,1] \qquad \mathbf{C}' = \begin{bmatrix} 1 & 0 \\ 0 & 1 \end{bmatrix}$$

so that $c = 1$ and $g = 2$. Then

$$\mathbf{F}\hat{\boldsymbol{\xi}}\mathbf{C}' = [0.4762, 0.8255]$$

$$\begin{bmatrix} \bar{\mathbf{x}}_1' \\ \bar{\mathbf{x}}_2' \end{bmatrix} \mathbf{D}^{-1}[\bar{\mathbf{x}}_1 \quad \bar{\mathbf{x}}_2] = \begin{bmatrix} 5.7536 & 6.3336 \\ 6.3336 & 7.0147 \end{bmatrix}$$

$$\hat{\boldsymbol{\xi}}'(\mathbf{B}'\mathbf{D}^{-1}\mathbf{B})\hat{\boldsymbol{\xi}} = \begin{bmatrix} 5.7543 & 6.3351 \\ 6.3351 & 7.0084 \end{bmatrix}$$

$$\mathbf{R} = 10^{-2}\begin{bmatrix} 9.024 & -0.152 \\ -0.152 & 6.879 \end{bmatrix} \qquad \mathbf{R}^{-1} = \begin{bmatrix} 11.086 & 0.245 \\ 0.245 & 14.542 \end{bmatrix}$$

$\mathbf{H} = 12.617$ and $\mathbf{E} = 2.7246$. Because r and q are both two, $s = 1$, $m = 0$; and $n = 2\frac{1}{2}$, and the test statistic

$$F = \frac{n+1}{m+1}\mathbf{HE}^{-1}$$
$$= 53.25$$

can be referred to critical values of the F distribution with 2 and 23 degrees of freedom. The hypothesis of zero slopes would of course be rejected at any reasonable level.

Next we might consider the hypothesis

$$H_0: \quad \xi_{21} = \xi_{22}$$

of equal slopes for boys and girls. Here

$$\mathbf{F} = [0,1] \qquad \mathbf{C}' = \begin{bmatrix} 1 \\ -1 \end{bmatrix}$$

and $c = g = 1$, $r = q = 2$. $\mathbf{CRC}' = 0.15599$, $\mathbf{F}\hat{\boldsymbol{\xi}}\mathbf{C}' = -0.3493$, $\mathbf{E} = 2.7246$ as before, and $\mathbf{H} = 0.7822$. The parameters are $s = 1$, $m = -\frac{1}{2}$, and $n = 2\frac{1}{2}$, so that the F statistic has the value 6.60 with degrees of freedom 1 and 23. We should reject the hypothesis of equal slopes at the 0.025 level. Finally, we note that the 95 percent simultaneous confidence interval for the difference of the slopes can be computed from (10) with $a = b = 1$ to be

$$-0.630 \leq \xi_{21} - \xi_{22} \leq -0.068$$

It is interesting to note that a computational shortcut can be obtained by fitting all p orthogonal polynomials at once. Let $\mathbf{B}$ be the $p \times p$ matrix of the values of the zero-order, . . . , $(p - 1)$th degree functions obtained from a standard table (e.g., Pearson and Hartley, 1966) for equally spaced points. By definition, the columns of $\mathbf{B}$ are orthogonal, so that $\mathbf{B}'\mathbf{B} = \mathbf{K}$ is a diagonal matrix. Since $\mathbf{B}^{-1} = \mathbf{K}^{-1}\mathbf{B}'$,

$$(\mathbf{B}'\mathbf{D}^{-1}\mathbf{B})^{-1} = \mathbf{K}^{-1}\mathbf{B}'\mathbf{D}\mathbf{B}\mathbf{K}^{-1}$$

and

$$\hat{\boldsymbol{\xi}} = \mathbf{K}^{-1}\mathbf{B}'[\bar{\mathbf{x}}_1 \quad \bar{\mathbf{x}}_2]$$

with a similar result for the estimate (4) with a general design matrix $\mathbf{A}_1$. Through the completion of $\mathbf{B}$ to all polynomials we have avoided the inversion of the matrix $\mathbf{D}$.

In the present example

$$\mathbf{B} = \begin{bmatrix} 1 & -3 & 1 & -1 \\ 1 & -1 & -1 & 3 \\ 1 & 1 & -1 & -3 \\ 1 & 3 & 1 & 1 \end{bmatrix} \qquad \mathbf{K} = \begin{bmatrix} 4 & 0 & 0 & 0 \\ & 20 & 0 & 0 \\ & & 4 & 0 \\ & & & 20 \end{bmatrix}$$

so that

$$\hat{\boldsymbol{\xi}} = \begin{bmatrix} 22.65 & 24.97 \\ 0.48 & 0.78 \\ -0.01 & 0.20 \\ 0.02 & -0.06 \end{bmatrix}$$

and

$$(\mathbf{B}'\mathbf{D}^{-1}\mathbf{B})^{-1} = \begin{bmatrix} 94.4162 & 1.6910 & -1.0518 & -3.9134 \\ & 2.9659 & -1.2299 & 0.2164 \\ & & 6.5728 & -0.1558 \\ & & & 3.1385 \end{bmatrix}$$

The estimates of the constant and linear coefficients are similar to the previous ones, but they are not identical because of the effect of $\mathbf{D}$. The earlier hypothesis tests would undoubtedly be sustained by inferences from the extended model.

In closing, let us use the results from the complete model to test the hypothesis that its quadratic and cubic coefficients are zero. For that test

$$\mathbf{F} = \begin{bmatrix} 0 & 0 & 1 & 0 \\ 0 & 0 & 0 & 1 \end{bmatrix} \qquad \mathbf{C}' = \begin{bmatrix} 1 & 0 \\ 0 & 1 \end{bmatrix} \qquad \mathbf{R} = \begin{bmatrix} 1/11 & 0 \\ 0 & 1/16 \end{bmatrix}$$

$$\mathbf{E} = \begin{bmatrix} 6.5728 & -0.1558 \\ -0.1558 & 3.1385 \end{bmatrix} \qquad \mathbf{H} = \begin{bmatrix} 0.6741 & -0.1892 \\ -0.1892 & 0.0550 \end{bmatrix}$$

$c = g = r = 2$, and of course $q = p = 4$. The greatest characteristic root of $\mathbf{HE}^{-1}$ is 0.1168, and $\theta = 0.104$. Since this is well below the 0.05 critical value based on $s = 2$, $m = -\frac{1}{2}$, and $n = 11$, the hypothesis of no quadratic or cubic components should be accepted.

5.8 OTHER TEST CRITERIA

While we have preferred to develop the multivariate analysis of variance through the union-intersection principle because it leads directly to simultaneous tests and confidence intervals, any number of functions of the roots of the determinantal equation $|\mathbf{H} - \lambda\mathbf{E}| = 0$ could have been chosen as test criteria. We shall describe two common alternatives whose exact or approximate percentage points are easily accessible. Two other statistics will be mentioned later in the comparison of the power probabilities of the tests.

The Wilks Λ *Criterion.* Multivariate analysis of variance was originally developed by Wilks (1932a) through the generalized likelihood-ratio principle. That approach led to the test statistic

$$\Lambda = \frac{|\mathbf{E}|}{|\mathbf{H} + \mathbf{E}|} = \frac{1}{|\mathbf{HE}^{-1} + \mathbf{I}|} \qquad (1)$$

Unlike the Roy statistic, Λ is the reciprocal of the product of *all* characteristic roots of $\mathbf{HE}^{-1} + \mathbf{I}$. When the general linear hypothesis (25) of Sec. 5.2 is true, the large-sample distribution theory of likelihood statistics implies that

$$\chi^2 = -[N - r - \tfrac{1}{2}(u - g + 1)] \ln \Lambda \qquad (2)$$

is distributed as a chi-squared variate with gu degrees of freedom as N tends to infinity. Closer asymptotic approximations have been given by Box (1949) and discussed by Anderson (1958, chap. 8). Transformations to exact chi-squared distributions and tables of the necessary multiplying factors have been given by Schatzoff (1966), Pillai and Gupta (1969), and Lee (1972) for a wide range of values of u and $s = \min(g,u)$.

For certain small values of g or u it is also possible to transform Λ into an exact F variate. If $s = 1$, it follows immediately from expression (41) of Sec. 5.2 that

$$F = \frac{1 - \Lambda}{\Lambda} \frac{n + 1}{m + 1} \tag{3}$$

has the F distribution with degrees of freedom $2m + 2$ and $2n + 2$ when the appropriate null hypothesis is true. $m = (|g - u| - 1)/2$ and $n = (N - r - u - 1)/2$ are the usual parameters introduced in Sec. 5.2. For $s = 2$,

$$F = \frac{1 - \Lambda^{1/2}}{\Lambda^{1/2}} \frac{2n + 2}{2m + 3} \tag{4}$$

is distributed as an F variate with $4m + 6$ and $4(n + 1)$ degrees of freedom under the null hypothesis.

The Lawley-Hotelling Trace Statistic. Lawley (1938) and Hotelling (1947, 1951) have proposed the sum of the roots of $\mathbf{HE}^{-1}$ as a test criterion. The exact distribution of

$$T_0^2 = \text{tr } \mathbf{HE}^{-1} \tag{5}$$

is complicated, but when the null hypothesis is true NT_0^2 tends to a chi-squared variate with gu degrees of freedom as the number of independent sampling units N becomes large. Small-sample F approximations to T_0^2 and tables for their use have been found by Hughes and Saw (1972), while Davis (1970) has computed tables of approximate percentage points by fitting appropriate Pearson-type distributions.

Power Comparisons of the Test Criteria. Ito (1962) has compared the large-sample power probabilities of the T_0^2 and Λ statistics for a simple class of alternative hypotheses, and has shown that for error degrees of freedom $N - r$ equal to 100 or 200 the powers differ at most in the second decimal place. Pillai and Jayachandran (1967) have computed the powers of the Roy, Wilks, and T_0^2 tests, as well as the statistic

$$V = \text{tr } \mathbf{H}(\mathbf{H} + \mathbf{E})^{-1} \tag{6}$$

proposed by Pillai (1955). For $p = 2$ responses and $\alpha = 0.05$ the powers of Λ, T_0^2, and V differed only in the third decimal place for small departures from the null hypothesis of zero population characteristic roots. For larger deviations the powers differed at most by 0.02. When the population roots were very different T_0^2 tended to have highest power, while for equal roots the power of V was highest. The power of the

Roy test was lowest for the alternatives considered, although the third-place differences were too small to have practical consequences.

Roy, Gnanadesikan, and Srivastava (1971) have described an earlier simulation of the power functions of the previous four criteria and a new statistic U equal to the product of the nonzero roots of $\mathbf{H}(\mathbf{H} + \mathbf{E})^{-1}$ for $p = 2$, small values of $N - r$, and $\alpha = 0.05$. Unlike the exact powers, large differences obtained among the tests. For equal population roots the Pillai statistic V tended to have highest power, followed closely by the U criterion. Λ and T_0^2 came next, while the greatest-root test had lowest power. However, for the case of a single large population root the Roy statistic tended to have the highest empirical power.

5.9 EXERCISES

1. The observations on the responses pigment creatinine (X_1), chloride (X_2), and choline (X_3), and the concomitant variable (specific gravity − 1) × 10^3 (U) of the analysis by Smith *et al.* described in Examples 5.3 and 5.6 are shown below for the four weight groups:

Group 1				Group 2				Group 3				Group 4			
U	X_1	X_2	X_3	U	X_1	X_2	X_3	U	X_1	X_2	X_3	U	X_1	X_2	X_3
24	17.6	5.15	7.5	31	18.1	9.00	14.5	18	17.0	4.55	1.9	32	12.5	2.90	22.5
32	13.4	5.75	7.1	23	19.7	5.30	12.5	10	12.5	2.65	0.7	25	8.7	3.00	19.5
17	20.3	4.35	2.3	32	16.9	9.85	8.0	33	21.5	6.50	8.3	28	9.4	3.40	1.3
30	22.3	7.55	4.0	20	23.7	3.60	4.9	25	22.2	4.85	9.3	27	15.0	5.40	20.0
30	20.5	8.50	2.0	18	19.2	4.05	0.2	35	13.0	8.75	13.0	23	12.9	4.45	1.0
27	18.5	10.25	2.0	23	18.0	4.40	3.6	33	13.0	5.20	18.3	25	12.1	4.30	5.0
25	12.1	5.95	16.8	31	14.8	7.15	12.0	31	10.9	4.75	10.5	26	13.2	5.00	3.0
30	12.0	6.30	14.5	28	15.6	7.25	5.2	34	12.0	5.85	14.5	34	11.5	3.40	5.1
28	10.1	5.45	0.9	21	16.2	5.30	10.2	16	22.8	2.85	3.3				
24	14.7	3.75	2.0	20	14.1	3.10	8.5	31	16.5	6.55	6.3				
26	14.8	5.10	0.4	15	17.5	2.40	9.6	28	18.4	6.60	4.9				
27	14.4	4.05	3.8	26	14.1	4.25	6.9								
				24	19.1	5.80	4.7								
				16	22.5	1.55	3.5								

a. Test the hypothesis of equal group means with and without adjustment for the specific gravity values.

b. Compute the 95 percent simultaneous confidence intervals for the six group differences of each of the response effects under the one-way multivariate analysis of variance and covariance models. Which variates appear to have contributed to any significance of the overall tests? To what extent does the concomitant variable affect the individual comparisons?

2. In an investigation of the effects of whole-body radiation Dutton (1954) subjected twelve rats to 500 r and twelve to 600 r. Each treatment group contained three young and three adult males, as well as three young and three adult females. The four responses in this $2 \times 2 \times 2$ factorial with $n = 3$ observations in each combination were the total weight losses in grams after 1, 3, 6, and 7 days following irradiation. The initial weight of each rat was used as a concomitant variable (U). The hypotheses and error matrices of the responses and the covariate had these values:

Radiation:

$$\mathbf{H}_{xx}^{(1)} = \begin{bmatrix} 1.50 & 8.75 & -7.75 & -10.50 \\ & 51.04 & -45.21 & -61.25 \\ & & 40.04 & 54.25 \\ & & & 73.50 \end{bmatrix} \qquad \mathbf{H}_{xu}^{(1)} = \begin{bmatrix} 10.25 \\ 59.79 \\ -52.96 \\ -71.75 \end{bmatrix}$$

$$\mathbf{H}_{uu}^{(1)} = 70.04$$

Age:

$$\mathbf{H}_{xx}^{(2)} = \begin{bmatrix} 60.17 & 4.75 & 106.08 & 114.00 \\ & 0.38 & 8.38 & 9.00 \\ & & 187.04 & 201.00 \\ & & & 216.00 \end{bmatrix} \qquad \mathbf{H}_{xu}^{(2)} = \begin{bmatrix} 847.08 \\ 66.88 \\ 1{,}493.54 \\ 1{,}605.00 \end{bmatrix}$$

$$\mathbf{H}_{uu}^{(2)} = 11{,}926.04$$

Sex:

$$\mathbf{H}_{xx}^{(3)} = \begin{bmatrix} 28.17 & 102.92 & 150.58 & 88.83 \\ & 376.04 & 550.21 & 324.58 \\ & & 805.04 & 474.92 \\ & & & 280.17 \end{bmatrix} \qquad \mathbf{H}_{xu}^{(3)} = \begin{bmatrix} 1{,}210.08 \\ 4{,}421.46 \\ 6{,}469.29 \\ 3{,}816.42 \end{bmatrix}$$

$$\mathbf{H}_{uu}^{(3)} = 51{,}987.04$$

Total interaction:

$$\mathbf{H}_{xx}^{(4)} = \begin{bmatrix} 86.01 & -52.33 & -54.33 & -61.16 \\ & 161.16 & 98.16 & 71.76 \\ & & 128.16 & 89.76 \\ & & & 73.51 \end{bmatrix} \qquad \mathbf{H}_{xu}^{(4)} = \begin{bmatrix} -204.66 \\ 124.84 \\ 32.34 \\ 172.75 \end{bmatrix}$$

$$\mathbf{H}_{uu}^{(4)} = 3{,}669.50$$

Error:

$$\mathbf{E}_{xx} = \begin{bmatrix} 95.98 & 59.33 & 53.34 & 49.33 \\ & 411.34 & 117.67 & 57.66 \\ & & 342.68 & 276.32 \\ & & & 385.32 \end{bmatrix} \qquad \mathbf{E}_{xu} = \begin{bmatrix} 36.00 \\ 284.65 \\ 66.67 \\ -90.67 \end{bmatrix}$$

$$\mathbf{E}_{uu} = 1{,}326.00$$

a. Under the assumption that the two radiation levels constitute fixed effects, set up the parameter and design matrices for this experiment. Show that the rank of the design matrix is $rcd = 8$.

b. Test the hypotheses of no radiation, age, and sex effects upon the four weight-loss responses.

c. Repeat the tests of part b after an adjustment for different values of the initial weight concomitant variable by the multivariate analysis of covariance. Compare the results with the unadjusted analysis.

3. Construct 95 percent simultaneous confidence intervals for the drug-effect differences $(\tau_{i1} + \tau_{i2}) - (\tau_{j1} + \tau_{j2})$ of total weight loss for the data of Examples 5.2 and 5.5.

4. In an experiment described by Dutton (1954) a Latin-square design was used to study the effect of radiation upon wound healing in rats. The six rows and columns of the square corresponded to the arrangement of the individual cages of 36 rats, while the six treatments consisted of the combinations of three radiation dosages and two times of wounding. The area of the wound was measured by a photoplanimeter at the end of each of eleven 3-day periods. An orthogonal polynomial was fitted to each animal's time series of logarithms of wound areas, and the two responses for the multivariate analysis of variance were the linear and quadratic coefficients of each rat's polynomial. The hypothesis and error matrices (with elements multiplied by 10^6 for convenience) follow:

$$\mathbf{H}_{\text{row}} = \begin{bmatrix} 27{,}033 & -51{,}317 \\ -51{,}317 & 114{,}147 \end{bmatrix} \qquad \mathbf{H}_{\text{col}} = \begin{bmatrix} 3{,}982 & -9{,}308 \\ -9{,}308 & 41{,}610 \end{bmatrix}$$

$$\mathbf{H}_{\text{treat}} = \begin{bmatrix} 15{,}730 & -52{,}378 \\ -52{,}378 & 225{,}535 \end{bmatrix} \qquad \mathbf{E} = \begin{bmatrix} 32{,}035 & -100{,}535 \\ -100{,}535 & 424{,}573 \end{bmatrix}$$

Dutton also introduced the logarithm of the initial area of the wound as a concomitant variable, although this did not alter the conclusions about the hypothesis of equal treatment effects.

a. Use the greatest-root statistic to test the hypotheses of equal row, column, and treatment effects.

b. Compare the conclusions of part a with those implied by the likelihood-ratio and T_0^2 statistics.

5. As an example of the multivariate analysis of variance, Bartlett (1934, 1947) employed the yields of grain and straw in an agricultural field trial. The eight fertilizer treatments were applied to plots in eight randomized blocks. The treatments, blocks, and error sums of squares and products matrices were

$$\mathbf{H}_t = \begin{bmatrix} 12{,}496.8 & -6{,}786.8 \\ -6{,}786.8 & 32{,}985.0 \end{bmatrix} \qquad \mathbf{H}_b = \begin{bmatrix} 86{,}045.8 & 56{,}073.6 \\ 56{,}073.6 & 75{,}841.5 \end{bmatrix}$$

$$\mathbf{E} = \begin{bmatrix} 136{,}972.6 & 58{,}549.0 \\ 58{,}549.0 & 71{,}496.1 \end{bmatrix}$$

Test the hypothesis of equal-treatment-effect vectors.

6. The following "two-arm bandit" device is frequently employed in learning and preference behavior experiments: a subject is seated before a panel with two lamps and matching buttons and is directed to try to predict which of the lamps will light at each of a sequence of trials. In a hypothetical experiment 30 subjects were assigned randomly to three versions of the game: (1) no monetary payoff for correct predictions, (2) payoff for correct predictions, and (3) payoff and loss for correct or incorrect choices. Each participant made 100 successive predictions. Under the assumption that predictive strategies would stabilize in the latter trials, the proportions of choices of the more frequent lamp on the blocks of trials 41–60, 61–80, and 81–100 were chosen as responses for the investigation of payoff condition differences. Those fictitious proportions (with decimal points omitted) are shown below:

No payoff			Payoff			Payoff and loss		
41–60	61–80	81–100	41–60	61–80	81–100	41–60	61–80	81–100
54	52	60	58	70	74	70	75	82
70	72	73	61	68	73	75	81	92
59	65	66	63	70	76	68	78	83
60	61	62	56	65	69	54	70	76
50	53	60	70	72	74	60	67	72
67	65	71	60	68	71	63	71	77
62	63	64	57	60	68	71	76	90
61	62	63	60	61	65	65	69	75
63	62	65	55	62	72	67	73	80
58	59	58	62	71	76	63	66	71

a. Carry out a complete profile analysis of these data by the appropriate multivariate techniques. Where those procedures reject the various hypotheses construct simultaneous confidence intervals to determine the significance of condition or trial-block differences.

b. Make the profile analysis by the approximate and conservative analysis-of-variance tests described in Sec. 5.6. Compare the conclusions about profile parallelism and trial-block differences with those obtained by the exact multivariate tests.

7. The following experiment and data were reported by Danford, Hughes, and McNee (1960) in an illustration of their proposed models for the analysis of repeated measurements. In an investigation of the effect of radiation therapy 45 patients suffering from cancerous lesions were trained to operate a psychomotor testing device. Four trials were conducted on the day preceding radiation and on each of the 10 days after the therapy, and the patient's average daily score was taken as the response. Six patients were not given radiation

and served as controls, while the remainder were treated with dosages of 25 to 50, 75 to 100, or 125 to 250 r. The average scores for the first three days following radiation are given below.

Controls, $N = 6$			25–50 r, $N = 14$			75–100 r, $N = 15$			125–250 r, $N = 10$		
1	2	3	1	2	3	1	2	3	1	2	3
223	242	248	53	102	104	206	199	237	202	229	232
72	81	66	45	50	54	208	222	237	126	159	157
172	214	239	47	45	34	224	224	261	54	75	75
171	191	203	167	188	209	119	149	196	158	168	175
138	204	213	193	206	210	144	169	164	175	217	235
22	24	24	91	154	152	170	202	181	147	183	181
			115	133	136	93	122	145	105	107	92
			32	97	86	237	243	281	213	263	260
			38	37	40	208	235	249	258	248	257
			66	131	148	187	199	205	257	269	270
			210	221	251	95	102	96			
			167	172	212	46	67	28			
			23	18	30	95	137	99			
			234	260	269	59	76	101			
						186	198	201			

Conduct a profile analysis of these data under the usual multinormal assumptions by (1) the exact multivariate method and (2) the approximate analysis of variance.

8. For profile data with only $p = 2$ responses show that the multivariate and analysis-of-variance approaches are equivalent.

9. Cole and Grizzle (1966) gave a statistical analysis of an experiment by Morris and Zeppa (1963) on the effects of morphine and trimethaphan on blood histamine levels in dogs. Sixteen mongrel dogs were divided into four treatment groups of four dogs apiece. Groups 1 and 2 received morphine sulphate intravenously; groups 3 and 4 received intravenous trimethaphan. The dogs in groups 2 and 4 had their supplies of histamine depleted at the times of innoculation with the treatment drugs. The observed levels of blood histamine at four times are shown in the table.*

   a. Carry out a complete profile analysis of these data under the two-way multivariate analysis-of-variance model.

   b. Use the methods of Sec. 5.8 to fit appropriate polynomials to blood histamine level as a function of time. Use a one- or two-way model for the four groups as suggested by any preliminary analyses.

* These data are reprinted by permission of the editor of *Biometrics*.

Treatment group	Blood histamine μg/ml			
	Control	1 min	3 min	5 min
1	0.04	0.20	0.10	0.08
	0.02	0.06	0.02	0.02
	0.07	1.40	0.48	0.24
	0.17	0.57	0.35	0.24
2	0.10	0.09	0.13	0.14
	0.12	0.11	0.10	0.11*
	0.07	0.07	0.07	0.07
	0.05	0.07	0.06	0.07
3	0.03	0.62	0.31	0.22
	0.03	1.05	0.73	0.60
	0.07	0.83	1.07	0.80
	0.09	3.13	2.06	1.23
4	0.10	0.09	0.09	0.08
	0.08	0.09	0.09	0.10
	0.13	0.10	0.12	0.12
	0.06	0.05	0.05	0.05

* Fictitious observation replacing a missing entry.

10. Consider a single random sample of equally spaced growth measurements on N individuals. Under the usual multinormal model show how the Hotelling T^2 statistic can be computed from the first, second, . . . , differences of the sample means to test the hypotheses of linear, quadratic, . . . , population growth functions (Hills, 1968).

11. Show that the estimator (2) of Sec. 5.7 reduces to the ordinary least-squares estimator

$$\hat{\boldsymbol{\xi}} = (\mathbf{B}'\mathbf{B})^{-1}\mathbf{B}'\mathbf{x}$$

when $\boldsymbol{\Sigma}$ has the equal-variance, equal-covariance pattern and $\mathbf{B}$ consists of a first column of ones and $q - 1$ mutually orthogonal columns.

12. Express the nonzero characteristic roots of the matrices $\mathbf{H}(\mathbf{H} + \mathbf{E})^{-1}$, $\mathbf{E}(\mathbf{H} + \mathbf{E})^{-1}$, and $\mathbf{I} + \mathbf{H}\mathbf{E}^{-1}$ in terms of the roots of $\mathbf{H}\mathbf{E}^{-1}$.

13. Compute the Λ and T_0^2 statistics for the data of Example 5.1 and Exercise 1, and determine their large-sample significance probabilities.

6 CLASSIFICATION BY THE LINEAR DISCRIMINANT FUNCTION

6.1 INTRODUCTION. The statistical methods developed in the preceding chapters for estimation, hypothesis testing, and confidence statements were based upon exact specifications of the populations of the response variates. In the applied sciences another kind of multivariate problem frequently occurs in which an observation must be assigned in some optimum fashion to one of several populations. For example, in plant taxonomy a botanist may wish to classify a new specimen as one of several recognized species of a flower. In educational psychology a candidate for admission to a school or program must be assigned to categories of the sort "admit," "admit conditionally," or "admission denied" on the basis of a vector of test scores, grades, and ratings. In routine banking or commercial finance an officer or analyst may wish to classify loan applicants as low or high credit risks on the basis of the elements of certain accounting statements. In each case the decision maker wishes to classify from simple functions of the observation vector rather than complicated regions in the higher-dimensional space of the original vector.

In this chapter we shall consider classification rules based on an index called the linear discriminant function. When the populations are multinormal with different mean vectors but a common covariance matrix such rules seem reasonable, and they are "optimum" in certain senses when the parameters are known. In the case of two populations the single linear discriminant function follows from the union-intersection construction of the two-sample T^2 test. That function is analogous to the population one given by the likelihood-ratio rule with known mean vectors and covariance matrix. While the misclassification probabilities follow immediately in that case, their evaluation in the more general one is very complicated, and some methods will be discussed for their estimation. In the final section classification rules will be given for several multinormal populations in terms of the one-way multivariate analysis of variance and the Mahalanobis distance statistic.

6.2 THE LINEAR DISCRIMINANT FUNCTION FOR TWO GROUPS

Let us begin by recalling from Sec. 4.4 the assumptions for the two-sample T^2 statistic: random samples of N_1 and N_2 observation vectors have been drawn independently from respective p-dimensional multinormal populations with mean vectors $\boldsymbol{\mu}_1$ and $\boldsymbol{\mu}_2$, and a common covariance matrix $\boldsymbol{\Sigma}$. Rather than test the usual hypothesis of equal mean vectors we wish to construct a linear compound or index for summarizing observations from the groups on a one-dimensional scale that *discriminates* between the populations by some measure of maximal separation. If $\bar{\mathbf{x}}_1$ and $\bar{\mathbf{x}}_2$ are the sample mean vectors and $\mathbf{S}$ is the pooled estimate of $\boldsymbol{\Sigma}$, we shall determine the coefficient vector $\mathbf{a}$ of the index $\mathbf{a}'\mathbf{x}$ as that which gives the greatest squared critical ratio

$$(1) \qquad t^2(\mathbf{a}) = \frac{[\mathbf{a}'(\bar{\mathbf{x}}_1 - \bar{\mathbf{x}}_2)]^2 N_1 N_2/(N_1 + N_2)}{\mathbf{a}'\mathbf{S}\mathbf{a}}$$

or, equivalently, which maximizes the absolute difference $|\mathbf{a}'(\bar{\mathbf{x}}_1 - \bar{\mathbf{x}}_2)|$ in the average values of the index for the two groups subject to the constraint $\mathbf{a}'\mathbf{S}\mathbf{a} = 1$. As in the union-intersection construction of the single-sample T^2 in Sec. 4.2, the coefficient vector $\mathbf{a}$ is given by the homogeneous system of equations

$$(2) \qquad \left[\frac{N_1 N_2}{N_1 + N_2}(\bar{\mathbf{x}}_1 - \bar{\mathbf{x}}_2)(\bar{\mathbf{x}}_1 - \bar{\mathbf{x}}_2)' - \lambda\mathbf{S}\right]\mathbf{a} = \mathbf{0}$$

where

$$(3) \qquad \begin{aligned} \lambda &= \max_{\mathbf{a}} t^2(\mathbf{a}) \\ &= \frac{N_1 N_2}{N_1 + N_2}(\bar{\mathbf{x}}_1 - \bar{\mathbf{x}}_2)'\mathbf{S}^{-1}(\bar{\mathbf{x}}_1 - \bar{\mathbf{x}}_2) \\ &= T^2 \end{aligned}$$

The rank of the coefficient matrix is $p - 1$, so that the system has only the single solution

$$(4) \qquad \mathbf{a} = \mathbf{S}^{-1}(\bar{\mathbf{x}}_1 - \bar{\mathbf{x}}_2)$$

and the linear discriminant function is

(5) $$y = (\bar{\mathbf{x}}_1 - \bar{\mathbf{x}}_2)'\mathbf{S}^{-1}\mathbf{x}$$

If the variances of the responses are nearly equal, the elements of **a** give the relative importance of the contribution of each response to the T^2 statistic; otherwise multiplication of the ith element by the standard deviation of its response will give comparable discriminant coefficients.

To use the linear discriminant function for classifying an observation of unknown population we begin by computing the mean values of the scores for the two samples:

$$\bar{y}_1 = (\bar{\mathbf{x}}_1 - \bar{\mathbf{x}}_2)'\mathbf{S}^{-1}\bar{\mathbf{x}}_1 \qquad \bar{y}_2 = (\bar{\mathbf{x}}_1 - \bar{\mathbf{x}}_2)'\mathbf{S}^{-1}\bar{\mathbf{x}}_2$$

The midpoint of these means on the discriminant function scale is

$$\tfrac{1}{2}(\bar{\mathbf{x}}_1 - \bar{\mathbf{x}}_2)'\mathbf{S}^{-1}(\bar{\mathbf{x}}_1 + \bar{\mathbf{x}}_2)$$

and we might adopt the classification rule

Assign the individual with observation **x** *to population* 1 *if*

$$(\bar{\mathbf{x}}_1 - \bar{\mathbf{x}}_2)'\mathbf{S}^{-1}\mathbf{x} > \tfrac{1}{2}(\bar{\mathbf{x}}_1 - \bar{\mathbf{x}}_2)'\mathbf{S}^{-1}(\bar{\mathbf{x}}_1 + \bar{\mathbf{x}}_2)$$

and to population 2 *if*

$$(\bar{\mathbf{x}}_1 - \bar{\mathbf{x}}_2)'\mathbf{S}^{-1}\mathbf{x} \le \tfrac{1}{2}(\bar{\mathbf{x}}_1 - \bar{\mathbf{x}}_2)'\mathbf{S}^{-1}(\bar{\mathbf{x}}_1 + \bar{\mathbf{x}}_2)$$

That is, sampling units are assigned to the group with the closer discriminant mean score.

Because the mean midpoint is a value of a random variable it would seem more appropriate to state the rule in terms of the single statistic

(6) $$W = \mathbf{x}'\mathbf{S}^{-1}(\bar{\mathbf{x}}_1 - \bar{\mathbf{x}}_2) - \tfrac{1}{2}(\bar{\mathbf{x}}_1 + \bar{\mathbf{x}}_2)'\mathbf{S}^{-1}(\bar{\mathbf{x}}_1 - \bar{\mathbf{x}}_2)$$

as

Assign **x** *to population* 1 *if* $W > 0$ *and otherwise to population* 2.

W is called the Wald-Anderson classification statistic in recognition of its origin (Wald, 1944) and results on its sampling properties (Anderson, 1951b). The distribution of W is exceedingly complex for small samples; some early investigations of it can be found in the collection edited by Solomon (1961, chaps. 15–19). Some later references will be cited in the context of the estimation of misclassification probabilities in Sec. 6.4.

Example 6.1. Let us compute the linear discriminant function for the data presented in Example 4.3. The vector of the WAIS mean differences for the two diagnostic groups is

$$(\bar{\mathbf{x}}_1 - \bar{\mathbf{x}}_2)' = [3.82,\ 4.24,\ 2.99,\ 3.22]$$

and the discriminant function is

$$y = 0.030x_1 + 0.204x_2 + 0.010x_3 + 0.443x_4$$

We see that the second test, similarities, and the fourth, picture completion, dominate the function, while the verbal subtests information and arithmetic make only negligible contributions to it. In subsequent investigations attention might be concentrated on the similarities and picture completion tests as indicators of the senile factor quality.

The mean discriminant scores for the non–senile factor and senile factor groups are $\bar{y}_1 = 5.97$ and $\bar{y}_2 = 3.54$. We may give the classification rule as

Assign the ith individual to the senile factor category if $y_i \leq 4.76$, *and to the non–senile factor diagnostic group if* $y_i > 4.76$.

Classification of the individual subjects by the rule gave these results:

Classification by discriminant function	Psychiatric diagnosis		
	Non-SF	SF	Total
Non-SF	29	4	33
SF	8	8	16
Total	37	12	49

While the table gives some measure of the ability of the linear discriminant function to reproduce the diagnoses, the sample misclassification proportions are badly biased estimates of the population error rates. We shall consider some alternative estimates of those probabilities in Sec. 6.4.

Frequency plots of the values of the discriminant functions for the two groups are shown in Fig. 6.1. The overlap of the frequency functions is rather large, as the high rates of misclassification in the table would suggest.

6.3 CLASSIFICATION WITH KNOWN PARAMETERS

Because the orientation of this text is toward the analysis of multivariate data we have preferred to introduce classification through the sample linear discriminant function. In this section we shall treat classification into one of two multinormal populations with known parameters. This model will be required for the definition of the misclassification probabilities given in Sec. 6.4 and is of some interest in its own right when very large samples are available.

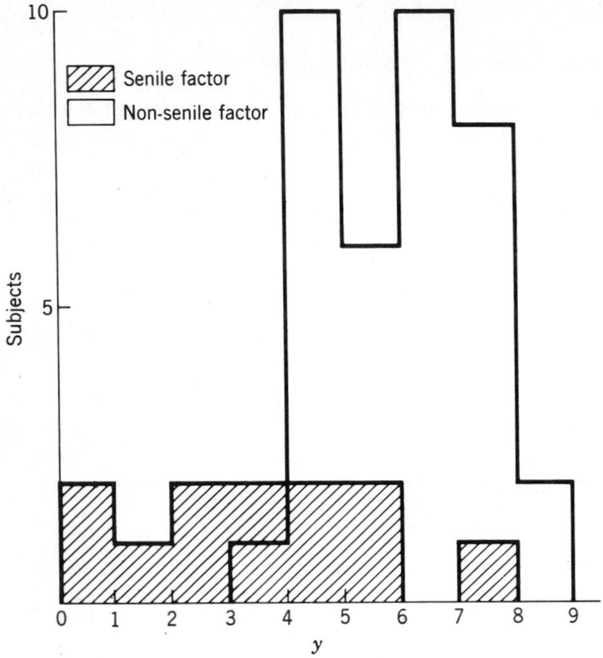

figure 6.1 Frequencies of discriminant function values.

Let us begin by developing the linear discriminant function by the likelihood-ratio approach of Welch (1939). The random variable $\mathbf{X}$ has either the density function $f_1(\mathbf{x};\boldsymbol{\theta}_1)$ or $f_2(\mathbf{x};\boldsymbol{\theta}_2)$, where the mathematical forms of the densities are specified and the parameter vectors $\boldsymbol{\theta}_i$ are known. Now assume that a single observation $\mathbf{x}$ has been taken on $\mathbf{X}$. If the likelihood $f_1(\mathbf{x};\boldsymbol{\theta}_1)$ is large relative to $f_2(\mathbf{x};\boldsymbol{\theta}_2)$, we would be inclined to believe that $\mathbf{x}$ came from the first population; if $f_2(\mathbf{x};\boldsymbol{\theta}_2)$ has the larger value, the second population would seem more likely. This rule may be written in terms of the likelihood ratio

$$\lambda = \frac{f_1(\mathbf{x};\boldsymbol{\theta}_1)}{f_2(\mathbf{x};\boldsymbol{\theta}_2)} \tag{1}$$

as

Classify $\mathbf{x}$ *as from population* 1 *if* $\lambda > 1$ *and from population* 2 *if* $\lambda \leq 1$.

Now let the functions be p-dimensional multinormal densities with respective mean vectors $\boldsymbol{\mu}_1$ and $\boldsymbol{\mu}_2$, and a common covariance matrix $\boldsymbol{\Sigma}$. Then

$$\lambda = \exp\left[(\boldsymbol{\mu}_1 - \boldsymbol{\mu}_2)'\boldsymbol{\Sigma}^{-1}\mathbf{x} - \tfrac{1}{2}(\boldsymbol{\mu}_1 - \boldsymbol{\mu}_2)'\boldsymbol{\Sigma}^{-1}(\boldsymbol{\mu}_1 + \boldsymbol{\mu}_2)\right] \tag{2}$$

and the multinormal rule is

Classify $\mathbf{x}$ *as an observation from population* 1 *if*

$$(\boldsymbol{\mu}_1 - \boldsymbol{\mu}_2)'\boldsymbol{\Sigma}^{-1}\mathbf{x} - \tfrac{1}{2}(\boldsymbol{\mu}_1 - \boldsymbol{\mu}_2)'\boldsymbol{\Sigma}^{-1}(\boldsymbol{\mu}_1 + \boldsymbol{\mu}_2) > 0$$

and from population 2 *otherwise.*

This rule is the population analogue of that for the linear discriminant in Sec. 6.2, that is, the mean vectors and covariance matrix have been replaced by their known parameters.

The misclassification probabilities of the rule can be easily calculated. The linear discriminant variate

$$(3) \qquad Y = (\boldsymbol{\mu}_1 - \boldsymbol{\mu}_2)'\boldsymbol{\Sigma}^{-1}\mathbf{X}$$

is normally distributed with parameters

$$(4) \qquad \mathrm{E}(Y) = (\boldsymbol{\mu}_1 - \boldsymbol{\mu}_2)'\boldsymbol{\Sigma}^{-1}\boldsymbol{\mu}_i$$

$$(5) \qquad \mathrm{var}\,(Y) = (\boldsymbol{\mu}_1 - \boldsymbol{\mu}_2)'\boldsymbol{\Sigma}^{-1}(\boldsymbol{\mu}_1 - \boldsymbol{\mu}_2) = \delta^2$$

when $\mathbf{X}$ is from the ith population. The variance δ^2 is known as the *Mahalanobis distance* of the two multinormal populations, after its originator (Mahalanobis, 1936). The probability of misclassification is

$$(6) \qquad \begin{aligned} P_{21} &= P(\text{classify } \mathbf{x} \text{ from } 2|\mathbf{x} \text{ from } 1) \\ &= P[Y \le \tfrac{1}{2}(\boldsymbol{\mu}_1 - \boldsymbol{\mu}_2)'\boldsymbol{\Sigma}^{-1}(\boldsymbol{\mu}_1 + \boldsymbol{\mu}_2)] \\ &= \Phi(-\tfrac{1}{2}\delta) \end{aligned}$$

where $\Phi(z)$ is the standard normal distribution function defined by (2), Sec. 1.3. From the symmetry of the classification rule and the normal density it follows that

$$(7) \qquad \begin{aligned} P_{12} &= P(\text{classify } x \text{ from } 1|x \text{ from } 2) \\ &= \Phi(-\tfrac{1}{2}\delta) \end{aligned}$$

Anderson (1951b; 1958, chap. 12) has extended the likelihood-ratio rule to the case of *a priori* probabilities h and $1 - h$ for populations 1 and 2 and losses $l(i;j)$ of classifying an observation from the jth population as one from the ith. For the previous multinormal model the Bayes, or minimum expected loss, rule states that the vector $\mathbf{x}$ should be assigned to population 1 if

$$(8) \quad (\boldsymbol{\mu}_1 - \boldsymbol{\mu}_2)'\boldsymbol{\Sigma}^{-1}\mathbf{x} - \tfrac{1}{2}(\boldsymbol{\mu}_1 - \boldsymbol{\mu}_2)'\boldsymbol{\Sigma}^{-1}(\boldsymbol{\mu}_1 + \boldsymbol{\mu}_2) > \ln \frac{(1-h)l(1;2)}{hl(2;1)}$$

and to population 2 otherwise. If the probability h is unknown, its value may be chosen to minimize the maximum expected cost of misclassification; details of the determination of the decision rule have been given by Anderson (1951b, 1958) and by Blackwell and Girshick (1954). If the costs of misclassification are equal, the minimax solution is given by $h = \frac{1}{2}$.

Example 6.2. The information about the relative frequencies with which observations from the two populations are encountered can be useful in discriminant classification. Let us reclassify the individuals in Example 6.1 by a sample analogue of the Bayes rule (8). The costs of misclassification will be assumed equal, but because the total sample was originally drawn without reference to overt senility we shall take as an estimate of the *a priori* probability of nonsenility the relative frequency $h = 37/49$. Then $\ln(12/37) = -1.260$, and the classification rule is

Assign the ith individual to the senile factor category if $Y_i \leq 3.50$, *and otherwise to the non-senile factor group.*

This rule led to the following classifications:

Classification by discriminant function	Psychiatric diagnosis		
	Non-SF	SF	Total
Non-SF	37	5	42
SF	0	7	7
Total	37	12	49

The use of the prior probabilities has reduced the number of misclassified individuals to less than half the number given by the symmetric rule of Example 6.1.

Throughout our treatment of classification we have assumed that both populations have a common covariance matrix. Development of the rule by the likelihood ratio of multinormal densities with unequal covariance matrices will lead to a quadratic discriminant function (Smith, 1947). Some alternative discriminant functions for the unequal-covariance-matrix case have been found by Kullback (1968) from statistical information principles.

6.4 ESTIMATION OF THE MISCLASSIFICATION PROBABILITIES

If the linear discriminant statistic W defined by (6) of Sec. 6.2 is to be used for the classification of future observations, some estimate of its error rates would be essential. The probability of misclassifying the observation $\mathbf{x}$ from the population $N(\boldsymbol{\mu}_1,\boldsymbol{\Sigma})$ as one from $N(\boldsymbol{\mu}_2,\boldsymbol{\Sigma})$ is

$$(1)\qquad P_1 = P(W \le 0) = P\left[\frac{\mathbf{x}'\mathbf{a} - \boldsymbol{\mu}_1'\mathbf{a}}{\sqrt{\mathbf{a}'\boldsymbol{\Sigma}\mathbf{a}}} \le \frac{\frac{1}{2}(\bar{\mathbf{x}}_1 + \bar{\mathbf{x}}_2)'\mathbf{a} - \boldsymbol{\mu}_1'\mathbf{a}}{\sqrt{\mathbf{a}'\boldsymbol{\Sigma}\mathbf{a}}}\right] = \Phi\left[\frac{\frac{1}{2}(\bar{\mathbf{x}}_1 + \bar{\mathbf{x}}_2)'\mathbf{a} - \boldsymbol{\mu}_1'\mathbf{a}}{\sqrt{\mathbf{a}'\boldsymbol{\Sigma}\mathbf{a}}}\right]$$

where $\mathbf{a} = \mathbf{S}^{-1}(\bar{\mathbf{x}}_1 - \bar{\mathbf{x}}_2)$. Similarly, the opposite misclassification probability can be expressed as

$$(2)\qquad P_2 = \Phi\left[\frac{\boldsymbol{\mu}_2'\mathbf{a} - \frac{1}{2}(\bar{\mathbf{x}}_1 + \bar{\mathbf{x}}_2)'\mathbf{a}}{\sqrt{\mathbf{a}'\boldsymbol{\Sigma}\mathbf{a}}}\right]$$

The unconditional evaluation of these probabilities is a difficult task, although Okamoto (1963) has given an asymptotic expansion for them and tables of coefficients for its use.

As an alternative approach we shall describe two procedures due to Lachenbruch and Mickey (1968) which appear to have good properties but avoid the complexities of the distribution of W. In both methods the W statistics are computed for each of the $N = N_1 + N_2$ sampling units, but with the particular individual's observation omitted in the calculation of the linear discriminant coefficients, means, and covariance matrix. While a greater amount of computation is necessary, the inversion of N matrices can be avoided by the use of the Bartlett identity (Bartlett, 1951). The statistic for the observation $\mathbf{x}_i$ from the kth sample is

$$(3)\qquad W_i = \left\{\mathbf{x}_i - \frac{1}{2}\left[\bar{\mathbf{x}}_1 + \bar{\mathbf{x}}_2 - \frac{1}{N_k - 1}(\mathbf{x}_i - \bar{\mathbf{x}}_k)\right]\right\}' \mathbf{S}_i^{-1} \cdot \left[\bar{\mathbf{x}}_1 - \bar{\mathbf{x}}_2 + \frac{(-1)^k}{N_k - 1}(\mathbf{x}_i - \bar{\mathbf{x}}_k)\right]$$

where

$$(4)\qquad \mathbf{S}_i^{-1} = \frac{N-3}{N-2}\left(\mathbf{S}^{-1} + \frac{c_k}{1 - c_k(\mathbf{x}_i - \bar{\mathbf{x}}_k)'\mathbf{S}^{-1}(\mathbf{x}_i - \bar{\mathbf{x}}_k)} \cdot \mathbf{S}^{-1}(\mathbf{x}_i - \bar{\mathbf{x}}_k)(\mathbf{x}_i - \bar{\mathbf{x}}_k)'\mathbf{S}^{-1}\right)$$

and

$$(5)\qquad c_k = \frac{N_k}{(N_k - 1)(N - 2)}$$

A simpler form for computational purposes is

$$(6)\quad W_i = \frac{N-3}{N-2}\Bigg(\mathbf{x}_i'\mathbf{S}^{-1}(\bar{\mathbf{x}}_1 - \bar{\mathbf{x}}_2) - \tfrac{1}{2}(\bar{\mathbf{x}}_1 + \bar{\mathbf{x}}_2)'\mathbf{S}^{-1}(\bar{\mathbf{x}}_1 - \bar{\mathbf{x}}_2)$$

$$+ \frac{1}{1 - c_k(\mathbf{x}_i - \bar{\mathbf{x}}_k)'\mathbf{S}^{-1}(\mathbf{x}_i - \bar{\mathbf{x}}_k)}$$

$$\cdot \Big\{c_k[(\mathbf{x}_i - \bar{\mathbf{x}}_k)'\mathbf{S}^{-1}(\mathbf{x}_i - \bar{\mathbf{x}}_k)][(\mathbf{x}_i - \bar{\mathbf{x}}_k)'\mathbf{S}^{-1}(\bar{\mathbf{x}}_1 - \bar{\mathbf{x}}_2)]$$

$$+ c_k[(\mathbf{x}_i - \bar{\mathbf{x}}_k)'\mathbf{S}^{-1}(\bar{\mathbf{x}}_1 - \bar{\mathbf{x}}_2)]^2$$

$$+ (-1)^k \frac{2N_k - 1}{2(N_k - 1)^2}[(\mathbf{x}_i - \bar{\mathbf{x}}_k)'\mathbf{S}^{-1}(\mathbf{x}_i - \bar{\mathbf{x}}_k)]^2\Big\}\Bigg)$$

As in Sec. 6.2 individuals with positive W_i scores would be assigned to population 1, and the remainder to the second population. The probabilities P_1 and P_2 are estimated in the first, or U, method by the proportions of misclassified cases for each group. For the second, or $\bar{U}$, method Lachenbruch and Mickey compute the statistics

$$(7)\quad \bar{W}_k = \frac{1}{N_k}\sum_{i=1}^{N_k} W_{ik} \qquad s_k^2 = \frac{1}{N_k - 1}\sum_{i=1}^{N_k}(W_{ik} - \bar{W}_k)^2 \qquad k = 1, 2$$

from the discriminant scores defined by (3) to (6). Then the estimates of the misclassification probabilities are

$$(8)\qquad \hat{P}_1 = \Phi\left(\frac{-\bar{W}_1}{s_1}\right) \qquad \hat{P}_2 = \Phi\left(\frac{\bar{W}_2}{s_2}\right)$$

Unlike the more general U method these estimates depend explicitly on the assumption that the linear discriminant variate is normally distributed.

Example 6.3. We shall estimate the misclassification probabilities for the data of Example 6.1 by the U and $\bar{U}$ methods. The values of the new discriminant function scores computed from (6) are summarized in this table:

	Psychiatric diagnosis	
	Non-SF	SF
Mean	1.126	−0.821
Standard deviation	1.494	2.679
Negative scores	11	8
Positive scores	26	4

The U estimate of the probability P_1 of misclassifying a non-senile factor patient as one from the senile factor population is now $^{11}\!/_{37} = 0.2973$. The U estimate of P_2 remains $\frac{1}{3}$. The $\bar{U}$ estimates are

$$\hat{P}_1 = \Phi(-0.7825) = 0.217$$

$$\hat{P}_2 = \Phi(-0.3064) = 0.397$$

The increase of the U estimate over that given by Example 6.1 for the non-senile factor sample is due to the additional misclassification of three cases with very small discriminant scores.

Lachenbruch and Mickey (1968) have compared the properties of several estimates of the misclassification probabilities, including U, $\bar{U}$, two variations using the Okamoto expansion, and the resubstitution method of Example 6.1. In general, resubstitution fared poorest, while the Okamoto procedure with a special estimate of the distance δ^2 was best. $\bar{U}$ was a close second, while the U technique appeared to be of middle quality. Recommendations for the choice of method for different numbers of responses and the sample distance measure have been given by the authors; further discussion of their results has been provided by Cochran (1968). Recent analytical results by McLachlan (1974b) have corroborated the Lachenbruch-Mickey conclusions.

Other aspects of the error rate problem have been investigated by a number of workers. Hills (1966) has surveyed the general problem and has found exact expressions in the univariate normal case. Dunn and Varady (1966) have simulated the sampling properties of the estimate $\Phi(\hat{\delta}/2)$, where $\hat{\delta}^2$ is the sample analogue of the Mahalanobis distance defined by (5), Sec. 6.3. Lachenbruch (1968) and Dunn (1971) have studied the expectations of certain estimated misclassification probabilities. Lachenbruch (1967) and others have found confidence intervals for the misclassification probability. Sorum (1971) has considered the error rate problem under the restrictive assumption of a known covariance matrix. An asymptotically unbiased estimate has been proposed and developed by McLachlan (1974c), who has also suggested (1974a) an improved expansion for the probability (1).

6.5 CLASSIFICATION FOR SEVERAL GROUPS

Now let us consider methods for classification of an unknown observation to one of k populations. The responses of the independent groups are described by multinormal random variables with mean vectors $\boldsymbol{\mu}_1, \ldots, \boldsymbol{\mu}_k$ and a common covariance matrix $\boldsymbol{\Sigma}$. If those parameters are known, Anderson (1958, chap. 6) has shown that the Bayes rule for given prior probabilities of the populations and a specified matrix of misclassification costs is based upon the likelihood ratios (2) of Sec. 6.3 for all pairs of groups. In the more common situation of unknown parameters the mean vectors and covariance matrix are replaced by their usual estimates

$$\hat{\mathbf{u}}_j = \bar{\mathbf{x}}_j$$

$$\text{(1)} \qquad \mathbf{S} = \frac{1}{N-k} \sum_{j=1}^{k} \mathbf{A}_j$$

defined in terms of the sample mean vector $\bar{\mathbf{x}}_j$ and the sums of squares and products matrix $\mathbf{A}_j$ for the jth group. If $\mathbf{x}$ is the new observation of unknown origin, we compute the linear discriminant scores

$$\text{(2)} \qquad W_{ij} = \mathbf{x}'\mathbf{S}^{-1}(\bar{\mathbf{x}}_i - \bar{\mathbf{x}}_j) - \tfrac{1}{2}(\bar{\mathbf{x}}_i + \bar{\mathbf{x}}_j)'\mathbf{S}^{-1}(\bar{\mathbf{x}}_i - \bar{\mathbf{x}}_j)$$

and follow the classification rule

(3) Assign $\mathbf{x}$ to population i if $W_{ij} > 0$ for all $j \neq i$

We note immediately that $W_{ij} = -W_{ji}$ and that any $k - 1$ linearly independent W_{ij} form a basis for the complete set of the statistics if $k - 1 \leq p$. If $p < k - 1$ the space of the W_{ij} will have rank p, and the classification rule can be defined in terms of p scores.

Let us consider some simple examples of classification rules. For the first, let $k = 3$ and let p be two or more. The distinct discriminant statistics are

$$\text{(4)} \qquad \begin{aligned} W_{12} &= \mathbf{x}'\mathbf{S}^{-1}(\bar{\mathbf{x}}_1 - \bar{\mathbf{x}}_2) - \tfrac{1}{2}(\bar{\mathbf{x}}_1 + \bar{\mathbf{x}}_2)'\mathbf{S}^{-1}(\bar{\mathbf{x}}_1 - \bar{\mathbf{x}}_2) \\ W_{13} &= \mathbf{x}'\mathbf{S}^{-1}(\bar{\mathbf{x}}_1 - \bar{\mathbf{x}}_3) - \tfrac{1}{2}(\bar{\mathbf{x}}_1 + \bar{\mathbf{x}}_3)'\mathbf{S}^{-1}(\bar{\mathbf{x}}_1 - \bar{\mathbf{x}}_3) \\ W_{23} &= \mathbf{x}'\mathbf{S}^{-1}(\bar{\mathbf{x}}_2 - \bar{\mathbf{x}}_3) - \tfrac{1}{2}(\bar{\mathbf{x}}_2 + \bar{\mathbf{x}}_3)'\mathbf{S}^{-1}(\bar{\mathbf{x}}_2 - \bar{\mathbf{x}}_3) \end{aligned}$$

Because $W_{23} = W_{13} - W_{12}$, it is only necessary to use the statistics W_{12} and W_{13}. The classification rule is defined in this way:

Classify $\mathbf{x}$ *as from*
Population 1 if $W_{12} > 0$ and $W_{13} > 0$
Population 2 if $W_{12} < 0$ and $W_{13} > W_{12}$
Population 3 if $W_{13} < 0$ and $W_{12} > W_{13}$

The three classification regions are illustrated in Fig. 6.2.

As a second example, let $k = 3$ and $p = 1$. For convenience the three populations have been labeled so that $\bar{x}_1 < \bar{x}_2 < \bar{x}_3$. The classification rules are those suggested by simple intuition:

Classify x *as from*
Population 1 if $x < \tfrac{1}{2}(\bar{x}_1 + \bar{x}_2)$
Population 2 if $\tfrac{1}{2}(\bar{x}_1 + \bar{x}_2) \leq x \leq \tfrac{1}{2}(\bar{x}_2 + \bar{x}_3)$
Population 3 if $x > \tfrac{1}{2}(\bar{x}_2 + \bar{x}_3)$

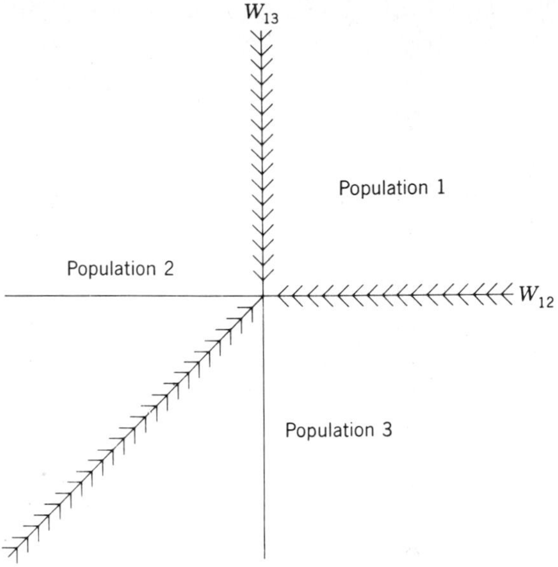

figure 6.2 Classification regions for $k = 3$, $p \geq 2$.

The number of classification statistics is now min $(k - 1, p) = 1$ rather than 2.

Classification by Minimum Distance. The classification rule (3) can be expressed in terms of the sample Mahalanobis distance

$$D_i^2 = (\mathbf{x} - \bar{\mathbf{x}}_i)'\mathbf{S}^{-1}(\mathbf{x} - \bar{\mathbf{x}}_i) \tag{5}$$

of the unknown observation from the mean of the ith sample as

(6) Assign $\mathbf{x}$ to population i if $D_i^2 = \min \{D_1^2, \ldots, D_k^2\}$

The equivalence of the rules follows immediately from the relation

$$W_{hj} = -\tfrac{1}{2}D_h^2 + \tfrac{1}{2}D_j^2 \tag{7}$$

so that (3) is satisfied if $D_i^2 < D_j^2$ for all $j \neq i$.

Kshirsagar and Arseven (1975) have expressed the multiple-group classification rule (3) in terms of the minimum Mahalanobis distance calculated from uncorrelated linear compounds of the original variates. The compound coefficient vectors $\mathbf{l}_i$ are the solutions of the equations

$$[\mathbf{H} - \lambda_i(\mathbf{H} + \mathbf{E})]\mathbf{l}_i = \mathbf{0} \qquad i = 1, \ldots, \min(p, k - 1) \tag{8}$$

where $\mathbf{H}$ and $\mathbf{E}$ are the one-way multivariate analysis-of-variance hypothesis and error matrices defined by (34), Sec. 5.2. The λ_i are the respective roots of the determinantal equation $|\mathbf{H} - \lambda(\mathbf{H} + \mathbf{E})| = 0$. Kshirsagar and Arseven have used all p roots and vectors, although it is easily shown that the $s = \min(p, k-1)$ vectors corresponding to the positive roots will give the same classifications with fewer variates. The uncorrelated nature of the derived variables and their smaller number when $p > k - 1$ are in keeping with the spirit of the "canonical reduction" methods of the next three chapters.

Example 6.4. Let us find the linear discriminant functions of the iris data given in Example 5.1 of Sec. 5.3. Since there are three species and four variates the two linearly independent functions are

$$W_{12} = -3.2456X_1 - 3.3907X_2 + 7.5530X_3 + 14.6358X_4 - 31.5226$$

$$W_{13} = -11.0759X_1 - 19.9161X_2 + 29.1874X_3 + 38.4608X_4 - 18.0933$$

The classification rule actions are

Classify an iris flower with observation x as
(1) *Virginica* if $W_{12} > 0$ and $W_{13} > 0$
(2) *Versicolor* if $W_{12} < 0$ and $W_{23} = W_{13} - W_{12} > 0$
(3) *Setosa* if $W_{13} < 0$ and $W_{23} < 0$

The rule is illustrated in Fig. 6.3. The mean coordinates of the three species are illustrated by the dots; their values are as follows:

	$\bar{W}_{12}$	$\bar{W}_{13}$
Virginica	8.5978	89.6782
Versicolor	−8.5979	36.3299
Setosa	−44.7502	−89.6783

The means fall nearly on a straight line, so that it would appear that a single linear discriminant function might be adequate for a classification rule. This is consistent with the very small size of the second positive characteristic root of $\mathbf{HE}^{-1}$ implied by the computations in Example 5.1. In the next example we shall consider another rule based on a reduced number of linear discriminant functions.

Example 6.5. As a second illustration of multiple discrimination let us use the three biochemical variables of Exercise 1, Sec. 5.9, without adjustment for the specific gravity values. Rather than finding the various W_{ij} scores or D^2 statistics let us approach the problem through the one-way multivariate analysis of variance. The mean values and hypothesis and error matrices are as follows:

Group	X_1	X_2	X_3
1	15.89	6.01	5.28
2	17.82	5.21	7.45
3	16.34	5.37	8.27
4	11.91	3.98	9.68

$$\mathbf{H} = \begin{bmatrix} 181.07 & 40.03 & -66.73 \\ & 20.03 & -41.37 \\ & & 103.80 \end{bmatrix}$$

$$\mathbf{E} = \begin{bmatrix} 496.77 & 4.51 & -225.02 \\ & 152.85 & 86.23 \\ & & 1{,}431.86 \end{bmatrix}$$

The characteristic roots of the matrix

$$\mathbf{E}^{-1}\mathbf{H} = \begin{bmatrix} 0.3597 & 0.0670 & -0.0971 \\ 0.2543 & 0.1443 & -0.3106 \\ -0.0054 & -0.0271 & 0.0759 \end{bmatrix}$$

are $l_1 = 0.4340$, $l_2 = 0.1410$, and $l_3 = 0.0049$. By the union-intersection development of Sec. 5.2 the characteristic vector corresponding to l_1 contains the coefficients

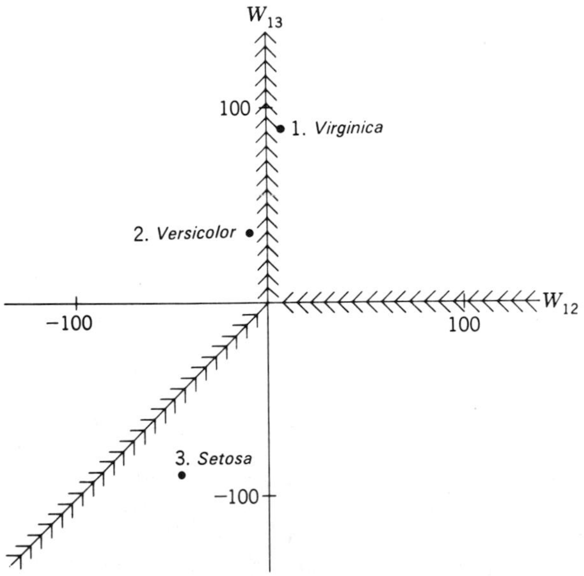

figure 6.3 Classification regions for the iris species.

of the linear compound of the observations which has a maximum univariate analysis-of-variance F statistic. Similarly, the next greatest F ratios for all linear compounds uncorrelated with the first are specified by the characteristic vectors associated with l_2 and l_3. However, the size of l_3 in relation to the other roots suggests that its linear compound has negligible discriminating power, and that two functions should suffice for defining the classification rule. That is, the four group means appear to lie nearly in a plane rather than in three-dimensional space. This effective rank of two of the matrix **H** seems due in part to the strong negative correlation of the X_2 and X_3 means.

The characteristic vectors of $\mathbf{E}^{-1}\mathbf{H}$ corresponding to l_1 and l_2 are

$$\mathbf{a}_1' = [0.038095, 0.037184, -0.003493]$$

$$\mathbf{a}_2' = [0.063154, -0.133265, 0.050281]$$

where the coefficients have been standardized so that $\mathbf{a}_1'\mathbf{E}\mathbf{a}_1 = \mathbf{a}_2'\mathbf{E}\mathbf{a}_2 = 1$. The new derived variates are

$$U_1 = 0.0381X_1 + 0.0372X_2 - 0.0035X_3$$

$$U_2 = 0.0632X_1 - 0.1333X_2 + 0.0503X_3$$

and their mean values for the four obesity groups are as follows:

$$\bar{\mathbf{u}}_1' = [0.8106, 0.8468, 0.7936, 0.5681]$$

$$\bar{\mathbf{u}}_2' = [0.4676, 0.8052, 0.7322, 0.7083]$$

The six basic linear discriminant functions

$$\mathbf{W}_{ij} = U_1(\bar{U}_{1i} - \bar{U}_{1j}) + U_2(\bar{U}_{2i} - \bar{U}_{2j}) - \tfrac{1}{2}(\bar{U}_{1i} - \bar{U}_{1j})(\bar{U}_{1i} + \bar{U}_{1j}) - \tfrac{1}{2}(\bar{U}_{2i} - \bar{U}_{2j})(\bar{U}_{2i} + \bar{U}_{2j})$$

are

$$W_{12} = -0.0362U_1 - 0.3377U_2 + 0.2449$$

$$W_{13} = 0.0170U_1 - 0.2646U_2 + 0.1451$$

$$W_{14} = 0.2425U_1 - 0.2407U_2 - 0.0256$$

$$W_{23} = 0.0532U_1 + 0.0731U_2 - 0.0998$$

$$W_{24} = 0.2787U_1 + 0.0969U_2 - 0.2705$$

$$W_{34} = 0.2255U_1 + 0.0239U_2 - 0.1707$$

where the other six are given by $W_{ji} = -W_{ij}$. Note that the coefficients are consistent; for example, $W_{23} = W_{13} - W_{12}$. From the twelve functions we can find the boundaries of the four classification regions. For example, the transformed observation $[u_1,u_2]$ should be classified as coming from the population of group 1 if

$$W_{12} > 0 \qquad W_{13} > 0 \qquad W_{14} > 0$$

so that its classification region would be the intersection of the half-planes defined by those inequalities. The other nine inequalities divide the U_1, U_2 space into unique regions for populations, 2, 3, and 4. These are shown in Fig. 6.4, where the

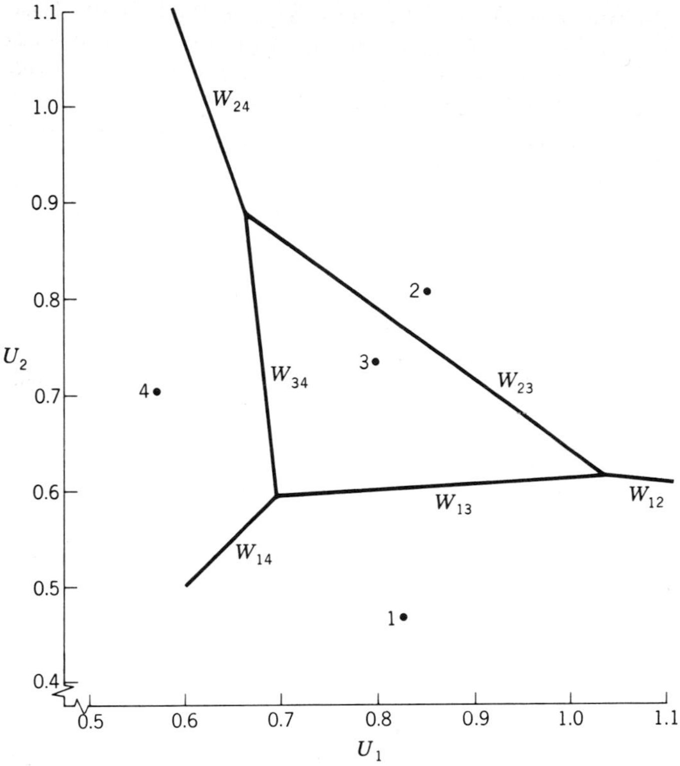

figure 6.4 Classification regions for the four obesity groups.

points indicate the group mean coordinates and the boundary segments are labeled by the appropriate W_{ij} functions.

For example, a subject with the observations $X_1 = 18$, $X_2 = 5$, and $X_3 = 8$ would have $U_1 = 0.8438$ and $U_2 = 0.8735$, and would be classified as coming from population 2. Each individual in the original four samples might be classified in this fashion, although the large amount of overlap of the data would undoubtedly lead to many misclassifications, particularly among groups 1 to 3.

Alternatively, the minimum D^2 classification rule applied to the scores U_1, U_2 might be more efficient if only a small number of individuals are to be classified. For the previous vector [18,5,8] the values of the Mahalanobis distances for groups 1 to 4 would be respectively 0.1658, 0.0047, 0.0225, and 0.1033, and assignment would be to population 2.

6.6 EXERCISES

1. From the growth data of Example 5.9, Sec. 5.7, find the linear discriminant

function of the four measurements. Can you make any empirical or approximate comparisons of classification with that rule and ones suggested by (a) distance at a single age and (b) distance predicted for each sex by the estimated growth function?

2. Use the profile data of Example 5.7, Sec. 5.6, for the following analyses:
   a. Find the multiple discriminant functions for the four groups and the classification regions defined by them.
   b. Classify the original observation vectors by the rules of (a), and calculate the misclassification proportions.
   c. Find the linear discriminant function for groups 1 and 4, and compare its ability to classify the original observations with that of the multiple-group rule.

3. Use the Bartlett identity to derive equation (3) of Sec. 6.4, and in turn verify that (6) follows from it.

4. Carry out a discriminant analysis on these three groups of contrived observations on the variates X_1, X_2, and X_3:

A			*B*			*C*		
X_1	X_2	X_3	X_1	X_2	X_3	X_1	X_2	X_3
6	7	4	11	13	30	18	16	25
7	5	1	15	16	42	24	22	30
9	10	11	22	20	50	20	21	32
8	8	6	17	16	45	19	20	27
8	9	6	12	11	38	22	25	34
10	9	8	13	14	35	17	16	32

7
INFERENCES FROM COVARIANCE MATRICES

7.1 INTRODUCTION.

The thrust of the last three chapters has been the analysis of the mean structure of multinormal variates. The covariance matrix was of interest only as part of the model, as in the requirements that it be a known positive definite matrix, that it be the same for all cells in an experimental design, or that it have a particular pattern with unknown elements. With this chapter we shall turn to the analysis and description of covariance structures. We shall begin with some likelihood-ratio tests of hypotheses for a single covariance matrix, for the equality of several matrices, and for the validity of certain important patterns. We shall then construct a test for the independence of two subsets of multinormal variates by the union-intersection principle, for that approach leads directly to the technique of *canonical correlation* for summarizing the dependence relations between the sets. That method is a generalization of multiple regression in which neither set of variables is restricted to the "dependent" or "predictor" roles. In Chaps. 8 and 9 we shall carry the generalization still further to models for covariance structure which do not require *a priori* partitionings of the response variates.

7.2 HYPOTHESIS TESTS FOR A SINGLE COVARIANCE MATRIX

Let us begin with the case of a single random sample of N observation vectors from the p-dimensional multinormal population with mean vector $\boldsymbol{\mu}$ and positive definite covariance matrix $\boldsymbol{\Sigma}$. We wish a test of the hypothesis

$$H_0\colon\ \boldsymbol{\Sigma} = \boldsymbol{\Sigma}_0 \tag{1}$$

and, possibly, a method of constructing a family of simultaneous confidence statements on certain quadratic forms $\mathbf{a}'\boldsymbol{\Sigma}\mathbf{a}$ in the elements of the matrix. Unlike the union-intersection developments of the tests in Chaps. 4 and 5 we shall prefer to find a test of (1) by the generalized likelihood ratio principle for these reasons: extensive and exact tables of the critical values of the union-intersection statistic are not presently available, nor does the simultaneous confidence interval adjunct of the tests appear to be as important for hypotheses on variances as for those on the mean vector. Nevertheless, the union-intersection statistic will be treated briefly at the end of this section.

The maximum-likelihood estimates of the multinormal parameters under H_0 are $\bar{\mathbf{x}}$ and $\boldsymbol{\Sigma}_0$, while those under the general parameter space are $\bar{\mathbf{x}}$ and $\mathbf{S}$. In keeping with the definition (7) of Sec. 1.5 the likelihood ratio is

$$(2)\qquad \lambda = \left[\left(\frac{N-1}{N}\right)^p \frac{|\mathbf{S}|}{|\boldsymbol{\Sigma}_0|}\right]^{\frac{1}{2}N} \exp\left[-\tfrac{1}{2}(N-1)(\operatorname{tr} \mathbf{S}\boldsymbol{\Sigma}_0^{-1} - Np)\right]$$

This ratio can be approximated by one in which $N - 1$ has been replaced by a general degrees-of-freedom parameter ν for $\mathbf{S}$; the test statistic is then

$$(3)\qquad L = \nu(\ln |\boldsymbol{\Sigma}_0| - \ln |\mathbf{S}| + \operatorname{tr} \mathbf{S}\boldsymbol{\Sigma}_0^{-1} - p)$$

When N is large, L is distributed as a chi-squared variate with $\frac{1}{2}p(p+1)$ degrees of freedom if the hypothesis (1) is true. For moderate N Bartlett (1954) has suggested the scaled statistic

$$(4)\qquad L' = \left\{1 - \frac{1}{6(N-1)}\,[2p + 1 - 2/(p+1)]\right\} L$$

as an improvement on the chi-squared approximation, and we shall adopt the decision rule

$$(5)\qquad \text{Reject } H_0 \text{ if } L' > \chi^2_{\alpha;\frac{1}{2}p(p+1)}$$

Korin (1968) has found more exact approximations to the distribution and percentage points of L, although for nearly all applications the critical values implied by (4) and (5) will be sufficiently accurate.

Example 7.1. Let us suppose that the reaction times in hundredths of a second after three preparatory intervals can be described by a trivariate normal random variable. Reaction times have been measured under the three conditions on a random sample of $N = 20$ normal subjects. From those data we wish to test the hypothesis

$$H_0\colon\ \boldsymbol{\Sigma} = \begin{bmatrix} 4 & 3 & 2 \\ & 6 & 5 \\ & & 10 \end{bmatrix}$$

that has been suggested by the effects of the lengthening preparatory intervals on the variances and correlations, as well as by previous experimental results. The sample covariance matrix is

$$\mathbf{S} = \begin{bmatrix} 3.42 & 2.60 & 1.89 \\ & 8.00 & 6.51 \\ & & 9.62 \end{bmatrix}$$

and

$$|\mathbf{\Sigma}_0| = 86 \qquad |\mathbf{S}| = 88.6355 \qquad \text{tr } \mathbf{S\Sigma}_0^{-1} = 3.2222$$
$$\nu = 19 \qquad L = 3.65 \qquad L' = 3.55$$

Since L' does not exceed any conventional upper percentage point of the chi-squared distribution with six degrees of freedom we conclude that the hypothesized covariance matrix is tenable.

The Union-Intersection Test. Roy (1957, p. 30) has shown that the union-intersection principle leads to the following decision rule for H_0: $\mathbf{\Sigma} = \mathbf{\Sigma}_0$:

$$\text{Accept } H_0 \text{ if } l \leq c_p \leq c_1 \leq u \tag{6}$$

where c_p and c_1 are the smallest and largest characteristic roots of $\mathbf{A\Sigma}_0^{-1}$, $\mathbf{A} = \nu\mathbf{S}$ is the sample sums of squares and products matrix based on ν degrees of freedom, and l and u are lower and upper percentage points of the characteristic root distribution such that

$$P(l \leq c_p \leq c_1 \leq u) = 1 - \alpha \tag{7}$$

Hanumara and Thompson (1968) have tabulated the approximate values of l and $u_1 \approx u$ for various marginal tail probabilities

$$\alpha = P(c_p \leq l) = P(c_1 \geq u_1)$$

They have indicated that the approximation of u by u_1 appears to be satisfactory for $\alpha \leq 0.05$ and $p \geq 2$. More extensive tables have been computed by Pillai and Chang (1968, 1970). An abbreviated version of the tables has been reproduced by Pearson and Hartley (1972, table 51). In these tables and expressions we note that contrary to the usage of Chap. 5, c_i will denote the ith largest characteristic root. We shall follow that notation in the later chapters as well.

The special quality of the union-intersection test is that it gives a family of simultaneous confidence intervals on all quadratic forms $\mathbf{a}'\mathbf{\Sigma}\mathbf{a}$ in the covariance matrix. The $100(1 - 2\alpha)$ percent family has the general interval

$$\frac{\mathbf{a}'\mathbf{A}\mathbf{a}}{u} \leq \mathbf{a}'\mathbf{\Sigma}\mathbf{a} \leq \frac{\mathbf{a}'\mathbf{A}\mathbf{a}}{l} \tag{8}$$

Alternatively, Bonferroni simultaneous confidence intervals might be obtained from (41), Sec. 1.5, for the k quadratic forms $\mathbf{a}_1'\mathbf{\Sigma}\mathbf{a}_1, \ldots,$ $\mathbf{a}_k'\mathbf{\Sigma}\mathbf{a}_k$ by dividing the tail probabilities by k:

(9)
$$\frac{\mathbf{a}_i'\mathbf{A}\mathbf{a}_i}{\chi^2_{\alpha_2/k;\nu}} \leq \mathbf{a}_i'\mathbf{\Sigma}\mathbf{a}_i \leq \frac{\mathbf{a}_i'\mathbf{A}\mathbf{a}_i}{\chi^2_{1-\alpha_1/k;\nu}}$$

In practice one usually takes $\alpha_1 = \alpha_2 = \alpha/2$ for a family confidence coefficient at least $1 - \alpha$.

7.3 TESTS FOR TWO SPECIAL PATTERNS

In Sec. 4.5 we indicated that the equal-variance, equal-covariance pattern was sufficient for the validity of the repeated-measurements equal-means test, while necessary and sufficient conditions on the matrix had also been obtained by Huynh and Feldt. In this section we shall give tests of hypotheses for those multinormal covariance models.

The test of

(1)
$$H_0\colon\ \mathbf{\Sigma} = \sigma^2 \begin{bmatrix} 1 & \rho & \cdots & \rho \\ \rho & 1 & \cdots & \rho \\ \cdots & \cdots & \cdots & \cdots \\ \rho & \rho & \cdots & 1 \end{bmatrix}$$

against some general alternative is due to Wilks (1946). Let us suppose that the usual unbiased estimate $\mathbf{S}$ of $\mathbf{\Sigma}$ based on ν degrees of freedom has been obtained, where $\nu = N - 1$ for a single random sample. Then the estimates of σ^2 and $\sigma^2\rho$ under the model (1) are

(2)
$$s^2 = \frac{1}{p}\sum_{i=1}^{p} s_{ii}$$
$$s^2r = \frac{1}{p(p-1)}\sum_{i \neq j}\sum s_{ij}$$

and the Wilks generalized likelihood-ratio statistic is

(3)
$$L = \frac{|S|}{(s^2)^p(1-r)^{p-1}[1+(p-1)r]}$$

Box (1949, 1950) has shown that

(4)
$$\chi^2 = -\left[\nu - \frac{p(p+1)^2(2p-3)}{6(p-1)(p^2+p-4)}\right]\ln L$$

is approximately distributed as a chi-squared variate with $f = \frac{1}{2}p(p+1) - 2$ degrees of freedom when (1) is true and ν is large.

We would of course reject H_0 in favor of a general positive definite matrix when the statistic (4) exceeds some upper critical value of the chi-squared distribution with f degrees of freedom.

Test for the Huynh and Feldt Type H Pattern. Huynh and Feldt (1970) have shown that their necessary and sufficient conditions on $\mathbf{\Sigma}$ given in Sec. 4.5 will hold if

$$\mathbf{C\Sigma C}' = \lambda \mathbf{I} \tag{5}$$

where $\mathbf{C}$ is any semiorthogonal $(p - 1) \times p$ submatrix of the orthogonal Helmert matrix

$$\begin{bmatrix} \dfrac{\mathbf{j}'}{\sqrt{p}} \\ \mathbf{C} \end{bmatrix} \tag{6}$$

and λ is one of $p + 1$ parameters defining the Type H pattern. $\mathbf{j}'$ is the $1 \times p$ vector of ones. The test for the validity of the H pattern is that for a spherical covariance matrix due to Mauchly (1940); its statistic is

$$W = \frac{(p-1)^{p-1}\,|\mathbf{CSC}'|}{(\text{tr}\,\mathbf{CSC}')^{p-1}} \tag{7}$$

and

$$\chi^2 = -\left[\nu - \frac{2p^2 - 3p + 3}{6(p-1)}\right] \ln W \tag{8}$$

has the chi-squared distribution with $f = \frac{1}{2}p(p - 1) - 1$ degrees of freedom when ν is large and $\mathbf{\Sigma}$ has the H pattern. Anderson (1958) has found the moments of W, and has shown in particular that for $p = 3$ the statistic (8) has an exact chi-squared distribution with $f = 2$ degrees of freedom. Consul (1967) has found exact distributions of W for other values of p.

Example 7.2. We shall now test the equal-variance, equal-covariance and Type H hypotheses for the reaction-time data of Example 4.7. For the first test,

$$s^2 = 2.4640 \qquad r = 0.7622 \qquad |\mathbf{S}| = 1.5061 \qquad L = 0.7052 \qquad \chi^2 = 5.588$$

The last statistic is slightly greater than $\chi^2_{0.25;4} = 5.39$, and we may conclude that the equal variance and covariance matrix is indeed a tenable model.

For the test of the H-pattern hypothesis we shall choose the semiorthogonal matrix

$$\mathbf{C} = \begin{bmatrix} \dfrac{1}{\sqrt{6}} & \dfrac{-2}{\sqrt{6}} & \dfrac{1}{\sqrt{6}} \\ \dfrac{1}{\sqrt{2}} & 0 & \dfrac{-1}{\sqrt{2}} \end{bmatrix}$$

Then

$$\mathbf{CSC}' = \begin{bmatrix} 0.44032 & -0.26434 \\ -0.26434 & 0.73155 \end{bmatrix}$$

$W = 0.73471$, and since $\chi^2 = 5.542$ does not exceed the exact upper 5 percent critical value of a chi-squared variate with two degrees of freedom we may consider the H pattern as a plausible model for $\mathbf{\Sigma}$.

7.4 TESTING THE EQUALITY OF SEVERAL COVARIANCE MATRICES

The hypothesis

$$H_0: \quad \mathbf{\Sigma}_1 = \cdots = \mathbf{\Sigma}_k \tag{1}$$

of the equality of the covariance matrices of k p-dimensional multinormal populations can be tested against the alternative of general positive definite matrices by a modified generalized likelihood-ratio statistic. Let $\mathbf{S}_i$ be the unbiased estimate of $\mathbf{\Sigma}_i$ based on n_i degrees of freedom, where $n_i = N_i - 1$ for the usual case of a random sample of N_i observation vectors from the ith population. When H_0 is true

$$\mathbf{S} = \frac{1}{\Sigma n_i} \sum_{i=1}^{k} n_i \mathbf{S}_i \tag{2}$$

is the pooled estimate of the common covariance matrix. The test statistic is

$$M = \Sigma n_i \ln |\mathbf{S}| - \sum_{i=1}^{k} n_i \ln \mathbf{S}_i \tag{3}$$

and Box (1949) has shown that if the scale factor

$$C^{-1} = 1 - \frac{2p^2 + 3p - 1}{6(p+1)(k-1)} \left(\sum_{i=1}^{k} \frac{1}{n_i} - \frac{1}{\Sigma n_i} \right) \tag{4}$$

is introduced the quantity MC^{-1} is approximately distributed as a chi-squared variate with degrees of freedom $\frac{1}{2}(k-1)p(p+1)$ as the n_i become large. If all the n_i are equal to n,

$$C^{-1} = 1 - \frac{(2p^2 + 3p - 1)(k+1)}{6(p+1)kn} \tag{5}$$

The chi-squared approximation appears to be good if k and p do not exceed four or five, and each n_i is perhaps twenty or more. For greater p and k and small n_i Box has proposed an F-distribution approximation. Further comments on the accuracy of the Box distributions have been given by Pearson (1969). Korin (1969) has prepared tables of the upper 0.05 critical values of M for the case of equal n_i; these have been reproduced by Pearson and Hartley (1972).

The statistic (3) is a generalization of the Bartlett test for homogeneity of variances, and in it the determinants assume the role of "generalized variances." As Box (1953) and others have shown, the Bartlett test is notably sensitive to nonnormality, and as one should expect, the distribution of the statistic M depends strongly on the multinormality assumption. Hopkins and Clay (1963) have shown by simulation that the probability of exceeding the nominal 0.05 critical value is greatly increased for leptokurtic, or highly peaked, nonnormal bivariate populations. The test should not be used indiscriminately with data for which the multinormal model seems unrealistic.

Example 7.3. In a reaction-time study* 32 male and 32 female young normal subjects reacted to visual stimuli preceded by warning intervals of different lengths. The sample covariance matrices of reaction times with preparatory intervals of 0.5 and 15 sec were

$$\mathbf{S}_M = \begin{bmatrix} 4.32 & 1.88 \\ 1.88 & 9.18 \end{bmatrix} \qquad \mathbf{S}_F = \begin{bmatrix} 2.52 & 1.90 \\ 1.90 & 10.06 \end{bmatrix} \qquad \mathbf{S} = \begin{bmatrix} 3.42 & 1.89 \\ 1.89 & 9.62 \end{bmatrix}$$

where the elements are in units of 10^{-4} sec^2. It is desired to test the hypothesis of a common covariance matrix in both sexes.

$$\begin{aligned} M &= 2(N-1)\ln|\mathbf{S}| - (N-1)(\ln|\mathbf{S}_M| + \ln|\mathbf{S}_F|) \\ &= 62 \ln 29.328 - 31(\ln 36.123 + \ln 21.741) \\ &= 2.82 \end{aligned}$$

$C^{-1} = 0.965$, and since $MC^{-1} = 2.72$ is much smaller than the percentage point $\chi^2_{0.05;3} = 7.81$, we conclude the null hypothesis is indeed tenable.

Greenstreet and Connor (1974) have simulated the power of the tests based on MC^{-1} and three other modifications of the likelihood criterion. A number of alternative tests have been proposed for the special case of $k = 2$ populations by Roy, Pillai, and others. The powers of four such criteria have been computed by Pillai and Jayachandran (1968), while Mardia (1971) has demonstrated the sensitivity of one test to nonnormality.

7.5 TESTING THE INDEPENDENCE OF SETS OF VARIATES

In Chap. 3 we saw that the independence of two responses with a joint bivariate normal distribution could be tested by their sample correlation and that the independence of one response with the remaining

* I am indebted to Jack Botwinick for kindly permitting the use of these data.

$p-1$ in a multinormal system could be tested through the multiple correlation coefficient. Now suppose that an investigator is concerned with a system of $p+q$ multinormal responses: the first p variates possess some common feature, while the remaining q are characterized in some other way. For example, if the responses were observations obtained during a stress situation, the first set might contain blood-pressure readings, galvanic skin resistance, and other physiological observations, while the second set might contain measures of anxiety and other affective features. These sets are always formed from the nature of the responses or other external means and never from an inspection of the data or the pattern of the sample correlations. The vector of the response variates is partitioned as

$$\mathbf{X}' = [\mathbf{X}_1' \quad \mathbf{X}_2']$$

according to the grouping of the responses, and the parameters of its $(p+q)$-dimensional multinormal distribution are similarly partitioned as

$$(1) \qquad \boldsymbol{\mu} = \begin{bmatrix} \boldsymbol{\mu}_1 \\ \boldsymbol{\mu}_2 \end{bmatrix} \qquad \boldsymbol{\Sigma} = \begin{bmatrix} \boldsymbol{\Sigma}_{11} & \boldsymbol{\Sigma}_{12} \\ \boldsymbol{\Sigma}_{12}' & \boldsymbol{\Sigma}_{22} \end{bmatrix}$$

$\boldsymbol{\Sigma}$ will be assumed to be nonsingular, and thereby we shall tacitly assume that its sample estimate $\mathbf{S}$ is nonsingular (apart from the usual collinear sample that occurs with probability zero). From this population N independent observation vectors have been drawn, and their covariance matrix has been partitioned in conformance with (1) as

$$(2) \qquad \mathbf{S} = \begin{bmatrix} \mathbf{S}_{11} & \mathbf{S}_{12} \\ \mathbf{S}_{12}' & \mathbf{S}_{22} \end{bmatrix}$$

We shall assume throughout that $N \geq p + q + 1$. Through these sample estimates we wish to test the hypothesis

$$(3) \qquad H_0\colon \quad \boldsymbol{\Sigma}_{12} = \mathbf{0}$$

of independence of the variates in the first and second sets, as opposed to the alternative

$$(4) \qquad H_1\colon \quad \boldsymbol{\Sigma}_{12} \neq \mathbf{0}$$

of some degree of dependence between the sets. We recall from Property 3.2 of Sec. 3.4 that $\boldsymbol{\Sigma}_{12} = \mathbf{0}$ is a necessary and sufficient condition for the independence of multinormal sets.

Unlike the tests of the previous sections we shall first construct that of (3) by the union-intersection principle. From the response variates form the linear compounds

$$x = \mathbf{a}'\mathbf{X}_1 \qquad y = \mathbf{b}'\mathbf{X}_2 \tag{5}$$

The covariance matrix of these new scalar variates is

$$\begin{bmatrix} \mathbf{a}' & \mathbf{0}' \\ \mathbf{0}' & \mathbf{b}' \end{bmatrix} \boldsymbol{\Sigma} \begin{bmatrix} \mathbf{a} & \mathbf{0} \\ \mathbf{0} & \mathbf{b} \end{bmatrix} = \begin{bmatrix} \mathbf{a}'\boldsymbol{\Sigma}_{11}\mathbf{a} & \mathbf{a}'\boldsymbol{\Sigma}_{12}\mathbf{b} \\ \mathbf{a}'\boldsymbol{\Sigma}_{12}\mathbf{b} & \mathbf{b}'\boldsymbol{\Sigma}_{22}\mathbf{b} \end{bmatrix} \tag{6}$$

and their correlation is

$$\rho(\mathbf{a},\mathbf{b}) = \frac{\mathbf{a}'\boldsymbol{\Sigma}_{12}\mathbf{b}}{\sqrt{\mathbf{a}'\boldsymbol{\Sigma}_{11}\mathbf{a}\mathbf{b}'\boldsymbol{\Sigma}_{22}\mathbf{b}}} \tag{7}$$

Now the matrix statement $\boldsymbol{\Sigma}_{12} = \mathbf{0}$ is equivalent to the infinite set of scalar equations $\mathbf{a}'\boldsymbol{\Sigma}_{12}\mathbf{b} = 0$ for all nonnull p- and q-component vectors $\mathbf{a}$ and $\mathbf{b}$, and in this manner acceptance of our hypothesis (3) is equivalent to accepting all hypotheses

$$H_0\colon \quad \rho(\mathbf{a},\mathbf{b}) = 0 \tag{8}$$

for different $\mathbf{a}$ and $\mathbf{b}$. The estimate of $\rho(\mathbf{a},\mathbf{b})$ is

$$r(\mathbf{a},\mathbf{b}) = \frac{\mathbf{a}'\mathbf{S}_{12}\mathbf{b}}{\sqrt{\mathbf{a}'\mathbf{S}_{11}\mathbf{a}\mathbf{b}'\mathbf{S}_{22}\mathbf{b}}} \tag{9}$$

and the acceptance region for the hypothesis (8) would be

$$r^2(\mathbf{a},\mathbf{b}) \le r_\beta^2 \tag{10}$$

where the right-hand term is the appropriate critical value of the distribution of r. Therefore, the acceptance region for the original hypothesis is the *intersection*

$$\bigcap_{\mathbf{a},\mathbf{b}} [r^2(\mathbf{a},\mathbf{b}) \le r_\beta^2] \tag{11}$$

of all regions of the sort specified by (10) for the totality of nonnull vectors $\mathbf{a}$ and $\mathbf{b}$. But, as we saw in the derivation of the Hotelling T^2 statistic in Chap. 4, such a region is equivalent to that defined by

$$\max_{\mathbf{a},\mathbf{b}} r^2(\mathbf{a},\mathbf{b}) \le r_\beta^2 \tag{12}$$

for if the maximum sample correlation of the pairs of linear compounds is in the acceptance region, all smaller correlations must also lie therein. The test statistic for the independence hypothesis is this maximum correlation.

To determine the maximum value of $r^2(\mathbf{a},\mathbf{b})$ we begin by imposing the constraints

$$\mathbf{a}'\mathbf{S}_{11}\mathbf{a} = 1 \qquad \mathbf{b}'\mathbf{S}_{22}\mathbf{b} = 1 \tag{13}$$

on the scales of the linear compounds, for as we have frequently observed in Chap. 3, correlation coefficients are invariant under changes of scale. If these constraints are introduced through the Lagrangian multipliers λ and μ, $r^2(\mathbf{a},\mathbf{b})$ is at a maximum when the derivatives of

$$f(\mathbf{a},\mathbf{b}) = (\mathbf{a}'\mathbf{S}_{12}\mathbf{b})^2 - \lambda(\mathbf{a}'\mathbf{S}_{11}\mathbf{a} - 1) - \mu(\mathbf{b}'\mathbf{S}_{22}\mathbf{b} - 1) \tag{14}$$

with respect to the elements of **a** and **b** are equal to zero. Those equations are

$$\begin{aligned} -\lambda\mathbf{S}_{11}\mathbf{a} + (\mathbf{a}'\mathbf{S}_{12}\mathbf{b})\mathbf{S}_{12}\mathbf{b} &= \mathbf{0} \\ (\mathbf{a}'\mathbf{S}_{12}\mathbf{b})\mathbf{S}'_{12}\mathbf{a} - \mu\mathbf{S}_{22}\mathbf{b} &= \mathbf{0} \end{aligned} \tag{15}$$

Premultiplication of the first equation by $\mathbf{a}'$ and the second by $\mathbf{b}'$ shows that

$$\lambda = \mu = (\mathbf{a}'\mathbf{S}_{12}\mathbf{b})^2 \tag{16}$$

and under the constraints (13) the single Lagrangian multiplier λ is the maximum value of $r^2(\mathbf{a},\mathbf{b})$ over all choices of **a** and **b**.

For the equations (15) to have a nonnull solution it is necessary that their determinant vanish. With the aid of (16) that condition can be written as

$$\begin{vmatrix} -\lambda^{1/2}\mathbf{S}_{11} & \mathbf{S}_{12} \\ \mathbf{S}'_{12} & -\lambda^{1/2}\mathbf{S}_{22} \end{vmatrix} = 0 \tag{17}$$

If $\mathbf{S}_{12}$ contains some nonzero elements, the maximum correlation cannot equal zero, and division by $\lambda^{1/2}$ is permissible. By the results on partitioned matrices of Sec. 2.11, the determinantal equation can be written as

$$\begin{aligned} |-\lambda\mathbf{S}_{11} + \mathbf{S}_{12}\mathbf{S}_{22}^{-1}\mathbf{S}'_{12}| &= 0 \\ |-\lambda\mathbf{S}_{22} + \mathbf{S}'_{12}\mathbf{S}_{11}^{-1}\mathbf{S}_{12}| &= 0 \end{aligned} \tag{18}$$

for the positive definiteness of **S** implies that $|\mathbf{S}_{11}|$ and $|\mathbf{S}_{22}|$ cannot be zero. Since the union-intersection principle states that the test statistic should be the greatest squared correlation, that statistic is the greatest root of (17) or its equivalent equations in (18). It is easily seen that the roots of the latter are the characteristic roots of the matrices

$$\mathbf{S}_{11}^{-1}\mathbf{S}_{12}\mathbf{S}_{22}^{-1}\mathbf{S}'_{12} \qquad \mathbf{S}_{22}^{-1}\mathbf{S}'_{12}\mathbf{S}_{11}^{-1}\mathbf{S}_{12} \tag{19}$$

or their other cyclic permutations.

Now let us formalize the test of the independence hypothesis. Denote the greatest characteristic root of the matrices (19) by c_1. When the two sets of variates are independent, c_1 has the greatest-characteristic-root distribution introduced in our preceding treatment of the multivariate analysis of variance. The hypothesis is accepted if

$$c_1 \leq x_{\alpha;s,m,n} \tag{20}$$

and rejected otherwise. $x_{\alpha;s,m,n}$ is the upper 100α percentage point of the greatest-root distribution with parameters

$$s = \min(p,q)$$

$$m = \frac{|p - q| - 1}{2} \tag{21}$$

$$n = \frac{N - p - q - 2}{2}$$

For a test of level α we compute c_1, determine $x_{\alpha;s,m,n}$ from the appropriate Appendix table or chart, and accept or reject the hypothesis according to the decision rule (20).

Before turning to an example let us investigate some essential properties of the test statistic c_1. We note first that its value is unchanged by separate nonsingular transformations on the response variates. Let such transformed responses be $\mathbf{Y}_1 = \mathbf{AX}_1$ and $\mathbf{Y}_2 = \mathbf{BX}_2$, where the determinants of the $p \times p$ matrix $\mathbf{A}$ and the $q \times q$ matrix $\mathbf{B}$ do not vanish. The first matrix of (19) becomes

$$(\mathbf{AS}_{11}\mathbf{A}')^{-1}\mathbf{AS}_{12}\mathbf{B}'(\mathbf{BS}_{22}\mathbf{B}')^{-1}(\mathbf{AS}_{12}\mathbf{B}')' = \mathbf{A}'^{-1}\mathbf{S}_{11}^{-1}\mathbf{S}_{12}\mathbf{S}_{22}^{-1}\mathbf{S}_{12}'\mathbf{A}' \tag{22}$$

Since the nonzero characteristic roots of the matrices $\mathbf{AB}$ and $\mathbf{BA}$ are equal, it follows immediately that the greatest root of (22) is c_1. In particular, if $\mathbf{A}$ and $\mathbf{B}$ are diagonal matrices containing the reciprocals of the sample standard deviations, the transformed version of the first matrix of (19) is the $p \times p$ matrix

$$R_{11}^{-1}R_{12}R_{22}^{-1}R_{12}' \tag{23}$$

computed from the appropriate submatrices of the partitioned sample correlation matrix. The correlations alone contain the information for the test.

Examination of the matrices (19) or (23) shows that c_1 is the ordinary squared sample correlation when $p = q = 1$. For $p = 1$ and general q, c_1 is the squared multiple correlation of the first variate with the second set. We shall see in the next section that the smaller characteristic roots are also interpretable as squared correlations.

Example 7.4. In an investigation of the relation of the Wechsler Adult Intelligence Scale to age Birren and Morrison (1961) obtained this matrix of correlations among the digit span and vocabulary subtests, chronological age, and years of formal education:

$$\mathbf{R} = \begin{bmatrix} 1 & 0.45 & -0.19 & 0.43 \\ & 1 & -0.02 & 0.62 \\ & & 1 & -0.29 \\ & & & 1 \end{bmatrix}$$

The sample consisted of $N = 933$ white, native-born men and women aged 25 to 64. From these data we wish to test at level $\alpha = 0.01$ the hypothesis that the pair of WAIS subtest variates is distributed independently of the age and education variates. Here $p = q = 2$, and

$$\mathbf{R}_{11} = \begin{bmatrix} 1 & 0.45 \\ 0.45 & 1 \end{bmatrix} \qquad \mathbf{R}_{12} = \begin{bmatrix} -0.19 & 0.43 \\ -0.02 & 0.62 \end{bmatrix} \qquad \mathbf{R}_{22} = \begin{bmatrix} 1 & -0.29 \\ -0.29 & 1 \end{bmatrix}$$

$$\mathbf{R}_{11}^{-1} = \begin{bmatrix} 1.254 & -0.564 \\ -0.564 & 1.254 \end{bmatrix} \qquad \mathbf{R}_{22}^{-1} = \begin{bmatrix} 1.092 & 0.317 \\ 0.317 & 1.092 \end{bmatrix}$$

$$\mathbf{R}_{12}\mathbf{R}_{22}^{-1}\mathbf{R}_{12}'\mathbf{R}_{11}^{-1} = \begin{bmatrix} 0.0937 & 0.2130 \\ 0.0873 & 0.3730 \end{bmatrix}$$

The characteristic equation of the last matrix is

$$\lambda^2 - 0.4667\lambda + 0.0164 = 0$$

and its roots are $c_1 = 0.4285$ and $c_2 = 0.0381$. The parameters of the distribution of c_1 under the independence hypothesis are $s = 2$, $m = -\frac{1}{2}$, and $n = 463.5$, and the 1 percent critical value can be read from Chart 10 as approximately

$$x_{0.01;\,2,-\frac{1}{2},463.5} = 0.02$$

Since c_1 greatly exceeds this value, we reject the hypothesis and accept one of some sort of dependence between the subtest and concomitant groups of variates.

Independence of k Sets of Variates. We have adopted the union-intersection test of this section for its direct connection with the notion of canonical correlation. For more than two subsets, however, that approach presents some complications. Wilks (1935) originally proposed a likelihood-ratio test for the independence of several sets of multinormal variates. If the ith of k sets contains p_i variates, so that the covariance matrix can be partitioned into submatrices $\mathbf{\Sigma}_{ij}$ of dimensions $p_i \times p_j$, the hypothesis is

$$H_0: \quad \mathbf{\Sigma}_{ij} = \mathbf{0} \tag{24}$$

for all $i \neq j$. Within the ith set the covariance matrix $\mathbf{\Sigma}_{ii}$ need only be positive definite. If a sample of N independent observation vectors has been drawn from this population and its covariance and correlation matrices have been partitioned in conformity with the population covariance matrix, Wilks's test statistic is the determinantal ratio

$$\begin{aligned} V &= \frac{|\mathbf{S}|}{|\mathbf{S}_{11}| \cdots |\mathbf{S}_{kk}|} \\ &= \frac{|\mathbf{R}|}{|\mathbf{R}_{11}| \cdots |\mathbf{R}_{kk}|} \end{aligned} \tag{25}$$

Although the exact distribution of V is complicated, Wilks and other workers have obtained good approximations to it in terms of tabulated functions. In particular Box (1949) has shown that

$$\chi^2 = -\frac{N-1}{C}\ln V \tag{26}$$

is approximately distributed as a chi-squared variate with f degrees of freedom when the hypothesis (24) is true. For this statistic

$$C^{-1} = 1 - \frac{1}{12f(N-1)}(2\Sigma_3 + 3\Sigma_2) \tag{27}$$

$$f = \tfrac{1}{2}\Sigma_2$$

where

$$\Sigma_s = \left(\sum_{i=1}^{k} p_i\right)^s - \sum_{i=1}^{k} p_i^s \qquad s = 2, 3$$

For a test of level α we would reject H_0 if $\chi^2 > \chi^2_{\alpha;f}$. The accuracy of this and other approximations has been investigated by Box.

Example 7.5. For a simple illustration of the Wilks statistic we shall use the WAIS correlations of Example 7.4. Then

$$\begin{aligned} V &= \frac{|\mathbf{R}_{11} - \mathbf{R}_{12}\mathbf{R}_{22}^{-1}\mathbf{R}'_{12}|}{|\mathbf{R}_{11}|} \\ &= |\mathbf{I} - \mathbf{R}_{11}^{-1}\mathbf{R}_{12}\mathbf{R}_{22}^{-1}\mathbf{R}'_{12}| \\ &= 0.5497 \end{aligned}$$

$\Sigma_2 = 8$, $\Sigma_3 = 48$, and $C^{-1} = 0.9973$, so that $\chi^2 = 556$. Since that would exceed any reasonable critical value we must conclude again that the subtests are dependent upon age and education.

7.6 CANONICAL CORRELATION

In the construction of the test of the preceding section we never explicitly produced the coefficients of the linear compounds, nor did we choose to discuss the test statistic in its squared-correlation sense. Now we shall consider the use of those quantities for describing the dependencies between the two sets. This technique is due to Hotelling (1935, 1936b), and is known as the method of *canonical correlation.* Although the problem can be formulated as one of estimating the canonical correlations and variates of a partitioned multinormal population, we shall prefer to develop it in descriptive terms from the sample.

Suppose that the $p + q$ variates $[\mathbf{X}'_1 \quad \mathbf{X}'_2]$ of some multidimensional population have been divided so that their covariance matrix $\mathbf{\Sigma}$ has the partitioned form of the last section. For the moment we shall drop the earlier requirement that the distribution be multinormal and merely specify that:

1. The elements of $\boldsymbol{\Sigma}$ are finite.
2. $\boldsymbol{\Sigma}$ is of full rank $p + q$.
3. The first $r \leq \min(p,q)$ characteristic roots of $\boldsymbol{\Sigma}_{11}^{-1}\boldsymbol{\Sigma}_{12}\boldsymbol{\Sigma}_{22}^{-1}\boldsymbol{\Sigma}_{12}'$ are distinct.

From this population N observation vectors have been randomly drawn, and their covariance matrix has been partitioned as

$$\mathbf{S} = \begin{bmatrix} \mathbf{S}_{11} & \mathbf{S}_{12} \\ \mathbf{S}_{12}' & \mathbf{S}_{22} \end{bmatrix} \tag{1}$$

in conformance with $\boldsymbol{\Sigma}$. The correlations computable from $\mathbf{S}_{12}$ appear to be substantial and significant for a sample of size N. However, the study of the dependence between the sets involves not only those pq correlations but the $\frac{1}{2}p(p-1) + \frac{1}{2}q(q-1)$ intraset correlations. In an attempt to reduce the number of variates and their nonzero correlations to more parsimonious degrees we propose the following question:

What are the linear compounds

$$\begin{array}{ll} u_1 = \mathbf{a}_1'\mathbf{X}_1 & v_1 = \mathbf{b}_1'\mathbf{X}_2 \\ \cdots\cdots & \cdots\cdots \\ u_s = \mathbf{a}_s'\mathbf{X}_1 & v_s = \mathbf{b}_s'\mathbf{X}_2 \end{array} \tag{2}$$

with the property that the sample correlation of u_1 and v_1 is greatest, the sample correlation of u_2 and v_2 greatest among all linear compounds uncorrelated with u_1 and v_1, and so on, for all $s = \min(p,q)$ possible pairs?

Clearly, the first pair has the coefficients determined implicitly in the union-intersection test of set independence, and its squared correlation is the test statistic. If we introduce the new constraints

$$\begin{aligned} \mathbf{a}_i'\mathbf{S}_{11}\mathbf{a}_j &= 0 \\ \mathbf{b}_i'\mathbf{S}_{22}\mathbf{b}_j &= 0 \\ \mathbf{a}_i'\mathbf{S}_{12}\mathbf{b}_j &= 0 \\ \mathbf{a}_j'\mathbf{S}_{12}\mathbf{b}_i &= 0 \end{aligned} \qquad i \neq j \tag{3}$$

into the earlier maximization problem, it can be shown that the coefficients of the ith pair are given by the homogeneous linear equations

(4)
$$\begin{aligned}(\mathbf{S}_{12}\mathbf{S}_{22}^{-1}\mathbf{S}_{12}' - c_i\mathbf{S}_{11})\mathbf{a}_i &= \mathbf{0}\\ (\mathbf{S}_{12}'\mathbf{S}_{11}^{-1}\mathbf{S}_{12} - c_i\mathbf{S}_{22})\mathbf{b}_i &= \mathbf{0}\end{aligned}$$

where c_i is the ith largest root of the determinantal equations

(5)
$$|\mathbf{S}_{12}\mathbf{S}_{22}^{-1}\mathbf{S}_{12}' - \lambda\mathbf{S}_{11}| = 0 \qquad \text{or} \qquad |\mathbf{S}_{12}'\mathbf{S}_{11}^{-1}\mathbf{S}_{12} - \lambda\mathbf{S}_{22}| = 0$$

Recall that c_i is the squared product-moment correlation of the ith linear compounds:

(6)
$$\begin{aligned}c_i &= r^2_{ui,vi}\\ &= \frac{(\mathbf{a}_i'\mathbf{S}_{12}\mathbf{b}_i)^2}{\mathbf{a}_i'\mathbf{S}_{11}\mathbf{a}_i\mathbf{b}_i'\mathbf{S}_{22}\mathbf{b}_i}\end{aligned}$$

If c_i is a distinct root, the coefficient vectors $\mathbf{a}_i$ and $\mathbf{b}_i$ will be unique, and their linear compounds will be uncorrelated with the other canonical variates. To demonstrate this property we assume that c_i and c_j are distinct roots and write the homogeneous equations for their respective pairs of coefficients as

$$\begin{aligned}(\mathbf{S}_{12}\mathbf{S}_{22}^{-1}\mathbf{S}_{12}' - c_i\mathbf{S}_{11})\mathbf{a}_i &= \mathbf{0}\\ (\mathbf{S}_{12}'\mathbf{S}_{11}^{-1}\mathbf{S}_{12} - c_i\mathbf{S}_{22})\mathbf{b}_i &= \mathbf{0}\\ (\mathbf{S}_{12}\mathbf{S}_{22}^{-1}\mathbf{S}_{12}' - c_j\mathbf{S}_{11})\mathbf{a}_j &= \mathbf{0}\\ (\mathbf{S}_{12}'\mathbf{S}_{11}^{-1}\mathbf{S}_{12} - c_j\mathbf{S}_{22})\mathbf{b}_j &= \mathbf{0}\end{aligned}$$

Premultiply these equations by $\mathbf{a}_j'$, $\mathbf{b}_j'$, $\mathbf{a}_i'$, and $\mathbf{b}_i'$, respectively. Then

$$\begin{aligned}(c_i - c_j)\mathbf{a}_i'\mathbf{S}_{11}\mathbf{a}_j &= 0\\ (c_i - c_j)\mathbf{b}_i'\mathbf{S}_{22}\mathbf{b}_j &= 0\end{aligned}$$

and since $c_i \neq c_j$, it must follow that the bilinear forms equal to the covariances of u_i, u_j and v_i, v_j are zero. Similarly, premultiplication of the first and fourth equations by $\mathbf{b}_j'\mathbf{S}_{12}'\mathbf{S}_{11}^{-1}$ and $\mathbf{a}_i'\mathbf{S}_{12}\mathbf{S}_{22}^{-1}$, respectively, and subtraction of the resulting first scalar equation from the second leads to the condition

$$(c_i - c_j)\mathbf{b}_j'\mathbf{S}_{12}'\mathbf{a}_i = 0$$

of zero correlation of u_i and v_j when c_i and c_j are distinct. If, as in the population, the r largest c_i are distinct, r pairs of unique canonical variates can be formed. Unlike similar results that we shall encounter in the next chapter, it does *not* necessarily follow that coefficient vectors with different subscripts are orthogonal.

Let us summarize these results. We began with a sample whose observation vectors had the covariance matrix (1). Through the canonical variate transformation we have passed to the new scores $u_1, \ldots, u_s, v_1, \ldots, v_s$ with correlation matrix

$$\begin{bmatrix} 1 & \cdots & 0 & c_1^{1/2} & \cdots & 0 \\ \cdot\;\cdot & \cdots & \cdot\;\cdot & \cdot\;\cdot & \cdots & \cdot\;\cdot \\ 0 & \cdots & 1 & 0 & \cdots & c_s^{1/2} \\ c_1^{1/2} & \cdots & 0 & 1 & \cdots & 0 \\ \cdot\;\cdot & \cdots & \cdot\;\cdot & \cdot\;\cdot & \cdots & \cdot\;\cdot \\ 0 & \cdots & c_s^{1/2} & 0 & \cdots & 1 \end{bmatrix} \tag{7}$$

All the correlation *between* the sets of the original variates has been channeled through the s canonical correlations.

Identical canonical correlations will be obtained from

$$\mathbf{S}_{11}^{-1}\mathbf{S}_{12}\mathbf{S}_{22}^{-1}\mathbf{S}_{12}'$$

or its counterpart

$$\mathbf{R}_{11}^{-1}\mathbf{R}_{12}\mathbf{R}_{22}^{-1}\mathbf{R}_{12}' \tag{8}$$

In the first case the elements of $\mathbf{a}_i$ and $\mathbf{b}_i$ will be in units proportional to those of the respective responses in the sets, and the dimensions of u_i and v_i will be meaningless. Canonical variates based on the correlation matrix are dimensionless and should be evaluated in terms of the standard scores

$$z_{ij} = \frac{x_{ij} - \bar{x}_j}{s_j}$$

of the original observations. It is essential to keep these distinctions in mind if the canonical variates are to be evaluated for each subject or sampling unit as part of a data-reduction scheme or as derived variates for later analyses.

Example 7.6. Let us find the canonical correlations and variates of the data treated in Example 7.4. The first characteristic root is $c_1 = 0.4285$, and the larger canonical correlation is 0.654. The smaller root is $c_2 = 0.0381$, and the second canonical correlation is approximately 0.196. The coefficients of the first WAIS canonical variate are given by the equations

$$\begin{bmatrix} 0.1896 & 0.2552 \\ 0.2552 & 0.4123 \end{bmatrix} \begin{bmatrix} a_{11} \\ a_{12} \end{bmatrix} = 0.4285 \begin{bmatrix} 1 & 0.45 \\ 0.45 & 1 \end{bmatrix} \begin{bmatrix} a_{11} \\ a_{12} \end{bmatrix}$$

or

$$-0.2389a_{11} + 0.0624a_{12} = 0$$

$$0.0624a_{11} - 0.0162a_{12} = 0$$

Their solution is proportional to $a_{11} = 0.26$, $a_{12} = 1$, where these coefficients can be scaled at the investigator's convenience. The coefficients of the first age-educational canonical variate are given by the equations

$$\begin{bmatrix} 0.0415 & -0.0467 \\ -0.0467 & 0.4130 \end{bmatrix} \begin{bmatrix} b_{11} \\ b_{12} \end{bmatrix} = 0.4285 \begin{bmatrix} 1 & -0.29 \\ -0.29 & 1 \end{bmatrix} \begin{bmatrix} b_{11} \\ b_{12} \end{bmatrix}$$

whose solution is proportional to $b_{11} = 0.20$, $b_{12} = 1.00$. If the greater coefficient in each variate is scaled to unity, the canonical responses are

$$u_1 = 0.26(\text{WAIS digit span}) + (\text{WAIS vocabulary})$$

$$v_1 = 0.20(\text{age}) + (\text{education})$$

We see that the canonical variate for the WAIS subtest set places nearly four times as much weight on vocabulary as on the digit-span score. Similarly, its correlative in the concomitant variates weights education five times that of chronological age. The major link between the sets appears to be a verbal-experiential one. Since the coefficients were computed from the correlation matrix, these weights refer to the *standard scores* of the four responses, and the actual evaluation of u_1 and v_1 would of course be computed from the subjects' transformed observations z_{ij}.

Replacement of 0.4285 by the smaller root 0.0381 in the preceding systems of equations leads to these coefficients for the second canonical variates:

$$a_{21} = 1 \qquad a_{22} = -0.64 \qquad b_{21} = 1 \qquad b_{22} = 0.10$$

These variates have an interesting interpretation: the first appears to be a weighted comparison of the "performance" and "verbal" subtests digit span and vocabulary with chronological age. If the negligible coefficient of education is disregarded, the pair of scores would reflect the widening gap with advancing age between accumulated knowledge and performance skills.

The reader may verify that the derived variates corresponding to different canonical correlations are uncorrelated. For such computations it should be noted that most of the eight canonical coefficients are accurate only to their given two places.

Some Further References. Kettenring (1971) has developed and compared extensions of canonical correlation to three or more sets of variates, and has given iterative schemes for the computation of the correlations and coefficients. McKeon (1965) has prepared a useful overview of canonical analysis and its relationships with the multivariate analysis of variance, classification, and the scaling of categorical data.

7.7 EXERCISES

1. Test the hypothesis

$$\mathbf{\Sigma} = \begin{bmatrix} 10 & 6 & 6 \\ & 10 & 6 \\ & & 10 \end{bmatrix}$$

on the covariance structure of the first three variates (information, similarities, and arithmetic) of Example 4.3, Sec. 4.4.

2. Test the hypothesis of Example 7.1 by the union-intersection statistic and the appropriate critical value from the Hanumara and Thompson (1968) or Pearson and Hartley (1972) tables.

3. Test the (a) equal variance, equal covariance and (b) Type H hypotheses for the covariance matrix of the growth measurements of Example 5.9, Sec. 5.7.

4. Test the hypothesis of equal population covariance matrices for the samples of Exercise 5, Chap. 4.

5. Jolicoeur and Mosimann (1960) measured the lengths, widths, and heights of the carapaces of 24 male and 24 female painted turtles. The covariance matrices of those observations were

$$\mathbf{S}_M = \begin{bmatrix} 138.77 & 79.15 & 37.38 \\ & 50.04 & 21.65 \\ & & 11.26 \end{bmatrix} \qquad \mathbf{S}_F = \begin{bmatrix} 451.39 & 271.17 & 168.70 \\ & 171.73 & 103.29 \\ & & 66.65 \end{bmatrix}$$

Test the hypothesis of equal population matrices for each sex at the 0.05 level.

6. In an investigation of the growth of fruit trees Pearce and Holland (1960) obtained this matrix of correlations of the logarithms of the weight of the mature tree at grubbing, basal trunk girth of the mature tree, total shoot growth during the first four years, and basal trunk girth at 4 years of age:

$$\mathbf{R} = \begin{bmatrix} 1.000 & 0.939 & 0.266 & 0.178 \\ & 1.000 & 0.424 & 0.358 \\ & & 1.000 & 0.835 \\ & & & 1.000 \end{bmatrix}$$

The trees were set out in a randomized-block design that led to 28 degrees of freedom for each sum of squares and products of the measurements.

a. Under the usual assumption of multinormality test the hypothesis of independence of the 4-year and maturity sets of responses. In (21) of Sec. 7.5, N is now equal to 29.

b. Extract the two canonical correlations and their associated coefficient vectors.

7. Show that the canonical variate vectors and characteristic roots for the WAIS scores and age-education sets of Example 3.3 are

$$\mathbf{a}_1' = [0.70, 1] \qquad \mathbf{b}_1' = [-0.12, 1] \qquad c_1 = 0.49$$

$$\mathbf{a}_2' = [1, -0.94] \qquad \mathbf{b}_2' = [1, 0.41] \qquad c_2 = 0.22$$

Compare this analysis with that of Example 7.2 and the multiple-regression solutions of Example 3.4.

8. Birren and Morrison (1961) reported the following canonical-correlation analysis for the complete set of eleven WAIS subtests and the age and education variates:

Response	Canonical variate	
	1	2
Set 1. WAIS Subtests:		
Information	0.91	0.50
Comprehension	0.12	0.04
Arithmetic	0.01	0.24
Similarities	0.24	−0.18
Digit span	0.14	−0.04
Vocabulary	0.07	1.00
Digit symbol	1.00	−0.69
Picture completion	−0.11	−0.23
Block design	0.04	−0.19
Picture arrangement	0.33	−0.48
Object assembly	0.14	−0.10
Set 2:		
Years of education	1.00	0.45
Age	−0.19	1.00
Canonical correlation	0.74	0.56

a. Compare these coefficients and correlations with those obtained in Example 7.6 and Exercise 7.

b. What substantive interpretations would you give to the canonical variates?

8
THE STRUCTURE OF MULTIVARIATE OBSERVATIONS: I. Principal Components

8.1 INTRODUCTION. Earlier we discussed the use of partial, multiple, and canonical correlation for analyzing the dependence structure of a multinormal population. The proper use of those methods required that certain roles be assigned to some of the responses. For a partial correlation analysis it is necessary to decide which variables are to be correlated and which of the remaining responses must be held constant. Multiple correlation demands that one response be dependent upon some or all of the remaining variates. Similarly, for a canonical-correlation analysis the responses must be collected into two or more sets. All these choices depend upon the nature of the responses and other information external to the mere values of their correlations. The conclusions we may draw about the dependence structure will in turn depend upon those choices. Furthermore, if the analyses are repeated for different choices of the dependent or constant variates, the successive findings will hardly be independent or contain mutually exclusive bits of information about the structure.

It would seem clear that a new class of techniques will be required for picking apart the dependence structure when the responses are symmetric in nature or no *a priori* patterns of causality are available. Those methods fall under the general heading of *factor analysis*, for by them one attempts to descry those hidden factors which have generated the dependence or variation in the responses. That is, the observable, or *manifest*, variates are represented as functions of a smaller number of *latent* factor variates. The mathematical form of the functions must be one which will generate the covariances or correlations among the responses. If that form is simple, and if the latent variates are few in number, a more parsimonious description of the dependence structure can be obtained. Now for simplicity linear functions are difficult to surpass, and in the two principal techniques of the sequel we shall always think of the responses and their observations as linear compounds of the latent variates. The analysis of the dependence structure amounts to the statistical estimation of the coefficients of the functions.

We shall begin our study by developing in this chapter the Hotelling principal-component technique. That methodology originated with K. Pearson (1901) as a means of fitting planes by orthogonal least squares, but was later proposed by Hotelling (1933, 1936a) for the particular purpose of analyzing correlation structures. We shall initially define the principal components of a multivariate sample statistically and algebraically and then in terms of the geometry of the scatter swarm of the observations. Some numerical methods for extracting components will be treated, and the problem of interpreting component coefficients will be illustrated by some examples from biology and cognitive psychology, and by some special patterned correlation matrices. Some results of Anderson and Lawley on the sampling properties of principal components will be discussed and illustrated in the last section.

8.2 THE PRINCIPAL COMPONENTS OF MULTIVARIATE OBSERVATIONS

Suppose that the random variables $X_1, \ldots, X_p$ of interest have a certain multivariate distribution with mean vector $\boldsymbol{\mu}$ and covariance matrix $\boldsymbol{\Sigma}$. We assume, of course, that the elements of $\boldsymbol{\mu}$ and $\boldsymbol{\Sigma}$ are finite. The rank of $\boldsymbol{\Sigma}$ is $r \leq p$, and the q largest characteristic roots

$$\lambda_1 > \cdots > \lambda_q$$

of $\boldsymbol{\Sigma}$ are all distinct. For the present we shall not require a multinormal distribution of the X_i.

From this population a sample of N independent observation vectors has been drawn. The observations can be written as the usual $N \times p$ data matrix

$$\mathbf{X} = \begin{bmatrix} x_{11} & \cdots & x_{1p} \\ \cdots & \cdots & \cdots \\ x_{N1} & \cdots & x_{Np} \end{bmatrix} \tag{1}$$

Here a cautionary note on the ranks of $\boldsymbol{\Sigma}$ and $\mathbf{X}$ is in order. Mathematically, those matrices need not be of full rank p, nor need $\boldsymbol{\Sigma}$ contain more than one distinct characteristic root. However, the exigency of simplicity in our description of the latent structure of the X_i calls for a data matrix of full rank. We do not wish to confound the problem by including as responses total scores, weighted averages suggested by earlier studies, or other linear compounds which will reduce the rank of $\mathbf{X}$ and obscure whatever latent structure may be present.

The estimate of $\mathbf{\Sigma}$ will be the usual sample covariance matrix $\mathbf{S}$ defined by (9) of Sec. 3.5. The information we shall need for our principal-component analysis will be contained in $\mathbf{S}$. However, it will be necessary to make a choice of measures of dependence: should we work with the variances and covariances of the observations, and carry out our analyses in the original units of the responses, or would a more accurate picture of the dependence pattern be obtained if each x_{ij} were transformed to a standard score

$$z_{ij} = \frac{x_{ij} - \bar{x}_j}{s_j} \tag{2}$$

and the correlation matrix employed? The components obtained from $\mathbf{S}$ and $\mathbf{R}$ are in general not the same, nor is it possible to pass from one solution to the other by a simple scaling of the coefficients. Most applications of the technique have involved the correlation matrix, as if in keeping with the usage established by factor analysts. If the responses are in widely different units (age in years, weight in kilograms, and biochemical excretions in a variety of units, to cite one plausible case), linear compounds of the original quantities would have little meaning and the standardized variates and correlation matrix should be employed. Conversely, if the responses are reasonably commensurable, the covariance form has a greater statistical appeal, for, as we shall presently see, the ith principal component is that linear compound of the responses which explains the ith largest portion of the total response variance, and maximization of such total variance of standard scores has a rather artificial quality (Anderson, 1963a, p. 139). Furthermore, as Anderson has shown, the sampling theory of components extracted from correlation matrices is exceedingly more complex than that of covariance-matrix components.

The first principal component of the observations $\mathbf{X}$ is that linear compound

$$\begin{aligned} Y_1 &= a_{11}X_1 + \cdots + a_{p1}X_p \\ &= \mathbf{a}_1'\mathbf{x} \end{aligned} \tag{3}$$

of the responses whose sample variance

$$\begin{aligned} s_{Y_1}^2 &= \sum_{i=1}^{p} \sum_{j=1}^{p} a_{i1}a_{j1}s_{ij} \\ &= \mathbf{a}_1'\mathbf{S}\mathbf{a}_1 \end{aligned} \tag{4}$$

is greatest for all coefficient vectors normalized so that $\mathbf{a}_1'\mathbf{a}_1 = 1$. To determine the coefficients we introduce the normalization constraint by means of the Lagrange multiplier l_1 and differentiate with respect to $\mathbf{a}_1$:

$$(5) \qquad \frac{\partial}{\partial \mathbf{a}_1}[s_{Y_1}^2 + l_1(1 - \mathbf{a}_1'\mathbf{a}_1)] = \frac{\partial}{\partial \mathbf{a}_1}[\mathbf{a}_1'\mathbf{S}\mathbf{a}_1 + l_1(1 - \mathbf{a}_1'\mathbf{a}_1)]$$
$$= 2(\mathbf{S} - l_1\mathbf{I})\mathbf{a}_1$$

The coefficients must satisfy the p simultaneous linear equations

$$(6) \qquad (\mathbf{S} - l_1\mathbf{I})\mathbf{a}_1 = \mathbf{0}$$

If the solution to these equations is to be other than the null vector, the value of l_1 must be chosen so that

$$(7) \qquad |\mathbf{S} - l_1\mathbf{I}| = 0$$

l_1 is thus a characteristic root of the covariance matrix, and $\mathbf{a}_1$ is its associated characteristic vector. To determine which of the p roots should be used, premultiply the system of equations (6) by $\mathbf{a}_1'$. Since $\mathbf{a}_1'\mathbf{a}_1 = 1$, it follows that

$$(8) \qquad l_1 = \mathbf{a}_1'\mathbf{S}\mathbf{a}_1$$
$$= s_{Y_1}^2$$

But the coefficient vector was chosen to maximize this variance, and l_1 must be the *greatest* characteristic root of $\mathbf{S}$. Let us summarize these results in this form:

Definition 8.1. *The first principal component of the complex of sample values of the responses $X_1, \ldots, X_p$ is the linear compound*

$$(9) \qquad Y_1 = a_{11}X_1 + \cdots + a_{p1}X_p$$

whose coefficients a_{i1} are the elements of the characteristic vector associated with the greatest characteristic root l_1 of the sample covariance matrix of the responses. The a_{i1} are unique up to multiplication by a scale factor, and if they are scaled so that $\mathbf{a}_1'\mathbf{a}_1 = 1$, *the characteristic root l_1 is interpretable as the sample variance of Y_1.*

But what is the utility of this artificial variate constructed from the original responses? In the extreme case of $\mathbf{X}$ of rank one the first principal component would explain all the variation in the multivariate system. In the more usual case of the data matrix of full rank the importance and usefulness of the component would be measured by the proportion of the total variance attributable to it. If 87 percent of the variation in a system of six responses could be accounted for by a simple weighted average of the response values, it would appear that almost all the variation could be expressed along a single continuum rather than in six-dimensional space. Not only would this appeal to our sense of parsimony, but the coefficients of the six responses would indicate the relative importance of each original variate in the new derived component.

The second principal component is that linear compound

$$Y_2 = a_{12}X_1 + \cdots + a_{p2}X_p \tag{10}$$

whose coefficients have been chosen, subject to the constraints

$$\begin{aligned} \mathbf{a}_2'\mathbf{a}_2 &= 1 \\ \mathbf{a}_1'\mathbf{a}_2 &= 0 \end{aligned} \tag{11}$$

so that the variance of Y_2 is a maximum. The first constraint is merely a scaling to assure the uniqueness of the coefficients, while the second requires that $\mathbf{a}_1$ and $\mathbf{a}_2$ be orthogonal. The immediate consequence of the orthogonality is that the variances of the successive components sum to the total variance of the responses. The geometric implications will become clear in the next section. The coefficients of the second component are found by introducing the constraints (11) by the Lagrange multipliers l_2 and μ and differentiating with respect to $\mathbf{a}_2$:

$$\frac{\partial}{\partial \mathbf{a}_2}[\mathbf{a}_2'\mathbf{S}\mathbf{a}_2 + l_2(1 - \mathbf{a}_2'\mathbf{a}_2) + \mu\mathbf{a}_1'\mathbf{a}_2] = 2(\mathbf{S} - l_2\mathbf{I})\mathbf{a}_2 + \mu\mathbf{a}_1 \tag{12}$$

If the right-hand side is set equal to $\mathbf{0}$ and premultiplied by $\mathbf{a}_1'$, it follows from the normalization and orthogonality conditions that

$$2\mathbf{a}_1'\mathbf{S}\mathbf{a}_2 + \mu = 0 \tag{13}$$

Similar premultiplication of the equations (6) by $\mathbf{a}_2'$ implies that

$$\mathbf{a}_1'\mathbf{S}\mathbf{a}_2 = 0 \tag{14}$$

and hence $\mu = 0$. The second vector must satisfy

$$(\mathbf{S} - l_2\mathbf{I})\mathbf{a}_2 = \mathbf{0} \tag{15}$$

and it follows that the coefficients of the second component are thus the elements of the characteristic vector corresponding to the second greatest characteristic root. The remaining principal components are found in their turn from the other characteristic vectors. Let us summarize the process in this formal definition:

Definition 8.2. *The jth principal component of the sample of p-variate observations is the linear compound*

$$Y_j = a_{1j}X_1 + \cdots + a_{pj}X_p \tag{16}$$

whose coefficients are the elements of the characteristic vector of the sample covariance matrix $\mathbf{S}$ *corresponding to the jth largest characteristic root* l_j. *If* $l_i \neq l_j$, *the coefficients of the ith and jth components are necessarily orthogonal; if* $l_i = l_j$, *the elements can be chosen to be orthogonal, although an infinity of such orthogonal vectors exists. The sample variance of the jth component is* l_j, *and the total system variance is thus*

$$l_1 + \cdot \cdot \cdot + l_p = \text{tr } \mathbf{S} \tag{17}$$

The importance of the jth component in a more parsimonious description of the system is measured by

$$\frac{l_j}{\text{tr } \mathbf{S}} \tag{18}$$

The algebraic sign and magnitude of a_{ij} indicate the direction and importance of the contribution of the ith response to the jth component. A more precise and widely used statistical interpretation is also available. The sample covariances of the responses with the jth component are given by the column vector

$$\mathbf{S}\mathbf{a}_j \tag{19}$$

By the definition $(\mathbf{S} - l_j\mathbf{I})\mathbf{a}_j = \mathbf{0}$ of $\mathbf{a}_j$,

$$\mathbf{S}\mathbf{a}_j = l_j\mathbf{a}_j \tag{20}$$

and the covariance of the ith response with Y_j is merely $l_j a_{ij}$. If we divide by the component and response standard deviations, it follows that

$$\frac{a_{ij}\sqrt{l_j}}{s_i} \tag{21}$$

is the product moment correlation of the ith response and the jth component. If the components have been extracted from the correlation matrix, the correlations of the responses with the jth component are given by the vector $\sqrt{l_j}\,\mathbf{a}_j$. In presenting components in the sequel we shall usually adopt that form of weight.

The vectors $\sqrt{l_j}\,\mathbf{a}_j$ bear an important relation to the correlation or covariance matrix from which they were extracted. The diagonalization theorem stated in Sec. 2.10 implies that every real symmetric matrix $\mathbf{S}$ can be written as

$$\mathbf{S} = \mathbf{P}\mathbf{D}(l_i)\mathbf{P}'$$

where $\mathbf{P}$ is an orthogonal matrix and $\mathbf{D}(l_i)$ is the diagonal matrix of the characteristic roots of $\mathbf{S}$. If we take as columns of $\mathbf{P}$ the characteristic vectors of $\mathbf{S}$, it follows that

$$\mathbf{S} = \mathbf{P}\mathbf{D}(\sqrt{l_i})\mathbf{D}(\sqrt{l_i})\mathbf{P}' \tag{22}$$

Let

$$\mathbf{L} = \mathbf{P}\mathbf{D}(\sqrt{l_i})$$

Then the columns of **L** *reproduce* **S** by the relation

$$\mathbf{S} = l_1\mathbf{a}_1\mathbf{a}_1' + \cdots + l_r\mathbf{a}_r\mathbf{a}_r' \tag{23}$$

$$= \mathbf{L}\mathbf{L}'$$

The rank r of **S** may be less than p. As successive components are extracted from **S**, the matrices $l_i\mathbf{a}_i\mathbf{a}_i'$ can be formed and their running sum compared with **S** to determine how well that matrix is being generated by a smaller number of variates.

By the relation (23) principal-component analysis is equivalent to a factorization of **S** into the product of a matrix **L** and its transpose. As we shall see in the next chapter, this is also the purpose of factor analysis, wherein "factorization" of a matrix has precisely that algebraic meaning. *However*, in component analysis this factorization is unique up to the coefficient signs, for the component coefficients have been chosen to partition the total variance orthogonally into successively smaller portions, and if the portions are distinct, only one set of coefficient vectors will accomplish this purpose. This uniqueness of component coefficients is frequently overlooked by some investigators, who subject every component matrix to a series of postmultiplications by orthogonal matrices to see which transformed set of weights has the simplest subject-matter interpretation. While the ability of the vectors to generate the original matrix **S** is unimpaired, their components no longer have the maximum-variance property.

If the components have been extracted from the correlation matrix rather than **S**, the sum of the characteristic roots will be

$$\operatorname{tr} \mathbf{R} = p \tag{24}$$

and the proportion of the total "variance" in the scatter of dimensionless standard scores attributable to the jth component will be l_j/p. The sum of the squared correlations $a_{ij}\sqrt{l_j}$ of the responses on that component will of course be the component variance l_j.

If the first r components explain a large amount of the total sample variance, they may be evaluated for each subject or sampling unit and used in later analyses in place of the original responses. For components extracted from the covariance matrix the component scores of the ith subject are

$$y_{i1} = \mathbf{a}_1'(\mathbf{x}_i - \bar{\mathbf{x}}), \quad \ldots, \quad y_{ir} = \mathbf{a}_r'(\mathbf{x}_i - \bar{\mathbf{x}}) \tag{25}$$

where $\mathbf{x}_i$ is the ith observation vector and $\bar{\mathbf{x}}$ is the sample mean vector.

The scores can be written as the $N \times r$ matrix

$$\mathbf{Y} = \left(\mathbf{I} - \frac{1}{N}\mathbf{E}\right)\mathbf{XA} \tag{26}$$

where $\mathbf{X}$ is the data matrix (1), $\mathbf{E}$ is the $N \times N$ matrix of ones in every position, and $\mathbf{A}$ is the $p \times r$ matrix whose columns are the first r characteristic vectors. Had the $\mathbf{a}_i$ been extracted from the correlation matrix, the scores would be computed from the standardized observations. Thus, the component values of the ith subject would be

$$y_{i1} = \mathbf{a}_1'\mathbf{z}_i, \quad . \; . \; . \; , \quad y_{ir} = \mathbf{a}_r'\mathbf{z}_i \tag{27}$$

where $\mathbf{z}_i$ is the vector of standard scores with jth element given by equation (2).

If the ith and jth principal components correspond to distinct characteristic roots, their sample values will be uncorrelated. We may verify this by premultiplying the matrix equation (20) by the ith characteristic vector:

$$\mathbf{a}_i'\mathbf{S}\mathbf{a}_j = l_j\mathbf{a}_i'\mathbf{a}_j \tag{28}$$

But the left-hand side is merely the sample covariance of the component values $y_{hi} = \mathbf{a}_i'\mathbf{x}_h$, $y_{hj} = \mathbf{a}_j'\mathbf{x}_h$, and if $l_i \neq l_j$, it follows from the orthogonality of the vectors that

$$\mathbf{a}_i'\mathbf{S}\mathbf{a}_j = 0$$

We have stated that one important use of the principal-component technique is that of summarizing most of the variation in a multivariate system in fewer variables. Unless the system is of less than full rank, some variance will always be unexplained if fewer than p components are taken to describe the system. How, then, should one decide that the first m components provide a parsimonious, yet fairly adequate, description of the complex, given that they account for K percent of the total variance? In practice one usually knows from earlier studies, the subject-matter nature of the data, or even the pattern of the covariances in $\mathbf{S}$ that a certain minimum number of components with large and distinct variances should be extracted. Beyond that number components might be computed until some arbitrarily large proportion (perhaps 75 percent or more) of the variances has been explained. It has been the author's experience that if that proportion cannot be explained by the first four or five components, it is usually fruitless to persist in extracting vectors, for even if the later characteristic roots are sufficiently distinct to allow easy computation of the components, the interpretation of the components may be difficult if not impossible. Frequently it is better to summarize the complex in terms of the first components with large and markedly distinct variances and include as highly specific and unique variates those responses which are generally independent in the system. Such unique responses would probably be represented by high loadings in the later components but only in the presence of considerable "noise" from the other unrelated variates. Anderson (1963a) has developed tests of hypotheses and confidence intervals for determining whether the remaining component variances are identical, and we shall discuss this approach at length in Sec. 8.7.

Example 8.1. Before turning to the geometric interpretation of principal components and some practical computation techniques let us fix the ideas of this section with an example from biometry. Jolicoeur and Mosimann (1960) have investigated the principal components of carapace length, width, and height of painted turtles in an effort to give meanings to the concepts of "size" and "shape." The covariance matrix of the lengths, widths, and heights in millimeters of the carapaces of 24 female turtles was

$$\mathbf{S} = \begin{bmatrix} 451.39 & 271.17 & 168.70 \\ & 171.73 & 103.29 \\ & & 66.65 \end{bmatrix}$$

The coefficients and variances of the three components extracted from this matrix are summarized in Table 8.1.

The first principal component accounts for nearly all the variance in the three dimensions. It is the new weighted mean of the carapace measurements

$$Y_1 = 0.81\ (\text{length}) + 0.50\ (\text{width}) + 0.31\ (\text{height})$$

The size of the turtle shells could be characterized by this single variable with little loss of information. Had the dimensions been expressed in logarithms of units, Y_1 would indeed be the logarithm of the volume of a box whose sides were powers of the actual carapace dimensions. Jolicoeur and Mosimann call the second and third components measures of carapace "shape," for they appear to be comparisons of length versus width and height, and height versus length and width, respectively.

Table 8.1. **Carapace Component Coefficients**

Dimension	Component		
	1	2	3
Length	0.8126	−0.5454	−0.2054
Width	0.4955	0.8321	−0.2491
Height	0.3068	0.1006	0.9465
Variance	680.40	6.50	2.86
Percentage of total variance	98.64	0.94	0.41

Table 8.2. **Correlations of Carapace Dimensions and Components**

Dimension	Component		
	1	2	3
Length	1.00	−0.07	−0.02
Width	0.99	0.16	−0.03
Height	0.98	0.03	0.20

We shall consider a test of the uniqueness of variances of these components in Sec. 8.7. The component correlation coefficients are given in Table 8.2. Component 1 appears to be almost equally correlated with the three dimensions. Components 2 and 3 are correlated with the width and height dimensions, but then only to a negligible degree.

8.3 THE GEOMETRICAL MEANING OF PRINCIPAL COMPONENTS

We have introduced principal components analytically as those linear combinations of the responses which explain progressively smaller portions of the total sample variance. We shall now discuss the geometrical interpretation of components as the variates corresponding to

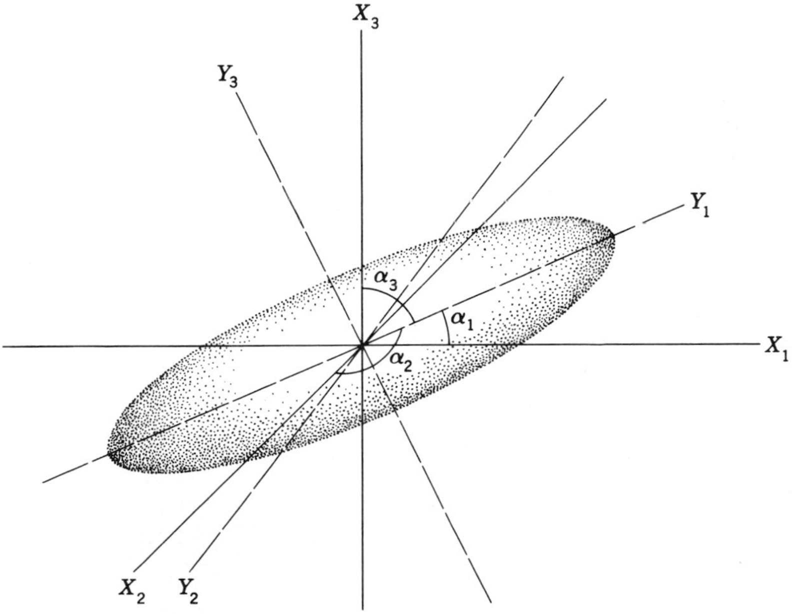

figure 8.1 Principal axes of trivariate observations.

the principal axes of the scatter of the observations in space. Imagine that a sample of N trivariate observations has the scatter plot shown in Fig. 8.1, where the origin of the response axes has been taken at the sample means. The swarm of points seems to have a generally ellipsoidal shape, with a major axis Y_1 and less well defined minor axes Y_2 and Y_3. Let us confine our attention for the moment to the major axis and denote its angles with the original response axes as α_1, α_2, α_3. If Y_1 passes through the sample mean point, its orientation is completely determined by the direction cosines

$$(1) \qquad a_{11} = \cos \alpha_1 \qquad a_{21} = \cos \alpha_2 \qquad a_{31} = \cos \alpha_3$$

where $a_{11}^2 + a_{21}^2 + a_{31}^2 = 1$. It is known from analytic geometry (Eisenhart, 1960; Somerville, 1958) that the value of the observation $[x_{i1},x_{i2},x_{i3}]$ on the new coordinate axis Y_1 will be

$$(2) \qquad y_{i1} = a_{11}(x_{i1} - \bar{x}_1) + a_{21}(x_{i2} - \bar{x}_2) + a_{31}(x_{i3} - \bar{x}_3)$$

Note that the mean of the Y_1 variate is

$$(3) \qquad \begin{aligned} \bar{y}_1 &= \frac{1}{N} \sum_{i=1}^{N} y_{i1} \\ &= \frac{1}{N} \sum_{i=1}^{N} \sum_{j=1}^{3} a_{j1}(x_{ij} - \bar{x}_j) \\ &= 0 \end{aligned}$$

Let us explicate the notion of the *major* axis of a discrete swarm of points by defining that axis as passing through the direction of *maximum variance* in the points. In the present case of three responses that variance is

$$(4) \qquad \frac{1}{N-1} \sum_{i=1}^{N} y_{i1}^2 = \frac{1}{N-1} \sum_{i=1}^{N} \left[\sum_{j=1}^{3} a_{j1}(x_{ij} - \bar{x}_j) \right]^2$$

and the angles of Y_1 would be found by differentiating this expression with respect to the a_{j1} (with suitable provision for the constraint) and solving for the a_{j1} which make the derivatives zero. The solution would be the characteristic vector of the greatest root of the sample covariance matrix of the x_{ij}, and Y_1 would be the continuum of the first *principal component* of the system.

Let us prove this statement for the general case of p responses. Write the direction cosines of the first principal-component axis as $\mathbf{a}_1' = [a_{11}, \ldots, a_{p1}]$, where the constraint

$$(5) \qquad \mathbf{a}_1'\mathbf{a}_1 = 1$$

must always be satisfied. The variance of the projections on the Y_1 axis is

$$s_{Y_1}^2 = \frac{1}{N-1}\sum_{i=1}^{N} y_{i1}^2 \tag{6}$$

$$= \frac{1}{N-1}\sum_{i=1}^{N}\left[\sum_{j=1}^{p} a_{j1}(x_{ij} - \bar{x}_j)\right]^2$$

$$= \frac{1}{N-1}\sum_{i=1}^{N} [(\mathbf{x}_i - \bar{\mathbf{x}})'\mathbf{a}_1]^2$$

$$= \frac{1}{N-1}\sum_{i=1}^{N} \mathbf{a}_1'(\mathbf{x}_i - \bar{\mathbf{x}})(\mathbf{x}_i - \bar{\mathbf{x}})'\mathbf{a}_1$$

$$= \mathbf{a}_1'\mathbf{S}\mathbf{a}_1$$

Introduce the constraint (5) by the Lagrange multiplier l_1. Then the maximand

$$\mathbf{a}_1'\mathbf{S}\mathbf{a}_1 + l_1(1 - \mathbf{a}_1'\mathbf{a}_1) \tag{7}$$

is precisely the same as that of equation (5) in Sec. 8.2. The direction cosines of the first principal axis are the elements of the first characteristic vector of **S**, and the maximized variance is the greatest characteristic root.

The remaining characteristic roots and vectors of **S** determine the lengths and orientations of the second and higher component axes. If two successive roots l_i and l_{i+1} are equal, the scatter configuration has no unique major axis in the plane of the axes of the roots and the appearance of the points is more circular than elliptical. Such dispersion is called *isotropic* or *spherical* in those dimensions with equal l_i. These results may be summarized in this geometrical definition of principal components:

Definition 8.3. *The principal components of the sample of N p-dimensional observations are the new variates specified by the axes of a rigid rotation of the original response coordinate system into an orientation corresponding to the directions of maximum variance in the sample scatter configuration. The direction cosines of the new axes are the normalized characteristic vectors corresponding to the successively smaller characteristic roots of the sample covariance matrix. If two or more roots are equal, the directions of the associated axes are not unique and may be chosen in an infinity of orthogonal positions. If the components are instead computed from the correlation matrix, the same geometrical interpretation holds, although the response coordinate system is expressed in standard units of zero means and unit variances.*

A second property of component axes was first considered by K. Pearson (1901) and is implicit in the preceding geometrical derivation: the choice of the new coordinate axes is such that the sums of squared distances of each point to its projections on the successive axes are minimized. Figure 8.2 shows one projection and distance of an observation in three-dimensional space. In the general case the squared ith distance is

$$(8) \qquad (P_i'P_i)^2 = (OP_i)^2 - (OP_i')^2$$
$$= \sum_{j=1}^{p} (x_{ij} - \bar{x}_j)^2 - \left[\sum_{j=1}^{p} a_{j1}(x_{ij} - \bar{x}_j)\right]^2$$

Minimization of the sum $\sum_{i=1}^{N} (P_i'P_i)^2$ of all squared distances to the new axis is equivalent to maximization of the second term of (8), which is of course proportional to the component variance (6). The orthogonal least-squares solution for the best-fitting line leads immediately to the first principal axis, and the other component axes would follow from the successive orthogonal least-squares solutions.

If the first characteristic root is large, an excellent approximation to the coefficients of the first component can be obtained from the line connecting the two extreme points in the scatter configuration. Denote by $\mathbf{x}_{(1)}$ and $\mathbf{x}_{(N)}$ the observations separated by the greatest distance

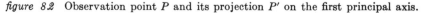

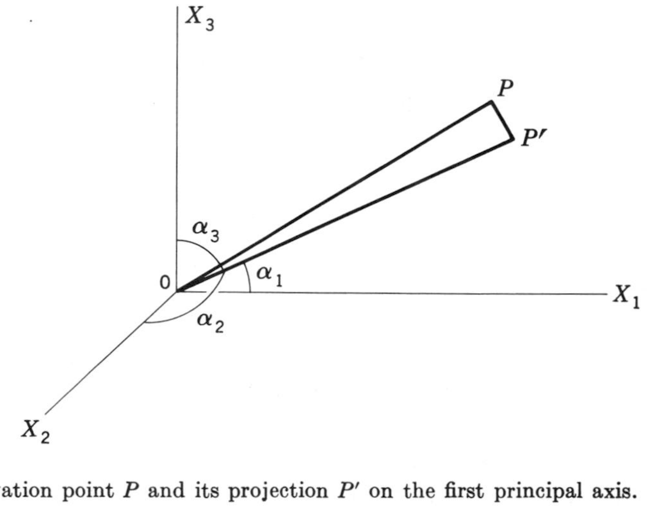

figure 8.2 Observation point P and its projection P' on the first principal axis.

in the sample space, or those for which

$$d^2 = \sum_{j=1}^{p} (x_{(N)j} - x_{(1)j})^2 \tag{9}$$

is a maximum for all pairs of observation vectors. The direction cosines of the line are

$$a_j = \frac{x_{(N)j} - x_{(1)j}}{d} \tag{10}$$

where d is the positive square root of the distance (9). These cosines or simply the differences $x_{(N)j} - x_{(1)j}$ can be used as the initial vector $\mathbf{x}_0$ for the iterative process of the next section.

Somc indication of the usefulness of this approximation follows from its application to the carapace data of Example 8.1. The approximate direction cosines of the greatest distance are [0.807,0.519,0.295], and these agree with the true values to nearly two places.

8.4 THE COMPUTATION OF PRINCIPAL COMPONENTS

Characteristic roots and vectors play an essential role in many problems of applied mathematics, dynamics, and statistical theory, and the numerical-analysis literature contains a number of techniques for their computation. It is of course possible to expand the characteristic equation determinant by equation (2) of Sec. 2.10 and reduce the problem to one of finding the roots of a polynomial and the solutions of certain homogeneous linear equations, but this approach is inefficient even if the matrix is small. Instead, an iterative procedure is usually employed to correct successively the guessed elements of the vector until they converge with sufficient accuracy to the true values. We shall present one common iterative method for calculating the greatest distinct characteristic root and its vector for any square matrix with real elements.

Let $\mathbf{A}$ be the $p \times p$ matrix of real elements. It is not necessary that $\mathbf{A}$ be symmetric. Order the characteristic roots λ_i of $\mathbf{A}$ by their absolute values:

$$|\lambda_1| > |\lambda_2| \geq \cdots \geq |\lambda_p| \tag{1}$$

and denote their respective vectors as

$$\mathbf{a}_1, \ldots, \mathbf{a}_p \tag{2}$$

Initially we shall require that only $|\lambda_1| > |\lambda_2|$. Let $\mathbf{x}_0$ be any vector of

p real components, and form the sequence

$$
\begin{aligned}
\mathbf{x}_1 &= \mathbf{A}\mathbf{x}_0 \\
&\cdots \\
\mathbf{x}_n &= \mathbf{A}\mathbf{x}_{n-1} \\
&= \mathbf{A}^n\mathbf{x}_0
\end{aligned}
\tag{3}
$$

of vectors. *Then if the successive* $\mathbf{x}_i$ *are scaled in some fashion, the sequence of standardized vectors will converge to the characteristic vector* $\mathbf{a}_1$. Probably the most convenient scaling is performed by dividing the elements by their maximum, with normalization to unit length merely reserved for the last, or exact, vector. Since

$$\mathbf{A}\mathbf{a}_1 = \lambda_1\mathbf{a}_1 \tag{4}$$

the characteristic root itself can be found by dividing any element of $\mathbf{A}\mathbf{a}_1$ by the corresponding element of $\mathbf{a}_1$.

The demonstration of this convergence is simple. Since characteristic vectors are unique only up to multiplication by a scale factor, it is always possible to choose their elements so that any vector $\mathbf{x}_0$ with p real components can be expressed as

$$\mathbf{x}_0 = \mathbf{a}_1 + \cdots + \mathbf{a}_p \tag{5}$$

If $\mathbf{x}_0$ is the trial vector in the iterative process, it follows that

$$\mathbf{x}_1 = \mathbf{A}\mathbf{a}_1 + \cdots + \mathbf{A}\mathbf{a}_p \tag{6}$$

and by (4)

$$\mathbf{x}_1 = \lambda_1\mathbf{a}_1 + \cdots + \lambda_p\mathbf{a}_p \tag{7}$$

After the nth iteration

$$\mathbf{x}_n = \lambda_1{}^n\mathbf{a}_1 + \cdots + \lambda_p{}^n\mathbf{a}_p \tag{8}$$

or

$$\lambda_1{}^{-n}\mathbf{x}_n = \mathbf{a}_1 + \left(\frac{\lambda_2}{\lambda_1}\right)^n \mathbf{a}_2 + \cdots + \left(\frac{\lambda_p}{\lambda_1}\right)^n \mathbf{a}_p \tag{9}$$

Since $|\lambda_1| > |\lambda_2|$, as the number of iterations increases without bound, the second and higher terms of the right-hand side of (9) tend to zero. The normalized vectors $\mathbf{x}_n$ converge to $\mathbf{a}_1$ regardless of the form of the starting guess $\mathbf{x}_0$. We note in passing that the ratio of corresponding elements of $\mathbf{x}_n$ and $\mathbf{x}_{n+1}$ is approximately $|\lambda_1|$ for large n.

In practice we ordinarily begin by computing the powers

$$\mathbf{A}^2, \mathbf{A}^4, \mathbf{A}^8, \ldots \tag{10}$$

of the matrix and start the iterative process with $\mathbf{A}^{16}$ or a higher power if p is large. The resulting vector is then compared with the vector obtained from $\mathbf{A}^{32}$ or the next similar power. If $|\lambda_1|$ is well separated from $|\lambda_2|$ and p is not large, convergence should follow with one or more iterations with such powers of $\mathbf{A}$.

The same iterative procedure can be used to compute any distinct characteristic root of $\mathbf{A}$. To extract the second largest root and its vector we normalize the first characteristic vector $\mathbf{a}_1$ to unit length, form the $p \times p$ matrix

$$\lambda_1 \mathbf{a}_1 \mathbf{a}_1' \tag{11}$$

and subtract it from $\mathbf{A}$ to give the residual matrix

$$\mathbf{A}_1 = \mathbf{A} - \lambda_1 \mathbf{a}_1 \mathbf{a}_1' \tag{12}$$

Note that $\sqrt{\lambda_1}\, \mathbf{a}_1$ is the vector of component correlations if $\mathbf{A}$ is a correlation matrix. Then the greatest root of

$$|\mathbf{A}_1 - \lambda \mathbf{I}| = 0 \tag{13}$$

will be the second largest characteristic root of the original matrix, and in place of the root λ_1 the characteristic equation will have a zero root. The iterative process is applied to $\mathbf{A}_1$, and if $|\lambda_2| > |\lambda_3|$, the successive approximations will converge to $\mathbf{a}_2$. If the first i characteristic roots are distinct in absolute value, the ith vector is found by applying the iterative algorithm to

$$\mathbf{A} - (\lambda_1 \mathbf{a}_1 \mathbf{a}_1' + \cdots + \lambda_{i-1} \mathbf{a}_{i-1} \mathbf{a}_{i-1}') \tag{14}$$

However, unless a substantial number of decimal places are carried along in the computations, the accumulation of roundoff error tends to make extraction of smaller roots and their vectors difficult. This is particularly true if the roots are close in value, for the iterative process may alternately approximate one vector and then the other without converging.

Occasionally it is necessary to obtain the smallest characteristic root, as for example in investigations of the rank of observations on a large number of responses. If $\mathbf{A}$ has the positive roots λ_i ordered as in (1), the matrix

$$(\text{tr}\, \mathbf{A})\mathbf{I} - \mathbf{A} \tag{15}$$

now has largest root

$$\text{tr}\, \mathbf{A} - \lambda_p \tag{16}$$

and λ_p can be found by applying the iterative process to it.

For the extraction of principal components when $\mathbf{A}$ is the covariance matrix it is usually convenient to scale the elements of $\mathbf{S}$ to values between -1 and 1. The characteristic vectors are unaffected by this change, and the characteristic roots can be returned to the original magnitude by multiplication by the scaling factor. The elements of the correlation matrix are already in such a convenient scale. It is also helpful to change the directions of the scales of some variates to make as many as possible of the covariances or correlations positive. Changing the signs of the covariances of the ith variate is equivalent to making the transformation

$$\mathbf{D}_i\mathbf{S}\mathbf{D}_i \tag{17}$$

where $\mathbf{D}_i$ is the diagonal matrix with unity in all positions but the ith, which contains -1. The characteristic vectors of the new matrix are $\mathbf{D}_i\mathbf{a}_1, \ldots, \mathbf{D}_i\mathbf{a}_p$ and merely have the signs of their ith elements reversed from the vectors of $\mathbf{S}$. The utility of this change is twofold: computations on a desk calculator are easier, and the interpretation of the components is facilitated. However, if some correlations are close to zero or otherwise beneath the "noise threshold" of the experiment, it may not be possible to make all signs positive.

Example 8.2. We shall illustrate the iterative procedure by extracting the first two components from a correlation matrix of fowl bone measurements analyzed by Wright (1954). The lengths of the humerus, ulna, tibia, and femur bones of 276 leghorn fowl were found to have the correlation matrix

$$\mathbf{R} = \begin{bmatrix} 1 & 0.940 & 0.875 & 0.878 \\ & 1 & 0.877 & 0.886 \\ & & 1 & 0.924 \\ & & & 1 \end{bmatrix}$$

The second power of this matrix is

$$\mathbf{R}^2 = \begin{bmatrix} 3.4201 & 3.4253 & 3.3856 & 3.3973 \\ & 3.4377 & 3.3952 & 3.4077 \\ & & 3.3885 & 3.3933 \\ & & & 3.4096 \end{bmatrix}$$

Since all correlations of $\mathbf{R}$ are nearly equal, a reasonable initial vector for the iterative process is $\mathbf{a}_0' = [1,1,1,1]$. Thus, $\mathbf{a}_0'\mathbf{R}^2 = [13.6283, 13.6659, 13.5626, 13.6079]$, or if we standardize by dividing by the second element, the first solution is proportional to

$$[0.99725, 1.00000, 0.99244, 0.99576]$$

Finally,

$$10^{-1} \times \mathbf{R}^4 = \begin{bmatrix} 4.6434 & 4.6562 & 4.6209 & 4.6364 \\ & 4.6689 & 4.6336 & 4.6491 \\ & & 4.5986 & 4.6140 \\ & & & 4.6294 \end{bmatrix}$$

The vector of column sums of this matrix is

$$[18.5569, 18.6078, 18.4671, 18.5289]$$

while the standardized vector is

$$[0.99726, 1.00000, 0.99244, 0.99576]$$

Since this vector is virtually identical to its predecessor, we conclude that the iterative process has converged to the first characteristic vector in just two steps. The vector of direction cosines is

$$\mathbf{a}_1' = [0.5004, 0.5018, 0.4980, 0.4997]$$

The characteristic vector must satisfy the linear equation

$$0.5004(1 - \lambda_1) + 0.5018r_{12} + 0.4980r_{13} + 0.4997r_{14} = 0$$

so that its characteristic root is $\lambda_1 = 3.6902$. We conclude that $^{369}\!/_{4} = 92.25$ percent of the total variance in the original bone lengths is accounted for by the new variate

$$Y_1 = \tfrac{1}{2}(\text{humerus length} + \text{ulna length} + \text{femur length} + \text{tibia length})$$

The residual matrix $\mathbf{R}_1$ required in the extraction of the second principal component is found by subtracting

$$\lambda_1\mathbf{a}_1\mathbf{a}_1' = \begin{bmatrix} 0.9241 & 0.9266 & 0.9196 & 0.9228 \\ & 0.9292 & 0.9222 & 0.9253 \\ & & 0.9152 & 0.9183 \\ & & & 0.9214 \end{bmatrix}$$

from $\mathbf{R}$:

$$\mathbf{R}_1 = \begin{bmatrix} 0.0759 & 0.0134 & -0.0446 & -0.0448 \\ & 0.0708 & -0.0452 & -0.0393 \\ & & 0.0848 & 0.0057 \\ & & & 0.0786 \end{bmatrix}$$

We shall operate as before upon the powers of $\mathbf{R}_1$ with some trial vector $\mathbf{a}_0$. The elements of this vector are suggested by the pattern of the original correlations among the four measurements. As one might expect, the humerus and ulna lengths are more correlated with each other than with those of the femur and tibia; similarly, the femur-tibia correlation is higher than any of the correlations of the lengths of those bones with the other members.

Such a partitioning of $\mathbf{R}_1$ seems to suggest that the next important source of variance might be due to some measure of difference between the limbs, so that the characteristic vector might be of the form $[-1,-1,1,1]$. If we compute $\mathbf{R}_1^2$, $\mathbf{R}_1^4$, $\mathbf{R}_1^8$, and $\mathbf{R}_1^{16}$ and postmultiply each by this initial vector, the successive approximations follow:

Power of $\mathbf{R}_1$	*Standardized iterate for* $\mathbf{a}_2'$
1	[−0.99113, −0.93566, 1, 0.93400]
2	[−0.98634, −0.91401, 1, 0.90752]
4	[−0.98183, −0.90289, 1, 0.89187]
8	[−0.98040, −0.90096, 1, 0.88850]
16	[−0.98033, −0.90090, 1, 0.88838]
32	[−0.98033, −0.90090, 1, 0.88838]

The second largest characteristic root can be computed from the inner product of $\mathbf{a}_2$ and the first row of $\mathbf{R}_1$ as

$$0.98033\lambda_2 = 0.0759(0.98033) + 0.0134(0.90090) + 0.0446 + 0.0448(0.88838)$$

or

$$\lambda_2 = 0.174307$$

where we have retained the spurious last two places for some subsequent comparisons. The vector in normalized direction-cosine form is

$$[-0.5195, \ -0.4774, \ 0.5299, \ 0.4707]$$

and we may think of the second principal component as the linear compound

$$Y_2 = -0.52 \text{ (femur length)} - 0.48 \text{ (tibia length)} + 0.53 \text{ (humerus length)} + 0.47 \text{ (ulna length)}$$

or a measure of the difference of leg and wing lengths.

The first two components of the bone dimensions may be summarized in this manner:

Bone length	Component correlation	
	1	2
Humerus	0.961	−0.217
Ulna	0.964	−0.199
Femur	0.957	0.221
Tibia	0.960	0.196
Component variance	3.690	0.174
Percentage of total variance	92.25	4.35

For presentation the correlation coefficients have been rounded to the number of places in the original sample correlations.

Aitken (1937) has proposed a technique known as the δ^2 process for speeding up the convergence of the iterative scheme. Let x_t be some function of the elements of the tth power of the matrix $\mathbf{A}$ with real characteristic roots $\lambda_1, \lambda_2, \lambda_3$ subject to the condition $|\lambda_1| > |\lambda_2|$. x_t can be an element of $\mathbf{x}_t = \mathbf{A}^t\mathbf{x}_0$, the tth approximation $\mathbf{x}_{i,t+1}/\mathbf{x}_{it}$ to λ_1 based upon the ith elements of $\mathbf{x}_{t+1}$ and $\mathbf{x}_t$, or some other quantity derivable from $\mathbf{A}$. From three consecutive values of $\mathbf{x}_t$ compute the ratio

$$\phi_2(t) = \frac{\begin{vmatrix} x_{t-1} & x_t \\ x_t & x_{t+1} \end{vmatrix}}{x_{t+1} - 2x_t + x_{t-1}} \tag{18}$$

Aitken has shown that the sequence of the $\phi_2(t)$ terms will converge more rapidly than that of the original x_t iterates. However, as in the case of the matrix-power iterative procedure, it is essential that a sufficient number of significant figures be retained in the x_t. The effect of lost accuracy is particularly severe in the denominator term, for errors in the seventh or eighth place of the x_t become greatly magnified by the subtraction and division operations.

Some idea of the acceleration in convergence can be gained by comparing the conventional iterative and δ^2 processes for the residual matrix $\mathbf{R}_1$ of Example 8.2. The first five standardized iterates obtained from the sequence $\mathbf{x}_{t+1} = \mathbf{R}_1\mathbf{x}_t$ are

$$\mathbf{x}_1' = [-0.99112590, \ -0.93566278, \ 1, \ 0.93399889]$$

$$\mathbf{x}_2' = [-0.98634084, \ -0.91400860, \ 1, \ 0.90752084]$$

$$\mathbf{x}_3' = [-0.98340373, \ -0.90596293, \ 1, \ 0.89652301]$$

$$\mathbf{x}_4' = [-0.98182782, \ -0.90289366, \ 1, \ 0.89187256]$$

$$\mathbf{x}_5' = [-0.98103785, \ -0.90169853, \ 1, \ 0.88988561]$$

The many decimal places are essential for the computation of the $\phi_2(t)$ elements. The iterates given by that operation are

$$\phi_2(2): [-0.97874, \ -0.90114, \ 1, \ 0.88871]$$

$$\phi_2(3): [-0.98000, \ -0.90100, \ 1, \ 0.88846]$$

$$\phi_2(4): [-0.98024, \ -0.90094, \ 1, \ 0.88840]$$

The second and fourth elements of the last iterate agree to four places with those obtained from the sixteenth power of $\mathbf{R}_1$. Had one more figure been carried in the $\mathbf{x}_t$ iterates, it would seem likely that another application of the δ^2 process to $\mathbf{x}_4$, $\mathbf{x}_5$, $\mathbf{x}_6$ would have produced a value of $\mathbf{a}_2$ whose elements agreed with those of $\mathbf{R}_1^{16}\mathbf{x}_0$ to at least four decimal places.

The ratios of the first elements in the initial five successive *unstandardized* $\mathbf{x}_t$ iterates gave this sequence of approximations to the second characteristic root of $\mathbf{R}$:

$$0.175767 \qquad 0.174755 \qquad 0.174440 \qquad 0.174344$$

Application of the δ^2 process to these values yields

$$\phi_2(3) = \phi_2(4) = 0.17430$$

or a good approximation to the value 0.174307 obtained by the power iterative procedure.

Useful discussions of the Aitken δ^2 technique and other methods for extracting characteristic roots and vectors have been given by Bodewig (1956) and Faddeeva (1959).

8.5 THE INTERPRETATION OF PRINCIPAL COMPONENTS

In this section we shall discuss some examples of principal-component solutions in an attempt to give subject-matter identities to the new variables.

Example 8.3. *Alternate Analyses of the Turtle Carapace Dimensions.* It is evident from the covariance matrix of Example 8.1 that the three shell dimensions have very unequal variances. To stabilize the variances the observations on each shell were transformed to common logarithms. The new means and covariance matrix had these values:

Mean log dimensions			Covariance matrix × 10^3
Length	Width	Height	
2.128	2.008	1.710	$\begin{bmatrix} 4.9810 & 3.8063 & 4.7740 \\ & 3.0680 & 3.7183 \\ & & 4.8264 \end{bmatrix}$

The characteristic roots of the matrix are 12.6372, 0.1386, and 0.0997. Since 98.15 percent of the total variance of the log-transformed observations is attributable to the first component and the scatter in the remaining two dimensions about that axis is nearly circular, it is reasonable to conclude that the new variables lie about a new dimension with direction cosines

$$\mathbf{a}_1' = [0.6235, 0.4860, 0.6124]$$

The new component variable is

$$\begin{aligned} Y_1 &= 0.62 \log (\text{length}) + 0.49 \log (\text{width}) + 0.61 \log (\text{height}) \\ &= \log (X_1^{0.62} X_2^{0.49} X_3^{0.61}) \end{aligned}$$

or the logarithm of a kind of carapace volume.

Alternatively the dimensions might have been scaled to standard scores with unit variances. Then the component analysis would be carried out upon the correlation matrix of

$$\begin{bmatrix} 1 & 0.974 & 0.973 \\ & 1 & 0.966 \\ & & 1 \end{bmatrix}$$

and as we might expect from such nearly equal correlations the direction-cosine vector

$$\mathbf{a}_1' = [0.5783, 0.5770, 0.5768]$$

of the first principal axis of the standard scores is almost exactly that of the equiangular line in three-space. This component accounts for 98.0 percent of the "variance" of the standard scores. The remaining two dimensions account for nearly equal parts of the total variance, and we conclude as before that removal of the inequalities of the variances produces a scatter with a single long axis surrounded by fairly isotropic variation.

Example 8.4. *Wechsler Adult Intelligence Scale Subtest Scores.* The correlation coefficients of principal components extracted from the correlation matrix given by Birren and Morrison (1961) for the 11 WAIS subtest scores, age, and years of formal education of 933 white, native-born male and female participants in a community testing program are shown in Table 8.3. The first component appears to be a measure of *general intellectual performance,* for all WAIS subtests have nearly equal high positive correlations with this dimension. The coefficient of education is also positive and of exactly the order of size as the subtest coefficients. Age is negatively represented in this component, and with a smaller absolute weight than those of the other variables. More than half the total variance in the 13 original dimensions is accounted for by component 1.

Component 2 explained a much smaller percentage (10.90 percent) of the total variance. This dimension may be thought of as an *experiential* or *age* factor, for that variable dominates the coefficients. In addition, the verbal subtests (1 to 6) have positive coefficients, while the performance tests (7 to 11) correlate negatively. In that sense component 2 is a bipolar dimension comparing one set of verbal or infor-

Table 8.3. **Correlation Coefficients of the WAIS Principal Components**

	Component			
	1	2	3	4
WAIS subtest:				
Information	0.83	0.33	−0.04	−0.10
Comprehension	0.75	0.31	0.07	−0.17
Arithmetic	0.72	0.25	−0.08	0.35
Similarities	0.78	0.14	0.00	−0.21
Digit span	0.62	0.00	−0.38	0.58
Vocabulary	0.83	0.38	−0.03	−0.16
Digit symbol	0.72	−0.36	−0.26	−0.08
Picture completion	0.78	−0.10	−0.25	−0.01
Block design	0.72	−0.26	0.36	0.18
Picture arrangement	0.72	−0.23	0.04	−0.05
Object assembly	0.65	−0.30	0.47	0.13
Age	−0.34	0.80	0.26	0.18
Years of education	0.75	0.01	−0.30	−0.23
Latent root (variance)	6.69	1.42	0.80	0.71
Percentage of total variance	51.47	10.90	6.15	5.48
Cumulative percentage of variance	51.47	62.37	68.52	74.01

mational skills known to increase with advancing age to subtests measuring spatial-perceptual qualities and other cognitive abilities known to decrease with age.

The third and fourth components explain small and nearly equal proportions of the total variance. Component 3 is highly correlated with block design and object assembly, though negatively with digit span, and may be thought of as a *spatial imagery or perception* dimension. Component 4 appears to be a measure of *numerical facility*.

Example 8.5. *Another Biometric Application: Bone Lengths of White Leghorn Fowl.* The correlation matrix of the complete set of six fowl bone measurements considered in part in Sec. 8.4 had the following form:

Skull length	1.000	0.584	0.615	0.601	0.570	0.600
Skull breadth		1.000	0.576	0.530	0.526	0.555
Humerus			1.000	0.940	0.875	0.878
Ulna				1.000	0.877	0.886
Femur					1.000	0.924
Tibia						1.000

These anatomical measurements have a *hierarchical* pattern: the skull contributes two dimensions, while two wing bones (humerus and ulna) and two leg bones (femur and tibia) are also represented. This partitioning of the matrix is apparent in the values of the correlation coefficients. It is even more striking in the patterns of the component correlations obtained by Wright (1954) and reproduced here in Table 8.4.

As a practical problem in principal axes or data reduction one would probably stop with the first two or three components, for no parsimony is gained by merely making an orthogonal transformation to six new dimensions. However, it is a rare

Table 8.4. **Component Correlations of the Fowl Bone Dimensions**

Dimension	Component					
	1	2	3	4	5	6
Skull:						
Length	0.74	0.45	0.49	−0.02	0.01	0.00
Breadth	0.70	0.59	−0.41	0.00	0.00	−0.01
Wing:						
Humerus	0.95	−0.16	−0.03	0.22	0.05	0.16
Ulna	0.94	−0.21	0.01	0.20	−0.04	−0.17
Leg:						
Femur	0.93	−0.24	−0.04	−0.21	0.18	−0.03
Tibia	0.94	−0.19	−0.03	−0.20	−0.19	0.04
Characteristic root	4.568	0.714	0.412	0.173	0.076	0.057
Percentage of total variance	76.1	11.9	6.9	2.9	1.3	0.9

matrix that permits straightforward identifications of all its components:

Component	*Name*
1	General average of all bone dimensions
2	Comparison of skull size with wing and leg lengths
3	Comparison of skull length and breadth; measure of skull shape
4	Comparison of wing and leg lengths
5	Comparison of femur and tibia
6	Comparison of humerus and ulna

We shall consider the case of hierarchically ordered variables in greater generality in the next section.

8.6 SOME PATTERNED MATRICES AND THEIR PRINCIPAL COMPONENTS

The component structure of a covariance or correlation matrix can sometimes be approximated rather well by inspection of the elements and a knowledge of the characteristic roots and vectors of certain patterned matrices. We shall now give the components of two such special matrices, and a general upper bound on the greatest characteristic root of any square matrix.

Equicorrelation Matrix. Here the $p \times p$ covariance matrix is

$$\mathbf{\Sigma} = \sigma^2 \begin{bmatrix} 1 & \rho & \cdots & \rho \\ \rho & 1 & \cdots & \rho \\ \cdot & \cdot & \cdots & \cdot \\ \rho & \rho & \cdots & 1 \end{bmatrix} \tag{1}$$

We shall require that $0 < \rho \leq 1$. The greatest characteristic root of this matrix is

$$\lambda_1 = \sigma^2[1 + (p - 1)\rho] \tag{2}$$

and its normalized characteristic vector is

$$\mathbf{a}_1' = \left[\frac{1}{\sqrt{p}}, \ldots, \frac{1}{\sqrt{p}}\right] \tag{3}$$

The first principal component

$$Y_1 = \frac{1}{\sqrt{p}} \sum_{j=1}^{p} X_j \tag{4}$$

is merely proportional to the mean of the original p responses; it accounts for $100[1 + (p - 1)\rho]/p$ percent of the total variance in the set. The remaining $p - 1$ characteristic roots are all equal to

$$\sigma^2(1 - \rho) \tag{5}$$

and their vectors are any of the $p - 1$ linearly independent solutions of the equation

$$\sigma^2\rho(a_{12} + a_{22} + \cdots + a_{p2}) = 0 \tag{6}$$

Note that this equation amounts to the requirement that each of the $p - 1$ characteristic vectors be orthogonal to the first vector, or

$$\mathbf{a}_1'\mathbf{a}_2 = 0 \tag{7}$$

Example 8.6. The largest characteristic root of the correlation matrix

$$\mathbf{P} = \begin{bmatrix} 1 & 0.6 & 0.6 & 0.6 \\ & 1 & 0.6 & 0.6 \\ & & 1 & 0.6 \\ & & & 1 \end{bmatrix}$$

is $\lambda_1 = 2.8$. Since the characteristic vector of that root is $\mathbf{a}_1' = [\frac{1}{2},\frac{1}{2},\frac{1}{2},\frac{1}{2}]$, 70 percent of the total variance in such an idealized system would be explained by the first principal component

$$Y_1 = \tfrac{1}{2}(X_1 + X_2 + X_3 + X_4)$$

The remaining 30 percent of the variance is equally attributable to three new variates symmetrically distributed about the Y_1 axis.

The equal variance and covariance structure thus always contains a single principal component explaining a greater proportion of the total variance as the common correlation becomes large. This dimension has an *equiangular* orientation in the midst of the axes of the original variates. The other independent variates have isotropic variation in the remaining $p - 1$ dimensions orthogonal to the major axis.

The Equipredictability Covariance Pattern. Bargmann (1957) has considered the properties of the patterned covariance matrix

$$\mathbf{\Sigma} = \begin{bmatrix} \sigma^2 & \sigma_{12} & \sigma_{13} & \sigma_{14} \\ & \sigma^2 & \sigma_{14} & \sigma_{13} \\ & & \sigma^2 & \sigma_{12} \\ & & & \sigma^2 \end{bmatrix} \tag{8}$$

Such matrices have the property that the four multiple correlation coefficients of one variable with the remaining three are equal. The matrix has this principal-component structure:

Variance	*Component direction cosines*
$\sigma^2 + \sigma_{12} + \sigma_{13} + \sigma_{14}$	$[\frac{1}{2}, \frac{1}{2}, \frac{1}{2}, \frac{1}{2}]$
$\sigma^2 + \sigma_{12} - \sigma_{13} - \sigma_{14}$	$[\frac{1}{2}, \frac{1}{2}, -\frac{1}{2}, -\frac{1}{2}]$
$\sigma^2 - \sigma_{12} + \sigma_{13} - \sigma_{14}$	$[\frac{1}{2}, -\frac{1}{2}, \frac{1}{2}, -\frac{1}{2}]$
$\sigma^2 - \sigma_{12} - \sigma_{13} + \sigma_{14}$	$[\frac{1}{2}, -\frac{1}{2}, -\frac{1}{2}, \frac{1}{2}]$

For example, the correlation matrix

$$\begin{bmatrix} 1 & 0.7 & 0.6 & 0.4 \\ & 1 & 0.4 & 0.6 \\ & & 1 & 0.7 \\ & & & 1 \end{bmatrix}$$

has those characteristic vectors with roots 2.7, 0.7, 0.5, and 0.1, respectively. Using the fact that the characteristic vectors give the four orthogonal contrasts in the 2^2 factorial experimental design, Bock (1960) has established interesting connections between covariance structures and the model II analysis of variance.

An Upper Bound on the Greatest Characteristic Root of a Matrix. Brauer (1953) and other algebraists (e.g., Marcus and Minc, 1964) have investigated the regions in which the characteristic roots of an arbitrary square matrix must lie. For the positive semidefinite covariance matrix $\mathbf{\Sigma}$ it can be shown that the maximum characteristic root cannot exceed the greatest of the row sums of absolute values of the elements of $\mathbf{\Sigma}$:

(9) $$\lambda_1 \leq \max_i \left(\sum_{j=1}^{p} |\sigma_{ij}| \right) \qquad i = 1, \ldots, p$$

If $\mathbf{\Sigma}$ is the equal-variance, equal-covariance matrix (1) this bound is equal to the greatest root. As an example, the row sums of the fowl bone dimensions correlation matrix of Example 7.5 are

3.970	3.771	4.884	4.834	4.772	4.843

The largest characteristic root must be less than 4.884, so that the first principal component can explain at most 81.4 percent of the total variance. Since this is a substantial proportion, the arithmetic labor might be begun with the assurance that the first component could possibly explain an appreciable amount of variance.

The use of this result for estimating λ_1 appears to be best reserved for correlation matrices, for the proximity of the greatest root to the upper bound diminishes as the matrix departs from the equal-variance, equal-covariance pattern. For example, the bound on the true root of 3.690 in Example 8.2 is 3.703, while the bound on the largest root 6.69

of the WAIS correlation matrix of Example 8.4 is 7.44. Even less precision is obtained if the inequality is applied to a covariance matrix with widely different diagonal elements, as for example, the carapace-dimension matrix of Example 8.1.

8.7 THE SAMPLING PROPERTIES OF PRINCIPAL COMPONENTS

Throughout the preceding sections we have treated principal-component analysis as a descriptive technique for studying the dependence or correlational structure of multivariate samples. Now we shall require that the sample has been drawn from a multinormal population whose covariance matrix has a specified covariance structure, and on the basis of that assumption we shall be able to state a number of large-sample distributional properties of the component coefficients and characteristic roots. In addition to providing knowledge of the stability of those quantities through their variances and covariances, these asymptotic distributions will also permit the construction of tests of hypotheses and confidence intervals for the population component structure. The initial derivation of these large-sample properties and tests was begun by Girshick (1936, 1939); subsequent extensions have been made by Anderson (1951a, 1963a), Bartlett (1954), and Lawley (1956, 1963). Since the mathematics of that research is well beyond the level of this text, the actual derivations will be omitted in the sequel.

Suppose that N independent observations have been taken on the p-dimensional random variable with distribution $N(\boldsymbol{\mu},\boldsymbol{\Sigma})$. $\boldsymbol{\Sigma}$ has the *distinct* characteristic roots

$$\lambda_1 > \cdots > \lambda_p > 0 \tag{1}$$

with corresponding vectors

$$\boldsymbol{\alpha}_1, \ldots, \boldsymbol{\alpha}_p \tag{2}$$

The sample estimate of $\boldsymbol{\Sigma}$ is the matrix $\mathbf{S}$ with elements based upon n degrees of freedom, and if $\mathbf{S}$ has the usual single-sample form of Sec. 3.5, $n = N - 1$. Denote the roots and vectors of $\mathbf{S}$ by the Latin counterparts

$$\begin{matrix} l_1 > \cdots > l_p > 0 \\ \mathbf{a}_1, \ldots, \mathbf{a}_p \end{matrix} \tag{3}$$

Both the population and sample characteristic vectors have the direction-cosine form of length one. Girshick and Anderson have demonstrated that the following results hold as n becomes very large:

1. l_i is distributed independently of the elements of its associated vector $\mathbf{a}_i$.
2. $\sqrt{n}\,(l_i - \lambda_i)$ is normally distributed with mean zero and variance $2\lambda_i^2$ as n tends to infinity and is independently distributed of the other sample characteristic roots.
3. The elements of $\sqrt{n}\,(\mathbf{a}_i - \boldsymbol{\alpha}_i)$ are distributed according to the multinormal distribution with null mean vector and covariance matrix

$$\lambda_i \sum_{\substack{h=1 \\ h \neq i}}^{p} \frac{\lambda_h}{(\lambda_h - \lambda_i)^2} \boldsymbol{\alpha}_h \boldsymbol{\alpha}_h' \tag{4}$$

4. The covariance of the rth element of $\mathbf{a}_i$ and the sth element of $\mathbf{a}_j$ is

$$-\frac{\lambda_i \lambda_j \alpha_{si} \alpha_{rj}}{n(\lambda_i - \lambda_j)^2} \qquad i \neq j \tag{5}$$

It is essential to bear in mind that these are *large* sample distributional properties for the case of p *distinct* population characteristic roots. However, as Anderson has indicated, the result (3) requires that only λ_i be distinct from the other $p - 1$ roots which may have any multiplicities.

The preceding general results can be used to construct tests of hypotheses and confidence intervals for the population roots and direction cosines. Perhaps a better idea of the sampling variation and dependence of component coefficients can be gained if the expressions are evaluated for a simple covariance matrix. If

$$\boldsymbol{\Sigma} = \sigma^2 \begin{bmatrix} 1 & \cdots & \rho \\ \cdot & \cdots & \cdot \\ \rho & \cdots & 1 \end{bmatrix}$$

its roots and vectors have the forms given by (2) to (3) and (5) to (7) of Sec. 8.6. The variances and covariances of the direction cosines of the first principal component have the values

$$\begin{aligned} \operatorname{var}(a_{r1}) &= \frac{[1 + (p-1)\rho](p-1)(1-\rho)}{np^3\rho^2} \\ \operatorname{cov}(a_{r1}, a_{s1}) &= -\frac{[1 + (p-1)\rho](1-\rho)}{np^3\rho^2} \end{aligned} \tag{6}$$

The correlation of any two coefficients is thus $-1/(p-1)$. Some further numerical examples of the correlations among characteristic vector elements have been computed by Jackson and Hearne (1973).

Asymptotic confidence intervals for the ith covariance population root follow from result (2). For large n, $\sqrt{\tfrac{1}{2}n}\,(l_i - \lambda_i)/\lambda_i$ is a standard normal variate, and the probability is $1 - \alpha$ that the inequality

$$\frac{n}{2\lambda_i^2}(l_i - \lambda_i)^2 \leq z_{\frac{1}{2}\alpha}^2 \tag{7}$$

holds, where $z_{\frac{1}{2}\alpha}$ denotes the upper 50α percentage point of the standard normal distribution. If the quadratic expression in λ_i is simplified, we have the more explicit $100(1 - \alpha)$ percent asymptotic confidence interval:

$$\frac{l_i}{1 + z_{\frac{1}{2}\alpha}\sqrt{2/n}} \leq \lambda_i \leq \frac{l_i}{1 - z_{\frac{1}{2}\alpha}\sqrt{2/n}} \tag{8}$$

Of course n and α must always be such that the right-hand side of (8) is greater than zero.

Example 8.7. Let us find the 95 percent confidence interval for the smallest characteristic root of the carapace dimensions in Example 8.1. Here $n = 23$, $l_3 = 2.86$, $z_{\frac{1}{2}\alpha} = 1.96$, and the confidence interval is

$$1.81 \leq \lambda_3 \leq 6.78$$

This interval is of course in the original millimeters squared units of the covariance matrix. The sample size of 24 observations is hardly of the "asymptotic" order of magnitude, and the true interval is probably considerably wider.

More General Inferences about the Dependence Structure. Anderson (1963a) has treated a test of the hypothesis

$$H_0: \quad \lambda_{q+1} = \cdots = \lambda_{q+r} \tag{9}$$

that r of the intermediate characteristic roots of $\mathbf{\Sigma}$ are equal. The q larger and $p - q - r$ smaller characteristic roots are unrestricted as to their values or multiplicities. The alternative hypothesis to H_0 is that some of the roots in the middle set are distinct. The likelihood-ratio criterion leads to the statistic

$$\chi^2 = -n \sum \ln l_j + nr \ln \frac{\Sigma l_j}{r} \tag{10}$$

where $n = N - 1$ for the single-sample covariance matrix and the summations extend over the values $j = q + 1, \ldots, q + r$. When the hypothesis (9) is true, the statistic has the chi-squared distribution with degrees of freedom $\frac{1}{2}r(r + 1) - 1$ for large n, and the hypothesis is of course rejected for large values of (10). An important special case of the hypothesis occurs when $q + r = p$ or when the variation in the last r dimensions is spherical.

Example 8.8. The hypothesis of equality of the second and third roots of the carapace data can be tested, albeit approximately because of the small sample size, by the statistic (10). Here $n = 23$, $q = 1$, $r = 2$, and

$$\chi^2 = -23(\ln 6.50 + \ln 2.86) + 46 \ln \frac{6.50 + 2.86}{2}$$

$$= 3.77$$

$\chi^2_{0.05;2} = 5.99$, and at the $\alpha = 0.05$ level we would not reject the hypothesis that the normal population ellipsoid has minor axes of equal length. It is thus reasonable to think of the three shell dimensions as distributed about a single principal axis and two minor axes of isotropic variation. We note that acceptance of H_0 on the basis of the asymptotic chi-squared criterion also implies that the hypothesis would be accepted if the exact distribution of the statistic were known for the small sample with 23 degrees of freedom.

It is also possible to generalize the asymptotic confidence interval (8) to the case of multiple roots. If λ_i is of multiplicity r, the $100(1 - \alpha)$ percent asymptotic confidence interval is

$$\frac{\bar{l}_i}{1 + z_{\frac{1}{2}\alpha}\sqrt{2/nr}} \leq \lambda_i \leq \frac{\bar{l}_i}{1 - z_{\frac{1}{2}\alpha}\sqrt{2/nr}} \tag{11}$$

where

$$\bar{l}_i = \frac{1}{r}(l_{q+1} + \cdots + l_{q+r}) \tag{12}$$

A one-sided $100(1 - \frac{1}{2}\alpha)$ percent statement can be obtained from the right-hand inequality for determining, in the case of $q + r = p$, whether the last r roots are probably negligible. Anderson (1963a) has also given a more conservative confidence interval than (11), for that expression tends to be narrow when the population roots are in fact not equal.

Finally let us state an asymptotic test of the important hypothesis

$$H_0: \quad \boldsymbol{\alpha}_i = \boldsymbol{\alpha}_{i0} \tag{13}$$

that the characteristic vector associated with the *distinct* root λ_i of $\boldsymbol{\Sigma}$ is equal to some specified vector $\boldsymbol{\alpha}_{i0}$. As in the case of the preceding tests and interval estimates, the result is due to Anderson (1963a). Starting from the result (3) and its covariance matrix (4) it can be shown that the test statistic

$$\chi^2 = n\left(l_i\boldsymbol{\alpha}'_{i0}\mathbf{S}^{-1}\boldsymbol{\alpha}_{i0} + \frac{1}{l_i}\boldsymbol{\alpha}'_{i0}\mathbf{S}\boldsymbol{\alpha}_{i0} - 2\right) \tag{14}$$

is asymptotically distributed as a chi-squared variate with $p - 1$ degrees of freedom when H_0 is true.

Example 8.9. *Allometry* is the study of the relationship of the size and shape of an organism, as well as the relative growth of its various dimensions. Its origins extend back to Galileo and his investigations of the limiting sizes of organisms and structures. The concept was introduced in modern biology by Huxley (1924, 1932). Gould (1966) has surveyed its biological role, while its statistical and mathematical properties have been described by a number of workers, including Jolicoeur (1963b), Mosimann (1970), and Sprent (1972). Jolicoeur has proposed this generalization of allometry to the growth relations of p dimensions of an organism: the growth of the dimensions is isometric, or in constant proportion with increasing size, if the first principal component of the covariance matrix of the logarithms of the dimensions has the equiangular direction-cosine vector

$$\boldsymbol{\alpha}_1' = \left[\frac{1}{\sqrt{p}}, \ldots, \frac{1}{\sqrt{p}}\right]$$

Let us test the hypothesis of isometric growth for the turtle carapace data introduced in Example 8.1. The covariance matrix of the logarithms of the three measurements for the 24 turtles and its first characteristic vector are given in Example 8.3. The inverse matrix is

$$\mathbf{S}^{-1} = \begin{bmatrix} 5.62396 & -3.54956 & -2.82829 \\ & 7.15698 & -2.00277 \\ & & 4.54773 \end{bmatrix}$$

and the statistic (14) has the value 31.71. Since this is much greater than the upper 1 percent critical value of the chi-squared distribution with two degrees of freedom, the isometric growth model is untenable for the turtle carapaces.

Principal Components Extracted from Correlation Matrices. The asymptotic distribution theory of the characteristic roots and vectors of correlation matrices is exceedingly more complicated than the preceding results, and will not be presented for want of space. However, two special cases lead to tractable results, and their importance merits consideration.

Lawley (1963) has proposed a test of the equality of the last $p - 1$ characteristic roots of the population correlation matrix. Such a hypothesis is precisely equivalent to that of the equality of all $\frac{1}{2}p(p - 1)$ correlations. Let us begin with our usual assumption of this section that a random sample of N p-element observation vectors has been drawn from some multinormal population $N(\boldsymbol{\mu},\boldsymbol{\Sigma})$. If the ijth element of $\boldsymbol{\Sigma}$ is $\sigma_{ij} = \rho_{ij}\sigma_i\sigma_j$, the null hypothesis is

$$H_0: \quad \rho_{ij} = \rho$$

for all subscripts $i \neq j$. Let the sample correlation of the ith and jth responses be denoted by r_{ij}. Then, following Lawley's notation, define these quantities:

$$n = N - 1$$

$$\lambda = 1 - \rho$$

$$\mu = \frac{(p - 1)^2(1 - \lambda^2)}{p - (p - 2)\lambda^2}$$

$$\bar{r}_k = \frac{1}{p-1} \sum_{\substack{i=1 \\ i \neq k}}^{p} r_{ik}$$

$$\bar{r} = \frac{2}{p(p-1)} \sum_{i<j} \sum r_{ij}$$

$\bar{r}_k$ is the average correlation of the kth response with the other variates, and $\bar{r}$ is the grand mean of the correlations. λ is of course the second characteristic root of the population correlation matrix when H_0 is true, and is estimated by $\hat{\lambda} = 1 - \bar{r}$. Lawley has shown that as n tends to infinity, the test statistic

$$\chi^2 = \frac{n}{\hat{\lambda}^2} \left[\sum_{i<j} \sum (r_{ij} - \bar{r})^2 - \hat{\mu} \sum_{k=1}^{p} (\bar{r}_k - \bar{r})^2 \right]$$

for H_0 is distributed according to the chi-squared distribution with $\frac{1}{2}(p+1)(p-2)$ degrees of freedom.

Example 8.10. Lawley's statistic will be used to test the hypothesis of a single principal axis in the standard scores of the four bone lengths of Example 8.2. The mean correlations of the four variates are 0.8977, 0.9010, 0.8920, and 0.8960, and the grand mean correlation is $\bar{r} = 0.8967$; $n = 275$, $\hat{\lambda} = 0.10333$, and $\hat{\mu} = 2.2379$. The total and between-variates sums of squares are

$$\sum_{i<j} \sum (r_{ij} - \bar{r})^2 = 0.003943$$

$$\sum_{k=1}^{4} (\bar{r}_k - \bar{r})^2 = 0.0000420$$

and the test statistic is

$$\chi^2 = \frac{275}{0.01067} [0.003943 - (2.2379)(0.0000420)]$$

$$= 99.20$$

The $\alpha = 0.001$ critical value of the chi-squared distribution with five degrees of freedom is only 20.52, and we conclude that the hypothesis of a single major axis among the four skeletal measurements should be rejected.

The second case is that of a population correlation matrix with the two different roots λ_1 and λ_2 of respective multiplicities q_1 and q_2. The unit diagonal elements of the matrix imply that $\lambda_1 q_1 + \lambda_2 q_2 = p$. Let the characteristic roots of the sample matrix $\mathbf{R}$ be $l_1 > \cdots > l_{q_1} > \cdots > l_p$, and let the estimates of λ_1 and λ_2 be the averages

$$\bar{l}_1 = \frac{1}{q_1} \sum_{i=1}^{q_1} l_i \tag{15}$$
$$\bar{l}_2 = \frac{1}{q_2} \sum_{i=q_1+1}^{p} l_i$$

As in the case of the population characteristic roots, $\bar{l}_2 = (p - q_1\bar{l}_1)/q_2$, and it will suffice to consider only one estimate, say $\bar{l}_2$, in the sequel. Anderson has shown that the asymptotic expectation of $\bar{l}_2$ is λ_2, its asymptotic variance is

$$2\lambda_2{}^2 \frac{(p - q_2\lambda_2)^2}{npq_1q_2} \tag{16}$$

and that as n becomes large, the standardized quantity

$$\frac{\bar{l}_2 - \lambda_2}{\lambda_2(p - q_2\lambda_2)} \sqrt{\frac{npq_1q_2}{2}} \tag{17}$$

tends to be distributed as a unit normal variate. From these results we can construct confidence intervals for either population root. For the $100(1 - \alpha)$ percent confidence interval for λ_2 we require that the square of the standardized estimate (17) be less than $z^2_{\frac{1}{2}\alpha}$, the square of the appropriate unit normal percentage point. This inequality leads to two quadratic equations in λ_2. The appropriate roots give the confidence interval

$$\frac{Kp + 1 - \sqrt{(Kp + 1)^2 - 4Kq_2\bar{l}_2}}{2Kq_2} \leq \lambda_2 \tag{18}$$
$$\leq \frac{Kp - 1 + \sqrt{(Kp - 1)^2 + 4Kq_2\bar{l}_2}}{2Kq_2}$$

where we have introduced the substitution $K = z_{\frac{1}{2}\alpha} \sqrt{2/npq_1q_2}$ in the interests of simplicity. A similar interval can be obtained for the larger characteristic root, although once again the reader is cautioned that the roots and their estimates are linearly dependent. That interval is given by expression (18) with q_2 replaced by q_1 and $\bar{l}_2$ replaced by $\bar{l}_1$.

Example 8.11. In the preceding discussion if $q_1 = 1$, $q_2 = p - 1$, we have the especially important case of the equicorrelation matrix. The matrix

$$\begin{bmatrix} 1 & 0.9740 & 0.9726 \\ & 1 & 0.9655 \\ & & 1 \end{bmatrix}$$

of the correlations of the carapace data of Example 8.1 certainly appears to have come from such a population, and it will be of interest to compute the 95 percent confidence interval for the double characteristic root. Here $p = 3$, $q_2 = 2$, $n = 23$, $z_{\frac{1}{2}\alpha} = 1.96$, and it can be shown that the greatest root $l_1 = 2.9414$. Thus $\bar{l}_2 = \frac{1}{2}(3 - 2.9414) = 0.0293$. $K = 0.236$, and the confidence interval is

$$0.017 \leq \lambda_2 \leq 0.088$$

If we had preferred, we might have obtained instead the 95 percent confidence interval for the largest root:

$$2.824 \leq \lambda_1 \leq 2.966$$

The true 95 percent confidence intervals for a small sample of 24 observations would be somewhat wider.

8.8 EXERCISES

1. Extract the first two principal components from the matrix **S** of Example 4.3. Interpret and name the components, and compare their coefficients with those of the linear discriminant function found in Example 6.1.

2. In a psychological experiment* the reaction times of 64 normal men and women to visual stimuli were recorded when warning intervals of 0.5, 1, 3, 6, and 15 sec preceded the stimulus. The correlations of the median reaction times of several replications of each preparatory interval for a subject formed this matrix:

$$\begin{bmatrix} 1 & 0.71 & 0.58 & 0.56 & 0.65 \\ & 1 & 0.71 & 0.60 & 0.69 \\ & & 1 & 0.75 & 0.71 \\ & & & 1 & 0.74 \\ & & & & 1 \end{bmatrix}$$

   a. What is the greatest amount of variance that a single component could possibly explain?

   b. Choose a trial vector consistent with the matrix pattern, and extract the first principal component.

   c. Compute the matrix reproduced by the first component and the residual matrix.

   d. From the pattern of the residual matrix select a trial vector orthogonal to the coefficients of the first principal component, and extract the second characteristic vector.

   e. Test the hypothesis of equal correlations in the population out of which the subjects were drawn, and if it is tenable, construct the 99 percent confidence interval for the multiple smaller characteristic root.

3. In a study of the relations among the optical elements of the eye van Alphen (1961)

* These correlations have been presented with the kind permission of Jack Botwinick and Joseph Brinley.

computed the correlations of ocular refraction, axial length, anterior chamber depth, corneal power, and lens power of the right eyes of 886 young men and women whose axial lengths did not exceed ± 3 diopters. The five elements had this matrix of correlations:

$$\begin{bmatrix} 1 & -0.45 & -0.40 & -0.21 & 0.13 \\ & 1 & 0.45 & -0.52 & -0.60 \\ & & 1 & 0.09 & -0.32 \\ & & & 1 & -0.09 \\ & & & & 1 \end{bmatrix}$$

a. Can the scales of the variates be reflected so that all correlations are positive?

b. Extract the first principal component. Does the conjecture of a single dimension among the five elements appear to be justified?

c. Compute the residual matrix of the first component. On the basis of its elements and the amount of variance explained by the first component would it appear that the second component can be computed easily?

4. Burton (1963) reexamined the data of Hartshorne and May (1928) in an investigation of the unidimensionality or generality of moral behavior. The correlations* of inventories measuring dishonesty in situations involving copying, speed, peeping, faking, athletic events, and lying had these values for one set of school children:

$$\begin{bmatrix} 1 & 0.450 & 0.400 & 0.400 & 0.288 & 0.350 \\ & 1 & 0.374 & 0.425 & 0.345 & 0.248 \\ & & 1 & 0.300 & 0.100 & 0.108 \\ & & & 1 & 0.300 & 0.256 \\ & & & & 1 & 0.230 \\ & & & & & 1 \end{bmatrix}$$

a. Verify that the correlation coefficients of the measures with the first two components are given by the vectors

$$\mathbf{a}_1' = [0.764, 0.754, 0.581, 0.703, 0.555, 0.526]$$

$$\mathbf{a}_2' = [0.092, 0.106, 0.660, 0.017, -0.504, -0.504]$$

b. What percentages of the total standard scores variance do these components explain?

5. Use the representation

$$\mathbf{\Sigma} = \sigma^2[1 + (p - 1)\rho]\boldsymbol{\alpha}_1\boldsymbol{\alpha}_1' + \sigma^2(1 - \rho)\sum_{h=2}^{p} \boldsymbol{\alpha}_h\boldsymbol{\alpha}_h'$$

of the equal-variance, equal-correlation matrix to verify the variance and covariance of expression (6), Sec. 8.7.

6. In an allometric investigation of the North American marten Jolicoeur (1963a) measured the following dimensions of the humerus and femur bones of 92 males and 47 females:

X_1 = humerus length from head to medial condyle
X_2 = maximum epicondylar width of the distal end of the humerus
X_3 = femur length from head to medial condyle
X_4 = maximum width of the femur distal end

The diagonal matrix of the characteristic roots of the within-sex covariance matrix **S** of the logarithms of those observations was

$$\hat{\mathbf{\Lambda}} = 10^{-4}\begin{bmatrix} 4.2813 & 0 & 0 & 0 \\ & 1.2603 & 0 & 0 \\ & & 0.5738 & 0 \\ & & & 0.1272 \end{bmatrix}$$

and the matrix formed from the column vectors of direction cosines was

$$\mathbf{A} = \begin{bmatrix} 0.39410 & 0.50972 & 0.22651 & 0.73046 \\ 0.57374 & -0.52333 & 0.61536 & -0.13518 \\ 0.39568 & 0.63427 & 0.02469 & -0.66373 \\ 0.59913 & -0.25301 & -0.75459 & 0.08731 \end{bmatrix}$$

Use the relationships $\mathbf{S} = \mathbf{A\Lambda A}'$ and $\mathbf{S}^{-1} = \mathbf{A\Lambda}^{-1}\mathbf{A}'$ to test the hypothesis

$$H_0\colon\ \boldsymbol{\alpha}_1' = [\tfrac{1}{2},\tfrac{1}{2},\tfrac{1}{2},\tfrac{1}{2}]$$

of equal direction cosines, or the same relative growth rates, for the dimensions in the *Martes americana* population.

7. Weiss (1963) reported this matrix of correlations among the hearing threshold measures left-ear audiogram hearing loss, right-ear audiogram hearing loss, 2,000 clicks/sec threshold, and 1 click/sec threshold:

$$\begin{bmatrix} 1 & 0.81 & -0.82 & -0.80 \\ & 1 & -0.91 & -0.80 \\ & & 1 & 0.86 \\ & & & 1 \end{bmatrix}$$

The sample consisted of $N = 43$ elderly men.

a. Extract the first principal component from this matrix.

b. Change the signs of the last two responses so that all correlations are positive. Carry out Lawley's test for the hypothesis of equality of the six population correlations among the four measures.

c. Construct the 95 percent confidence interval for the greatest population characteristic root.

d. If the hypothesis tested in part b can be accepted, construct the 95 percent confidence interval for the smaller characteristic root of the matrix of positive correlations.

9
THE STRUCTURE OF MULTIVARIATE OBSERVATIONS: II. Factor Analysis

9.1 INTRODUCTION.

In the preceding chapter the Hotelling principal-component technique was shown to consist of an orthogonal transformation of the coordinate axes of a multivariate system to new orientations through the natural shape of the scatter swarm of the observation points. By partitioning the total variance of all responses into successively smaller portions we arrived at an objective means of determining the new coordinates. If the first few axes accounted for most of the total variance, and if their positions could be interpreted meaningfully, the system could be described more parsimoniously by them. The values of each sampling unit on the first few "major" axes might be computed and used in later statistical analyses as part of a plan for reducing the data to a smaller number of uncorrelated scores.

For the study of multinormal *dependence* structures we would prefer a technique for explaining the *covariances* of the responses. Although principal-component analysis accomplishes this in some degree through its factorization of the covariance matrix, it is still merely a transformation rather than the result of a fundamental model for covariance structure. The method possesses other shortcomings: the forms of the components are not invariant under changes in the scales of the responses, no rational criteria exist for deciding when a sufficient proportion of the variance has been accounted for by the "principal" components, nor can provision be made for variance components that are attributable only to the unreliability or sampling variation of the individual responses.

In this chapter we shall discuss a mathematical model that leads to the psychometric and statistical technique of *factor analysis*. Through this approach we can avoid the deficiencies of the principal-component solution, although we shall also acquire a new sort of indeterminacy in the estimates of the factor parameters. Under the factor model each response variate will be represented as a linear function of a small number of unobservable *common-factor* variates and a single latent *specific* variate. The common factors generate the covariances among the observable responses, while the specific terms contribute only to the variances of their particular responses. The coefficients of the common factors will not be restricted to be orthogonal, and in fact their matrix will be unique only up to postmultiplication by an orthogonal matrix. In return for this greater parsimony and generality we must assume that the observations arose from a multinormal population of full rank and that the exact number of common factors can be specified before the analysis.

In principle the concept of latent factors seems to have been first suggested by Galton (1888). The actual mathematical model for factor structure as distinct from the primordial principal-axis solution of K. Pearson originated with Spearman (1904), who hypothesized that the correlations among a set of intelligence-test scores could be generated by a single latent factor of general intellective ability and a second set of factors reflecting the unique qualities of the individual tests. Later workers—particularly L. L. Thurstone—extended this model to include many common factors. Thurstone also proposed his centroid method for extracting the factor coefficients from a correlation matrix. In both periods divergent schools developed around other models and their computational problems. The various approaches are discussed by Harman (1967) in his scholarly and comprehensive text. While many of the models included "error" terms reflecting the sampling variation of the observed correlations, none actually used the results of the new discipline of statistical inference. It was not until 1940 that D. N. Lawley reduced the extraction of factor parameters to a problem in maximum-likelihood estimation and by so doing eliminated the indeterminacies and subjective decisions required by the centroid method. Furthermore, the goodness of fit of a solution with just m factors could now be tested rigorously by the generalized likelihood-ratio principle. For those reasons our treatment of factor analysis will be restricted to its modern explication as a problem in statistical estimation.

We shall begin our development with a description of the model and the maximum-likelihood estimation of its parameters. The likelihood equations will be expressed in characteristic vector form and solved iteratively. We shall briefly treat graphical and analytical solutions of the "rotation" problem, as well as an alternative derivation of the estimates which does not assume multinormality. Unlike principal components, the sampling properties of the estimates and the evaluation of factor scores are not so straightforward, and we shall consider some methods for their determination. Finally we shall treat a different model for covariance structure that links the observed responses to a latent stochastic process.

9.2 THE MATHEMATICAL MODEL FOR FACTOR STRUCTURE

Suppose that the multivariate system consists of p responses described by the *observable* random variables $X_1, \ldots, X_p$. The X_i have a nonsingular multinormal distribution. Since only the covariance structure will be of interest, we can assume without loss of generality that the population means of the X_i are zero. Let us begin our quest for a more parsimonious explanation of the covariance structure with the following model for the responses:

$$\begin{aligned} X_1 &= \lambda_{11}Y_1 + \cdots + \lambda_{1m}Y_m + e_1 \\ &\cdots\cdots\cdots\cdots\cdots\cdots \\ X_p &= \lambda_{p1}Y_1 + \cdots + \lambda_{pm}Y_m + e_p \end{aligned} \tag{1}$$

where Y_j = jth common-factor variate
λ_{ij} = parameter reflecting importance of jth factor in composition of ith response
e_i = ith specific-factor variate

In the usage of factor analysts λ_{ij} is called the *loading* of the ith response on the jth common factor. For the matrix version of the model let

$$\mathbf{x}' = [X_1, \ldots, X_p]$$

$$\mathbf{y}' = [Y_1, \ldots, Y_m]$$

$$\boldsymbol{\varepsilon}' = [e_1, \ldots, e_p]$$

and

$$\boldsymbol{\Lambda} = \begin{bmatrix} \lambda_{11} & \cdots & \lambda_{1m} \\ \cdot & \cdots & \cdot \\ \lambda_{p1} & \cdots & \lambda_{pm} \end{bmatrix} \tag{2}$$

Then the factor model can be written as

$$\mathbf{x} = \boldsymbol{\Lambda}\mathbf{y} + \boldsymbol{\varepsilon} \tag{3}$$

Let the m common-factor variates in $\mathbf{y}$ be independently and normally distributed with zero means and unit variances. Similarly, assume that the elements of $\boldsymbol{\varepsilon}$ are normally and independently distributed with mean zero and variances

$$\text{(4)} \qquad \operatorname{var}(e_i) = \Psi_i$$

Ψ_i is called the *specific variance* or *specificity* of the ith response. In the sequel it will be convenient to write those parameters as the diagonal matrix

$$\text{(5)} \qquad \mathbf{\Psi} = \begin{bmatrix} \Psi_1 & \cdots & 0 \\ \cdot\cdot & \cdots & \cdot\cdot \\ 0 & \cdots & \Psi_p \end{bmatrix}$$

Thurstone (1945) and other factor analysts have proposed more general linear models with several specific or residual terms in the representations of some variates. Our Ψ_i terms will account for variation from all such sources and will be the equivalent of Thurstone's *uniqueness* variance component. Finally, we shall require that the variates $\mathbf{y}$ and $\boldsymbol{\varepsilon}$ be independently distributed.

From the properties of the latent variates it follows that the variance of the ith response variate can be written as

$$\text{(6)} \qquad \sigma_i^2 = \lambda_{i1}^2 + \cdots + \lambda_{im}^2 + \Psi_i$$

and the covariance of the ith and jth as

$$\text{(7)} \qquad \sigma_{ij} = \lambda_{i1}\lambda_{j1} + \cdots + \lambda_{im}\lambda_{jm}$$

These relations can be expressed concisely in matrix form as

$$\text{(8)} \qquad \mathbf{\Sigma} = \mathbf{\Lambda\Lambda}' + \mathbf{\Psi}$$

In this fundamental representation of $\mathbf{\Sigma}$ under the factor model the diagonal elements of $\mathbf{\Lambda\Lambda}'$

$$\text{(9)} \qquad \sigma_i^2 - \Psi_i = \sum_{j=1}^{m} \lambda_{ij}^2$$

are called the *communalities* of the responses.

The parameter λ_{ij} is the covariance of the ith response with the jth common factor. This is easily verified by writing the covariance matrix of the observable and common-factor variates as

$$\text{(10)} \qquad \operatorname{cov}(\mathbf{x},\mathbf{y}') = E[(\mathbf{\Lambda y} + \boldsymbol{\varepsilon})\mathbf{y}'] = \mathbf{\Lambda}$$

from the mutual independence of the latent variates. If $\mathbf{\Sigma}$ is the population correlation matrix, the λ_{ij} are the correlations of the responses and common factors. In the literature of the behavioral sciences the term *loading* is almost invariably used in that sense of a correlation.

The purpose of factor analysis is the determination of the elements of the loading matrix $\mathbf{\Lambda}$, with the elements of $\mathbf{\Psi}$ following from the constraint (6) imposed upon the communalities. For this reason factor analysis is outwardly similar to the extraction of the principal components of $\mathbf{\Sigma}$, for both techniques begin with linear models and end with matrix factorization. However, as Bartlett (1953) has pointed out, the model for component analysis must be linear by the very fact that it refers to a rigid rotation of the response coordinate axes, while the model (1) of factor analysis is as much a part of our hypothesis about the dependence structure as the choice of exactly m common factors. Selection of a linear model was due to the convenient interpretation of its parameters as correlations and the fact that normal variates are still normally distributed after a linear transformation. If the covariances reproduced by the m-factor linear model fail to fit the sample values adequately, it is as proper to reject linearity as it is to advance the more usual finding that m common factors are inadequate for explaining the sample correlations.

Now let us examine this common feature of matrix factorization. In our discussion of principal components as the factorization of the covariance matrix it was clear that the factor matrix was unique if the component variances were all different. This is not true for the factor-loading matrix, for no condition has been imposed that the sums of squares of the loadings become successively smaller as one passes from the first to the mth factor. If we multiply $\mathbf{\Lambda}$ by the $m \times m$ orthogonal matrix $\mathbf{T}$, the representation (8) of $\mathbf{\Sigma}$ becomes

$$\begin{aligned} \mathbf{\Lambda T}(\mathbf{\Lambda T})' + \mathbf{\Psi} &= \mathbf{\Lambda T T'\Lambda}' + \mathbf{\Psi} \\ &= \mathbf{\Lambda\Lambda}' + \mathbf{\Psi} \\ &= \mathbf{\Sigma} \end{aligned} \tag{11}$$

and although the elements of $\mathbf{\Lambda T}$ are different from the original loadings, their ability to generate the covariances is unchanged. By choosing different orthogonal transformations an infinity of loading matrices can be computed from $\mathbf{\Lambda}$ which would lead to the same covariance matrix. Thurstone (1945) proposed the concept of *simple structure* as a means of selecting the loadings most meaningful and interpretable in the subject-matter sense of the responses. Howe (1955) and others have suggested that the transformation should be carried out to make the loadings conform as closely as possible with some *a priori* notion of the pattern of the dependence structure. Factor rotation will be treated in detail in Sec. 9.7.

Example 9.1. It will be necessary to defer the consideration of real data examples of factor structure until we have developed the statistical and computational procedures of the next three sections. However, perhaps a simple contrived example will help at this point to fix the notion of the latent-factor model. The matrix

$$\mathbf{\Sigma} = \begin{bmatrix} 1 & 0.66 & 0.39 & 0.15 & 0.68 & 0.72 \\ & 1 & 0.17 & 0.05 & 0.65 & 0.56 \\ & & 1 & 0.28 & 0.05 & 0.40 \\ & & & 1 & -0.01 & 0.16 \\ & & & & 1 & 0.56 \\ & & & & & 1 \end{bmatrix}$$

can be shown to have specific variances 0.18, 0.42, 0.39, 0.87, 0.26, and 0.36 for its respective variates and factor-loading matrix

$$\mathbf{\Lambda}' = \begin{bmatrix} 0.9 & 0.7 & 0.5 & 0.2 & 0.7 & 0.8 \\ -0.1 & -0.3 & 0.6 & 0.3 & -0.5 & 0 \end{bmatrix}$$

An infinity of such factorizations can be generated by postmultiplying $\mathbf{\Lambda}$ by the 2×2 orthogonal matrix

$$\mathbf{T} = \begin{bmatrix} \cos\theta & -\sin\theta \\ \sin\theta & \cos\theta \end{bmatrix}$$

The reader should verify that this is the case for some simple value of θ, e.g., $\theta = 45°$ and $\cos\theta = \sin\theta = \sqrt{2}/2$.

Since $\mathbf{\Sigma}$ is a correlation matrix, the elements of $\mathbf{\Lambda}$ are interpretable as correlation coefficients. For example, the first response variate has correlation 0.9 with the first factor and a negligible correlation of -0.1 with the second. The generally small correlations of the fourth variate with its neighbors are reflected in the small loadings on both factors, and in its large specific variance component.

9.3 ESTIMATION OF THE FACTOR LOADINGS

In practice the parameters of the factor model are never known and must be estimated from sample observations. Lawley (1940, 1942, 1943) attacked this estimation problem by the method of maximum likelihood and showed that the estimates of the elements of $\mathbf{\Lambda}$ and $\mathbf{\Psi}$ could be expressed as the solutions of certain implicit matrix equations. For the development of those equations let us suppose that a sample of $N = n + 1$ independent observation vectors has been drawn from the multinormal population with mean vector $\boldsymbol{\mu}$ and covariance matrix $\mathbf{\Sigma} = \mathbf{\Psi} + \mathbf{\Lambda}\mathbf{\Lambda}'$. It will be essential for our development to require that $\mathbf{\Sigma}$ be of full rank p. The loading matrix $\mathbf{\Lambda}$ has m common-factor columns, where m has been specified before the extraction of the factor estimates. The information in the sample covariance matrix $\mathbf{S}$ is sufficient for the estimation of the factor parameters, and we shall choose to express the fundamental likelihood function for **S** in terms of the Wishart density

$$(1) \qquad f(\mathbf{S}) = C|\mathbf{S}|^{\frac{1}{2}(n-p-1)}|\mathbf{\Sigma}|^{-\frac{1}{2}n} \exp\left(-\tfrac{1}{2}n \operatorname{tr} \mathbf{\Sigma}^{-1}\mathbf{S}\right)$$

Under the m-factor model the logarithm of the likelihood is

$$(2) \qquad l(\mathbf{\Lambda},\mathbf{\Psi}) = \ln C + \tfrac{1}{2}(n-p-1)\ln|\mathbf{S}| - \tfrac{1}{2}n \ln|\mathbf{\Psi} + \mathbf{\Lambda}\mathbf{\Lambda}'| - \tfrac{1}{2}n \operatorname{tr}\,[\mathbf{\Psi} + \mathbf{\Lambda}\mathbf{\Lambda}']^{-1}\mathbf{S}$$

The maximum-likelihood equations follow from setting the derivatives of $l(\mathbf{\Lambda},\mathbf{\Psi})$ with respect to the $p(m+1)$ elements of $\mathbf{\Psi}$ and $\mathbf{\Lambda}$ equal to zero.

Let us begin by differentiating with respect to the specific variances. Then

$$(3) \qquad \frac{\partial l(\mathbf{\Lambda},\mathbf{\Psi})}{\partial \Psi_i} = -\tfrac{1}{2}n \frac{1}{|\mathbf{\Psi} + \mathbf{\Lambda}\mathbf{\Lambda}'|}\frac{\partial|\mathbf{\Psi} + \mathbf{\Lambda}\mathbf{\Lambda}'|}{\partial \Psi_i} + \tfrac{1}{2}n \operatorname{tr}\,[\mathbf{\Psi} + \mathbf{\Lambda}\mathbf{\Lambda}']^{-1}\frac{\partial \mathbf{\Psi}}{\partial \Psi_i}[\mathbf{\Psi} + \mathbf{\Lambda}\mathbf{\Lambda}']^{-1}\mathbf{S}$$

$$= -\tfrac{1}{2}n \frac{1}{|\mathbf{\Psi} + \mathbf{\Lambda}\mathbf{\Lambda}'|}|\mathbf{\Psi} + \mathbf{\Lambda}\mathbf{\Lambda}'|_{ii} + \tfrac{1}{2}n \operatorname{tr}\,[\mathbf{\Psi} + \mathbf{\Lambda}\mathbf{\Lambda}']^{-1}\mathbf{S}[\mathbf{\Psi} + \mathbf{\Lambda}\mathbf{\Lambda}']^{-1}\frac{\partial \mathbf{\Psi}}{\partial \Psi_i}$$

$\partial\mathbf{\Psi}/\partial\Psi_i$ is the $p \times p$ matrix with unity in its ith diagonal position and zeros everywhere else, while $|\mathbf{\Psi} + \mathbf{\Lambda}\mathbf{\Lambda}'|_{ii}$ is the cofactor of the ith diagonal element of $\mathbf{\Psi} + \mathbf{\Lambda}\mathbf{\Lambda}'$. For the likelihood to be a maximum it is necessary that each of the p derivatives defined by (3) equals zero. That condition can be written in matrix form as

$$\operatorname{diag}\{[\mathbf{\Psi} + \mathbf{\Lambda}\mathbf{\Lambda}']^{-1}(\mathbf{I} - \mathbf{S}[\mathbf{\Psi} + \mathbf{\Lambda}\mathbf{\Lambda}']^{-1})\} = \mathbf{0}$$

or

$$(4) \qquad \operatorname{diag}(\hat{\mathbf{\Sigma}}^{-1}) = \operatorname{diag}(\hat{\mathbf{\Sigma}}^{-1}\mathbf{S}\hat{\mathbf{\Sigma}}^{-1})$$

The quantity

$$(5) \qquad \hat{\mathbf{\Sigma}} = \hat{\mathbf{\Psi}} + \hat{\mathbf{\Lambda}}\hat{\mathbf{\Lambda}}'$$

is the *matrix of variances and covariances reproduced from the m-factor model.* It will be essential in the sequel not to confuse $\hat{\mathbf{\Sigma}}$ with the usual maximum-likelihood estimate of the covariance matrix that was introduced in the more general context of Chap. 3. After we have obtained the second set of equations, we shall see that the set (4) can be reduced to a considerably simpler form.

Now let us obtain the other set of equations by differentiating $l(\mathbf{\Lambda},\mathbf{\Psi})$ with respect to the loading parameters. The ijth derivative is

$$(6)\qquad \frac{\partial l(\mathbf{\Lambda},\mathbf{\Psi})}{\partial \lambda_{ij}} = -\tfrac{1}{2}n \frac{1}{|\mathbf{\Psi}+\mathbf{\Lambda\Lambda}'|} \sum_{g=1}^{p}\sum_{h=1}^{p} |\mathbf{\Psi}+\mathbf{\Lambda\Lambda}'|_{gh} \frac{\partial \sigma_{gh}}{\partial \lambda_{ij}} + \tfrac{1}{2}n \operatorname{tr} [\mathbf{\Psi}+\mathbf{\Lambda\Lambda}']^{-1} \frac{\partial \mathbf{\Sigma}}{\partial \lambda_{ij}} [\mathbf{\Psi}+\mathbf{\Lambda\Lambda}']^{-1}\mathbf{S}$$

The first term of the right-hand side is equal to

$$(7)\qquad -\tfrac{1}{2}n \operatorname{tr} \mathbf{\Sigma}^{-1} \frac{\partial \mathbf{\Sigma}}{\partial \lambda_{ij}}$$

where

$$(8)\qquad \frac{\partial \mathbf{\Sigma}}{\partial \lambda_{ij}} = \left[\frac{\partial \sigma_{gh}}{\partial \lambda_{ij}}\right] = \begin{bmatrix} 0 & \cdots & 0 & \lambda_{1j} & 0 & \cdots & 0 \\ \cdot & \cdots & \cdot & \cdot & \cdot & \cdots & \cdot \\ 0 & \cdots & 0 & \lambda_{i-1,j} & 0 & \cdots & 0 \\ \lambda_{1j} & \cdots & \lambda_{i-1,j} & 2\lambda_{ij} & \lambda_{i+1,j} & \cdots & \lambda_{pj} \\ 0 & \cdots & 0 & \lambda_{i+1,j} & 0 & \cdots & 0 \\ \cdot & \cdots & \cdot & \cdot & \cdot & \cdots & \cdot \\ 0 & \cdots & 0 & \lambda_{pj} & 0 & \cdots & 0 \end{bmatrix}$$

is a symmetric matrix with zeros in all positions except those of the ith row and column. If we denote the ijth element of $\mathbf{\Sigma}^{-1}$ by σ^{ij}, it follows that

$$(9)\qquad \begin{aligned} \operatorname{tr} \mathbf{\Sigma}^{-1} \frac{\partial \mathbf{\Sigma}}{\partial \lambda_{ij}} &= \sigma^{1i}\lambda_{1j} + \sigma^{2i}\lambda_{2j} + \cdots + \sigma^{1i}\lambda_{1j} + \cdots \\ &\quad + 2\sigma^{ii}\lambda_{ij} + \cdots + \sigma^{ip}\lambda_{pj} + \cdots + \sigma^{ip}\lambda_{pj} \\ &= 2(\sigma^{1i}\lambda_{1j} + \cdots + \sigma^{ip}\lambda_{pj}) \\ &= 2\boldsymbol{\sigma}^{i\prime}\boldsymbol{\lambda}_j \end{aligned}$$

where $\boldsymbol{\sigma}^{i\prime}$ is the ith row of $\mathbf{\Sigma}^{-1}$, and $\boldsymbol{\lambda}_j$ is the jth column of $\mathbf{\Lambda}$. Hence the pm derivatives of the first term $-\tfrac{1}{2}n \ln |\mathbf{\Sigma}|$ are given by

$$(10)\qquad -n\mathbf{\Sigma}^{-1}\mathbf{\Lambda}$$

In the second term two cyclic permutations of the matrices within the trace symbol yield

$$(11)\qquad \tfrac{1}{2}n \operatorname{tr} \mathbf{\Sigma}^{-1} \frac{\partial \mathbf{\Sigma}}{\partial \lambda_{ij}} \mathbf{\Sigma}^{-1}\mathbf{S} = \tfrac{1}{2}n \operatorname{tr} \mathbf{\Sigma}^{-1}\mathbf{S}\mathbf{\Sigma}^{-1} \frac{\partial \mathbf{\Sigma}}{\partial \lambda_{ij}}$$

For convenience in the ensuing manipulations let us write $\mathbf{\Sigma}^{-1}\mathbf{S}\mathbf{\Sigma}^{-1}$ as the matrix $\mathbf{Z}$ with general element z_{gh}. Then, by the same argument employed to obtain equation (9),

$$\text{(12)} \qquad \operatorname{tr} \mathbf{Z} \frac{\partial \mathbf{\Sigma}}{\partial \lambda_{ij}} = 2\mathbf{z}_i'\boldsymbol{\lambda}_j$$

where $\mathbf{z}_i'$ is the ith row of $\mathbf{Z}$. All pm such derivatives of the second term can be written as the matrix $\mathbf{Z}\mathbf{\Lambda}$, and the second set of equations to be satisfied by the estimates is

$$\text{(13)} \qquad (\hat{\mathbf{\Sigma}}^{-1} - \hat{\mathbf{\Sigma}}^{-1}\mathbf{S}\hat{\mathbf{\Sigma}}^{-1})\,\hat{\mathbf{\Lambda}} = \mathbf{0}$$

or

$$\text{(14)} \qquad \mathbf{S}\hat{\mathbf{\Sigma}}^{-1}\hat{\mathbf{\Lambda}} = \hat{\mathbf{\Lambda}}$$

Now we are ready to determine the final simplified versions of both sets of equations. In the first set (4) pre- and postmultiply the matrices of the left- and right-hand sides by the diagonal matrix

$$\hat{\mathbf{\Psi}} = \hat{\mathbf{\Sigma}} - \hat{\mathbf{\Lambda}}\hat{\mathbf{\Lambda}}'$$

The equations become

$$\operatorname{diag}(\hat{\mathbf{\Sigma}} - 2\hat{\mathbf{\Lambda}}\hat{\mathbf{\Lambda}}' + \hat{\mathbf{\Lambda}}\hat{\mathbf{\Lambda}}'\hat{\mathbf{\Sigma}}^{-1}\hat{\mathbf{\Lambda}}\hat{\mathbf{\Lambda}}')$$
$$= \operatorname{diag}(\mathbf{S} - \mathbf{S}\hat{\mathbf{\Sigma}}^{-1}\hat{\mathbf{\Lambda}}\hat{\mathbf{\Lambda}}' - \hat{\mathbf{\Lambda}}\hat{\mathbf{\Lambda}}'\hat{\mathbf{\Sigma}}^{-1}\mathbf{S} + \hat{\mathbf{\Lambda}}\hat{\mathbf{\Lambda}}'\hat{\mathbf{\Sigma}}^{-1}\mathbf{S}\hat{\mathbf{\Sigma}}^{-1}\hat{\mathbf{\Lambda}}\hat{\mathbf{\Lambda}}')$$

Simplification of the terms in the right-hand matrix by the second maximum-likelihood equation (14) leads straightaway to the condition

$$\text{(15)} \qquad \operatorname{diag}(\hat{\mathbf{\Sigma}}) = \operatorname{diag}(\mathbf{S})$$

or the not-surprising constraint that the estimated specific variance and communality of each response must sum to the sample variance.

Inversion of a $p \times p$ matrix can be avoided by writing (14) in an alternate form. We begin by introducing the useful identity

$$\text{(16)} \qquad (\hat{\mathbf{\Psi}} + \hat{\mathbf{\Lambda}}\hat{\mathbf{\Lambda}}')^{-1}\hat{\mathbf{\Lambda}} = \hat{\mathbf{\Psi}}^{-1}\hat{\mathbf{\Lambda}}(\mathbf{I} + \hat{\mathbf{\Lambda}}'\hat{\mathbf{\Psi}}^{-1}\hat{\mathbf{\Lambda}})^{-1}$$

By that device equation (14) can be transformed into

$$\text{(17)} \qquad \mathbf{S}\hat{\mathbf{\Psi}}^{-1}\hat{\mathbf{\Lambda}}(\mathbf{I} + \hat{\mathbf{\Lambda}}'\hat{\mathbf{\Psi}}^{-1}\hat{\mathbf{\Lambda}})^{-1} = \hat{\mathbf{\Lambda}}$$

or

$$\text{(18)} \qquad \mathbf{S}\hat{\mathbf{\Psi}}^{-1}\hat{\mathbf{\Lambda}} = \hat{\mathbf{\Lambda}}(\mathbf{I} + \hat{\mathbf{\Lambda}}'\hat{\mathbf{\Psi}}^{-1}\hat{\mathbf{\Lambda}})$$

The maximum-likelihood loading estimates enjoy a powerful invariance property: *changes in the scales of the response variates only appear as scale changes of the loadings.* In particular, the loadings extracted from the correlation matrix differ from those of the covariance matrix by the factors $1/s_i$. The property may be verified by making the diagonal transformation

$$\mathbf{z}_i = \mathbf{D}\mathbf{x}_i \tag{19}$$

upon the observation vectors $\mathbf{x}_i$. The new sample covariance matrix is **DSD**, and its inverse is $\mathbf{D}^{-1}\mathbf{S}^{-1}\mathbf{D}^{-1}$. The factor loadings of the transformed observations follow from (14) as the solutions of the equation

$$[\hat{\boldsymbol{\Psi}}_z + \hat{\boldsymbol{\Lambda}}_z\hat{\boldsymbol{\Lambda}}_z']\mathbf{D}^{-1}\mathbf{S}^{-1}\mathbf{D}^{-1}\hat{\boldsymbol{\Lambda}}_z = \hat{\boldsymbol{\Lambda}}_z \tag{20}$$

where $\hat{\boldsymbol{\Lambda}}_z$ is the loading matrix of the new observations. Premultiplication of both sides of (20) by $\mathbf{D}^{-1}$ yields the estimation equations

$$[\mathbf{D}^{-1}\hat{\boldsymbol{\Psi}}_z\mathbf{D}^{-1} + (\mathbf{D}^{-1}\hat{\boldsymbol{\Lambda}}_z)(\mathbf{D}^{-1}\hat{\boldsymbol{\Lambda}}_z)']\mathbf{S}^{-1}(\mathbf{D}^{-1}\hat{\boldsymbol{\Lambda}}_z) = \mathbf{D}^{-1}\hat{\boldsymbol{\Lambda}}_z$$

so that the factor structure of the responses is related to the original loadings $\hat{\boldsymbol{\Lambda}}_x$ by the scale transformation

$$\hat{\boldsymbol{\Lambda}}_z = \mathbf{D}\hat{\boldsymbol{\Lambda}}_x \tag{21}$$

9.4 NUMERICAL SOLUTION OF THE ESTIMATION EQUATIONS

For the solution of the maximum-likelihood loading equations we shall adopt an approach first suggested by Rao (1955) and subsequently advocated by Maxwell (1964) which appears to have good convergence properties. We begin by writing the equation (18) of Sec. 9.3 as

$$\hat{\boldsymbol{\Lambda}}(\hat{\boldsymbol{\Lambda}}'\hat{\boldsymbol{\Psi}}^{-1}\hat{\boldsymbol{\Lambda}}) = (\mathbf{S} - \hat{\boldsymbol{\Psi}})\hat{\boldsymbol{\Psi}}^{-1}\hat{\boldsymbol{\Lambda}} \tag{1}$$

Premultiplication of both sides by $\hat{\boldsymbol{\Psi}}^{-1/2}$ and some rearrangement of terms yield

$$[\hat{\boldsymbol{\Psi}}^{-1/2}(\mathbf{S} - \hat{\boldsymbol{\Psi}})\hat{\boldsymbol{\Psi}}^{-1/2}]\hat{\boldsymbol{\Psi}}^{-1/2}\hat{\boldsymbol{\Lambda}} = \hat{\boldsymbol{\Psi}}^{-1/2}\hat{\boldsymbol{\Lambda}}\mathbf{J} \tag{2}$$

where $\mathbf{J} = \hat{\boldsymbol{\Lambda}}'\hat{\boldsymbol{\Psi}}^{-1}\hat{\boldsymbol{\Lambda}}$. In order that the elements of $\hat{\boldsymbol{\Lambda}}$ be uniquely defined we shall require in the sequel that $\mathbf{J}$ be a diagonal matrix. Then the first m characteristic roots of $(\mathbf{S} - \hat{\boldsymbol{\Psi}})\hat{\boldsymbol{\Psi}}^{-1}$, and hence of $\hat{\boldsymbol{\Psi}}^{-1/2}(\mathbf{S} - \hat{\boldsymbol{\Psi}})\hat{\boldsymbol{\Psi}}^{-1/2}$, are equal to the successive elements of $\mathbf{J}$, and the ith column of $\hat{\boldsymbol{\Psi}}^{-1/2}\hat{\boldsymbol{\Lambda}}$ is merely the characteristic vector corresponding to the ith largest root of $\hat{\boldsymbol{\Psi}}^{-1/2}(\mathbf{S} - \hat{\boldsymbol{\Psi}})\hat{\boldsymbol{\Psi}}^{-1/2}$. The computation of these vectors is still not the usual straightforward process, for the elements of $\hat{\boldsymbol{\Psi}}$ are also unknown, and must be found from the relation $\hat{\boldsymbol{\Psi}} = \operatorname{diag}(\mathbf{S} - \hat{\boldsymbol{\Lambda}}\hat{\boldsymbol{\Lambda}}')$. However, we can proceed iteratively by solving for the characteristic vectors corresponding to the guessed or approximate elements of $\hat{\boldsymbol{\Psi}}$, and then using that solution to compute a more precise set of specific variance estimates.

For the single-factor model the characteristic vector of the greatest root of $\mathbf{S}$ can be used as the starting vector for such an iterative process. The elements of the vector must be scaled so that the sum of their squares is equal to the characteristic root. The iterative process follows this plan:

1. Compute the greatest characteristic root l_{10} and its vector $\mathbf{a}_{10}$ of $\mathbf{S}$, where the elements of the vector have been scaled such that $\mathbf{a}'_{10}\mathbf{a}_{10} = l_{10}$.
2. Approximate the specific variances from

$$\hat{\mathbf{\Psi}}_{10} = \text{diag}\,(\mathbf{S} - \mathbf{a}_{10}\mathbf{a}'_{10}) \tag{3}$$

where in the sequel the subscripts i, j of $\hat{\mathbf{\Psi}}$ and $\mathbf{a}$ will denote the jth iterates of the i-factor solution.
3. Form the matrix

$$\hat{\mathbf{\Psi}}_{10}^{-1/2}(\mathbf{S} - \hat{\mathbf{\Psi}}_{10})\hat{\mathbf{\Psi}}_{10}^{-1/2} \tag{4}$$

and extract the vector $\mathbf{a}_{11}$ associated with its greatest characteristic root l_{11}. Scale the elements so that $\mathbf{a}'_{11}\mathbf{a}_{11} = l_{11}$ and premultiply the vector by $\hat{\mathbf{\Psi}}_{10}^{1/2}$ to obtain the first approximation $\hat{\boldsymbol{\lambda}}_{11}$ to the single column of $\hat{\mathbf{\Lambda}}_1$.
4. Compute

$$\hat{\mathbf{\Psi}}_{11} = \text{diag}\,(\mathbf{S} - \hat{\boldsymbol{\lambda}}_{11}\hat{\boldsymbol{\lambda}}'_{11}) \tag{5}$$

and repeat the process for the second approximation to $\hat{\mathbf{\Lambda}}_1$. Continue in this fashion until the corresponding elements of the successive iterates $\hat{\boldsymbol{\lambda}}_{1i}$ and $\hat{\boldsymbol{\lambda}}_{1,i+1}$ do not differ by more than some predetermined amount. The resulting column vector $\hat{\mathbf{\Lambda}}_1$ will contain the maximum-likelihood estimates of the loadings for the one-factor model.

Maxwell has cautioned that the loadings obtained from the principal-component and factor analyses may not be sufficiently close to permit the convergence of an iterative procedure starting from the first m principal components of $\mathbf{S}$. To obtain the estimated loadings of the second, third, . . . , mth factors Maxwell begins by computing the residual matrix

$$\mathbf{S}_1 = \mathbf{S} - \hat{\mathbf{\Lambda}}_1\hat{\mathbf{\Lambda}}'_1$$

of the single-factor solution. The characteristic vector $\mathbf{a}_{20}$ of the greatest root of $\mathbf{S}_1$ is then taken as the initial approximation to the loadings of the second factor. The elements of $\mathbf{a}_{20}$ have of course been scaled so that the sum of their squares is that greatest root. From the single-factor solution and the new initial vector form the $p \times 2$ matrix

$$\hat{\mathbf{\Lambda}}_{20} = [\hat{\mathbf{\Lambda}}_1 \quad \mathbf{a}_{20}]$$

for the zero-order approximation to the estimated loadings of the two-factor model. The iterative process for those loadings follows as before: the zero-order approximation to the specific variances is computed as

$$\hat{\mathbf{\Psi}}_{20} = \text{diag}\,(\mathbf{S} - \hat{\mathbf{\Lambda}}_{20}\hat{\mathbf{\Lambda}}'_{20})$$

and the symmetric matrix

$$\hat{\mathbf{\Psi}}_{20}^{-\frac{1}{2}}(\mathbf{S} - \hat{\mathbf{\Psi}}_{20})\hat{\mathbf{\Psi}}_{20}^{-\frac{1}{2}}$$

is formed. Its first two characteristic roots and their vectors are extracted, scaled to the usual loading form, and written as the $p \times 2$ matrix $[\mathbf{a}_{11} \quad \mathbf{a}_{21}]$. It is essential to note that the first column vector is *not* the same as the first iterate $\mathbf{a}_{11}$ of the single-factor solution. Premultiply $[\mathbf{a}_{11} \quad \mathbf{a}_{21}]$ by $\hat{\mathbf{\Psi}}_{20}^{\frac{1}{2}}$ to obtain the first approximation $\hat{\mathbf{\Lambda}}_{21}$ to the loading estimates of the two-factor model. This process is repeated until all elements of the iterates $\hat{\mathbf{\Lambda}}_{2i}$ have converged with specified accuracy to the two-factor solution $\hat{\mathbf{\Lambda}}_2$.

The solution of the m-factor likelihood equations begins in like manner from the $(m - 1)$-factor solution: those latter estimates provide the starting values for the first $m - 1$ factors of the new model, while the mth trial vector is found from the characteristic vector of the greatest root of

$$\mathbf{S}_{m-1} = \mathbf{S} - \hat{\mathbf{\Lambda}}_{m-1}\hat{\mathbf{\Lambda}}'_{m-1}$$

The iterative process is repeated until the elements of $\hat{\mathbf{\Lambda}} \equiv \hat{\mathbf{\Lambda}}_m$ have converged with appropriate accuracy. Such convergence may be difficult to attain if m is large, although in one computer program* written for the process convergence was greatly accelerated by the use of the Aitken δ^2 method described in Sec. 8.4. Some solutions obtained by that program will be discussed in Sec. 9.6.

Other numerical methods have been developed for solving the maximum-likelihood equations. Originally Lawley (1942) proposed an iterative scheme which has been shown by later workers (e.g., Howe, 1955) to have poor convergence properties. More recently, Jöreskog (1967a) has developed a direct maximization method and a computer program for its use (1967b). That approach will provide solutions in the improper, or Heywood, case (Harman, 1967, sec. 7.3) in which the sta-

* I am indebted to S. J. Farlow for supplying me with a copy of this program.

tionary maximum occurs for nonpositive specific variances. Jennrich and Robinson (1969) have constructed and illustrated an algorithm based upon a Newton-Raphson solution of the likelihood equations. Another Newton-Raphson solution for the maximum likelihood and two other factor estimates which makes provision for the Heywood case has been programmed by Jöreskog and van Thillo (1971); the algorithms have been described by Jöreskog (1975).

9.5 TESTING THE GOODNESS OF FIT OF THE FACTOR MODEL

The maximum-likelihood nature of the loading estimates leads to a formal test of the adequacy of the m-factor model for generating the observed covariances or correlations. The null hypothesis is

$$H_0\colon\ \boldsymbol{\Sigma} = \boldsymbol{\Psi} + \boldsymbol{\Lambda}\boldsymbol{\Lambda}' \tag{1}$$

where $\boldsymbol{\Lambda}$ has dimensions $p \times m$, and the alternative hypothesis is that $\boldsymbol{\Sigma}$ is any $p \times p$ symmetric positive definite matrix. Use of the likelihood-ratio principle gives the test statistic

$$\chi^2 = [N - 1 - \tfrac{1}{6}(2p + 5) - \tfrac{2}{3}m] \ln \frac{|\hat{\boldsymbol{\Psi}} + \hat{\boldsymbol{\Lambda}}\hat{\boldsymbol{\Lambda}}'|}{|\mathbf{S}|} \tag{2}$$

where $\hat{\boldsymbol{\Psi}}$, $\hat{\boldsymbol{\Lambda}}$ are the solutions of the maximum-likelihood equations, $\mathbf{S}$ is the sample covariance matrix, and the rather lengthy coefficient has been determined by Bartlett (1954) to improve the convergence of the large-sample distribution of the statistic to the chi-squared type. If the hypothesis (1) is in fact true, as N becomes large, the statistic tends to be distributed as a chi-squared variate with

$$\nu = \tfrac{1}{2}[(p - m)^2 - p - m] \tag{3}$$

degrees of freedom, and the null hypothesis of exactly m common factors would be rejected at the α level if

$$\chi^2 \geq \chi^2_{\alpha;\nu} \tag{4}$$

and accepted otherwise. From the invariance property of the estimated loadings and specific variances it follows that the same value of the test statistic would be obtained from a factor solution in terms of the correlation matrix. If $m = 0$, the hypothesis (1) states that the population covariance matrix is diagonal, and the test statistic is equivalent to that given earlier in Sec. 3.8 for testing that hypothesis of no correlation.

Lawley (1940) has shown that the logarithm of the determinantal ratio can be approximated exceedingly well by the sum of scaled squared residuals

$$\sum_{i<j}\sum \frac{(s_{ij} - \hat{\sigma}_{ij})^2}{\hat{\Psi}_i \hat{\Psi}_j}$$

where

(5) $$\hat{\sigma}_{ij} = \sum_{h=1}^{m} \hat{\lambda}_{ih}\hat{\lambda}_{jh} \qquad i \neq j$$

is the covariance reproduced by the model. The approximate test statistic is thus

(6) $$\chi^2 = [N - 1 - \tfrac{1}{6}(2p + 5) - \tfrac{2}{3}m] \sum_{i<j}\sum \frac{(s_{ij} - \hat{\sigma}_{ij})^2}{\hat{\Psi}_i \hat{\Psi}_j}$$

and its asymptotic distribution is identical with that of the exact expression (2).

The terms of the sum of expression (6) are the superdiagonal elements of the matrix $\hat{\mathbf{\Psi}}^{-\frac{1}{2}}[\mathbf{S} - \hat{\mathbf{\Lambda}}\hat{\mathbf{\Lambda}}']\hat{\mathbf{\Psi}}^{-\frac{1}{2}}$. It is easily shown that those quantities are invariant under scale changes of the original observations, so that the identical standardized residual matrix could be obtained from the factor analysis of the sample correlations. We shall also show in Sec. 9.8 that $(s_{ij} - \hat{\sigma}_{ij})\hat{\Psi}_i^{-\frac{1}{2}}\hat{\Psi}_j^{-\frac{1}{2}}$ is the sample partial correlation of the ith and jth responses with the m common factors held constant.

In most applications one does not know the number m of common factors, and frequently successively larger numbers are extracted until the goodness-of-fit hypothesis is accepted or the iterative computing scheme fails to converge. It should be emphasized that the chi-squared statistics obtained in this fashion are not independent, and the true significance level of the test of the hypothesized number of factors may be very different from the nominal value used at each stage of the extraction process.

For the application of the goodness-of-fit test it is of course necessary that the degrees of freedom of the chi-squared statistic be positive. This means that the number of common factors cannot exceed the largest integer satisfying

(7) $$m < \tfrac{1}{2}(2p + 1 - \sqrt{8p + 1})$$

for a fixed number of p responses. The bound also follows from algebraic restrictions on the number of common factors (Thurstone, 1945, p. 293).

9.6 EXAMPLES OF FACTOR ANALYSES

Now that we have set forth the machinery for the estimation, testing, and numerical-analysis procedures for the factor model, let us consider some complete factor analyses.

Example 9.2. One- and two-factor solutions were extracted from the 6×6 fowl bone correlation matrix given in Example 8.5. The loadings and specific variances for the two solutions are shown in Table 9.1. With the Aitken δ^2 process the one-factor solution converged to four decimal places in 11 iterations, although some oscillations were observed in the values of the second and third places between the fifth and final cycles. Twenty-one iterations were required for convergence of the two-factor solution with a maximum difference of 0.0002 between corresponding elements of the last two iterates. Without the δ^2 modification the one-factor estimates appeared to converge more monotonically in 13 iterations. The similar two-factor solution was terminated after 5 min of computer time and 138 iterations. In the last cycles steady changes of the order of -0.0002 were observed in some elements of the second factor and were reminiscent of the kind of slow divergence observed by the author in the solution by the original iterative algorithm proposed by Lawley. The loadings of the 138th iterate still differed from those obtained by the Aitken δ^2 solution by as much as 0.1, and it appears that the starting vectors were too different from the actual solution to converge readily without the aid of the δ^2 modification.

The residual matrices formed by subtracting the correlations generated by the respective factor models from the original sample values are given below.

Single-factor solution:

$$\mathbf{R}_1 = \begin{bmatrix} 0 & 0.2122 & 0.0044 & -0.0102 & -0.0216 & 0.0041 \\ & 0 & 0.0161 & -0.0305 & -0.0165 & 0.0086 \\ & & 0 & 0.0195 & -0.0160 & -0.0195 \\ & & & 0 & -0.0149 & -0.0123 \\ & & & & 0 & 0.0545 \\ & & & & & 0 \end{bmatrix}$$

Two-factor solution:

$$\mathbf{R}_2 = \begin{bmatrix} 0 & -0.0001 & 0.0014 & 0.0023 & -0.0123 & 0.0048 \\ & 0 & 0.0007 & -0.0006 & -0.0004 & 0.0003 \\ & & 0 & 0.0167 & -0.0148 & -0.0185 \\ & & & 0 & -0.0167 & -0.0125 \\ & & & & 0 & 0.0583 \\ & & & & & 0 \end{bmatrix}$$

Inclusion of the second factor has substantially improved the reproduction of the r_{12} correlation. It also appears that a third factor reflecting leg size might also be present in the multivariate system, for the 5, 6 residual element remains rather large in both solutions.

Details of the goodness-of-fit tests for the two solutions are summarized in

Table 9.1. **Factor Analyses of the Fowl Bone Dimensions**

Dimension	Single-factor solution		Two-factor solution		
	$\hat{\Psi}_i$	Factor loading	$\hat{\Psi}_i$	Factor loadings 1	Factor loadings 2
1. Skull length	0.594	0.637	0.521	0.674	0.158
2. Skull breadth	0.659	0.584	0.067	0.715	0.650
3. Humerus	0.080	0.959	0.078	0.947	−0.157
4. Ulna	0.079	0.960	0.071	0.939	−0.217
5. Femur	0.137	0.929	0.140	0.908	−0.189
6. Tibia	0.124	0.936	0.128	0.920	−0.159

Table 9.2. The value of the determinant of the sample correlation matrix is 0.001009. Note that the chi-squared statistics computed from the determinantal and residual definitions are scarcely different. The tests resoundingly reject the one- or two-factor models and suggest that a higher-dimensional or nonlinear mechanism has generated the observed correlations. Such significant results are in fact common in testing the fit of a small number of factors to correlations obtained from even moderately large samples.

The interpretations of the factors are similar to those given for the original principal components of Example 8.5. The loadings of the first component are very similar to those of both first factors, and in each case those factors can be thought of as a general size dimension, with the wing and leg bones dominant. The importance of skull length is markedly less in factor 2 than in component 2, and we may think of that factor as a measure of skull breadth and perhaps to a lesser degree a *shape* factor comparing the skull dimensions with the wing and leg lengths. However, this rough interpretation probably reflects in part the inadequacy of the linear two-factor model.

Finally, any factor analysis should always consider orthogonal transformations of the loadings which might have simpler interpretations than the original solutions. Rather than anticipate such notions of "simplicity" here, we shall defer this final touch to the next section.

Example 9.3. Now we shall consider a maximum-likelihood factor analysis of the 11 × 11 WAIS standardization sample correlations. From the earlier principal-

Table 9.2. **Goodness-of-fit Tests**

	One factor	Two factors
Determinant of the reproduced matrix	0.001628	0.001400
Chi-squared statistic	131.01	92.62
Degrees of freedom	9	4
Significance level	$P < 0.001$	$P < 0.001$
Approximate chi-squared statistic	129.60	92.86

component study discussed in Example 8.4 and from the nature of the subtests it was hypothesized that three common factors might suffice to describe the dependence structure parsimoniously, whereas four or more components appeared to be necessary in that *open-ended* analysis. Through the Maxwell iterative scheme solutions embodying one, two, three, and four factors were successively extracted. Some difficulty was experienced in achieving convergence in the third and fourth places of the four-factor solution, and the computations were stopped short of a completely satisfactory solution. It is interesting to examine the approximate chi-squared statistics for the solutions:

Number of factors	Goodness of fit χ^2	Degrees of freedom
1	900	44
2	160	34
3	86.3	25
4	48.8	17

All these statistics would lead to the rejection of the adequacy of the respective factor models at the $\alpha = 0.001$ level.

The three-factor model had the estimated loadings and specific variances shown in Table 9.3. Factor 1 is a so-called g factor of general intellective ability,

Table 9.3. **WAIS Factor Loadings**

Subtest	Specific variance	Factor		
		1	2	3
1. Information	0.23	0.84	−0.20	−0.15
2. Comprehension	0.39	0.75	−0.22	−0.09
3. Arithmetic	0.51	0.70	0.03	−0.03
4. Similarities	0.41	0.75	−0.18	0.01
5. Digit span	0.13	0.69	0.62	−0.10
6. Vocabulary	0.15	0.86	−0.27	−0.22
7. Digit symbol	0.55	0.63	0.05	0.22
8. Picture completion	0.40	0.72	−0.09	0.28
9. Block design	0.34	0.66	−0.02	0.48
10. Picture arrangement	0.49	0.66	−0.02	0.26
11. Object assembly	0.50	0.55	−0.05	0.44

with the verbal subtests 1 to 6 showing slightly higher loadings than the performance tests 7 to 11. The second factor is dominated by the large digit-span loading, although in contrast the verbal subtests information, comprehension, similarities, and vocabulary enter with smaller, though appreciable, negative loadings. Factor 3 appears to be a dimension of performance, although it also has a bipolar interpretation as a comparison of the performance tests 7 to 11 with the verbal tests information, comprehension, digit span, and vocabulary.

The largest element of the residual matrix occurred in the reproduction of the digit symbol–picture arrangement correlation and had the value 0.0456. It is difficult to discern a clear pattern of the inadequacy of the model from the residuals, although the correlations involving the digit-symbol and arithmetic subtests appeared to be more poorly reproduced. Orthogonal transformations of the loading matrix failed to yield factors more easily interpretable or of structure "simple" in the sense to be defined in Sec. 9.7.

Example 9.4. Sometimes a correlation matrix will be encountered for which the communality of a variable is nearly unity. This is a simple approximate example of the Heywood case cited in Sec. 9.4. For a single-factor solution the variable with the nearly unit loading is that factor, and the other factor loadings are its correlations with the remaining variables. For example, van Alphen (1961) obtained the following matrix of correlations among ocular refraction, axial length, anterior chamber depth, corneal power, and lens power of the right eyes of 96 adult subjects:

$$\mathbf{R} = \begin{bmatrix} 1 & -0.77 & -0.46 & -0.30 & 0.28 \\ & 1 & 0.46 & -0.28 & -0.49 \\ & & 1 & 0.19 & -0.46 \\ & & & 1 & -0.10 \\ & & & & 1 \end{bmatrix}$$

A solution for a single factor by the original Lawley iterative process gave the loadings shown in Table 9.4. The factor is closely identified with axial length, and its loadings are nearly equal to the correlations of that variable in the second row and column of the matrix.

Anderson (1963b) has discussed the effect of vanishing specific variances and has suggested approaches to the factor analysis of observations generated by such models.

Table 9.4. **Factor Solution**

Variable	Communality	Factor loading
Refraction	0.603	−0.776
Axial length	0.974	0.987
Anterior chamber	0.224	0.473
Corneal power	0.062	−0.249
Lens power	0.243	−0.493

9.7 FACTOR ROTATION

In Sec. 9.2 we saw that the classical factorization

$$\mathbf{\Sigma} = \mathbf{\Psi} + \mathbf{\Lambda\Lambda}'$$

of the response covariance matrix was not unique, for postmultiplication of the loading matrix by any conformable orthogonal matrix would yield an equally valid factorization. For the Rao-Maxwell iterative solution of the estimation equations this indeterminacy was removed by requiring that the characteristic roots associated with the m factors be distinct or, equivalently, that $\mathbf{J} = \hat{\mathbf{\Lambda}}'\hat{\mathbf{\Psi}}^{-1}\hat{\mathbf{\Lambda}}$ be a diagonal matrix containing the m roots in descending order of magnitude. Nevertheless, once this particular loading matrix had been extracted, could one or more orthogonal transformation matrices be found which would lead to a pattern of loadings more easily interpreted or identifiable with the subject-matter nature of the responses? Since such transformations amount to a rigid rotation or reflection of the coordinate axes of the m-dimensional factor space, they are commonly called *rotations* of the loadings. Rotations that are not rigid, in the sense of preserving the perpendicularity of the coordinate axes, are called *oblique*.

Perhaps a simple example will illustrate this arbitrariness of the positions of the axes and serve as motivation for the construction of some objective standards for "good" orientations. In Fig. 9.1 the loadings of the two factors of Example 9.1 have been plotted, with each point identified by its response number. The loadings are scattered rather symmetrically on both sides of the factor 1 axis. Responses 1 and 6 appear to be loading only on that factor, while the other responses are loaded highly, and frequently with mixed signs, on both factors. Such *bipolar* factors with large positive and negative loadings are usually more difficult to interpret, and we note that the negative loadings can be virtually removed by a rotation of the factor axes through an angle of about $\theta = -35°$. The new γ_1, γ_2 axes are indicated by the broken lines. The loadings of the rotated factors can be determined approximately from the figure, or directly by the transformation equations

$$[\gamma_{i1},\gamma_{i2}] = [\lambda_{i1},\lambda_{i2}]\begin{bmatrix} 0.8192 & 0.5736 \\ -0.5736 & 0.8192 \end{bmatrix} \tag{1}$$

where cos 35 = 0.8192. The transposed matrix of new loadings is

$$\mathbf{\Gamma}' = \begin{bmatrix} 0.79 & 0.75 & 0.07 & -0.01 & 0.86 & 0.66 \\ 0.43 & 0.16 & 0.78 & 0.36 & -0.01 & 0.46 \end{bmatrix} \tag{2}$$

The number of very small loadings has been increased, while the number of large loadings, e.g., greater than 0.6, has remained about the same. The negative loadings of the new factors are negligibly small. The rotated loadings would appear to permit an easier identification of the latent-factor dimensions with the six observed responses.

Positiveness of the loadings is but one criterion for choosing a rotation of the factor axes. As one resolution of the problem Thurstone

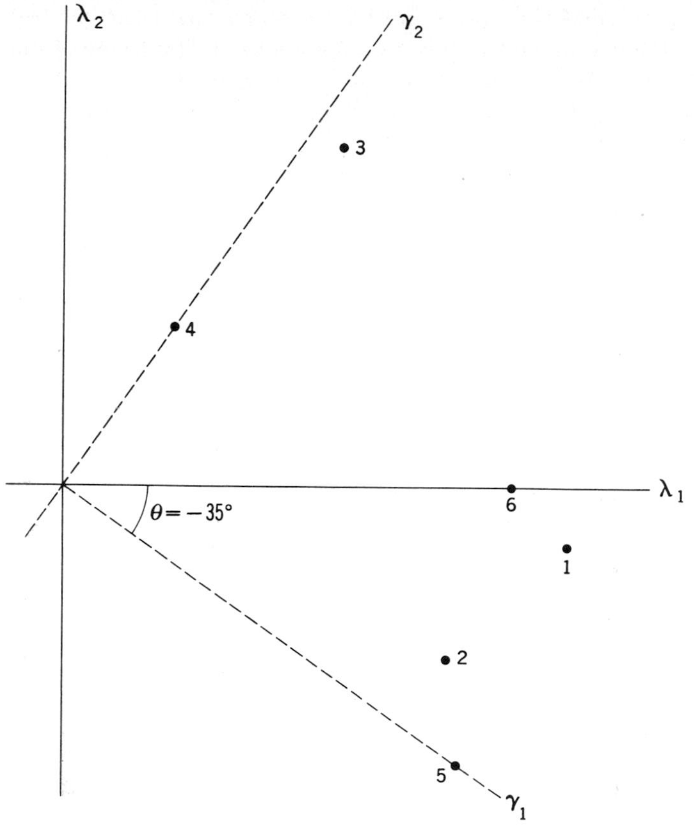

figure 9.1

(1945, chap. 14) has suggested that the $p \times m$ matrix $\mathbf{\Gamma}$ of rotated loadings should meet his criteria for *simple structure:*

1. Each row of $\mathbf{\Gamma}$ should contain at least one zero.
2. Each column of $\mathbf{\Gamma}$ should contain at least m zeros.
3. Every pair of columns of $\mathbf{\Gamma}$ should contain several responses whose loadings vanish in one column but not in the other.
4. If the number of factors m is four or more, every pair of columns of $\mathbf{\Gamma}$ should contain a large number of responses with zero loadings in both columns.
5. Conversely, for every pair of columns of $\mathbf{\Gamma}$ only a small number of responses should have nonzero loadings in both columns.

In essence these criteria say that under a simple structure the responses fall into generally mutually exclusive groups whose loadings are high on single factors, perhaps moderate to low on a few factors, and of negligible size on the remaining dimensions. We note in passing that Thurstone permitted nonorthogonal, or oblique, rotations in his definition and search for simple structure.

It has been the author's experience that simple structures meeting Thurstone's conditions 1 to 5 rarely exist in factor solutions extracted from real data. As anyone who has attempted to interpret large solutions can affirm, the location of that elusive state by graphically rotating successive pairs of factors is a tedious and frequently unrewarding task. The reader faced with that problem is referred to Harman's text (1967, chaps. 12–13) for a discussion of systematic methods for graphical rotation. It is also wise to work closely with someone familiar with the substantive nature of the response variables. In addition to passing on the reasonableness of the interpretations of the rotated factors, such collaborators can often suggest a few responses which might be constrained to load highly with certain factors and not at all with the others. For example, if a factor analysis were carried out on the 13-variate complex of Wechsler subtests, age and years of education of examinees discussed in Example 8.4, those latter responses might be used as concomitant variates for choosing rotations that maximize their loadings with different factors. Results of earlier factor analyses of the same responses are also useful for hypothesizing both the number m of factors and the forms of the rotated structures. In that connection Howe (1955) and Lawley and Maxwell (1971) have proposed least-squares methods for rotating as closely as possible to some prescribed pattern of zero and nonzero loadings. Jöreskog (1966, 1969) has given hypothesis tests for proposed simple-structure loading matrices and numerical methods for their execution.

Let us use this cut-and-try technique of graphical rotation to see whether simpler structures can be found for two of the earlier examples of factor solutions. For such repeated rotations the following notation will be adopted:

Y_j = jth factor variate, $j = 1, \ldots, m$

$Y_j^{(h)}$ = jth factor variate obtained after hth rotation of its original loadings

λ_j = factor space axis of jth original factor

$\gamma_j^{(h)}$ = factor space axis of loadings of jth factor after hth rotation

Example 9.5. The loadings of the two factors extracted from the six fowl bone dimensions in Example 9.2 have been plotted in Fig. 9.2. It does not appear that a simpler structure can be found, although the negative loadings can be virtually removed by a rotation of $-12°$. The transformation matrix is

$$\mathbf{T} = \begin{bmatrix} 0.9782 & 0.2079 \\ -0.2079 & 0.9782 \end{bmatrix}$$

and the new loadings have the values shown in Table 9.5. The new first factor is even more closely identified with wing and leg bone length than its unrotated predecessor. The second rotated factor appears to be a dimension measuring only skull size and, in particular, skull breadth.

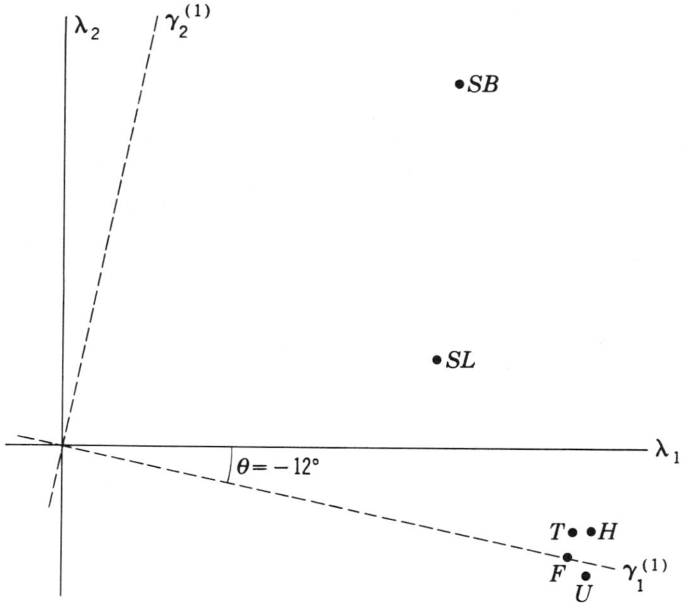

figure 9.2 Rotation of the fowl-bone dimensions factors.

Table 9.5. **Rotated Factor Loadings**

Bone dimension	Factor 1	Factor 2
Skull length	0.63	0.29
Skull breadth	0.56	0.78
Humerus	0.96	0.04
Ulna	0.96	−0.02
Femur	0.93	0.00
Tibia	0.93	0.04

Example 9.6. From the Wechsler scale data originally discussed in Example 8.4 the partial correlations of the subtests were computed with the age and education variates held constant. Maximum-likelihood estimates of the loadings for the one, two, and three common-factor models were computed by the Rao-Maxwell method. Results of an earlier principal-component analysis of the partial correlations, as well as convergence difficulties experienced with a four-factor solution, led to the choice of the three-factor solution for describing the dependence structure. Those loadings are shown in the first three columns of Table 9.6. The goodness-of-fit chi-squared statistic was equal to 78.1, and with its 25 degrees of freedom the hypothesis of the adequacy of the three-factor model would be rejected at the 0.001 level.

The search for a simpler factor structure began with the plotting of the loadings of the first and second factors as in Fig. 9.3. A rotation of about $-28°$ would remove the many negative loadings in the second factor and at the same time decrease the loading of the digit-span subtest on the first factor. The loading matrix was accordingly postmultiplied by the transformation matrix

$$\mathbf{T}_1 = \begin{bmatrix} 0.8830 & 0.4695 & 0 \\ -0.4695 & 0.8830 & 0 \\ 0 & 0 & 1 \end{bmatrix}$$

Plots of the rotated loadings against those of the original third factor indicated that a rotation of about $-30°$ of the new first factor $\gamma_1^{(1)}$ and original third-factor axes would lead to positive loadings on the third dimension, as well as some simplification of the other parts of the structure. Those loadings and their new axes are shown in Fig. 9.4. The loadings of the second rotation were computed by postmultiplying the matrix of loadings of the factors $\gamma_1^{(1)}$, $\gamma_2^{(1)}$, and λ_3 by the matrix

$$\mathbf{T}_2 = \begin{bmatrix} 0.8660 & 0 & 0.5000 \\ 0 & 1 & 0 \\ -0.5000 & 0 & 0.8660 \end{bmatrix}$$

Study of the plots of those loadings did not suggest further rotations to a simpler structure, and the process was terminated. The loadings obtained by the two graphical rotations are shown in the last three columns of Table 9.6. We shall presently see that those values are similar to ones obtained by another approach to the definition of simple structure.

Analytical Methods for Rotation. Reduction of the simple structure precepts to a convenient mathematical function eluded factor analysts until Carroll, Ferguson, Neuhaus and Wrigley, and Saunders arrived independently and nearly simultaneously at analytical descriptions of the notion. If the rotation is rigid, each explication leads to the same maximand and transformation matrix.

In our usual notation let us suppose that the $p \times m$ loading matrix $\mathbf{\Lambda}$ has been subjected to the rigid rotation specified by the orthogonal $m \times m$ matrix $\mathbf{T}$, so that the element γ_{hj} of $\mathbf{\Gamma} = \mathbf{\Lambda T}$ will denote the loading of the hth response on the jth rotated factor. The orthogonality of

Table 9.6. **WAIS Partial Correlation Factors**

Subtest	Original factors			Rotated factors		
	1	2	3	1	2	3
Information	0.74	−0.16	−0.14	0.70	0.21	0.24
Comprehension	0.65	−0.19	−0.11	0.62	0.14	0.24
Arithmetic	0.57	0.13	0.01	0.38	0.38	0.23
Similarities	0.62	−0.16	−0.05	0.56	0.15	0.27
Digit span	0.53	0.68	−0.10	0.18	0.85	−0.02
Vocabulary	0.80	−0.24	−0.26	0.84	0.16	0.18
Digit symbol	0.42	0.07	0.10	0.24	0.26	0.25
Picture completion	0.64	−0.08	0.24	0.40	0.23	0.51
Block design	0.57	−0.01	0.52	0.18	0.26	0.71
Picture arrangement	0.54	−0.02	0.16	0.34	0.23	0.38
Object assembly	0.44	−0.05	0.43	0.14	0.17	0.58

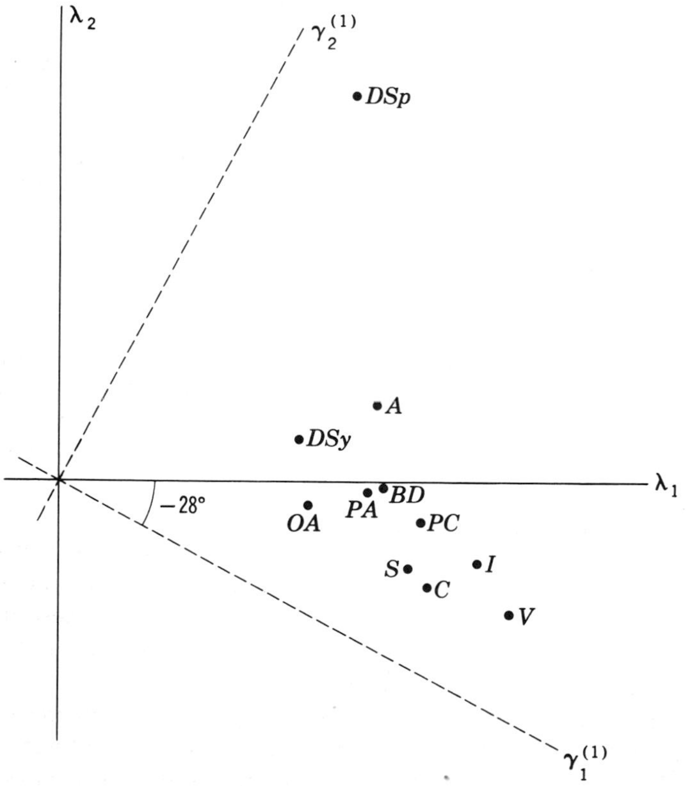

figure 9.3 **First rotation of the WAIS factors.**

T leaves the p communalities

$$\sum_{j=1}^{m} \lambda_{hj}^2 = \sum_{j=1}^{m} \gamma_{hj}^2 \tag{3}$$

unchanged, and of course the sum of their squares,

$$\sum_{h=1}^{p} \Big(\sum_{j=1}^{m} \gamma_{hj}^2\Big)^2 = \sum_{h=1}^{p} \sum_{j=1}^{m} \gamma_{hj}^4 + 2 \sum_{h=1}^{p} \sum_{\substack{i=1 \\ i>j}}^{m} \sum_{j=1}^{m} \gamma_{hi}^2 \gamma_{hj}^2 \tag{4}$$

is also invariant. With that result in mind let us consider the four measures of simplicity or parsimony proposed by the factor analysts. Ferguson (1954) suggested the second, or total cross-product, term of (4) as a measure of parsimony, and proposed that the rotation matrix **T** should be chosen so as to minimize it. At virtually the same time Carroll (1953) arrived at the same quantity as an objective measure of parsimony, although in his development oblique as well as rigid rotations were permitted.

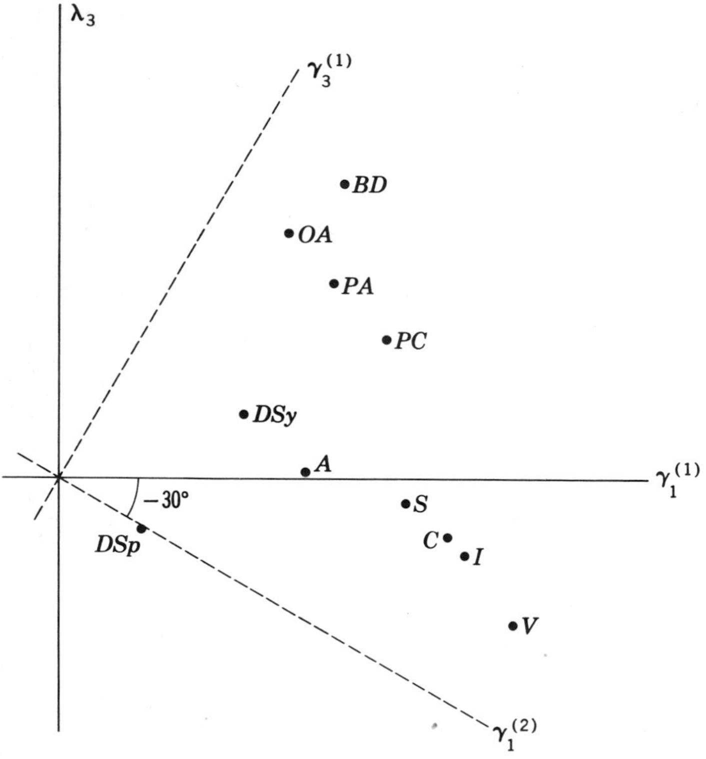

figure 9.4 Second rotation of the WAIS factors.

Neuhaus and Wrigley (1954) attempted to reduce the simple-structure criteria to a single index by defining simplicity as the variance of the squares of all pm loadings. Under this explication the rotation should be chosen so as to maximize

$$V = \sum_{h=1}^{p} \sum_{j=1}^{m} \gamma_{hj}^4 - \frac{1}{pm}\Big(\sum_{h=1}^{p} \sum_{j=1}^{m} \gamma_{hj}^2\Big)^2 \tag{5}$$

and since the correction term is merely the sum of the communalities, maximization of V is equivalent to maximizing the sum of the fourth powers of the loadings, or to minimization of the parsimony measure of Ferguson and Carroll. Finally, Saunders (1960) arrived at the objective function V by the following argument: in essence simple structure means that the factors contain many large and many zero loadings and only a minimum of intermediate values. If the negative of each loading is included as well, one measure of the simplicity of the doubled set of loadings would be the quantity

$$\frac{\sum_{h=1}^{p} \sum_{j=1}^{m} \gamma_{hj}^4}{\Big(\sum_{h=1}^{p} \sum_{j=1}^{m} \gamma_{hj}^2\Big)^2} - 3 \tag{6}$$

that is often taken as a measure of the shape of a sample frequency distribution. Inclusion of the negative value of each loading in the "sample" has caused the usual correction term for the second and fourth moments to vanish. The denominator is merely the square of the sum of the communalities and is the same for all rotations. Saunders's measure of simplicity is therefore equivalent to the other three indices. Rotations based upon those objective functions are called *quartimax*, for they seek to maximize the fourth powers of the loadings.

Kaiser (1956, 1958, 1959) has proposed as a measure of simple structure the sum of the variances of the squared loadings within each column of the factor matrix. His *raw varimax* criterion for rotation to a "simple" structure is

$$v^* = \frac{1}{p^2} \sum_{j=1}^{m} \Big[p \sum_{h=1}^{p} \gamma_{hj}^4 - \Big(\sum_{h=1}^{p} \gamma_{hj}^2\Big)^2\Big] \tag{7}$$

However, this criterion gives equal weight to responses with both large and negligible communalities, and Kaiser has suggested that a better maximand might be

(8) $$v = \frac{1}{p^2} \sum_{j=1}^{m} \left[p \sum_{h=1}^{p} x_{hj}^4 - \left(\sum_{h=1}^{p} x_{hj}^2 \right)^2 \right]$$

where

(9) $$x_{hj} = \frac{\gamma_{hj}}{\sqrt{\sum_{j=1}^{m} \lambda_{hj}^2}}$$

is the hjth loading divided by the square root of its communality. Following the rotation each loading x_{hj} is multiplied by the square root of its respective communality to restore its proper dimensionality. Kaiser has called this criterion the *normal varimax* or merely the *varimax*.

Let us sketch the computational process for a varimax rotation. We begin by restricting our attention to a single pair of factors, e.g., the rth and sth with normalized loadings x_{hr}, x_{hs}. Rotation of those factors involves the single angle ϕ, and by differentiating (8) with respect to ϕ Kaiser has shown that the angle must satisfy the relation

(10) $$\tan 4\phi = \frac{2[2p\Sigma(x_{hr}^2 - x_{hs}^2)x_{hr}x_{hs} - \Sigma(x_{hr}^2 - x_{hs}^2)(2\Sigma x_{hr}x_{hs})]}{p\Sigma[(x_{hr}^2 - x_{hs}^2)^2 - (2x_{hr}x_{hs})^2] - \{[\Sigma(x_{hr}^2 - x_{hs}^2)]^2 - 2(\Sigma x_{hs}x_{hr})^2\}}$$

where each of the summations extends over the p responses. For the second derivative of the criterion to be negative, it is necessary that 4ϕ be assigned to the appropriate quadrant. That choice is determined by the signs of the numerator and denominator of (10), and is specified by the cells of Table 9.7. The iterative solution for the rotation proceeds in this fashion: the first and second factors are rotated by an angle determined from expression (10) and Table 9.7, the new first factor is rotated with the original third factor, and so on, until all $\frac{1}{2}m(m-1)$ pairs of factors have been rotated. Such a sequence of rotations is called a *cycle*. The cycles are repeated until one is completed in which all angles are less

Table 9.7. **Quadrant of 4ϕ**

Sign of denominator	Sign of numerator: +	Sign of numerator: −
+	I: $0° \le 4\phi < 90°$	IV: $-90° \le 4\phi < 0°$
−	II: $90° \le 4\phi < 180°$	III: $-180° \le 4\phi < -90°$

than some specified convergence criterion ϵ. The normalized loadings achieved by that succession of transformations are those which maximize the function (8). Kaiser has shown that (8) cannot exceed $(m - 1)/m$, and since each varimax rotation must lead to either no change or an increase in that quantity, the iterative process must converge.

A detailed outline for writing a varimax computer program has been given by Kaiser (1959). Examples of varimax rotations and the comparison of their loadings with those obtained according to the quartimax method can be found in Harman's text (1967).

Example 9.7. As a simple illustration of the varimax method we shall rotate the two factors extracted from the six fowl bone measurements in Example 9.2. For those normalized loadings

$$\tan 4\phi = \frac{-10.16}{-7.072} = 1.437$$

Since both the numerator and denominator signs are negative, 4ϕ must be in the third quadrant, or $4\phi \approx -124°50'$. The angle of the normal varimax rotation is approximately $-31°12'$. This is a greater rotation than our original graphical one chosen to minimize the loadings of the wing and leg dimensions on the new second factor. The varimax criterion was impartial in locating the axes nearly equidistant from the skull and wing-leg loadings. The new loading matrix is

$$\mathbf{\Gamma}' = \begin{bmatrix} 0.495 & 0.274 & 0.891 & 0.916 & 0.875 & 0.869 \\ 0.484 & 0.926 & 0.356 & 0.301 & 0.309 & 0.341 \end{bmatrix}$$

The first factor is still essentially identified with the leg and wing dimensions, while the second is correlated highest with skull breadth.

Example 9.8. The three WAIS partial correlation factors described in Example 9.6 were subjected to a varimax rotation. This matrix of new loadings was obtained:

$$\mathbf{\Gamma}' = \begin{bmatrix} 0.70 & 0.63 & 0.37 & 0.57 & 0.15 & 0.84 & 0.24 & 0.41 & 0.20 & 0.35 & 0.16 \\ 0.18 & 0.11 & 0.35 & 0.12 & 0.85 & 0.16 & 0.22 & 0.15 & 0.14 & 0.17 & 0.06 \\ 0.25 & 0.24 & 0.29 & 0.28 & 0.13 & 0.18 & 0.29 & 0.53 & 0.74 & 0.41 & 0.59 \end{bmatrix}$$

These loadings are remarkably close to those found graphically in Example 9.6.

9.8 AN ALTERNATIVE MODEL FOR FACTOR ANALYSIS

Howe (1955) has proposed an alternative approach to factor analysis which leads directly to the model of Sec. 9.2, and which with suitable assumptions also results in estimates equivalent to those obtained by the maximum-likelihood principle. This formulation is especially powerful, for it does not require multivariate normality of the response and factor variates but only that their second-order moments be defined. Let the manifest responses be described by a p-dimensional random variable $\mathbf{x}$, where for convenience $E(\mathbf{x}) = \mathbf{0}$. The covariance matrix of these responses will be denoted by $E(\mathbf{xx}') = \mathbf{\Sigma}$. Howe poses the problem of factor analysis through this question: do m uncorrelated random variables exist, which, if held constant, would leave the manifest response variates uncorrelated? Let us denote these hypothesized variates by the $m \times 1$ vector $\mathbf{y}$, where for convenience $E(\mathbf{y}) = \mathbf{0}$, and from the uncorrelated nature of the elements of $\mathbf{y}$,

$$(1) \qquad E(\mathbf{yy}') = \mathbf{I}$$

By definition, the partial correlation of the ith and jth responses with the elements of $\mathbf{y}$ held constant must vanish for all i and j:

$$(2) \qquad \rho_{ij.\mathbf{y}} = 0$$

The entire matrix of partial correlations can be written as

$$(3) \qquad \mathbf{\Psi}^{-\frac{1}{2}}\{E[(\mathbf{x} - \mathbf{\Lambda y})(\mathbf{x} - \mathbf{\Lambda y})']\}\mathbf{\Psi}^{-\frac{1}{2}}$$

where

$$(4) \qquad \mathbf{\Psi} = \text{diag}\,\{E[(\mathbf{x} - \mathbf{\Lambda y})(\mathbf{x} - \mathbf{\Lambda y})']\}$$

and

$$(5) \qquad \mathbf{\Lambda} = E(\mathbf{xy}')$$

The hypothesis that all such partial correlations vanish can be written as

$$(6) \qquad E[(\mathbf{x} - \mathbf{\Lambda y})(\mathbf{x} - \mathbf{\Lambda y})'] = \mathbf{\Psi}$$

Evaluation of the expectations of the left-hand side of (6) shows that this is equivalent to requiring that

$$(7) \qquad \mathbf{\Sigma} = \mathbf{\Psi} + \mathbf{\Lambda\Lambda}'$$

The hypothesis of no partial correlation is then equivalent to stating that the population covariance or correlation matrix can be decomposed into the usual factor model introduced in Sec. 9.2.

Through the partial-correlation definition of factors Howe has arrived at the estimation equation (18) of Sec. 9.3. He has proposed that $\mathbf{\Psi}$ and $\mathbf{\Lambda}$ be estimated such that the sample counterparts of the partial correlations (3) should be as small as possible. Now as the elements of a correlation matrix decrease in absolute size, their determinant approaches its maximum value of one, and Howe has taken as maximand in the problem the determinant

$$\frac{|\mathbf{S} - \mathbf{\Lambda\Lambda}'|}{|\mathbf{\Psi}|} \tag{8}$$

Let us denote (8) as a function of the loading parameters by $f(\mathbf{\Lambda})$ and differentiate with respect to the ijth element of $\mathbf{\Lambda}$. By the results of Sec. 2.12,

$$\frac{\partial f(\mathbf{\Lambda})}{\partial \lambda_{ij}} = -\frac{1}{|\mathbf{\Psi}|^2}\frac{\partial|\mathbf{\Psi}|}{\partial \lambda_{ij}}|\mathbf{S} - \mathbf{\Lambda\Lambda}'| + \frac{1}{|\mathbf{\Psi}|}\frac{\partial|\mathbf{S} - \mathbf{\Lambda\Lambda}'|}{\partial \lambda_{ij}} \tag{9}$$

Now

$$\begin{aligned} \mathbf{\Psi} &= \operatorname{diag}(\mathbf{\Sigma} - \mathbf{\Lambda\Lambda}') \\ &= \mathbf{D}\left(\sigma_i^2 - \sum_{j=1}^{m} \lambda_{ij}^2\right) \end{aligned} \tag{10}$$

where $\mathbf{D}(a_i)$ denotes a diagonal matrix with a_i in the ith diagonal position, and since the determinant of $\mathbf{D}(a_i)$ is merely the product of its diagonal elements,

$$\begin{aligned} \frac{\partial|\mathbf{\Psi}|}{\partial \lambda_{ij}} &= -2\lambda_{ij} \prod_{\substack{g=1 \\ g\neq i}}^{p} \left(\sigma_g^2 - \sum_{h=1}^{m} \lambda_{gh}^2\right) \\ &= \frac{-2\lambda_{ij}|\mathbf{\Psi}|}{\Psi_i} \end{aligned} \tag{11}$$

Furthermore,

$$\begin{aligned} \frac{\partial|\mathbf{S} - \mathbf{\Lambda\Lambda}'|}{\partial \lambda_{ij}} &= \sum_{g=1}^{p}\sum_{h=1}^{p} |\mathbf{S} - \mathbf{\Lambda\Lambda}'|_{gh} \frac{\partial\left(s_{gh} - \sum_{k=1}^{m} \lambda_{gk}\lambda_{hk}\right)}{\partial \lambda_{ij}} \\ &= -2\sum_{g=1}^{p} |\mathbf{S} - \mathbf{\Lambda\Lambda}'|_{ig}\lambda_{gj} \end{aligned} \tag{12}$$

where $|\mathbf{S} - \mathbf{\Lambda\Lambda}'|_{ig}$ is the cofactor of the igth element of $\mathbf{S} - \mathbf{\Lambda\Lambda}'$. Substitution of the results of (11) and (12) into the right-hand side of equation (9) yields, upon simplification and equation of the expression to zero,

$$\frac{\lambda_{ij}}{\Psi_i} = \sum_{g=1}^{p} \frac{|\mathbf{S} - \mathbf{\Lambda\Lambda}'|_{ig}\lambda_{gj}}{|\mathbf{S} - \mathbf{\Lambda\Lambda}'|} \tag{13}$$

for all subscripts $i = 1, \ldots, p$, and $j = 1, \ldots, m$. In matrix form the last set of equations can be written as

$$\mathbf{\Psi}^{-1}\mathbf{\Lambda} = [\mathbf{S} - \mathbf{\Lambda\Lambda}']^{-1}\mathbf{\Lambda} \tag{14}$$

Premultiplication by $\mathbf{S} - \mathbf{\Lambda\Lambda}'$ and regrouping of the terms yields

$$\hat{\mathbf{\Lambda}}(\hat{\mathbf{\Lambda}}'\hat{\mathbf{\Psi}}^{-1}\hat{\mathbf{\Lambda}}) = \mathbf{S}\hat{\mathbf{\Psi}}^{-1}\hat{\mathbf{\Lambda}} - \hat{\mathbf{\Lambda}} \tag{15}$$

or equation (18) of Sec. 9.3. The Lawley maximum-likelihood and Howe minimum-partial-correlation approaches lead to the same estimates of the factor loadings and specificities.

9.9 SAMPLING VARIATION OF LOADING ESTIMATES

Lawley (1953) has found the approximate asymptotic covariance matrix of the estimates for the special factor model

$$\mathbf{\Sigma} = \sigma^2\mathbf{I} + \mathbf{\Lambda\Lambda}' \tag{1}$$

In this model Lawley has defined $\mathbf{\Lambda}$ uniquely by requiring its columns to be orthogonal and the first m characteristic roots of $\mathbf{\Sigma}$ to be distinct. The rth such root is equal to

$$\mu_r = \sum_{h=1}^{p} \lambda_{hr}^2 + \sigma^2 \qquad r = 1, \ldots, m \tag{2}$$

while the remaining $p - m$ characteristic roots are equal to σ^2. Although the restriction to equal specific variances is a severe one, some insight into the magnitude of the sampling variation of the estimates in the general case might be gained from its particular results. Using the Taylor-series approach employed originally by Girshick for the moments of principal components, Lawley found that the asymptotic covariance of the ith and jth loadings of the rth factor in the set of m estimated by maximum likelihood from the sample *covariance* matrix based upon N independent observation vectors from a multinormal population with covariance matrix (1) was

$$\begin{aligned}(3)\quad &\operatorname{cov}(\hat{\lambda}_{ir},\hat{\lambda}_{jr})\\ &= \frac{\mu_r}{(N-1)(\mu_r - \sigma^2)}\left\{\sigma_{ij} - \frac{\mu_r}{2(\mu_r - \sigma^2)}\lambda_{ir}\lambda_{jr} + \sum_{\substack{h=1\\h\neq r}}^{m}\frac{\mu_h}{\mu_h - \sigma^2}\right.\\ &\quad\left.\cdot\left[\left(\frac{\mu_r - \sigma^2}{\mu_r - \mu_h}\right)^2 - 1\right]\lambda_{ih}\lambda_{jh}\right\} + \frac{\sigma^4}{2(N-1)(p-m)(\mu_r - \sigma^2)^2}\lambda_{ir}\lambda_{jr}\end{aligned}$$

σ_{ij} is the ijth element of $\mathbf{\Sigma}$. The asymptotic variance of the ith loading in the rth factor estimate is found by setting $i = j$ in expression (3). The asymptotic covariance of the ith and jth estimated loadings of the rth and sth factors is

$$\text{(4)} \quad \operatorname{cov}(\hat{\lambda}_{ir},\hat{\lambda}_{js}) = -\frac{\mu_r\mu_s}{(N-1)(\mu_r-\mu_s)^2}\lambda_{is}\lambda_{jr} + \frac{\sigma^4}{2(N-1)(p-m)(\mu_r-\sigma^2)(\mu_s-\sigma^2)}\lambda_{ir}\lambda_{js}$$

The final terms in both covariance expressions are due to the estimation of the specific variance parameter; if σ^2 were known, those terms would vanish. The reader should note the similarity with the covariances of the principal-component loadings computed by Girshick and Anderson and presented in Sec. 8.7. The corresponding covariances for loadings extracted from the sample correlation matrix would have to reflect the scaling of the estimates by standard deviations and would of necessity be more complicated.

Perhaps a better sense of the size of such variances and covariances can be had by considering a simple example. Let $\sigma^2 = 0.2$, and let the factor-loading matrix be

$$\mathbf{\Lambda}' = \begin{bmatrix} 0.8 & 0.8 & 0.7 & 0.6 & 0.5 \\ -0.4 & -0.5 & 0.1 & 0.5 & 0.7 \end{bmatrix}$$

Then

$$\mathbf{\Sigma} = \begin{bmatrix} 1.00 & 0.84 & 0.52 & 0.28 & 0.12 \\ & 1.09 & 0.51 & 0.23 & 0.05 \\ & & 0.70 & 0.47 & 0.42 \\ & & & 0.81 & 0.65 \\ & & & & 0.94 \end{bmatrix}$$

The expressions $(N-1)\operatorname{var}(\hat{\lambda}_{ih})$ for the large-sample variances of the sample loadings for such a population are given in Table 9.8. These variances reflect sampling variation relative to the size of the population loadings, and the coefficients of variation provide a better indication of the stabilities of the individual loadings. As one might expect from intuition, the variances of the estimates of the small loadings (par-

Table 9.8. **Large-sample Variances and Coefficients of Variation of Estimated Loadings**

Variable	$(N-1)\operatorname{var}(\hat{\lambda}_{i1})$	$\sqrt{N-1}\,CV$	$(N-1)\operatorname{var}(\hat{\lambda}_{i2})$	$\sqrt{N-1}\,CV$
1	1.28	1.41	0.99	−2.48
2	1.70	1.63	1.03	−2.03
3	0.51	1.02	0.75	8.69
4	1.56	2.08	0.74	1.72
5	2.62	3.24	0.74	1.23

ticularly of factor 2) are relatively much larger than those corresponding to the high loadings. We note that a sample size of approximately 677 independent observations would be required to estimate the smallest loading $\hat{\lambda}_{32} = 0.10$ with an asymptotic standard deviation equal to one-third that loading. The correlations of equal or nearly equal loadings of the same factor in this example are very high. Similarly, the estimates of the larger loadings in different factors are highly correlated: the asymptotic correlation of $\hat{\lambda}_{41}$ and $\hat{\lambda}_{42}$ is equal to -0.66.

Lawley (1955) has applied these results to the calculation of the large-sample efficiency of the centroid estimates of loadings and has shown those approximations to be highly efficient for some simple population structures. Jöreskog (1963) has computed the moments (3) and (4) for a two-factor model and has compared their values with those obtained from sampling studies. He has also described extensive simulation studies of the estimates and their sampling properties for a variety of factor models and distributional assumptions. Further analytical and numerical results on the asymptotic standard errors have also been given by Lawley and Maxwell (1971).

9.10 THE EVALUATION OF FACTORS

Frequently factor analysis is used to reduce a large number of responses to a smaller set of uncorrelated scores. If those derived variates are to be used in further statistical analyses, it will be necessary to assign each subject or other sampling unit scores for each of the factors. In Chap. 8 we saw that this could be easily accomplished for principal-component variates, for by definition each component is a linear compound of the original responses. For the factor model evaluation of such scores is neither so simple nor so uniquely defined, and in this section we shall consider two intuitively appealing methods for computing the values of the latent factors of a subject.

The first of these is due to Thomson (1951). Recall that the linear model (1) of Sec. 9.2 related the response variate $\mathbf{x}$ to the factor and specific variates as

$$\mathbf{x} = \mathbf{\Lambda y} + \mathbf{e} \tag{1}$$

If $\mathbf{y}$ and $\mathbf{e}$ are independently distributed as m- and p-dimensional multinormal variates with respective covariance matrices $\mathbf{I}$ and $\mathbf{\Psi}$, the response and common-factor variates have the $(m + p)$-dimensional multinormal distribution with covariance matrix

$$\begin{bmatrix} \mathbf{\Lambda\Lambda}' + \mathbf{\Psi} & \mathbf{\Lambda} \\ \mathbf{\Lambda}' & \mathbf{I} \end{bmatrix} \tag{2}$$

If $\hat{\mathbf{\Lambda}}$ and $\hat{\mathbf{\Psi}}$ are the usual maximum-likelihood estimates obtained in Sec. 9.3, the estimate of the covariance matrix is

$$\begin{bmatrix} \hat{\mathbf{\Lambda}}\hat{\mathbf{\Lambda}}' + \hat{\mathbf{\Psi}} & \hat{\mathbf{\Lambda}} \\ \hat{\mathbf{\Lambda}}' & \mathbf{I} \end{bmatrix} \tag{3}$$

In consideration of this joint density Thomson has proposed that the unobservable factor score vectors should be predicted by the m regression functions connecting the two sets of variates. The vectors of coefficients of those functions are given by the rows of the matrix

$$\mathbf{B} = \mathbf{\Lambda}'(\mathbf{\Lambda}\mathbf{\Lambda}' + \mathbf{\Psi})^{-1} \tag{4}$$

as defined in Sec. 3.4. In terms of the estimated regression coefficients the vector of factor scores predicted from the observation vector $\mathbf{x}_i$ of the ith sampling unit is

$$\mathbf{y}_i = \hat{\mathbf{\Lambda}}'(\hat{\mathbf{\Lambda}}\hat{\mathbf{\Lambda}}' + \hat{\mathbf{\Psi}})^{-1}\mathbf{x}_i \tag{5}$$

and the $N \times m$ matrix of factor scores for the entire sample can be computed as

$$\mathbf{Y} = \mathbf{X}(\hat{\mathbf{\Lambda}}\hat{\mathbf{\Lambda}}' + \hat{\mathbf{\Psi}})^{-1}\hat{\mathbf{\Lambda}} \tag{6}$$

The origin of the factor scales is arbitrary, and the usual sample mean term of the regression equation has been omitted. However, it is implicit in the equations for the estimated factor scores that $\hat{\mathbf{\Lambda}}$ and $\hat{\mathbf{\Psi}}$ have been computed from the sample covariance matrix. If those parameters had been extracted from the correlation matrix, $\mathbf{X}$ would be replaced by the matrix of standard scores $\mathbf{Z}$.

The equations (5) and (6) can be simplified by the nature of the maximum-likelihood estimates. Recall that equation (14) of Sec. 9.3 required those quantities to satisfy

$$\mathbf{S}(\hat{\mathbf{\Lambda}}\hat{\mathbf{\Lambda}}' + \hat{\mathbf{\Psi}})^{-1}\hat{\mathbf{\Lambda}} = \hat{\mathbf{\Lambda}} \tag{7}$$

or

$$(\hat{\mathbf{\Lambda}}\hat{\mathbf{\Lambda}}' + \hat{\mathbf{\Psi}})^{-1}\hat{\mathbf{\Lambda}} = \mathbf{S}^{-1}\hat{\mathbf{\Lambda}} \tag{8}$$

Hence

$$\mathbf{Y} = \mathbf{X}\mathbf{S}^{-1}\hat{\mathbf{\Lambda}} \tag{9}$$

or the form of the estimation equations given by Harman (1967). Finally, through the identity (16) of Sec. 9.3 we need only invert a smaller $m \times m$ matrix in this version:

$$\mathbf{Y} = \mathbf{X}\hat{\mathbf{\Psi}}^{-1}\hat{\mathbf{\Lambda}}(\mathbf{I} + \hat{\mathbf{\Lambda}}'\hat{\mathbf{\Psi}}^{-1}\hat{\mathbf{\Lambda}})^{-1} \tag{10}$$

Example 9.9. Let us compute scores for the three WAIS factors discussed in Example 9.3 for two hypothetical subjects. The subtest scores for the first individual were taken to be the means for participants in the original standardization sample of ages 25 to 34 and 9 to 12 years of education, while the second subject's scores fell at the mean of examinees of the same ages but with 13 or more years of formal education. The standard score matrix (multiplied by 100 to eliminate decimal points) is

$$\mathbf{Z}' = \begin{bmatrix} 4 & 4 & 1 & 25 & 24 & -4 & 59 & 26 & 33 & 39 & 37 \\ 87 & 66 & 68 & 76 & 65 & 81 & 122 & 84 & 82 & 93 & 63 \end{bmatrix}$$

The diagonal matrix

$$\hat{\mathbf{\Lambda}}'\hat{\mathbf{\Psi}}^{-1}\hat{\mathbf{\Lambda}} = \begin{bmatrix} 20.23 & 0 & 0 \\ & 3.74 & 0 \\ & & 2.01 \end{bmatrix}$$

was originally computed as part of the Rao-Maxwell solution of the estimation equations. The vectors of factor scores are

$$\mathbf{y}_1' = [20,20,42]$$

and

$$\mathbf{y}_2' = [100,\ -2,\ 28]$$

Bartlett (1937, 1938) has proposed another method for evaluating factors. The scores for the ith sampling unit are computed so as to minimize the sum of squares

$$\sum_{j=1}^{p} \hat{\Psi}_j^{-1} e_{ij}^2 = (\mathbf{x}_i - \hat{\mathbf{\Lambda}}\mathbf{y}_i)'\hat{\mathbf{\Psi}}^{-1}(\mathbf{x}_i - \hat{\mathbf{\Lambda}}\mathbf{y}_i) \tag{11}$$

of the specific variates divided by their estimated variances. Differentiation of the right-hand expression with respect to $\mathbf{y}_i$ leads to the least-squares solution

$$\mathbf{y}_i = (\hat{\mathbf{\Lambda}}'\hat{\mathbf{\Psi}}^{-1}\hat{\mathbf{\Lambda}})^{-1}\hat{\mathbf{\Lambda}}'\hat{\mathbf{\Psi}}^{-1}\mathbf{x}_i \tag{12}$$

and to the complete matrix of factor scores

$$\mathbf{Y} = \mathbf{X}\hat{\mathbf{\Psi}}^{-1}\hat{\mathbf{\Lambda}}(\hat{\mathbf{\Lambda}}'\hat{\mathbf{\Psi}}^{-1}\hat{\mathbf{\Lambda}})^{-1} \tag{13}$$

If the elements of $\hat{\mathbf{\Lambda}}'\hat{\mathbf{\Psi}}^{-1}\hat{\mathbf{\Lambda}}$ are much larger than one, the two methods will give nearly the same factor scores.

9.11 MODELS FOR THE DEPENDENCE STRUCTURE OF ORDERED RESPONSES

Let us suppose that a natural ordering exists among the responses of some multivariate system. For example, growth curves or profiles are ordered by the times at which the successive measurements were recorded. Similarly, the nature of the tasks in a psychomotor experiment might suggest a ranking of their scores on a difficulty or complexity scale. In every case we shall require that these orderings follow from *external* criteria and never from the magnitudes of the observations themselves.

When time is the ordering index, the data from each sampling unit are called a *time series*. The analysis of time series has received considerable attention from economists and, more recently, from engineers and physicists. Those investigators have proposed statistical models for time-series phenomena, generally with a view toward predicting future values from the observations. The older *deterministic* models consisted of a simple mathematical function (linear trend, a polynomial of fairly low degree, sinusoids, or exponential growth or decay) merely disturbed by an additive random variable at each time. In many applications this representation proved to be inadequate, for memory, averaging or cumulative effects, and other persistent factors led to random fluctuations which cast doubt upon the validity of the deterministic component and upon the important assumption of *independence* of the successive random errors. Those time series appeared to be observation vectors from multidimensional distributions or realizations of *stochastic processes*.

For our purposes it will suffice to think of a stochastic, or random, process as a random variable $X(t)$ indexed by the continuous parameter t. In the time-series context t is the time scale. If the process is observed at the particular p points

$$t_1 < \cdots < t_p$$

the random vector

$$[X(t_1), \ldots, X(t_p)]$$

has some multidimensional distribution characterizing the process. Our attention will be restricted to normal processes, or those in which the variates $X(t_1), \ldots, X(t_p)$ have a p-dimensional multinormal distribution. This distribution will of course be specified by the vector of expectations $E[X(t_i)]$ and the covariance matrix with general element

$$\text{cov}\,[X(t_i),X(t_j)] = \sigma_i\sigma_j\rho(t_i,t_j) \tag{1}$$

The term $\rho(t_i,t_j)$ is called the *correlation function* of the stochastic process. If that function has a simple form, $X(t)$ might be taken as the model for the parsimonious generation of the correlations among the original responses.

Let us consider one such process. For convenience and without loss of generality we shall let

$$E[X(t)] = 0 \qquad \text{var}\,[X(t)] = 1 \tag{2}$$

and for a covariance function we shall take

(3) $$\rho(t_i,t_j) = \rho^{|t_i-t_j|} \qquad 0 \le \rho < 1$$

$X(t)$ is a *stationary* process: the joint distribution of its variates $X(t_i)$ is unchanged by shifts of the t scale. Clearly, the expectation and variance do not depend upon t, and the covariance or correlation function depends only upon the *difference* of the t_i indices. Let the response variates of the ordered set be $x_1, \ldots, x_p$. Since the assumptions of a common zero mean and unit variance are rather restrictive for most behavioral-sciences applications, we shall relate the responses to the stochastic process by

(4) $$x_i = \mu_i + \sigma_i X(t_i) \qquad i = 1, \ldots, p$$

The correlation structures of the x_i and $X(t_i)$ variates are identical, although for consistency in the sequel we shall prefer to describe the structure in terms of the process.

Now, in what way does the correlation function (3) lead to a parsimonious description of the correlation structure? *It follows from that function that the variate $X(t_i)$ can depend upon the preceding variates $X(t_1), \ldots, X(t_{i-1})$ only through its immediate antecedent $X(t_{i-1})$.* That is, *the partial correlation of $X(t_i)$ and $X(t_j)$ with any $X(t_h)$, $i < h < j$, held constant is zero.* By expression (17) of Sec. 3.4, that correlation is

$$\begin{aligned} \rho_{ij.h} &= \frac{\rho^{t_j-t_i} - \rho^{t_h-t_i}\rho^{t_j-t_h}}{\sqrt{1-\rho^{2(t_h-t_i)}}\sqrt{1-\rho^{2(t_j-t_h)}}} \\ &= 0 \end{aligned}$$

The higher-order partial correlations with two or more intermediate variates held constant must also vanish. From the symmetry of the correlation matrix, the similar partial correlations for $i > j$ must also vanish. The implications for the multiple regressions among the responses are clear: for the prediction of the ith response all regression coefficients except those of x_{i-1} and x_{i+1} must vanish. Only the immediate neighbors of x_i are useful for its estimation.

Stochastic processes with the correlation function (3) are called *Markovian* after the Russian probabilist A. A. Markov (1856–1922) responsible for their development. A Markov process can be thought of as the next step beyond the assumption of independence. The process can be generated by the *first-order autoregressive* model

(5) $$X(t+\tau) = \rho^{|\tau|}X(t) + e_{t+\tau}$$

$e_{t+\tau}$ is a random variable with zero mean and variance $\sigma^2(1 - \rho^{2|\tau|})$ and is distributed independently of $X(t)$. As one should expect, the variance of that random disturbance tends to zero as $t + \tau$ is chosen closer to t. The correlation of $X(t + \tau)$ and $X(t)$ is easily shown to be $\rho^{|\tau|}$, and the Markovian nature of the process is immediate.

Finally let us examine the pattern of the Markov correlation matrix and its inverse. For simplicity write the separation of the ith and $(i + 1)$th variates as

$$\tau_i = t_{i+1} - t_i \tag{6}$$

The $p \times p$ correlation matrix is

$$\mathbf{P} = \begin{bmatrix} 1 & \rho^{\tau_1} & \rho^{\tau_1+\tau_2} & \cdots & \rho^{\Sigma\tau_i} \\ & 1 & \rho^{\tau_2} & \cdots & \rho^{\Sigma\tau_i} \\ & & \cdots & \cdots & \cdots \\ & & & & \rho^{\tau_{p-1}} \\ & & & & 1 \end{bmatrix} \tag{7}$$

The summations in the last column run over the last $p - 1$, $p - 2$, . . . , 1 subscripts. The correlation structure is completely generated by the elements of the first sub- and superdiagonals. Of equal importance is this property: knowledge of those correlations will establish the spacings of the ordered responses on the t scale. However, those positions are determined only up to a linear transformation: the $p - 1$ increments τ_i do not set the origin of the scale, and because ρ is indeterminate, the units are arbitrary. The inverse of the correlation matrix is

$$\begin{bmatrix} a_1 & b_1 & 0 & \cdots & 0 & 0 \\ & a_2 & b_2 & \cdots & 0 & 0 \\ & & a_3 & \cdots & 0 & 0 \\ & & & \cdots & \cdots & \cdots \\ & & & & a_{p-1} & b_{p-1} \\ & & & & & a_p \end{bmatrix} \tag{8}$$

where

$$a_1 = \frac{1}{1 - \rho^{2\tau_1}}$$

$$a_i = \frac{1 - \rho^{2(\tau_{i-1}+\tau_i)}}{(1 - \rho^{2\tau_{i-1}})(1 - \rho^{2\tau_i})} \qquad i = 2, \ldots, p - 1$$

$$a_p = \frac{1}{1 - \rho^{2\tau_{p-1}}}$$

$$b_i = -\frac{\rho^{\tau_i}}{1 - \rho^{2\tau_i}} \qquad i = 1, \ldots, p - 1$$

All elements outside the main and first diagonals are zero. By the relation between the inverse matrix and the p multiple-regression functions stated in Exercise 5 of Sec. 3.10 the pattern of zero and nonzero coefficients is apparent. The multiple correlations can be computed by the expression of that exercise.

The connection between the correlations of ordered responses and positions on some latent continuum is due to Guttman (1954, 1955), while the explicit representation of the response variates by a Markov process is due to Anderson (1960). Guttman originally proposed the representation

$$r_{ij} = \frac{t_i}{t_j} \tag{9}$$

of the ijth sample correlation as the ratio of the positions of the ith and jth variates on some latent scale. Ordered responses with such a correlation structure are said to have the *perfect simplex* property. The simplex notion was suggested by the correlation matrices of some cognitive tests, and in that context the parameter t was interpreted as the *complexity scale* of the battery. As in the Markov model, the correlations in a perfect simplex matrix are generated by the elements of the first sub- and superdiagonals, the same partial correlations vanish, and the inverse contains the same pattern of zero and nonzero elements. For that reason Guttman proposed as a crude test of the presence of the simplex property the examination of the inverse of the correlation matrix. A statistical test has since been developed, and will be given presently.

As a third model one might employ the Winer stochastic process, or one whose successive increments

$$Y(t_i) - Y(t_{i+1})$$

are uncorrelated. That process is used as a model for the Brownian motion of a particle, since the successive displacements might be treated as independent normal variates. The Winer process is not stationary: its variance is proportional to the index parameter, and its correlation function has the form

$$\rho(t_i,t_j) = \sqrt{\frac{t_i}{t_j}} \qquad t_i \leq t_j$$

Although the Markov, simplex, and Winer models generate identical correlation structures, the scalings of the responses on their t continua need not be the same. In the Markov model distances of responses are proportional to the logarithms of their correlations, while the simplex leads to a scale whose ratios equal the correlations of the associated responses. As Anderson has indicated, an infinity of scalings can be produced through monotonic transformations of the t parameter in either process. The investigator must decide at the outset which stochastic model is more appropriate for a particular multivariate system. If the responses arose from a repeated-measurements situation, the presence of memory, order, or other persistent effects would seem to suggest a Markov model. The Wiener model might be more suitable for growth profiles or other cumulative variates whose increments could be considered independent.

Estimation of the Complexity Scale. Let us suppose that N observation vectors have been drawn from a multinormal population of ordered responses generated by the Markov stochastic process. The spacings τ_i of the responses on the t scale are unknown, and it is desired to estimate these from the sample data. From the form of the inverse matrix (8) it is easily shown that the sample variances and the correlations $r_{12}, \ldots, r_{p-1,p}$ constitute a set of sufficient statistics for $\sigma_1^2, \ldots, \sigma_p^2$ and $\rho^{\tau_1}, \ldots, \rho^{\tau_{p-1}}$, and it follows that the maximum-likelihood estimate of the distance of the ith and $(i+1)$st responses is proportional to

$$\hat{\tau}_i = -\ln r_{i,i+1} \qquad i = 1, \ldots, p-1 \tag{10}$$

The proportionality constant is the unestimatable quantity $-1/\ln \rho$.

If the simplex model had been chosen, the maximum-likelihood estimate of t_i/t_{i+1} would be

$$\frac{\hat{t}_i}{\hat{t}_{i+1}} = r_{i,i+1} \qquad i = 1, \ldots, p-1 \tag{11}$$

An arbitrary value must be assigned to one of the t_i. The remaining scale positions are in units of that value. Kaiser (1962) has proposed that a least-squares solution be used to estimate the simplex scale, but for simplicity as well as statistical rigor the maximum-likelihood approach would seem preferable.

Example 9.10. Botwinick and Brinley (1962) measured the reaction times of 64 normal subjects to visual stimuli presented after preparatory intervals of 0.5, 1, 3, 6, and 15 sec. This matrix of correlations was observed among the responses:

$$\mathbf{R} = \begin{bmatrix} 1 & 0.56 & 0.46 & 0.42 & 0.36 \\ & 1 & 0.64 & 0.61 & 0.52 \\ & & 1 & 0.73 & 0.61 \\ & & & 1 & 0.85 \\ & & & & 1 \end{bmatrix}$$

The t Markov scale standardized to zero origin and unit length had the following values:

Preparatory interval	0.5	1	3	6	15
Scale value	0	0.39	0.68	0.89	1

Shorter preparatory intervals tend to be more widely spaced on the scale.

Other Models for Ordered Responses. Guttman added specific variance components to the diagonal terms of the simplex correlation matrix to form his *quasi-simplex* model. Maximum-likelihood estimates and generalized likelihood-ratio tests have been constructed by Mukherjee (1966) for the quasi-simplex parameters expressed in terms of the covariance matrix. Schönemann (1970) has treated estimation schemes for the quasi-simplex based on correlations. Bock and Bargmann (1966) have developed estimates and tests for variance-components models which include the simplex and quasi-simplex cases. Jöreskog (1970) has considered maximum-likelihood estimation in a very general covariance-matrix model which includes the factor and ordered-responses forms of this chapter.

Guttman (1954) has proposed his *circumplex* model for describing the correlations among responses arranged on a circular scale. The correlation matrix of the variates has an approximate circulant appearance: the correlations of successive diagonals first decrease as one moves away from the main diagonal, then increase as the upper right- and lower left-hand corners are neared. The circumplex structure is based upon a circular moving-average model for the responses, while Anderson (1960) has proposed an alternative representation as a circular Markov process. Under that parsimonious model the positions of the responses on the circular scale can be estimated in much the same way as in the linear Markov process. Guttman has considered a two-dimensional structure called the *radex* by combining the circumplex and simplex models. Anderson has formalized the radex structure in terms of linear and circular stochastic processes in a plane.

A Test for Ordered Dependence. The assumption of merely first-order dependence among ordered responses is rather restrictive, and in many economic and physical stochastic models longer persistence of the effects is permitted. Such sth-degree antedependence has been defined by Gabriel (1962) in this manner: the multinormal system of p ordered responses is said to be *sth-antedependent* if the partial correlation of x_i and $x_{i-s-z-1}, \ldots, x_1$, for $x_{i-1}, \ldots, x_{i-s-z}$ held constant, is zero for all nonnegative z. This is equivalent to the condition that all elements in the $(s + 1)$th, . . . , last sub- and superdiagonals of the inverse covariance matrix must vanish. From the standpoint of the prediction of the ith variate this pattern implies that only the s adjacent responses have nonzero regression weights.

Gabriel has treated a likelihood-ratio test for the hypothesis H_s that the covariance structure of a multinormal population is sth-antedependent against the alternative H_{s+1} of the next higher order of antedependence. For a test of level α the large-sample decision rule is

$$\text{(12)} \quad \text{Accept } H_s \text{ if } -N \sum_{i=1}^{p-s-1} \ln\,(1 - r^2_{i,i+s+1.i+1,\ldots,i+s}) \leq \chi^2_{\alpha;p-s-1}$$

and reject H_s otherwise. The critical value is the upper 100α percentage point of the chi-squared distribution with $p - s - 1$ degrees of freedom. The asymptotic distribution of the statistic can be improved a bit by replacing the factor N by $N - s - 1$.

From the likelihood-ratio construction of the test it is possible to test H_s against any hypothesis H_{s+k} of higher antedependence by adding the statistics for $H_s, \ldots, H_{s+k-1}$. The sum is asymptotically distributed as a chi-squared variate with degrees of freedom equal to the total of those of the individual statistics when H_s is true. In this fashion the test described in Sec. 3.8 of H_0 (complete independence) against the alternative H_{p-1} of an arbitrary positive definite correlation matrix can be derived.

Example 9.11. Let us test the hypothesis of first-order antedependence in the reaction-time correlations of Example 9.10. We begin by testing the hypothesis of independence against that of first-order antedependence: the test statistic is

$$-(N-1) \sum_{i=1}^{p-1} \ln\,(1 - r^2_{i,i+1}) = 186$$

and independence would be rejected at any reasonable level. For the next test of first-order dependence the requisite partial correlations are

$$r_{13.2} = 0.160 \qquad r_{24.3} = 0.272 \qquad r_{35.4} = -0.029$$

and the test statistic (with corrected coefficient) is equal to 6.50. Since the 5 percent critical value of the chi-squared distribution with three degrees of freedom is 7.81, we can conclude that the hypothesis of first-order antedependence is tenable.

9.12 EXERCISES

1. From the residual matrices of the one- and two-factor solutions of Example 9.2 compute the matrices of partial correlations of the dimensions with the respective factors held constant.
2. Compute the reproduced correlation, residual, and partial correlation matrices for the factor solution of Example 9.4, and test by the approximate chi-squared statistic the hypothesis that a single common factor would account for the observed correlations.
3. Compute the determinants of the original and reproduced correlation matrices of Example 9.4, and test the single-factor hypothesis. Compare the value of the determinantal-ratio statistic with that of the approximate quantity com-

puted in Exercise 2. What are the implications of the size of the determinant of the observed correlation matrix for the single-factor hypothesis and solution?

4. Verify that the Wishart density approach of Sec. 9.3 and use of the multinormal likelihood (2) of Sec. 3.5 lead to identical estimation equations for the factor parameters.

5. From the reaction-time correlation matrix of Example 9.10 the following two-factor solution was extracted:

$$\hat{\boldsymbol{\Lambda}}' = \begin{bmatrix} 0.48 & 0.65 & 0.76 & 0.98 & 0.86 \\ 0.74 & 0.35 & 0.16 & -0.07 & -0.08 \end{bmatrix}$$

   a. The determinants of the original and reproduced correlation matrices were 0.0473 and 0.0494, respectively. Use those values to test at the $\alpha = 0.05$ level the adequacy of the two-factor linear model.

   b. Compute the residual matrix and repeat the goodness-of-fit test with the approximate test statistic. Can you account for the poor approximation to the determinantal statistic?

   c. What are the partial correlations of the five reaction times with the two factors held constant?

   d. Can the loadings be rotated to a simpler structure? Would an oblique rotation to correlated factors lead to greater simplicity?

   e. Compare the nature of the two-factor solution with the simplex scaling given in Example 9.10.

6. From the reaction-time correlations presented in Exercise 2, Chap. 8, this single factor was extracted:

$$\hat{\boldsymbol{\Lambda}}' = [0.74, 0.82, 0.85, 0.82, 0.86]$$

   a. An attempt to compute a two-factor solution by the Rao-Maxwell equations was unsuccessful. Could that failure have been predicted intuitively from the pattern of the correlations?

   b. Compute the residual matrix and test by the approximate method at the 0.01 level the adequacy of the single-factor model.

   c. Use the Lawley test of Sec. 8.7 to test the hypothesis of equal correlations among the five responses.

   d. The pattern of the correlations has the *circular* form described in Sec. 9.11. Would the interpretation of such a circular ordering of the responses be appropriate in light of the result of the test conducted in part b?

7. From the matrix of correlations of Exercise 4, Chap. 8, these factors were extracted:

$$\hat{\boldsymbol{\Lambda}}' = \begin{bmatrix} 0.669 & 0.653 & 0.689 & 0.570 & 0.383 & 0.360 \\ 0.148 & 0.175 & -0.442 & 0.209 & 0.368 & 0.302 \end{bmatrix}$$

   a. Attempt to rotate these loadings to "simple" structure.

   b. What names or interpretations could be given to the rotated factors?

c. How closely do the principal components of the earlier exercise approximate some rotated form of the two-factor solution?

8. Use expression (10) of Sec 9.7 to carry out the varimax rotation of the two factors of Example 9.1. Compare the varimax rotation with the graphical rotation of the factors illustrated in Fig. 9.1.

9. From the correlation matrix of Exercise 12, Sec. 3.10, Sinha and Lee (1970) extracted these factor loadings by the Jöreskog maximum-likelihood formulation and computer program:

Variate	Original loadings			Varimax-rotated loadings		
	1	2	3	1	2	3
Grade	0.24	0.61	−0.40	0.22	0.74	0.06
Moisture	0.53	0.53	0.17	0.37	0.37	0.57
Dockage	0.02	0.57	−0.16	−0.05	0.57	0.17
Acarus	0.29	0.14	0.18	0.21	0.03	0.31
Cheyletus	0.79	0.03	0.07	0.73	0.00	0.32
Glycyphagus	0.28	0.37	0.56	0.06	0.02	0.73
Tarsonemus	1.00	0.00	0.00	0.95	0.02	0.31
Cryptolestes	0.35	0.08	−0.32	0.39	0.25	−0.11
Psocoptera	−0.15	−0.39	−0.16	−0.03	−0.25	−0.38

a. What are the specific variances of the nine variates?

b. What interpretations can be made of the rotated factors?

c. Calculate the matrix of correlations reproduced from the factor loadings.

10. In the same experiment cited earlier Botwinick and Brinley found these correlations among the reaction times of the 64 subjects to auditory stimuli with the same sequence of preparatory intervals:

$$\begin{bmatrix} 1 & 0.49 & 0.39 & 0.28 & 0.17 \\ & 1 & 0.68 & 0.50 & 0.47 \\ & & 1 & 0.74 & 0.68 \\ & & & 1 & 0.73 \\ & & & & 1 \end{bmatrix}$$

a. Test the successive hypotheses of zero-, first-, and second-order antedependence.

b. If the first-order model is tenable, compute the scalings of the five preparatory intervals under the Markov and simplex models.

REFERENCES

SOME BIBLIOGRAPHIC SOURCES. Anderson *et al.* (1972) have compiled an extensive bibliography of books on multivariate analysis through 1970 and journal articles through 1966. Subrahmaniam and Subrahmaniam (1973) have prepared abstracts of nearly 1,200 papers dealing with multivariate normal theory and methods which appeared in the years 1957–1972. Their work complements the references of Anderson's fundamental text (1958) and his later bibliography. Johnson and Kotz (1972) have treated all aspects of continuous multivariate distributions in great detail, with extensive references to original sources.

Factor analysis has a vast literature, much of which is outside the major statistical journals. Harman's text (1967) contains a bibliography of more than 500 books and papers. Earlier overviews and literature surveys have been given by Wolfle (1940), by Carroll and Schweiker (1951) for the period 1939–1951, and by Solomon and Rosner (1954) for 1952–1954.

REFERENCES

Aitken, A. C. (1934): On least squares and linear combination of observations, *Proceedings of the Royal Society of Edinburgh,* vol. 55, pp. 42–47.

Aitken, A. C. (1937): Studies in practical mathematics, II: The evaluation of the latent roots and latent vectors of a matrix, *Proceedings of the Royal Society of Edinburgh,* vol. 57, pp. 269–304.

Aitkin, M. A. (1969): Some tests for correlation matrices, *Biometrika,* vol. 56, pp. 443–446.

Aitkin, M. A. (1971): Correction to "Some tests for correlation matrices," *Biometrika,* vol. 58, p. 245.

Albert, A. (1972): "Regression and the Moore-Penrose Pseudoinverse," Academic Press, New York.

Anderson, T. W. (1951a): The asymptotic distribution of certain characteristic roots and vectors, "Proceedings of the Second Berkeley Symposium on Mathematical Statistics and Probability," pp. 103–130, University of California Press, Berkeley.

Anderson, T. W. (1951b): Classification by multivariate analysis, *Psychometrika,* vol. 16, pp. 31–50.

Anderson, T. W. (1957): Maximum likelihood estimates for a multivariate normal distribution when some observations are missing, *Journal of the American Statis-*

tical Association, vol. 52, pp. 200–203.

Anderson, T. W. (1958): "An Introduction to Multivariate Statistical Analysis," John Wiley & Sons, Inc., New York.

Anderson, T. W. (1960): Some stochastic process models for intelligence test scores, in K. J. Arrow *et al.* (eds.), "Mathematical Methods in the Social Sciences," pp. 205–220. Stanford University Press, Stanford, CA.

Anderson, T. W. (1963a): Asymptotic theory for principal component analysis, *Annals of Mathematical Statistics,* vol. 34, pp. 122–148.

Anderson, T. W. (1963b): The use of factor analysis in the statistical analysis of multiple time series, *Psychometrika,* vol. 28, pp. 1–25.

Anderson, T. W., S. Das Gupta, and G. P. H. Styan (1972): "A Bibliography of Multivariate Statistical Analysis," Halsted Press, John Wiley & Sons, Inc., New York.

Bargmann, R. E. (1957): A study of independence and dependence in multivariate normal analysis, *University of North Carolina Institute of Statistics Mimeographed Series,* no. 186, Chapel Hill.

Bartlett, M. S. (1933): On the theory of statistical regression, *Proceedings of the Royal Society of Edinburgh,* vol. 53, pp. 260–283.

Bartlett, M. S. (1934): The vector representation of a sample, *Proceedings of the Cambridge Philosophical Society,* vol. 30, pp. 327–340.

Bartlett, M. S. (1937): The statistical conception of mental factors, *British Journal of Psychology,* vol. 28, pp. 97–104.

Bartlett, M. S. (1938): Methods of estimating mental factors, *Nature,* vol. 141, pp. 609–610.

Bartlett, M. S. (1947): Multivariate analysis, *Journal of the Royal Statistical Society, Series B,* vol. 9, pp. 176–197.

Bartlett, M. S. (1951): An inverse matrix adjustment arising in discriminant analysis, *Annals of Mathematical Statistics,* vol. 22, pp. 107–111.

Bartlett, M. S. (1953): Factor analysis in psychology as a statistician sees it, "Uppsala Symposium on Psychological Factor Analysis," pp. 23–34, Almqvist & Wiksell, Uppsala, Sweden.

Bartlett, M. S. (1954): A note on multiplying factors for various chi-squared approximations, *Journal of the Royal Statistical Society, Series B,* vol. 16, pp. 296–298.

Basickes, S. (1972): "On Parameter Estimation and Hypothesis Tests for Mean Vectors with Incomplete Multivariate Data," Ph.D. dissertation, University of Pennsylvania, Philadelphia.

Bellman, R. (1960): "Introduction to Matrix Analysis," McGraw-Hill Book Company, New York.

Bhargava, R. P. (1962): "Multivariate Tests of Hypotheses with Incomplete Data," Technical Report No. 3, Applied Mathematics and Statistics Laboratories, Stanford University, Stanford, CA.

Bhoj, D. S. (1971): "On Multivariate Tests of Hypotheses on Mean Vectors with Missing Observations," Ph.D. dissertation, University of Pennsylvania, Philadelphia.

Bhoj, D. S. (1973a): Percentage points of the statistics for testing hypotheses on mean vectors of multivariate normal distributions with missing observations,

Journal of Statistical Computation and Simulation, vol. 2, pp. 211–224.

Bhoj, D. S. (1973b): On the distribution of a statistic used for testing a multinormal mean vector with incomplete data, *Journal of Statistical Computation and Simulation,* vol. 2, pp. 309–316.

Birren, J. E., and D. F. Morrison (1961): Analysis of the WAIS subtests in relation to age and education, *Journal of Gerontology,* vol. 16, pp. 363–369.

Birren, J. E., J. Botwinick, A. D. Weiss, and D. F. Morrison (1963): Interrelations of mental and perceptual tests given to healthy elderly men, in J. E. Birren *et al.* (eds.), "Human Aging: A Biological and Behavioral Study," pp. 143–156, U.S. Government Printing Office, Washington.

Blackwell, D., and M. A. Girshick (1954): "Theory of Games and Statistical Decisions," John Wiley & Sons, Inc., New York.

Bock, R. D. (1960): Components of variance analysis as a structural and discriminal analysis for psychological tests, *British Journal of Statistical Psychology,* vol. 13, pp. 151–163.

Bock, R. D. (1963): Multivariate analysis of variance of repeated measurements, in C. W. Harris (ed.), "Problems in Measuring Change," pp. 85–103, University of Wisconsin Press, Madison.

Bock, R. D., and R. E. Bargmann (1966): Analysis of covariance structures, *Psychometrika,* vol. 31, pp. 507–534.

Bodewig, E. (1956): "Matrix Calculus," North-Holland Publishing Company, Amsterdam.

Bose, R. C. (1944): The fundamental theorem of linear estimation, *Proceedings of the 31st Indian Scientific Congress,* pp. 2–3.

Bose, R. C., and S. N. Roy (1938): The exact distribution of the studentized D^2-statistic, *Sankhyā,* vol. 4, pp. 19–38.

Botwinick, J., and J. F. Brinley (1962): An analysis of set in relation to reaction time, *Journal of Experimental Psychology,* vol. 63, pp. 568–574.

Boullion, T. L., and P. L. Odell (1971): "Generalized Inverse Matrices," John Wiley & Sons, Inc., New York.

Bowker, A. H. (1960): A representation of Hotelling's T^2 and Anderson's classification statistic W in terms of simple statistics, in I. Olkin *et al.* (eds.), "Contributions to Probability and Statistics: Essays in Honor of Harold Hotelling," pp. 142–149, Stanford University Press, Stanford, CA.

Box, G. E. P. (1949): A general distribution theory for a class of likelihood criteria, *Biometrika,* vol. 36, pp. 317–346.

Box, G. E. P. (1950): Problems in the analysis of growth and wear curves, *Biometrics,* vol. 6, pp. 362–389.

Box, G. E. P. (1953): Non-normality and tests on variances, *Biometrika,* vol. 40, pp. 318–335.

Brauer, A. T. (1953): Bounds for characteristic roots of matrices, in L. J. Paige *et al.* (eds.), "Simultaneous Linear Equations and the Determination of Eigenvalues," pp. 101–106, U.S. Department of Commerce, Washington.

Browne, E. T. (1958): "Introduction to the Theory of Determinants and Matrices," The University of North Carolina Press, Chapel Hill.

Burton, R. V. (1963): Generality of honesty reconsidered, *Psychological Review,* vol. 70, pp. 481–499.

Carroll, J. B. (1953): An analytical solution for approximating simple structure in factor analysis, *Psychometrika,* vol. 18, pp. 23–38.

Carroll, J. B., and R. F. Schweiker (1951): Factor analysis in educational research, *Review of Educational Research,* vol. 21, pp. 368–388.

Cochran, W. G. (1968): Commentary on "Estimation of error rates in discriminant analysis," *Technometrics,* vol. 10, pp. 204–205.

Cole, J. W. L. (1969): Multivariate analysis of variance using patterned covariance matrices, *University of North Carolina Institute of Statistics Mimeographed Series,* No. 640, Chapel Hill.

Cole, J. W. L., and J. E. Grizzle (1966): Applications of multivariate analysis of variance to repeated measurements experiments, *Biometrics,* vol. 22, pp. 810–828.

Collier, R. O., Jr., F. B. Baker, G. K. Mandeville, and T. F. Hayes (1967): Estimates of test size for several test procedures based on conventional variance ratios in the repeated measures design, *Psychometrika,* vol. 32, pp. 339–353.

Constantine, A. G. (1963): Some noncentral distribution problems in multivariate analysis, *Annals of Mathematical Statistics,* vol. 34, pp. 1270–1285.

Consul, P. C. (1967): On the exact distributions of the criterion W for testing sphericity in a p-variate normal distribution, *Annals of Mathematical Statistics,* vol. 38, pp. 1170–1174.

Courant, R. (1966): "Differential and Integral Calculus," vol. 2, Interscience Publishers, New York.

Danford, M. B., H. M. Hughes, and R. C. McNee (1960): On the analysis of repeated-measurements experiments, *Biometrics,* vol. 16, pp. 547–565.

David, F. N. (1938): "Tables of the Ordinates and Probability Integral of the Distribution of the Correlation Coefficient in Small Samples," Cambridge University Press, Cambridge.

David, H. A. (1956): On the application to statistics of an elementary theorem in probability, *Biometrika,* vol. 43, pp. 85–91.

Davis, A. W. (1970): Exact distributions of Hotelling's generalized T_0^2, *Biometrika,* vol. 57, pp. 187–191.

Dempster, A. P. (1969): "Elements of Continuous Multivariate Analysis," Addison-Wesley Publishing Company, Reading, MA.

Dixon, W. J., and F. J. Massey, Jr. (1969): "Introduction to Statistical Analysis," 3d ed., McGraw-Hill Book Company, New York.

Doppelt, J. E., and W. L. Wallace (1955): Standardization of the Wechsler Adult Intelligence Scale for older persons, *Journal of Abnormal and Social Psychology,* vol. 51, pp. 312–330.

Dunn, O. J. (1971): Some expected values for probabilities of correct classification in discriminant analysis, *Technometrics,* vol. 13, pp. 345–353.

Dunn, O. J., and P. J. Varady (1966): Probabilities of correct classification in discriminant analysis, *Biometrics,* vol. 22, pp. 908–924.

Dutton, A. M. (1954): Application of some multivariate analysis techniques to data from radiation experiments, in O. Kempthorne *et al.* (eds.), "Statistics and Mathematics in Biology," pp. 81–91, The Iowa State University Press, Ames.

Dwyer, P. S. (1951): "Linear Computations," John Wiley & Sons, Inc., New York.

Dwyer, P. S. (1967): Some applications of matrix derivatives in multivariate analysis, *Journal of the American Statistical Association,* vol. 62, pp. 607–625.

Dwyer, P. S., and M. S. MacPhail (1948): Symbolic matrix derivatives, *Annals of Mathematical Statistics,* vol. 19, pp. 517–534.

Eisenhart, C. (1947): The assumptions underlying the analysis of variance, *Biometrics,* vol. 3, pp. 1–21.

Eisenhart, L. P. (1960): "Coordinate Geometry," Dover Publications, Inc., New York.

Elashoff, R. M., and A. A. Afifi (1966): Missing values in multivariate statistics—I. Review of the literature, *Journal of the American Statistical Association,* vol. 61, pp. 595–604.

Faddeeva, V. N. (1959): "Computational Methods of Linear Algebra," Dover Publications, Inc., New York.

Feller, W. (1968): "An Introduction to Probability Theory and Its Applications," vol. 1, 3d ed., John Wiley & Sons, Inc., New York.

Feller, W. (1971): "An Introduction to Probability Theory and Its Applications," vol. 2, 2d ed., John Wiley & Sons, Inc., New York.

Ferguson, G. A. (1954): The concept of parsimony in factor analysis, *Psychometrika,* vol. 19, pp. 281–290.

Fisher, R. A. (1915): Frequency distribution of the values of the correlation coefficient in samples from an indefinitely large population, *Biometrika,* vol. 10, pp. 507–521.

Fisher, R. A. (1921): On the "probable error" of a coefficient of correlation deduced from a small sample, *Metron,* vol. 1, pp. 1–32.

Fisher, R. A. (1924): The distribution of the partial correlation coefficient, *Metron,* vol. 3, pp. 329–332.

Fisher, R. A. (1928): The general sampling distribution of the multiple correlation coefficient, *Proceedings of the Royal Society of London, Series A,* vol. 121, pp. 654–673.

Fisher, R. A. (1936): The use of multiple measurements in taxonomic problems, *Annals of Eugenics,* vol. 7, pp. 179–188.

Fisher, R. A. (1939): The sampling distribution of some statistics obtained from nonlinear equations, *Annals of Eugenics,* vol. 9, pp. 238–249.

Fisher, R. A. (1969): "Statistical Methods for Research Workers," 14th ed., Hafner Publishing Company, London.

Fisher, R. A., and G. Prance (1972): "The Design of Experiments," 8th ed., Hafner Publishing Company, London.

Frazer, R. A., W. J. Duncan, and A. R. Collar (1963): "Elementary Matrices and Some Applications to Dynamics and Differential Equations," Cambridge University Press, Cambridge.

Freund, J. E. (1971): "Mathematical Statistics," 2d ed., Prentice-Hall, Inc., Englewood Cliffs, NJ.

Gabriel, K. R. (1962): Ante-dependence analysis of an ordered set of variables, *Annals of Mathematical Statistics,* vol. 33, pp. 201–212.

Galton, F. (1888): Co-relations and their measurement, chiefly from anthropometric data, *Proceedings of the Royal Society,* vol. 45, pp. 135–140.

Galton, F. (1889): "Natural Inheritance," Macmillan & Co., London.

Geisser, S. (1963): Multivariate analysis of variance for a special covariance case, *Journal of the American Statistical Association,* vol. 58, pp. 660–669.

Geisser, S., and S. W. Greenhouse (1958): An extension of Box's results on the use of the F distribution in multivariate analysis, *Annals of Mathematical Statistics,* vol. 29, pp. 885–891.

Genizi, A. (1973): Some power comparisons between bivariate tests and a twofold univariate test, *Journal of the Royal Statistical Society, Series B,* vol. 35, pp. 86–96.

Girshick, M. A. (1936): Principal components, *Journal of the American Statistical Association,* vol. 31, pp. 519–528.

Girshick, M. A. (1939): On the sampling theory of roots of determinantal equations, *Annals of Mathematical Statistics,* vol. 10, pp. 203–224.

Goldberg, S. (1960): "Probability: An Introduction," Prentice-Hall, Inc., Englewood Cliffs, NJ.

Goldberger, A. S. (1964): "Econometric Theory," John Wiley & Sons, Inc., New York.

Gould, S. J. (1966): Allometry and size in ontogeny and phylogeny, *Biological Reviews,* vol. 41, pp. 587–640.

Graybill, F. A. (1961): "An Introduction to Linear Statistical Models," vol. 1, McGraw-Hill Book Company, New York.

Graybill, F. A. (1969): "Introduction to Matrices with Applications in Statistics," Wadsworth Publishing Company, Inc., Belmont, CA.

Greenhouse, S. W., and S. Geisser (1959): On methods in the analysis of profile data, *Psychometrika,* vol. 24, pp. 95–112.

Greenstreet, R. L., and R. J. Connor (1974): Power of tests for equality of covariance matrices, *Technometrics,* vol. 16, pp. 27–30.

Grizzle, J. E., and D. M. Allen (1969): Analysis of growth and dose response curves, *Biometrics,* vol. 25, pp. 357–381.

Gupta, S. S. (1963a): Probability integrals of multivariate normal and multivariate *t, Annals of Mathematical Statistics,* vol. 34, pp. 792–828.

Gupta, S. S. (1963b): Bibliography of the multivariate normal integrals and related topics, *Annals of Mathematical Statistics,* vol. 34, pp. 829–838.

Guttman, L. (1954): A new approach to factor analysis: the radex, in P. F. Lazarsfeld (ed.), "Mathematical Thinking in the Social Sciences," pp. 258–348, The Free Press of Glencoe, New York.

Guttman, L. (1955): A generalized simplex for factor analysis, *Psychometrika,* vol. 20, pp. 173–192.

Hadley, G. (1961): "Linear Algebra," Addison-Wesley Publishing Company, Inc., Reading, MA.

Hadley, G. (1964): "Nonlinear and Dynamic Programming," Addison-Wesley Publishing Company, Inc., Reading, MA.

Hancock, H. (1960): "Theory of Maxima and Minima," Dover Publications, Inc., New York.

Hanumara, R. C., and W. A. Thompson, Jr. (1968): Percentage points of the extreme roots of a Wishart matrix, *Biometrika,* vol. 55, pp. 505–512.

Harman, H. H. (1967): "Modern Factor Analysis," 2d ed., University of Chicago Press, Chicago.

Hartley, H. O., and R. R. Hocking (1971): The analysis of incomplete data (with discussion), *Biometrics,* vol. 27, pp. 783–823.

Hartshorne, H., and M. A. May (1928): "Studies in the Nature of Character," vol. 1, "Studies in Deceit," The Macmillan Company, New York.

Healy, M. J. R. (1969): Rao's paradox concerning multivariate tests of significance, *Biometrics,* vol. 25, pp. 411–413.

Heck, D. L. (1960): Charts of some upper percentage points of the distribution of the largest characteristic root, *Annals of Mathematical Statistics,* vol. 31, pp. 625–642.

Hills, M. (1966): Allocation rules and their error rates, *Journal of the Royal Statistical Society, Series B,* vol. 28, pp. 1–31.

Hills, M. (1968): A note on the analysis of growth curves, *Biometrics,* vol. 24, pp. 192–196.

Hills, M. (1969): On looking at large correlation matrices, *Biometrika,* vol. 56, pp. 249–253.

Hocking, R. R. (1973): A discussion of the two-way mixed model, *The American Statistician,* vol. 27, pp. 148–152.

Hodges, J. L., Jr., and E. L. Lehmann (1964): "Basic Concepts of Probability and Statistics," Holden-Day, Inc., San Francisco.

Hogg, R. V., and A. T. Craig (1959): "Introduction to Mathematical Statistics," The Macmillan Company, New York.

Hohn, F. E. (1964): "Elementary Matrix Algebra," 2d ed., The Macmillan Company, New York.

Hopkins, J. W., and P. P. F. Clay (1963): Some empirical distributions of bivariate T^2 and homoscedasticity criterion M under unequal variance and leptokurtosis, *Journal of the American Statistical Association,* vol. 58, pp. 1048–1053.

Horst, P. (1963): "Matrix Algebra for Social Scientists," Holt, Rinehart and Winston, Inc., New York.

Hotelling, H. (1931): The generalization of Student's ratio, *Annals of Mathematical Statistics,* vol. 2, pp. 360–378.

Hotelling, H. (1933): Analysis of a complex of statistical variables into principal components, *Journal of Educational Psychology,* vol. 24, pp. 417–441, 498–520.

Hotelling, H. (1935): The most predictable criterion, *Journal of Educational Psychology,* vol. 26, pp. 139–142.

Hotelling, H. (1936a): Simplified calculation of principal components, *Psychometrika,* vol. 1, pp. 27–35.

Hotelling, H. (1936b): Relations between two sets of variates, *Biometrika,* vol. 28, pp. 321–377.

Hotelling, H. (1943): Some new methods for matrix calculation, *Annals of Mathematical Statistics,* vol. 14, pp. 1–33.

Hotelling, H. (1947): Multivariate quality control, illustrated by the air testing of sample bombsights, in C. Eisenhart *et al.* (eds), "Selected Techniques of Statistical Analysis," pp. 111–184, McGraw-Hill Book Company, New York.

Hotelling, H. (1951): A generalized T test and measure of multivariate dispersion, "Proceedings of the Second Berkeley Symposium on Mathematical Statistics and Probability," pp. 23–41, University of California Press, Berkeley.

Hotelling, H. (1953): New light on the correlation coefficient and its transforms, *Journal of the Royal Statistical Society, Series B,* vol. 15, pp. 193–232.

Hotelling, H. (1954): Multivariate analysis, in O. Kempthorne *et al.* (eds.), "Statistics and Mathematics in Biology," pp. 67–80, The Iowa State University Press, Ames.

Householder, A. S. (1953): "Principles of Numerical Analysis," McGraw-Hill Book Company, New York.

Householder, A. S. (1964): "The Theory of Matrices in Numerical Analysis," Blaisdell Publishing Company, New York.

Howe, W. G. (1955): "Some Contributions to Factor Analysis," Oak Ridge National Laboratory, Oak Ridge, TN.

Hsu, P. L. (1938): Notes on Hotelling's generalized T^2, *Annals of Mathematical Statistics,* vol. 9, pp. 231–243.

Hsu, P. L. (1939): On the distribution of roots of certain determinantal equations, *Annals of Eugenics,* vol. 9, pp. 250–258.

Hughes, D. T., and J. G. Saw (1972): Approximating the percentage points of Hotelling's generalized T_0^2 statistic, *Biometrika,* vol. 59, pp. 224–226.

Huxley, J. S. (1924): Constant differential growth ratios and their significance, *Nature,* vol. 114, pp. 895–896.

Huxley, J. S. (1932): "Problems of Relative Growth," Methuen and Company, Ltd., London.

Huynh, H., and L. S. Feldt (1970): Conditions under which mean square ratios in repeated measurements designs have exact *F*-distributions, *Journal of the American Statistical Association,* vol. 65, pp. 1582–1589.

Ito, K. (1962): A comparison of the powers of two multivariate analysis of variance tests, *Biometrika,* vol. 49, pp. 455–462.

Ito, K., and W. J. Schull (1964): On the robustness of the T_0^2 test in multivariate analysis of variance when variance-covariance matrices are not equal, *Biometrika,* vol. 51, pp. 71–82.

Jackson, J. E. (1959): Some multivariate statistical techniques used in color matching data, *Journal of the Optical Society of America,* vol. 49, pp. 585–592.

Jackson, J. E. (1962): Some multivariate statistical techniques used in color matching data—addenda and errata, *Journal of the Optical Society of America,* vol. 52, pp. 835–836.

Jackson, J. E., and F. T. Hearne (1973): Relationships among coefficients of vectors used in principal components, *Technometrics,* vol. 15, pp. 601–610.

James, A. T. (1964): Distribution of matrix variates and latent roots derived from normal samples, *Annals of Mathematical Statistics,* vol. 35, pp. 475–501.

Jennrich, R. I., and S. M. Robinson (1969): A Newton-Raphson algorithm for maximum likelihood factor analysis, *Psychometrika,* vol. 34, pp. 111–123.

Johnson, N. L., and S. Kotz (1972): "Distributions in Statistics: Continuous Multivariate Distributions," John Wiley & Sons, Inc., New York.

Jolicoeur, P. (1963a): The degree of generality in *Martes americana, Growth,* vol. 27, pp. 1–27.

Jolicoeur, P. (1963b): The multivariate generalization of the allometry equation, *Biometrics,* vol. 19, pp. 497–499.

Jolicoeur, P., and J. E. Mosimann (1960): Size and shape variation in the painted turtle: A principal component analysis, *Growth,* vol. 24, pp. 339–354.

Jöreskog, K. G. (1963): "Statistical Estimation in Factor Analysis," Almqvist & Wiksell, Uppsala, Sweden.

Jöreskog, K. G. (1966): Testing a simple structure hypothesis in factor analysis, *Psychometrika,* vol. 31, pp. 165–178.

Jöreskog, K. G. (1967a): Some contributions to maximum likelihood factor analysis, *Psychometrika,* vol. 32, pp. 443–482.

Jöreskog, K. G. (1967b): "UMLFA: A Computer Program for Unrestricted Maximum Likelihood Factor Analysis," Research Memorandum RM-66-20, Educational Testing Service, Princeton, NJ.

Jöreskog, K. G. (1969): A general approach to confirmatory maximum likelihood factor analysis, *Psychometrika,* vol. 34, pp. 183–202.

Jöreskog, K. G. (1970): A general method for analysis of covariance structures, *Biometrika,* vol. 57, pp. 239–251.

Jöreskog, K. G. (1975): Factor analysis by least squares and maximum likelihood, in K. Enslein, A. Ralston, and H. S. Wilf (eds.), "Statistical Methods for Digital Computers," John Wiley & Sons, Inc., New York.

Jöreskog, K. G., and D. N. Lawley (1968): New methods in maximum likelihood factor analysis, *British Journal of Mathematical and Statistical Psychology,* vol. 21, pp. 85–96.

Jöreskog, K. G., and M. van Thillo (1971): "New Rapid Algorithms for Factor Analysis by Unweighted Least Squares, Generalized Least Squares and Maximum Likelihood," Research Memorandum RM-71-5, Educational Testing Service, Princeton, NJ.

Kaiser, H. F. (1956): "The Varimax Method of Factor Analysis," Ph.D. dissertation, University of California.

Kaiser, H. F. (1958): The varimax criterion for analytic rotation in factor analysis, *Psychometrika,* vol. 23, pp. 187–200.

Kaiser, H. F. (1959): Computer program for varimax rotation in factor analysis, *Journal of Educational and Psychological Measurement,* vol. 19, pp. 413–420.

Kaiser, H. F. (1962): Scaling a simplex, *Psychometrika,* vol. 27, pp. 155–162.

Kendall, M. G., and A. Stuart (1958): "The Advanced Theory of Statistics," vol. 1, Charles Griffin & Company, Ltd., London.

Kendall, M. G., and A. Stuart (1961): "The Advanced Theory of Statistics," vol. 2, Charles Griffin & Company, Ltd., London.

Kettenring, J. R. (1971): Canonical analysis of several sets of variables, *Biometrika,* vol. 58, pp. 433–451.

Khatri, C. G. (1966): A note on a MANOVA model applied to problems in growth curve, *Annals of the Institute of Statistical Mathematics,* vol. 18, pp. 75–86.

Korin, B. P. (1968): On the distribution of a statistic used for testing a covariance matrix, *Biometrika,* vol. 55, pp. 171–178.

Korin, B. P. (1969): On testing the equality of k covariance matrices, *Biometrika,* vol. 56, pp. 216–218.

Kowalski, C. J. (1973): Non-normal bivariate distributions with normal marginals, *The American Statistician,* vol. 27, pp. 103–106.

Kshirsagar, A. M., and E. Arseven (1975): A note on the equivalency of two discrimination procedures, *The American Statistician,* vol. 29, pp. 38–39.

Kullback, S. (1968): "Information Theory and Statistics," Dover Publications, Inc., New York.

Lachenbruch, P. A. (1967): An almost unbiased method of obtaining confidence intervals for the probability of misclassification in discriminant analysis, *Biometrics,* vol. 23, pp. 639–645.

Lachenbruch, P. A. (1968): On expected probabilities of misclassification in discriminant analysis, necessary sample size, and a relation with the multiple correlation coefficient, *Biometrics,* vol. 24, pp. 823–834.

Lachenbruch, P. A., and M. R. Mickey (1968): Estimation of error rates in discriminant analysis, *Technometrics,* vol. 10, pp. 1–11.

Lawley, D. N. (1938): "A generalization of Fisher's z-test," *Biometrika,* vol. 30, pp. 180–187.

Lawley, D. N. (1940): The estimation of factor loadings by the method of maximum likelihood, *Proceedings of the Royal Society of Edinburgh,* vol. 60, pp. 64–82.

Lawley, D. N. (1942): Further investigations in factor estimation, *Proceedings of the Royal Society of Edinburgh,* Section *A,* vol. 61, pp. 176–185.

Lawley, D. N. (1943): The application of the maximum likelihood method to factor analysis, *British Journal of Psychology,* vol. 33, pp. 172–175.

Lawley, D. N. (1953): A modified method of estimation in factor analysis and some large sample results, "Uppsala Symposium on Psychological Factor Analysis," pp. 35–42, Almqvist & Wiksell, Uppsala, Sweden.

Lawley, D. N. (1955): A statistical examination of the centroid method, *Proceedings of the Royal Society of Edinburgh,* Section *A,* vol. 64, pp. 175–189.

Lawley, D. N. (1956): Tests of significance for the latent roots of covariance and correlation matrices, *Biometrika,* vol. 43, pp. 128–136.

Lawley, D. N. (1963): On testing a set of correlation coefficients for equality, *Annals of Mathematical Statistics,* vol. 34, pp. 149–151.

Lawley, D. N., and A. E. Maxwell (1971): "Factor Analysis as a Statistical Method," 2d ed., American Elsevier Publishing Company, Inc., New York.

Lee, Y. S. (1972): Some results on the distribution of Wilks's likelihood-ratio criterion, *Biometrika,* vol. 59, pp. 649–664.

Lehmann, E. L. (1959): "Testing Statistical Hypotheses," John Wiley & Sons, Inc., New York.

Lin, P. I., and L. E. Stivers (1974): On difference of means with incomplete data, *Biometrika,* vol.61, pp. 325–334.

Lindgren, B. W. (1962): "Statistical Theory," The Macmillan Company, New York.

Mahalanobis, P. C. (1936): On the generalized distance in statistics, *Proceedings of the National Institute of Sciences of India,* vol. 12, pp. 49–55.

Marcus, M., and H. Minc (1964): "A Survey of Matrix Theory and Matrix Inequalities," Allyn and Bacon, Inc., Boston.

Mardia, K. V. (1971): The effect of nonnormality on some multivariate tests and robustness to nonnormality in the linear model, *Biometrika,* vol. 58, pp. 105–121.

Mauchly, J. W. (1940): Significance test for sphericity of a normal n-variate distribution, *Annals of Mathematical Statistics,* vol. 11, pp. 204–209.

Maxwell, A. E. (1964): Calculating maximum likelihood factor loadings, *Journal of the Royal Statistical Society, Series A,* vol. 127, pp. 238–241.

McHugh, R. B., G. Sivanich, and S. Geisser (1961): On the evaluation of personality changes as measured by psychometric test profiles, *Psychological Reports,* vol. 9, pp. 335–344.

McKeon, J. J. (1965): Canonical analysis: some relations between canonical correlation, factor analysis, discriminant function analysis, and scaling theory, *Psychometric Monographs,* no. 13, The Psychometric Society, University of Chicago Press, Chicago.

McLachlan, G. J. (1974a): The asymptotic distributions of the conditional error rate and risk in discriminant analysis, *Biometrika,* vol. 61, pp. 131–135.

McLachlan, G. J. (1974b): Estimation of the errors of misclassification on the criterion of asymptotic mean square error, *Technometrics,* vol. 16, pp. 255–260.

McLachlan, G. J. (1974c): An asymptotic unbiased technique for estimating the error rates in discriminant analysis, *Biometrics,* vol. 30, pp. 239–249.

Mehta, J. S., and J. Gurland (1969): Testing equality of means in the presence of correlation, *Biometrika,* vol. 56, pp. 119–126.

Mehta, J. S., and J. Gurland (1973): A test for equality of means in the presence of correlation and missing values, *Biometrika,* vol. 60, pp. 211–213.

Miller, R. G., Jr. (1966): "Simultaneous Statistical Inference," McGraw-Hill Book Company, New York.

Mood, A. M. (1951): On the distribution of the characteristic roots of normal second-moment matrices, *Annals of Mathematical Statistics,* vol. 22, pp. 266–273.

Mood, A. M., F. A. Graybill, and D. C. Boes (1974): "Introduction to the Theory of Statistics," 3d ed., McGraw-Hill Book Company, New York.

Morris, K. J., and R. Zeppa (1963): Histamine-induced hypotension due to morphine and Arfonad in the dog, *Journal of Surgical Research,* vol. 3, pp. 313–317.

Morrison, D. F. (1970): The optimal spacing of repeated measurements, *Biometrics,* vol. 26, pp. 281–290.

Morrison, D. F. (1971): Expectations and variances of maximum likelihood estimates of the multivariate normal distribution parameters with missing data, *Journal of the American Statistical Association,* vol. 66, pp. 602–604.

Morrison, D. F. (1972): The analysis of a single sample of repeated measurements, *Biometrics,* vol. 28, pp. 55–71.

Morrison, D. F. (1973): A test for equality of means of correlated variates with missing data on one response, *Biometrika,* vol. 60, pp. 101–105.

Morrison, D. F., and D. S. Bhoj (1973): Power of the likelihood ratio test on the mean vector of the multivariate normal distribution with missing observations, *Biometrika,* vol. 60, pp. 365–368.

Mosimann, J. E. (1968): "Elementary Probability for the Biological Sciences," Appleton-Century-Crofts, New York.

Mosimann, J. E. (1970): Size allometry: size and shape variables with characterizations of the lognormal and generalized gamma distributions, *Journal of the American Statistical Association,* vol. 65, pp. 930–945.

Mueller, P. S. (1963): Plasma free fatty acid concentrations (FFA) in chronic schizophrenia before and after insulin stimulation, *Psychiatric Research,* vol. 1, pp. 106–115.

Mukherjee, B. N. (1966): Derivation of likelihood-ratio tests for Guttman quasi-simplex covariance structures, *Psychometrika,* vol. 31, pp. 97–123.

Nanda, D. N. (1948): Distribution of a root of a determinantal equation, *Annals of Mathematical Statistics,* vol. 19, pp. 47–57.

Nanda, D. N. (1951): Probability distribution tables of the largest root of a determinantal equation with two roots, *Journal of the Indian Society of Agricultural Statistics,* vol. 3, pp. 175–177.

National Bureau of Standards (1959): "Tables of the Bivariate Normal Distribution Function and Related Functions," U.S. Department of Commerce, Washington.

Neuhaus, J., and C. Wrigley (1954): The quartimax method: an analytical approach to orthogonal simple structure, *British Journal of Statistical Psychology,* vol. 7, pp. 81–91.

Neyman, J., and E. S. Pearson (1928): On the use and interpretation of certain test criteria for purposes of statistical inference, *Biometrika,* vol. 20A, pp. 175–240, 263–294.

Neyman, J., and E. S. Pearson (1933): On the problem of the most efficient tests of statistical hypotheses, *Philosophical Transactions of the Royal Society of London, Series A,* vol. 231, pp. 289–337.

Okamoto, M. (1963): An asymptotic expansion for the distribution of the linear discriminant function, *Annals of Mathematical Statistics,* vol. 34, pp. 1286–1301.

Olkin, I., and J. W. Pratt (1958): Unbiased estimation of certain correlation coefficients, *Annals of Mathematical Statistics,* vol. 29, pp. 201–211.

Opatowski, I. (1964): Analysis of degree of dependence of pathological conditions in two different organs: sclerosis in aorta and in coronary arteries, in J. Gurland (ed.), "Stochastic Models in Medicine and Biology," pp. 85–96, The University of Wisconsin Press, Madison.

Parzen, E. (1960): "Modern Probability Theory and Its Applications," John Wiley & Sons, Inc., New York.

Pearce, S. C., and D. A. Holland (1960): Some applications of multivariate methods in botany, *Applied Statistics,* vol. 9, pp. 1–7.

Pearson, E. S. (1969): Some comments on the accuracy of Box's approximations to the distribution of *M, Biometrika,* vol. 56, pp. 219–220.

Pearson, E. S., and H. O. Hartley (1951): Charts of the power function of the analysis of variance tests, derived from the noncentral F-distribution, *Biometrika,* vol. 38, pp. 112–130.

Pearson, E. S., and H. O. Hartley (1966): "Biometrika Tables for Statisticians," vol. 1, 3d ed., Cambridge University Press, Cambridge.

Pearson, E. S., and H. O. Hartley (1972): "Biometrika Tables for Statisticians," vol. 2, Cambridge University Press, Cambridge.

Pearson, K. (1901): On lines and planes of closest fit to systems of points in space, *Philosophical Magazine,* ser. 6, vol. 2, pp. 559–572.

Pearson, K. (1968): "Tables of the Incomplete Beta-Function," 2d ed., Cambridge University Press for the Biometrika Trustees, Cambridge.

Penrose, R. A. (1955): A generalized inverse for matrices, *Proceedings of the Cambridge Philosophical Society,* vol. 51, pp. 406–413.

Perlis, S. (1952): "Theory of Matrices," Addison-Wesley Publishing Company, Inc., Reading, MA.

Pillai, K. C. S. (1955): Some new test criteria in multivariate analysis, *Annals of Mathematical Statistics,* vol. 26, pp. 117–121.

Pillai, K. C. S. (1956a): On the distribution of the largest or the smallest root of a matrix in multivariate analysis, *Biometrika,* vol. 43, pp. 122–127.

Pillai, K. C. S. (1956b): Some results useful in multivariate analysis, *Annals of*

Mathematical Statistics, vol. 27, pp. 1106–1114.

Pillai, K. C. S. (1964): On the distribution of the largest of seven roots of a matrix in multivariate analysis, *Biometrika,* vol. 51, pp. 270–275.

Pillai, K. C. S. (1965): On the distribution of the largest characteristic root of a matrix in multivariate analysis, *Biometrika,* vol. 52, pp. 405–414.

Pillai, K. C. S. (1967): Upper percentage points of the largest root of a matrix in multivariate analysis, *Biometrika,* vol. 54, pp. 189–194.

Pillai, K. C. S., and C. G. Bantegui (1959): On the distribution of the largest of six roots of a matrix in multivariate analysis, *Biometrika,* vol. 46, pp. 237–240.

Pillai, K. C. S., and T. C. Chang (1968): "An Approximation to the C.D.F. of the Largest Root of a Covariance Matrix," Purdue Department of Statistics Mimeographed Series, no. 184, Purdue University, Lafayette, IN.

Pillai, K. C. S., and T. C. Chang (1970): An approximation to the C.D.F. of the largest root of a covariance matrix, *Annals of the Institute of Statistical Mathematics, Supplement b,* pp. 115–124.

Pillai, K. C. S., and A. K. Gupta (1969): On the exact distribution of Wilks's criterion, *Biometrika,* vol. 56, pp. 109–118.

Pillai, K. C. S., and K. Jayachandran (1967): Power comparisons of tests of two multivariate hypotheses based on four criteria, *Biometrika,* vol. 54, pp. 195–210.

Pillai, K. C. S., and K. Jayachandran (1968): Power comparisons of tests of equality of two covariance matrices based on four criteria, *Biometrika,* vol. 55, pp. 335–342.

Potthoff, R. F., and S. N. Roy (1964): A generalized multivariate analysis of variance model useful especially for growth curve problems, *Biometrika,* vol. 51, pp. 313–326.

Press, S. J. (1972): "Applied Multivariate Analysis," Holt, Rinehart and Winston, Inc., New York.

Rao, C. R. (1955): Estimation and tests of significance in factor analysis, *Psychometrika,* vol. 20, pp. 93–111.

Rao, C. R. (1962): A note on a generalized inverse of a matrix with applications to problems in mathematical statistics, *Journal of the Royal Statistical Society, Series B,* vol. 24, pp. 152–158.

Rao, C. R. (1965): "Linear Statistical Inference and Its Applications," John Wiley & Sons, Inc., New York.

Rao, C. R. (1966): Covariance adjustment and related problems in multivariate analysis, in P. Krishnaiah (ed.), "Multivariate Analysis," pp. 87–103, Academic Press, New York.

Rao, C. R., and S. K. Mitra (1971): "Generalized Inverse of Matrices and Its Applications," John Wiley & Sons, Inc., New York.

Roy, S. N. (1939): p-statistics or some generalizations in analysis of variance appropriate to multivariate problems, *Sankhyā,* vol. 4, pp. 381–396.

Roy, S. N. (1942): The sampling distribution of p-statistics and certain allied statistics on the non-null hypothesis, *Sankhyā,* vol. 6, pp. 15–34.

Roy, S. N. (1945): The individual sampling distribution of the maximum, the minimum, and any intermediate one of the p-statistics on the null hypothesis, *Sankhyā,* vol. 7, pp. 133–158.

Roy, S. N. (1953): On a heuristic method of test construction and its use in multivariate analysis, *Annals of Mathematical Statistics,* vol. 24, pp. 220–238.

Roy, S. N. (1957): "Some Aspects of Multivariate Analysis," John Wiley & Sons, Inc., New York.

Roy, S. N., and R. C. Bose (1953): Simultaneous confidence interval estimation, *Annals of Mathematical Statistics,* vol. 24, pp. 513–536.

Roy, S. N., R. Gnanadesikan, and J. N. Srivastava (1971): "Analysis and Design of Certain Quantitative Multiresponse Experiments," Pergamon Press, Oxford.

Roy, S. N., and A. E. Sarhan (1956): On inverting a class of patterned matrices, *Biometrika,* vol. 43, pp. 227–231.

Saunders, D. R. (1960): A computer program to find the best-fitting orthogonal factors for a given hypothesis, *Psychometrika,* vol. 25, pp. 207–210.

Schatzoff, M. (1966): Exact distributions of Wilks's likelihood-ratio criterion, *Biometrika,* vol. 53, pp. 347–358.

Scheffé, H. (1953): A method for judging all contrasts in the analysis of variance, *Biometrika,* vol. 40, pp. 87–104.

Scheffé, H. (1959): "The Analysis of Variance," John Wiley & Sons, Inc., New York.

Schönemann, P. H. (1970): Fitting a simplex symmetrically, *Psychometrika,* vol. 35, pp. 1–21.

Searle, S. R. (1966): "Matrix Algebra for the Biological Sciences," John Wiley & Sons, Inc., New York.

Searle, S. R. (1971): "Linear Models," John Wiley & Sons, Inc., New York.

Searle, S. R., and W. H. Hausman (1970): "Matrix Algebra for Business and Economics," Interscience-Wiley, New York.

Sinha, R. N., and P. J. Lee (1970): Maximum likelihood factor analysis of natural arthropod infestations in stored grain bulks, *Researches on Population Ecology,* vol. 12, pp. 51–60.

Smith, C. A. B. (1947): Some examples of discrimination, *Annals of Eugenics,* vol. 13, pp. 272–282.

Smith, H., R. Gnanadesikan, and J. B. Hughes (1962): Multivariate analysis of variance (MANOVA), *Biometrics,* vol. 18, pp. 22–41.

Snedecor, G. W., and W. G. Cochran (1967): "Statistical Methods," 6th ed., The Iowa State University Press, Ames.

Solomon, H. (ed.) (1961): "Studies in Item Analysis and Prediction," Stanford University Press, Stanford, CA.

Solomon, H., and B. Rosner (1954): Factor analysis, *Review of Educational Research,* vol. 24, pp. 421–438.

Somerville, D. M. Y. (1958): "An Introduction to the Geometry of N Dimensions," Dover Publications, Inc., New York.

Sorum, M. J. (1971): Estimating the conditional probability of misclassification, *Technometrics,* vol. 13, pp. 333–343.

Spearman, C. (1904): General intelligence objectively determined and measured, *American Journal of Psychology,* vol. 15, pp. 201–293.

Sprent, P. (1972): The mathematics of size and shape, *Biometrics,* vol. 28, pp. 23–37.

Subrahmaniam, K., and K. Subrahmaniam (1973): "Multivariate Analysis: A

Selected and Abstracted Bibliography, 1957–1972," Marcel Dekker, Inc., New York.

Thomson, G. H. (1951): "The Factorial Analysis of Human Ability," University of London Press, Ltd., London.

Thurstone, L. L. (1945): "Multiple Factor Analysis," University of Chicago Press, Chicago.

Timm, N. H. (1970): The estimation of variance-covariance and correlation matrices from incomplete data, *Psychometrika,* vol. 35, pp. 417–437.

Tracy, D. S., and P. S. Dwyer (1969): Multivariate maxima and minima with matrix derivatives, *Journal of the American Statistical Association,* vol. 64, pp. 1576–1594.

Tukey, J. W. (1953): "The Problem of Multiple Comparisons," mimeographed manuscript, Princeton University.

van Alphen, G. W. H. M. (1961): "On Emmetropia and Ametropia," S. Karger, Basel.

Wald, A. (1944): On a statistical problem arising in the classification of an individual into one of two groups, *Annals of Mathematical Statistics,* vol. 15, pp. 145–162.

Weil, A. T., N. E. Zinberg, and J. M. Nelsen (1968): Clinical and psychological effects of marihuana in man, *Science,* vol. 162, pp. 1234–1242.

Weiss, A. (1963): Auditory perception in relation to age, in J. E. Birren *et al.* (eds.), "Human Aging: A Biological and Behavioral Study," pp. 111–140, U.S. Department of Health, Education, and Welfare, Washington.

Welch, B. L. (1939): Note on discriminant functions, *Biometrika,* vol. 31, pp. 218–220.

Wilks, S. S. (1932a): Certain generalizations in the analysis of variance, *Biometrika,* vol. 24, pp. 471–494.

Wilks, S. S. (1932b): Moments and distributions of estimates of population parameters from fragmentary samples, *Annals of Mathematical Statistics,* vol. 3, pp. 163–195.

Wilks, S. S. (1935): On the independence of *k* sets of normally distributed statistical variables, *Econometrica,* vol. 3, pp. 309–326.

Wilks, S. S. (1946): Sample criteria for testing equality of means, equality of variances, and equality of covariances in a normal multivariate distribution, *Annals of Mathematical Statistics,* vol. 17, pp. 257–281.

Wilks, S. S. (1962): "Mathematical Statistics," John Wiley & Sons, Inc., New York.

Williams, E. J. (1970): Comparing means of correlated variates, *Biometrika,* vol. 57, pp. 459–461.

Winer, B. J. (1971): "Statistical Principles in Experimental Design," 2d ed., McGraw-Hill Book Company, New York.

Wishart, J. (1928): The generalized product moment distribution in samples from a normal multivariate population, *Biometrika,* vol. 20, pp. 32–52.

Wishart, J. (1931): The mean and second moment coefficient of the multiple correlation coefficient in samples from a normal population, *Biometrika,* vol. 22, pp. 353–361.

Wolfle, D. (1940): Factor analysis to 1940, *Psychometric Monographs,* no. 3, University of Chicago Press, Chicago.

Wright, S. (1954): The interpretation of multivariate systems, in O. Kempthorne *et al.* (eds.), "Statistics and Mathematics in Biology," pp. 11–33, The Iowa State University Press, Ames.

Yule, G. U., and M. G. Kendall (1950): "An Introducton to the Theory of Statistics," 14th ed., Charles Griffin & Company, Ltd., London.

Zelen, M., and N. C. Severo (1965): Probability functions, in M. Abramowitz and I. A. Stegun (eds.), "Handbook of Mathematical Functions," pp. 925–995, U.S. Department of Commerce, Washington.

APPENDIX: Tables and Charts

Table 1. Cumulative Normal Distribution Function. The values of the standard normal distribution function

$$\Phi(z) = \frac{1}{\sqrt{2\pi}} \int_{-\infty}^{z} e^{-\frac{1}{2}x^2}\, dx$$

are given in the body of the upper part of Table 1. The lower table gives selected upper percentage points z_α, where

$$P(Z > z_\alpha) = \alpha$$

or

$$\Phi(z_\alpha) = 1 - \alpha$$

Table 2. Percentage Points of the Chi-squared Distribution. This table contains the lower and upper percentage points $\chi^2_{\alpha;n}$ of the chi-squared distribution with n degrees of freedom. The cumulative distribution function is

$$F(\chi^2) = \frac{1}{2^{n/2}\Gamma(n/2)} \int_0^{\chi^2} x^{(n-2)/2} e^{-x/2}\, dx$$

and the 100α percentage point is defined as

$$1 - \alpha = F(\chi^2_{\alpha;n})$$

Table 3. Upper Percentage Points of the t Distribution. The cumulative distribution function of the Student-Fisher t variate with n degrees of freedom is

$$F(t) = \frac{\Gamma[(n+1)/2]}{\sqrt{\pi n}\,\Gamma(n/2)} \int_{-\infty}^{t} \frac{dx}{(1 + x^2/n)^{(n+1)/2}}$$

The upper 100α percentage points $t_{\alpha;n}$ defined by

$$1 - \alpha = F(t_{\alpha;n})$$

are given in the body of the table.

Table 4. Upper Percentage Points of the F Distribution. The F variate with degrees of freedom m and n has the cumulative distribution function

$$G(F) = \frac{\Gamma[(m+n)/2]}{\Gamma(m/2)\Gamma(n/2)} \left(\frac{m}{n}\right)^{m/2} \int_0^F x^{(m-2)/2} \left(1 + \frac{m}{n}x\right)^{-(m+n)/2} dx$$

The upper 100α percentage points $F_{\alpha;m,n}$ defined by

$$1 - \alpha = G(F_{\alpha;m,n})$$

are given in the body of the table for various combinations of m and n and common values of $1 - \alpha$.

Table 5. The Fisher z Transformation. Values of the inverse hyperbolic tangent transformation

$$z = \tanh^{-1} r$$

$$= \tfrac{1}{2} \ln \frac{1+r}{1-r}$$

for stabilizing the large-sample variance of the correlation coefficient are given in this table.

Charts 1 to 8. Power Functions of the F Test. The Pearson-Hartley charts of the power functions of tests whose statistics have the F distribution are reproduced as Charts 1 to 8. Each chart corresponds to a value from one to eight of the first degrees-of-freedom parameter ν_1 of the F statistic and contains two families of curves for the values 0.05 and 0.01 of the α level. To determine the power of a test one selects the proper chart and locates the value of the noncentrality parameter ϕ given by the alternative hypothesis on the upper or lower horizontal scale in accordance with the value of α. Next one reads up to the curve associated with the second degrees-of-freedom parameter ν_2 in the appropriate family. The power of the test corresponds to that intersection and is read from the left-hand vertical scale.

Charts 9 to 16 and Tables 6 to 14. Upper Percentage Points of the Distribution of the Largest Characteristic Root. Charts 9 to 16 were prepared by D. Heck to give the upper 0.05 and 0.01 critical values $x_\alpha \equiv x_{\alpha;s,m,n}$ for the largest characteristic root of random matrices encountered in multivariate analysis. A separate chart is required for each combination of the two α levels and the four values $s = 2, \ldots, 5$ of the first parameter. Each chart contains a family of curves corresponding to the values $-\tfrac{1}{2}, 0, 1, 2, \ldots, 10$ of the parameter m. The parameter n is given logarithmically on the vertical axis. To find the percentage point x_α for a given α and some combination of the parameters one selects the chart corresponding to α and s, locates n on the left vertical axis, and reads across horizontally to the curve of the m parameter. If the curve is in the upper family, the critical value x_α is read from the upper scale. If the curve is in the continuation of the family in the lower left portion of the chart, x_α is read from the lower scale. Tables 6 to 14 were compiled by K. C. S. Pillai to give similar percentage points for $s = 6, \ldots, 10, 14, 16, 18, 20$, and various values of m and n.

Table 1. **Cumulative Normal Distribution Function***

z	0.00	0.01	0.02	0.03	0.04	0.05	0.06	0.07	0.08	0.09
0.0	0.5000	0.5040	0.5080	0.5120	0.5160	0.5199	0.5239	0.5279	0.5319	0.5359
0.1	0.5398	0.5438	0.5478	0.5517	0.5557	0.5596	0.5636	0.5675	0.5714	0.5753
0.2	0.5793	0.5832	0.5871	0.5910	0.5948	0.5987	0.6026	0.6064	0.6103	0.6141
0.3	0.6179	0.6217	0.6255	0.6293	0.6331	0.6368	0.6406	0.6443	0.6480	0.6517
0.4	0.6554	0.6591	0.6628	0.6664	0.6700	0.6736	0.6772	0.6808	0.6844	0.6879
0.5	0.6915	0.6950	0.6985	0.7019	0.7054	0.7088	0.7123	0.7157	0.7190	0.7224
0.6	0.7257	0.7291	0.7324	0.7357	0.7389	0.7422	0.7454	0.7486	0.7517	0.7549
0.7	0.7580	0.7611	0.7642	0.7673	0.7704	0.7734	0.7764	0.7794	0.7823	0.7852
0.8	0.7881	0.7910	0.7939	0.7967	0.7995	0.8023	0.8051	0.8078	0.8106	0.8133
0.9	0.8159	0.8186	0.8212	0.8238	0.8264	0.8289	0.8315	0.8340	0.8365	0.8389
1.0	0.8413	0.8438	0.8461	0.8485	0.8508	0.8531	0.8554	0.8577	0.8599	0.8621
1.1	0.8643	0.8665	0.8686	0.8708	0.8729	0.8749	0.8770	0.8790	0.8810	0.8830
1.2	0.8849	0.8869	0.8888	0.8907	0.8925	0.8944	0.8962	0.8980	0.8997	0.9015
1.3	0.9032	0.9049	0.9066	0.9082	0.9099	0.9115	0.9131	0.9147	0.9162	0.9177
1.4	0.9192	0.9207	0.9222	0.9236	0.9251	0.9265	0.9279	0.9292	0.9306	0.9319
1.5	0.9332	0.9345	0.9357	0.9370	0.9382	0.9394	0.9406	0.9418	0.9429	0.9441
1.6	0.9452	0.9463	0.9474	0.9484	0.9495	0.9505	0.9515	0.9525	0.9535	0.9545
1.7	0.9554	0.9564	0.9573	0.9582	0.9591	0.9599	0.9608	0.9616	0.9625	0.9633
1.8	0.9641	0.9649	0.9656	0.9664	0.9671	0.9678	0.9686	0.9693	0.9699	0.9706
1.9	0.9713	0.9719	0.9726	0.9732	0.9738	0.9744	0.9750	0.9756	0.9761	0.9767
2.0	0.9772	0.9778	0.9783	0.9788	0.9793	0.9798	0.9803	0.9808	0.9812	0.9817
2.1	0.9821	0.9826	0.9830	0.9834	0.9838	0.9842	0.9846	0.9850	0.9854	0.9857
2.2	0.9861	0.9864	0.9868	0.9871	0.9875	0.9878	0.9881	0.9884	0.9887	0.9890
2.3	0.9893	0.9896	0.9898	0.9901	0.9904	0.9906	0.9909	0.9911	0.9913	0.9916
2.4	0.9918	0.9920	0.9922	0.9925	0.9927	0.9929	0.9931	0.9932	0.9934	0.9936
2.5	0.9938	0.9940	0.9941	0.9943	0.9945	0.9946	0.9948	0.9949	0.9951	0.9952
2.6	0.9953	0.9955	0.9956	0.9957	0.9959	0.9960	0.9961	0.9962	0.9963	0.9964
2.7	0.9965	0.9966	0.9967	0.9968	0.9969	0.9970	0.9971	0.9972	0.9973	0.9974
2.8	0.9974	0.9975	0.9976	0.9977	0.9977	0.9978	0.9979	0.9979	0.9980	0.9981
2.9	0.9981	0.9982	0.9982	0.9983	0.9984	0.9984	0.9985	0.9985	0.9986	0.9986
3.0	0.9987	0.9987	0.9987	0.9988	0.9988	0.9989	0.9989	0.9989	0.9990	0.9990
3.1	0.9990	0.9991	0.9991	0.9991	0.9992	0.9992	0.9992	0.9992	0.9993	0.9993
3.2	0.9993	0.9993	0.9994	0.9994	0.9994	0.9994	0.9994	0.9995	0.9995	0.9995
3.3	0.9995	0.9995	0.9995	0.9996	0.9996	0.9996	0.9996	0.9996	0.9996	0.9997
3.4	0.9997	0.9997	0.9997	0.9997	0.9997	0.9997	0.9997	0.9997	0.9997	0.9998

z_α	1.282	1.645	1.960	2.326	2.576	3.090	3.291	3.891	4.417
$1 - \alpha = \Phi(z_\alpha)$	0.90	0.95	0.975	0.99	0.995	0.999	0.9995	0.99995	0.999995
2α	0.20	0.10	0.05	0.02	0.01	0.002	0.001	0.0001	0.00001

* Reproduced from A. M. Mood: "Introduction to the Theory of Statistics," McGraw-Hill Book Company, New York, 1950, with the kind permission of the author and publisher.

Table 2. **Percentage Points of the Chi-squared Distribution***

$1-\alpha$ / n	0.005	0.010	0.025	0.050	0.100	0.250	0.500	0.750	0.900	0.950	0.975	0.990	0.995
1	0.0^4393	0.0^3157	0.0^3982	0.0^2393	0.0158	0.102	0.455	1.32	2.71	3.84	5.02	6.63	7.88
2	0.0100	0.0201	0.0506	0.103	0.211	0.575	1.39	2.77	4.61	5.99	7.38	9.21	10.6
3	0.0717	0.115	0.216	0.352	0.584	1.21	2.37	4.11	6.25	7.81	9.35	11.3	12.8
4	0.207	0.297	0.484	0.711	1.06	1.92	3.36	5.39	7.78	9.49	11.1	13.3	14.9
5	0.412	0.554	0.831	1.15	1.61	2.67	4.35	6.63	9.24	11.1	12.8	15.1	16.7
6	0.676	0.872	1.24	1.64	2.20	3.45	5.35	7.84	10.6	12.6	14.4	16.8	18.5
7	0.989	1.24	1.69	2.17	2.83	4.25	6.35	9.04	12.0	14.1	16.0	18.5	20.3
8	1.34	1.65	2.18	2.73	3.49	5.07	7.34	10.2	13.4	15.5	17.5	20.1	22.0
9	1.73	2.09	2.70	3.33	4.17	5.90	8.34	11.4	14.7	16.9	19.0	21.7	23.6
10	2.16	2.56	3.25	3.94	4.87	6.74	9.34	12.5	16.0	18.3	20.5	23.2	25.2
11	2.60	3.05	3.82	4.57	5.58	7.58	10.3	13.7	17.3	19.7	21.9	24.7	26.8
12	3.07	3.57	4.40	5.23	6.30	8.44	11.3	14.8	18.5	21.0	23.3	26.2	28.3
13	3.57	4.11	5.01	5.89	7.04	9.30	12.3	16.0	19.8	22.4	24.7	27.7	29.8
14	4.07	4.66	5.63	6.57	7.79	10.2	13.3	17.1	21.1	23.7	26.1	29.1	31.3
15	4.60	5.23	6.26	7.26	8.55	11.0	14.3	18.2	22.3	25.0	27.5	30.6	32.8
16	5.14	5.81	6.91	7.96	9.31	11.9	15.3	19.4	23.5	26.3	28.8	32.0	34.3
17	5.70	6.41	7.56	8.67	10.1	12.8	16.3	20.5	24.8	27.6	30.2	33.4	35.7
18	6.26	7.01	8.23	9.39	10.9	13.7	17.3	21.6	26.0	28.9	31.5	34.8	37.2
19	6.84	7.63	8.91	10.1	11.7	14.6	18.3	22.7	27.2	30.1	32.9	36.2	38.6
20	7.43	8.26	9.59	10.9	12.4	15.5	19.3	23.8	28.4	31.4	34.2	37.6	40.0
21	8.03	8.90	10.3	11.6	13.2	16.3	20.3	24.9	29.6	32.7	35.5	38.9	41.4
22	8.64	9.54	11.0	12.3	14.0	17.2	21.3	26.0	30.8	33.9	36.8	40.3	42.8
23	9.26	10.2	11.7	13.1	14.8	18.1	22.3	27.1	32.0	35.2	38.1	41.6	44.2
24	9.89	10.9	12.4	13.8	15.7	19.0	23.3	28.2	33.2	36.4	39.4	43.0	45.6
25	10.5	11.5	13.1	14.6	16.5	19.9	24.3	29.3	34.4	37.7	40.6	44.3	46.9
26	11.2	12.2	13.8	15.4	17.3	20.8	25.3	30.4	35.6	38.9	41.9	45.6	48.3
27	11.8	12.9	14.6	16.2	18.1	21.7	26.3	31.5	36.7	40.1	43.2	47.0	49.6
28	12.5	13.6	15.3	16.9	18.9	22.7	27.3	32.6	37.9	41.3	44.5	48.3	51.0
29	13.1	14.3	16.0	17.7	19.8	23.6	28.3	33.7	39.1	42.6	45.7	49.6	52.3
30	13.8	15.0	16.8	18.5	20.6	24.5	29.3	34.8	40.3	43.8	47.0	50.9	53.7

* Abridged from Catherine M. Thompson: Tables of percentage points of the incomplete beta function and of the chi-square distribution, *Biometrika*, vol. 32 (1941), pp. 187–191, and published here with the kind permission of the editor of *Biometrika*.

Table 3. **Upper Percentage Points of the t Distribution***

$1-\alpha$ / n	0.75	0.90	0.95	0.975	0.99	0.995	0.9995
1	1.000	3.078	6.314	12.706	31.821	63.657	636.619
2	0.816	1.886	2.920	4.303	6.965	9.925	31.598
3	0.765	1.638	2.353	3.182	4.541	5.841	12.941
4	0.741	1.533	2.132	2.776	3.747	4.604	8.610
5	0.727	1.476	2.015	2.571	3.365	4.032	6.859
6	0.718	1.440	1.943	2.447	3.143	3.707	5.959
7	0.711	1.415	1.895	2.365	2.998	3.499	5.405
8	0.706	1.397	1.860	2.306	2.896	3.355	5.041
9	0.703	1.383	1.833	2.262	2.821	3.250	4.781
10	0.700	1.372	1.812	2.228	2.764	3.169	4.587
11	0.697	1.363	1.796	2.201	2.718	3.106	4.437
12	0.695	1.356	1.782	2.179	2.681	3.055	4.318
13	0.694	1.350	1.771	2.160	2.650	3.012	4.221
14	0.692	1.345	1.761	2.145	2.624	2.977	4.140
15	0.691	1.341	1.753	2.131	2.602	2.947	4.073
16	0.690	1.337	1.746	2.120	2.583	2.921	4.015
17	0.689	1.333	1.740	2.110	2.567	2.898	3.965
18	0.688	1.330	1.734	2.101	2.552	2.878	3.922
19	0.688	1.328	1.729	2.093	2.539	2.861	3.883
20	0.687	1.325	1.725	2.086	2.528	2.845	3.850
21	0.686	1.323	1.721	2.080	2.518	2.831	3.819
22	0.686	1.321	1.717	2.074	2.508	2.819	3.792
23	0.685	1.319	1.714	2.069	2.500	2.807	3.767
24	0.685	1.318	1.711	2.064	2.492	2.797	3.745
25	0.684	1.316	1.708	2.060	2.485	2.787	3.725
26	0.684	1.315	1.706	2.056	2.479	2.779	3.707
27	0.684	1.314	1.703	2.052	2.473	2.771	3.690
28	0.683	1.313	1.701	2.048	2.467	2.763	3.674
29	0.683	1.311	1.699	2.045	2.462	2.756	3.659
30	0.683	1.310	1.697	2.042	2.457	2.750	3.646
40	0.681	1.303	1.684	2.021	2.423	2.704	3.551
60	0.679	1.296	1.671	2.000	2.390	2.660	3.460
120	0.677	1.289	1.658	1.980	2.358	2.617	3.373
∞	0.674	1.282	1.645	1.960	2.326	2.576	3.291

* Taken from Table III of R. A. Fisher and F. Yates: "Statistical Tables for Biological, Agricultural, and Medical Research," published by Oliver & Boyd Ltd., Edinburgh, by permission of the authors and publishers.

Table 4. Upper Percentage Points of the F Distribution*

$1-\alpha$	n \ m	1	2	3	4	5	6	7	8	9	10	12	15	20	30	60	120	∞
0.90		39.9	49.5	53.6	55.8	57.2	58.2	58.9	59.4	59.9	60.2	60.7	61.2	61.7	62.3	62.8	63.1	63.3
0.95		161	200	216	225	230	234	237	239	241	242	244	246	248	250	252	253	254
0.975	1	648	800	864	900	922	937	948	957	963	969	977	985	993	1,000	1,010	1,010	1,020
0.99		4,050	5,000	5,400	5,620	5,760	5,860	5,930	5,980	6,020	6,060	6,110	6,160	6,210	6,260	6,310	6,340	6,370
0.995		16,200	20,000	21,600	22,500	23,100	23,400	23,700	23,900	24,100	24,200	24,400	24,600	24,800	25,000	25,200	25,400	25,500
0.90		8.53	9.00	9.16	9.24	9.29	9.33	9.35	9.37	9.38	9.39	9.41	9.42	9.44	9.46	9.47	9.48	9.49
0.95		18.5	19.0	19.2	19.2	19.3	19.3	19.4	19.4	19.4	19.4	19.4	19.4	19.5	19.5	19.5	19.5	19.5
0.975	2	38.5	39.0	39.2	39.2	39.3	39.3	39.4	39.4	39.4	39.4	39.4	39.4	39.4	39.5	39.5	39.5	39.5
0.99		98.5	99.0	99.2	99.2	99.3	99.3	99.4	99.4	99.4	99.4	99.4	99.4	99.4	99.5	99.5	99.5	99.5
0.995		199	199	199	199	199	199	199	199	199	199	199	199	199	199	199	199	199
0.90		5.54	5.46	5.39	5.34	5.31	5.28	5.27	5.25	5.24	5.23	5.22	5.20	5.18	5.17	5.15	5.14	5.13
0.95		10.1	9.55	9.28	9.12	9.01	8.94	8.89	8.85	8.81	8.79	8.74	8.70	8.66	8.62	8.57	8.55	8.53
0.975	3	17.4	16.0	15.4	15.1	14.9	14.7	14.6	14.5	14.5	14.4	14.3	14.3	14.2	14.1	14.0	13.9	13.9
0.99		34.1	30.8	29.5	28.7	28.2	27.9	27.7	27.5	27.3	27.2	27.1	26.9	26.7	26.5	26.3	26.2	26.1
0.995		55.6	49.8	47.5	46.2	45.4	44.8	44.4	44.1	43.9	43.7	43.4	43.1	42.8	42.5	42.1	42.0	41.8
0.90		4.54	4.32	4.19	4.11	4.05	4.01	3.98	3.95	3.93	3.92	3.90	3.87	3.84	3.82	3.79	3.78	3.76
0.95		7.71	6.94	6.59	6.39	6.26	6.16	6.09	6.04	6.00	5.96	5.91	5.86	5.80	5.75	5.69	5.66	5.63
0.975	4	12.2	10.6	9.98	9.60	9.36	9.20	9.07	8.98	8.90	8.84	8.75	8.66	8.56	8.46	8.36	8.31	8.26
0.99		21.2	18.0	16.7	16.0	15.5	15.2	15.0	14.8	14.7	14.5	14.4	14.2	14.0	13.8	13.7	13.6	13.5
0.995		31.3	26.3	24.3	23.2	22.5	22.0	21.6	21.4	21.1	21.0	20.7	20.4	20.2	19.9	19.6	19.5	19.3
0.90		4.06	3.78	3.62	3.52	3.45	3.40	3.37	3.34	3.32	3.30	3.27	3.24	3.21	3.17	3.14	3.12	3.11
0.95		6.61	5.79	5.41	5.19	5.05	4.95	4.88	4.82	4.77	4.74	4.68	4.62	4.56	4.50	4.43	4.40	4.37
0.975	5	10.0	8.43	7.76	7.39	7.15	6.98	6.85	6.76	6.68	6.62	6.52	6.43	6.33	6.23	6.12	6.07	6.02
0.99		16.3	13.3	12.1	11.4	11.0	10.7	10.5	10.3	10.2	10.1	9.89	9.72	9.55	9.38	9.20	9.11	9.02
0.995		22.8	18.3	16.5	15.6	14.9	14.5	14.2	14.0	13.8	13.6	13.4	13.1	12.9	12.7	12.4	12.3	12.1
0.90		3.78	3.46	3.29	3.18	3.11	3.05	3.01	2.98	2.96	2.94	2.90	2.87	2.84	2.80	2.76	2.74	2.72
0.95		5.99	5.14	4.76	4.53	4.39	4.28	4.21	4.15	4.10	4.06	4.00	3.94	3.87	3.81	3.74	3.70	3.67
0.975	6	8.81	7.26	6.60	6.23	5.99	5.82	5.70	5.60	5.52	5.46	5.37	5.27	5.17	5.07	4.96	4.90	4.85
0.99		13.7	10.9	9.78	9.15	8.75	8.47	8.26	8.10	7.98	7.87	7.72	7.56	7.40	7.23	7.06	6.97	6.88
0.995		18.6	14.5	12.9	12.0	11.5	11.1	10.8	10.6	10.4	10.2	10.0	9.81	9.59	9.36	9.12	9.00	8.88
0.90		3.59	3.26	3.07	2.96	2.88	2.83	2.78	2.75	2.72	2.70	2.67	2.63	2.59	2.56	2.51	2.49	2.47
0.95		5.59	4.74	4.35	4.12	3.97	3.87	3.79	3.73	3.68	3.64	3.57	3.51	3.44	3.38	3.30	3.27	3.23
0.975	7	8.07	6.54	5.89	5.52	5.29	5.12	4.99	4.90	4.82	4.76	4.67	4.57	4.47	4.36	4.25	4.20	4.14
0.99		12.2	9.55	8.45	7.85	7.46	7.19	6.99	6.84	6.72	6.62	6.47	6.31	6.16	5.99	5.82	5.74	5.65
0.995		16.2	12.4	10.9	10.1	9.52	9.16	8.89	8.68	8.51	8.38	8.18	7.97	7.75	7.53	7.31	7.19	7.08
0.90		3.46	3.11	2.92	2.81	2.73	2.67	2.62	2.59	2.56	2.54	2.50	2.46	2.42	2.38	2.34	2.31	2.29
0.95		5.32	4.46	4.07	3.84	3.69	3.58	3.50	3.44	3.39	3.35	3.28	3.22	3.15	3.08	3.01	2.97	2.93
0.975	8	7.57	6.06	5.42	5.05	4.82	4.65	4.53	4.43	4.36	4.30	4.20	4.10	4.00	3.89	3.78	3.73	3.67
0.99		11.3	8.65	7.59	7.01	6.63	6.37	6.18	6.03	5.91	5.81	5.67	5.52	5.36	5.20	5.03	4.95	4.86
0.995		14.7	11.0	9.60	8.81	8.30	7.95	7.69	7.50	7.34	7.21	7.01	6.81	6.61	6.40	6.18	6.06	5.95

0.90		3.36	3.01	2.81	2.69	2.61	2.55	2.51	2.47	2.44	2.42	2.38	2.34	2.30	2.25	2.21	2.18	2.16
0.95		5.12	4.26	3.86	3.63	3.48	3.37	3.29	3.23	3.18	3.14	3.07	3.01	2.94	2.86	2.79	2.75	2.71
0.975	9	7.21	5.71	5.08	4.72	4.48	4.32	4.20	4.10	4.03	3.96	3.87	3.77	3.67	3.56	3.45	3.39	3.33
0.99		10.6	8.02	6.99	6.42	6.06	5.80	5.61	5.47	5.35	5.26	5.11	4.96	4.81	4.65	4.48	4.40	4.31
0.995		13.6	10.1	8.72	7.96	7.47	7.13	6.88	6.69	6.54	6.42	6.23	6.03	5.83	5.62	5.41	5.30	5.19
0.90		3.29	2.92	2.73	2.61	2.52	2.46	2.41	2.38	2.35	2.32	2.28	2.24	2.20	2.15	2.11	2.08	2.06
0.95		4.96	4.10	3.71	3.48	3.33	3.22	3.14	3.07	3.02	2.98	2.91	2.84	2.77	2.70	2.62	2.58	2.54
0.975	10	6.94	5.46	4.83	4.47	4.24	4.07	3.95	3.85	3.78	3.72	3.62	3.52	3.42	3.31	3.20	3.14	3.08
0.99		10.0	7.56	6.55	5.99	5.64	5.39	5.20	5.06	4.94	4.85	4.71	4.56	4.41	4.25	4.08	4.00	3.91
0.995		12.8	9.43	8.08	7.34	6.87	6.54	6.30	6.12	5.97	5.85	5.66	5.47	5.27	5.07	4.86	4.75	4.64
0.90		3.18	2.81	2.61	2.48	2.39	2.33	2.28	2.24	2.21	2.19	2.15	2.10	2.06	2.01	1.96	1.93	1.90
0.95		4.75	3.89	3.49	3.26	3.11	3.00	2.91	2.85	2.80	2.75	2.69	2.62	2.54	2.47	2.38	2.34	2.30
0.975	12	6.55	5.10	4.47	4.12	3.89	3.73	3.61	3.51	3.44	3.37	3.28	3.18	3.07	2.96	2.85	2.79	2.72
0.99		9.33	6.93	5.95	5.41	5.06	4.82	4.64	4.50	4.39	4.30	4.16	4.01	3.86	3.70	3.54	3.45	3.36
0.995		11.8	8.51	7.23	6.52	6.07	5.76	5.52	5.35	5.20	5.09	4.91	4.72	4.53	4.33	4.12	4.01	3.90
0.90		3.07	2.70	2.49	2.36	2.27	2.21	2.16	2.12	2.09	2.06	2.02	1.97	1.92	1.87	1.82	1.79	1.76
0.95		4.54	3.68	3.29	3.06	2.90	2.79	2.71	2.64	2.59	2.54	2.48	2.40	2.33	2.25	2.16	2.11	2.07
0.975	15	6.20	4.77	4.15	3.80	3.58	3.41	3.29	3.20	3.12	3.06	2.96	2.86	2.76	2.64	2.52	2.46	2.40
0.99		8.68	6.36	5.42	4.89	4.56	4.32	4.14	4.00	3.89	3.80	3.67	3.52	3.37	3.21	3.05	2.96	2.87
0.995		10.8	7.70	6.48	5.80	5.37	5.07	4.85	4.67	4.54	4.42	4.25	4.07	3.88	3.69	3.48	3.37	3.26
0.90		2.97	2.59	2.38	2.25	2.16	2.09	2.04	2.00	1.96	1.94	1.89	1.84	1.79	1.74	1.68	1.64	1.61
0.95		4.35	3.49	3.10	2.87	2.71	2.60	2.51	2.45	2.39	2.35	2.28	2.20	2.12	2.04	1.95	1.90	1.84
0.975	20	5.87	4.46	3.86	3.51	3.29	3.13	3.01	2.91	2.84	2.77	2.68	2.57	2.46	2.35	2.22	2.16	2.09
0.99		8.10	5.85	4.94	4.43	4.10	3.87	3.70	3.56	3.46	3.37	3.23	3.09	2.94	2.78	2.61	2.52	2.42
0.995		9.94	6.99	5.82	5.17	4.76	4.47	4.26	4.09	3.96	3.85	3.68	3.50	3.32	3.12	2.92	2.81	2.69
0.90		2.88	2.49	2.28	2.14	2.05	1.98	1.93	1.88	1.85	1.82	1.77	1.72	1.67	1.61	1.54	1.50	1.46
0.95		4.17	3.32	2.92	2.69	2.53	2.42	2.33	2.27	2.21	2.16	2.09	2.01	1.93	1.84	1.74	1.68	1.62
0.975	30	5.57	4.18	3.59	3.25	3.03	2.87	2.75	2.65	2.57	2.51	2.41	2.31	2.20	2.07	1.94	1.87	1.79
0.99		7.56	5.39	4.51	4.02	3.70	3.47	3.30	3.17	3.07	2.98	2.84	2.70	2.55	2.39	2.21	2.11	2.01
0.995		9.18	6.35	5.24	4.62	4.23	3.95	3.74	3.58	3.45	3.34	3.18	3.01	2.82	2.63	2.42	2.30	2.18
0.90		2.79	2.39	2.18	2.04	1.95	1.87	1.82	1.77	1.74	1.71	1.66	1.60	1.54	1.48	1.40	1.35	1.29
0.95		4.00	3.15	2.76	2.53	2.37	2.25	2.17	2.10	2.04	1.99	1.92	1.84	1.75	1.65	1.53	1.47	1.39
0.975	60	5.29	3.93	3.34	3.01	2.79	2.63	2.51	2.41	2.33	2.27	2.17	2.06	1.94	1.82	1.67	1.58	1.48
0.99		7.08	4.98	4.13	3.65	3.34	3.12	2.95	2.82	2.72	2.63	2.50	2.35	2.20	2.03	1.84	1.73	1.60
0.995		8.49	5.80	4.73	4.14	3.76	3.49	3.29	3.13	3.01	2.90	2.74	2.57	2.39	2.19	1.96	1.83	1.69
0.90		2.75	2.35	2.13	1.99	1.90	1.82	1.77	1.72	1.68	1.65	1.60	1.54	1.48	1.41	1.32	1.26	1.19
0.95		3.92	3.07	2.68	2.45	2.29	2.18	2.09	2.02	1.96	1.91	1.83	1.75	1.66	1.55	1.43	1.35	1.25
0.975	120	5.15	3.80	3.23	2.89	2.67	2.52	2.39	2.30	2.22	2.16	2.05	1.94	1.82	1.69	1.53	1.43	1.31
0.99		6.85	4.79	3.95	3.48	3.17	2.96	2.79	2.66	2.56	2.47	2.34	2.19	2.03	1.86	1.66	1.53	1.38
0.995		8.18	5.54	4.50	3.92	3.55	3.28	3.09	2.93	2.81	2.71	2.54	2.37	2.19	1.98	1.75	1.61	1.43
0.90		2.71	2.30	2.08	1.94	1.85	1.77	1.72	1.67	1.63	1.60	1.55	1.49	1.42	1.34	1.24	1.17	1.00
0.95		3.84	3.00	2.60	2.37	2.21	2.10	2.01	1.94	1.88	1.83	1.75	1.67	1.57	1.46	1.32	1.22	1.00
0.975	∞	5.02	3.69	3.12	2.79	2.57	2.41	2.29	2.19	2.11	2.05	1.94	1.83	1.71	1.57	1.39	1.27	1.00
0.99		6.63	4.61	3.78	3.32	3.02	2.80	2.64	2.51	2.41	2.32	2.18	2.04	1.88	1.70	1.47	1.32	1.00
0.995		7.88	5.30	4.28	3.72	3.35	3.09	2.90	2.74	2.62	2.52	2.36	2.19	2.00	1.79	1.53	1.36	1.00

* Abridged from Maxine Merrington and Catherine M. Thompson: Tables of percentage points of the inverted beta distribution, *Biometrika*, vol. 33 (1943), pp. 73–88, and published here with the kind permission of the editor of *Biometrika*.

Table 5. **The Fisher z Transformation***

r	0.00	0.01	0.02	0.03	0.04	0.05	0.06	0.07	0.08	0.09
0.0	0.0000	0.0100	0.0200	0.0300	0.0400	0.0500	0.0601	0.0701	0.0802	0.0902
0.1	0.1003	0.1104	0.1206	0.1307	0.1409	0.1511	0.1614	0.1717	0.1820	0.1923
0.2	0.2027	0.2132	0.2237	0.2342	0.2448	0.2554	0.2661	0.2769	0.2877	0.2986
0.3	0.3095	0.3205	0.3316	0.3428	0.3541	0.3654	0.3769	0.3884	0.4001	0.4118
0.4	0.4236	0.4356	0.4477	0.4599	0.4722	0.4847	0.4973	0.5101	0.5230	0.5361
0.5	0.5493	0.5627	0.5763	0.5901	0.6042	0.6184	0.6328	0.6475	0.6625	0.6777
0.6	0.6931	0.7089	0.7250	0.7414	0.7582	0.7753	0.7928	0.8107	0.8291	0.8480
0.7	0.8673	0.8872	0.9076	0.9287	0.9505	0.9730	0.9962	1.0203	1.0454	1.0714
0.8	1.0986	1.1270	1.1568	1.1881	1.2212	1.2562	1.2933	1.3331	1.3758	1.4219
0.9	1.4722	1.5275	1.5890	1.6584	1.7380	1.8318	1.9459	2.0923	2.2976	2.6467

* Reproduced from M. Abramowitz and I. A. Stegun (eds.): "Handbook of Mathematical Functions," p. 221, U.S. Department of Commerce, Washington, D.C., 1964. The table was compiled from Harvard Computation Laboratory, "Tables of Inverse Hyperbolic Functions," Harvard University Press, Cambridge, MA, 1949, and is reprinted with the permission of the Harvard University Press.

Chart 1

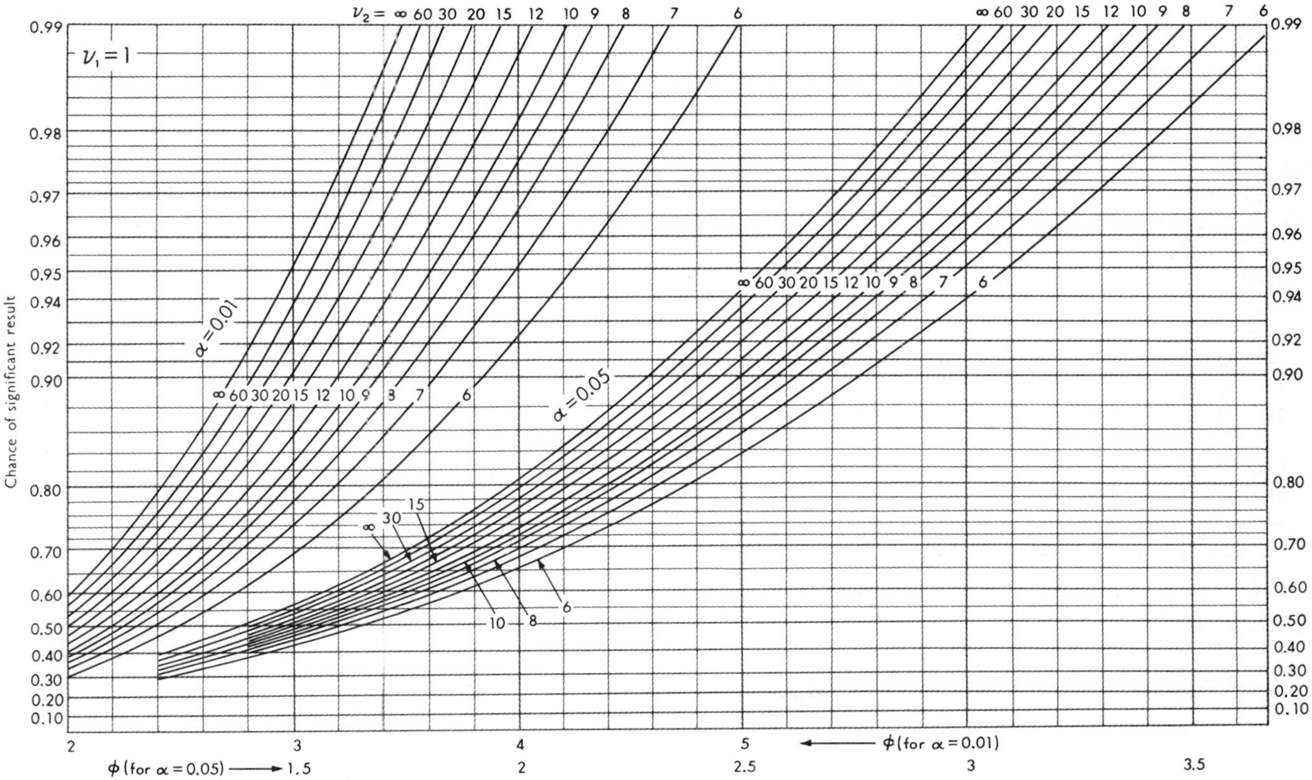

Reprinted from E. S. Pearson and H. O. Hartley: Charts of the power function of the analysis of variance tests, derived from the non-central *F*-distribution, *Biometrika*, vol. 38 (1951), pp. 112–130, with the permission of the authors and the publisher.

Chart 2

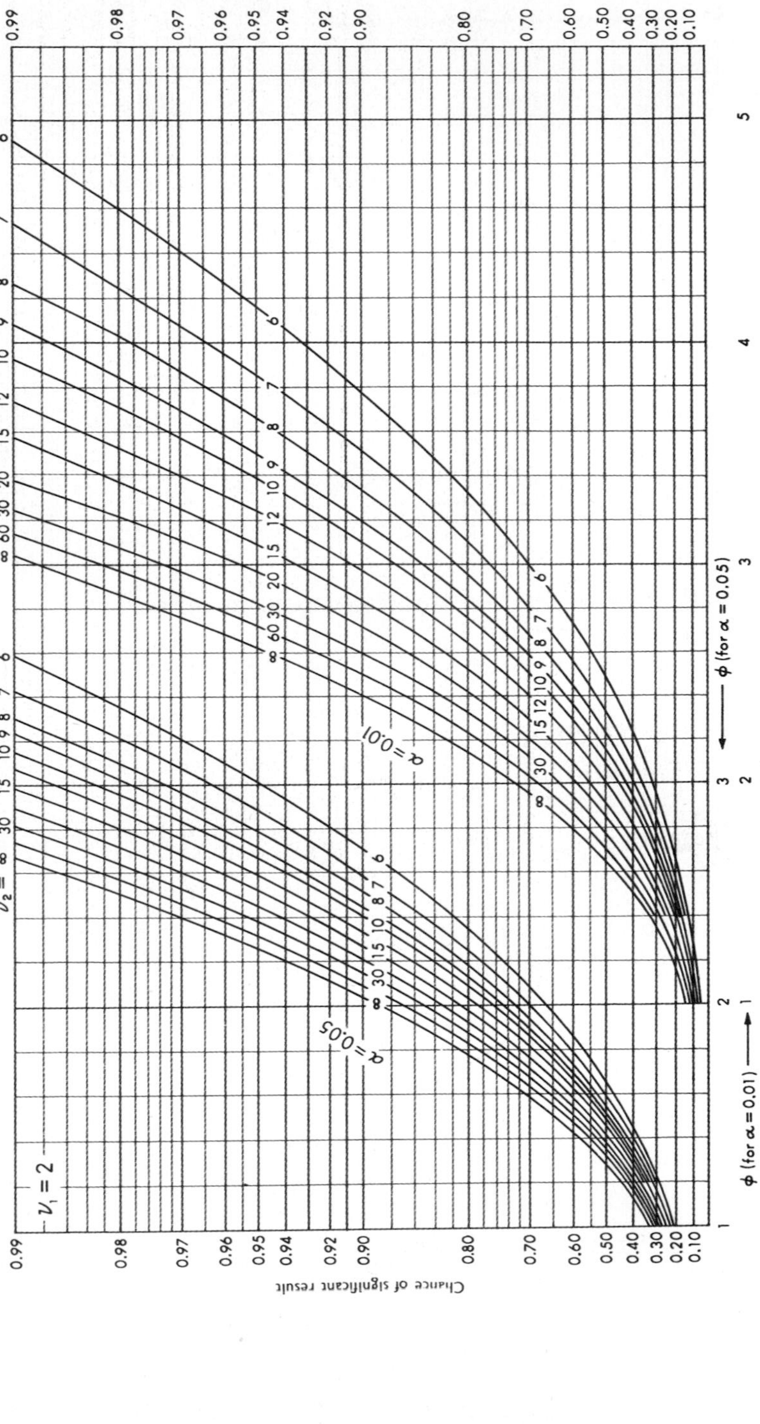

$\nu_1 = 2$
$\nu_2 = \infty$ 60 30 20 15 12 10 9 8 7 6
$\alpha = 0.05$
$\alpha = 0.01$
Chance of significant result
0.99 0.98 0.97 0.96 0.95 0.94 0.92 0.90 0.80 0.70 0.60 0.50 0.40 0.30 0.20 0.10
ϕ (for $\alpha = 0.01$) ⟶
⟵ ϕ (for $\alpha = 0.05$)
1 2 3 4 5
1 2 3

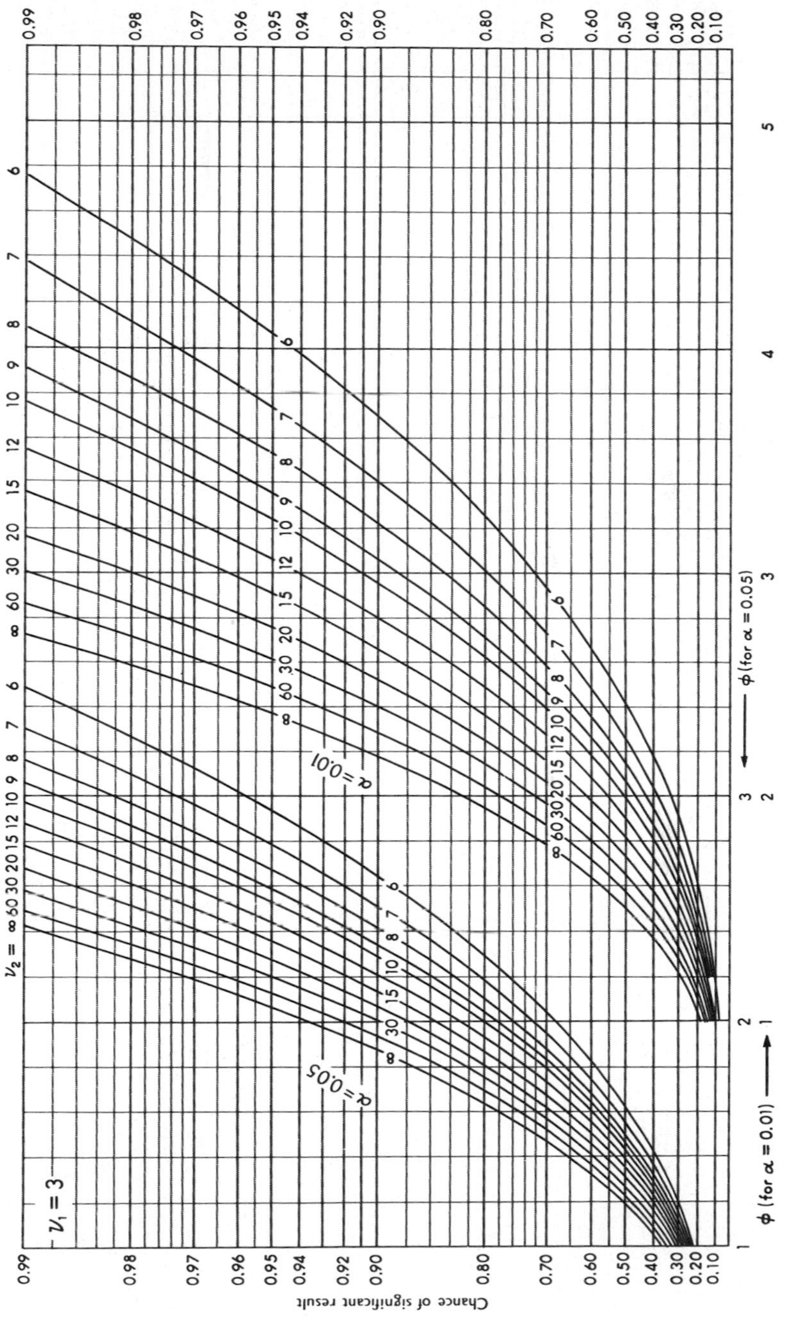

$\nu_1 = 3$
$\nu_2 = \infty$ 60 30 20 15 12 10 9 8 7 6
∞ 60 30 20 15 12 10 9 8 7 6
α = 0.05
α = 0.01
Chance of significant result
0.99 0.98 0.97 0.96 0.95 0.94 0.92 0.90 0.80 0.70 0.60 0.50 0.40 0.30 0.20 0.10
φ (for α = 0.01) ⟶
⟵ φ (for α = 0.05)
1 2 3
1 2 3 4 5

Chart 3

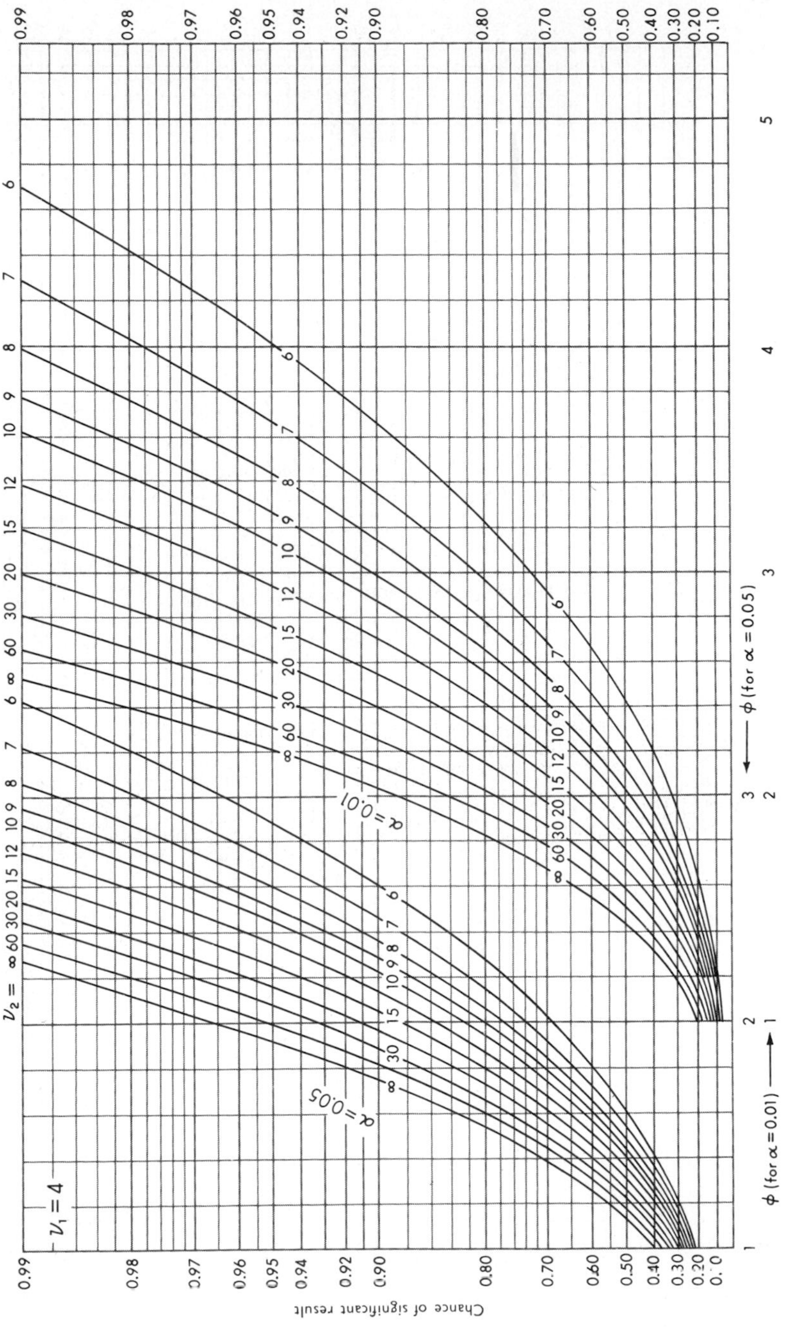
$\nu_1 = 4$
$\nu_2 =$ ∞ 60 30 20 15 12 10 9 8 7 6 ∞ 60 30 20 15 12 10 9 8 7 6
α = 0.01
α = 0.05
Chance of significant result
0.99 0.98 0.97 0.96 0.95 0.94 0.92 0.90 0.80 0.70 0.60 0.50 0.40 0.30 0.20 0.10
φ (for α = 0.01) ⟶
1 2 3 4 5
⟵ φ (for α = 0.05)
1 2 3

Chart 4

Chart 5

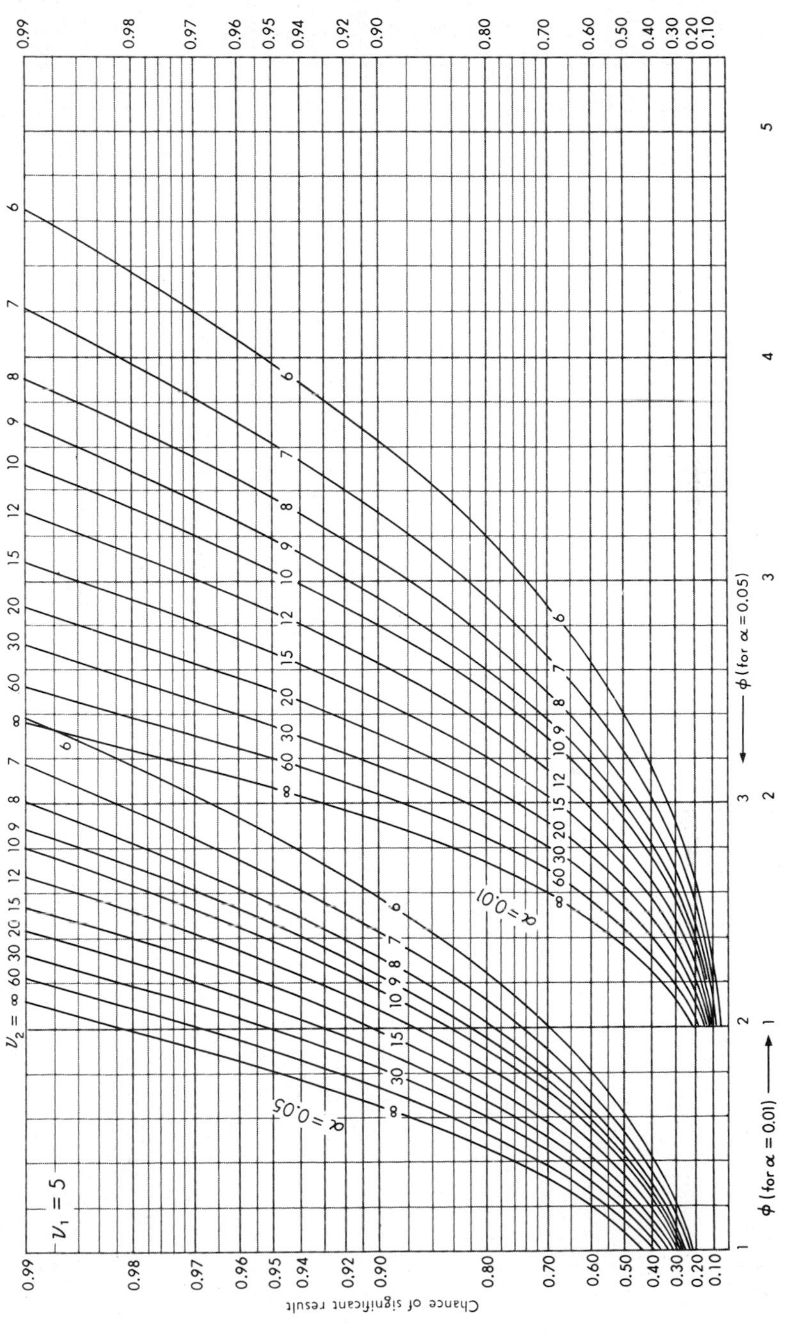
$\nu_1 = 5$
$\nu_2 = \infty$ 60 30 20 15 12 10 9 8 7 6
$\alpha = 0.05$
$\alpha = 0.01$
Chance of significant result
0.99 0.98 0.97 0.96 0.95 0.94 0.92 0.90 0.80 0.70 0.60 0.50 0.40 0.30 0.20 0.10
ϕ (for $\alpha = 0.01$)
ϕ (for $\alpha = 0.05$)
1 2 3 4 5

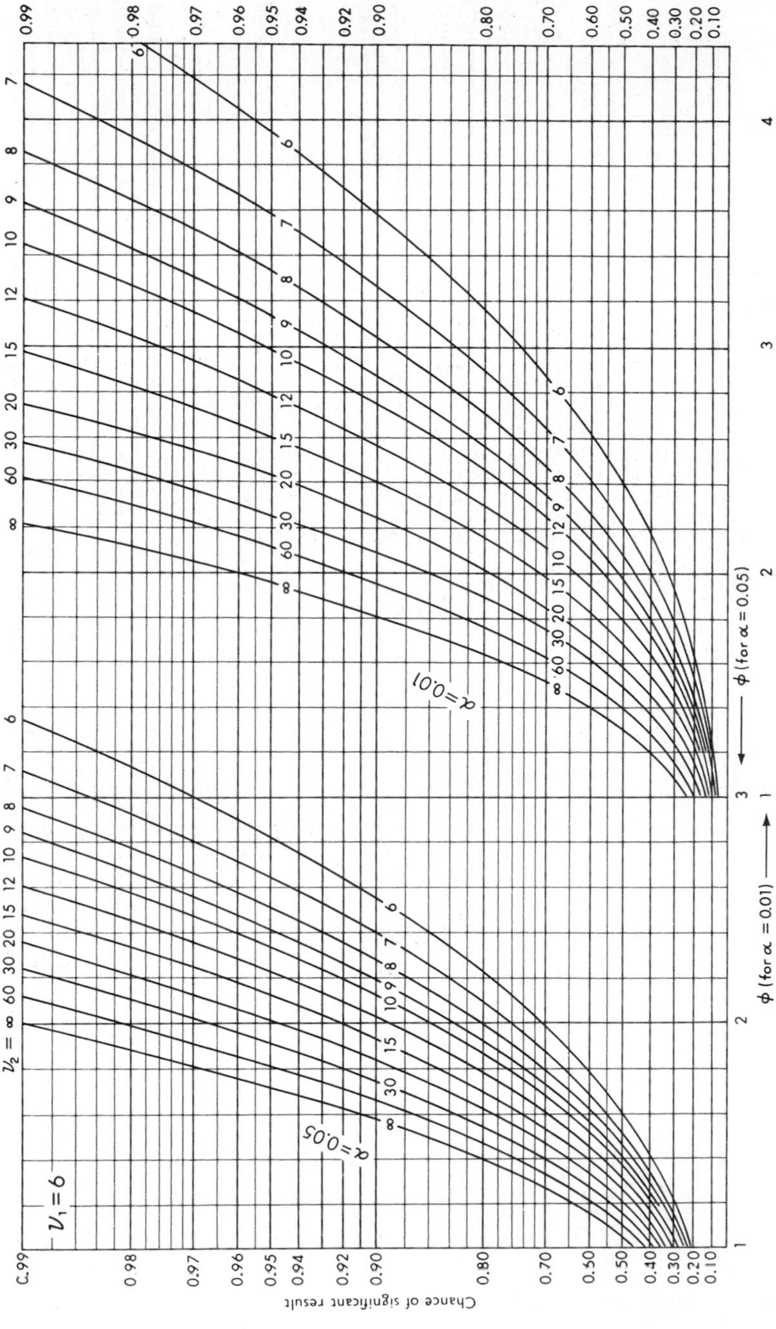

$\nu_1 = 6$
$\nu_2 = \infty$ 60 30 20 15 12 10 9 8 7 6
$\alpha = 0.05$
$\alpha = 0.01$
ϕ (for $\alpha = 0.01$)
ϕ (for $\alpha = 0.05$)
Chance of significant result

Chart 6

Chart 7

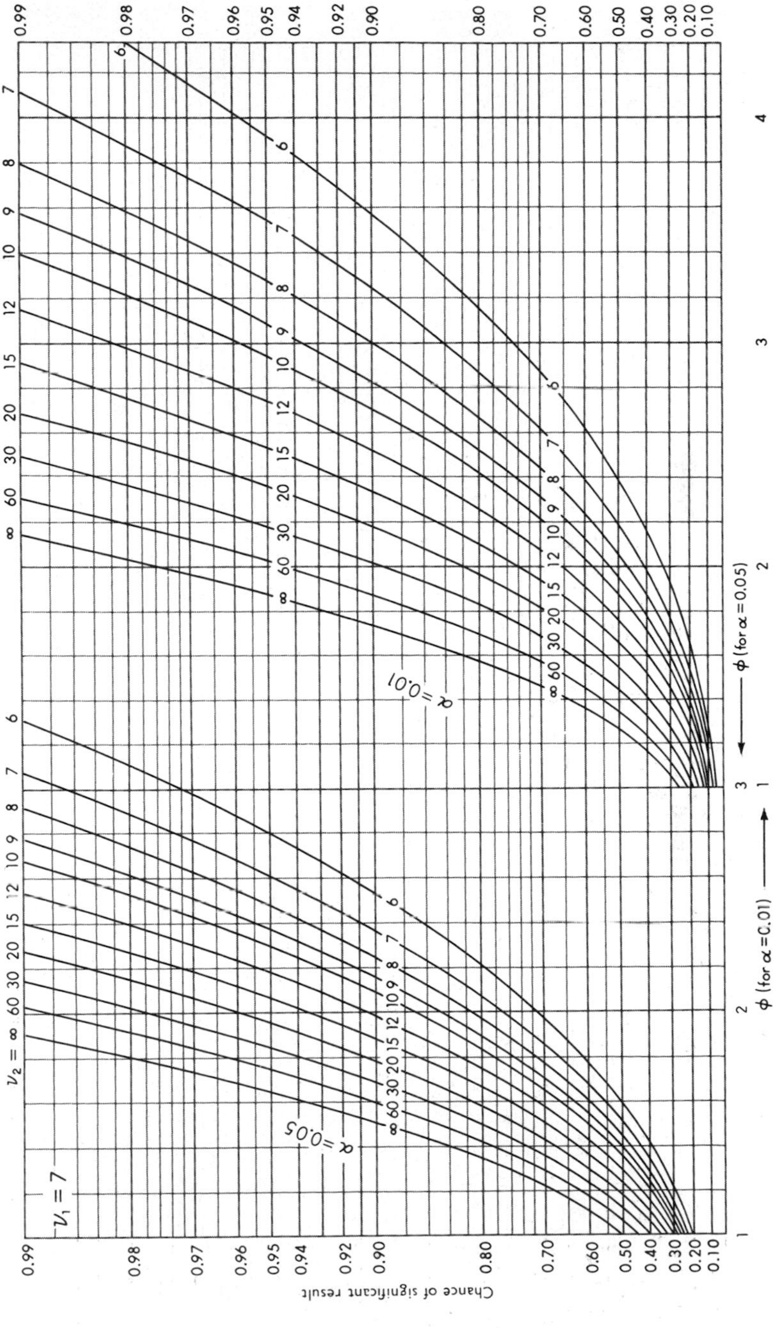
$\nu_1 = 7$
$\nu_2 = \infty$ 60 30 20 15 12 10 9 8 7 6
$\alpha = 0.05$
$\alpha = 0.01$
ϕ (for $\alpha = 0.01$)
ϕ (for $\alpha = 0.05$)
Chance of significant result
0.99 0.98 0.97 0.96 0.95 0.94 0.92 0.90 0.80 0.70 0.60 0.50 0.40 0.30 0.20 0.10

Chart 8

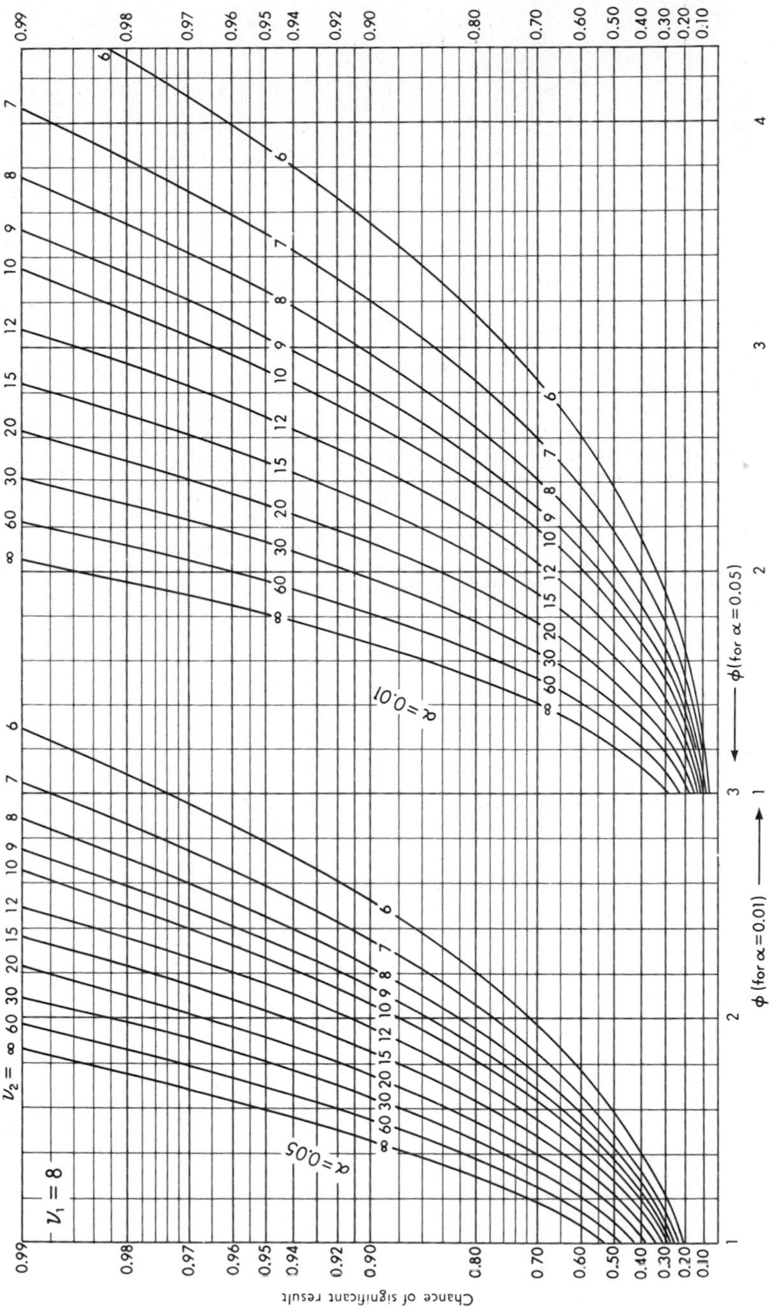

$\nu_1 = 8$
$\nu_2 =$
$\alpha = 0.05$
$\alpha = 0.01$
ϕ (for $\alpha = 0.01$)
ϕ (for $\alpha = 0.05$)
Chance of significant result

Chart 9

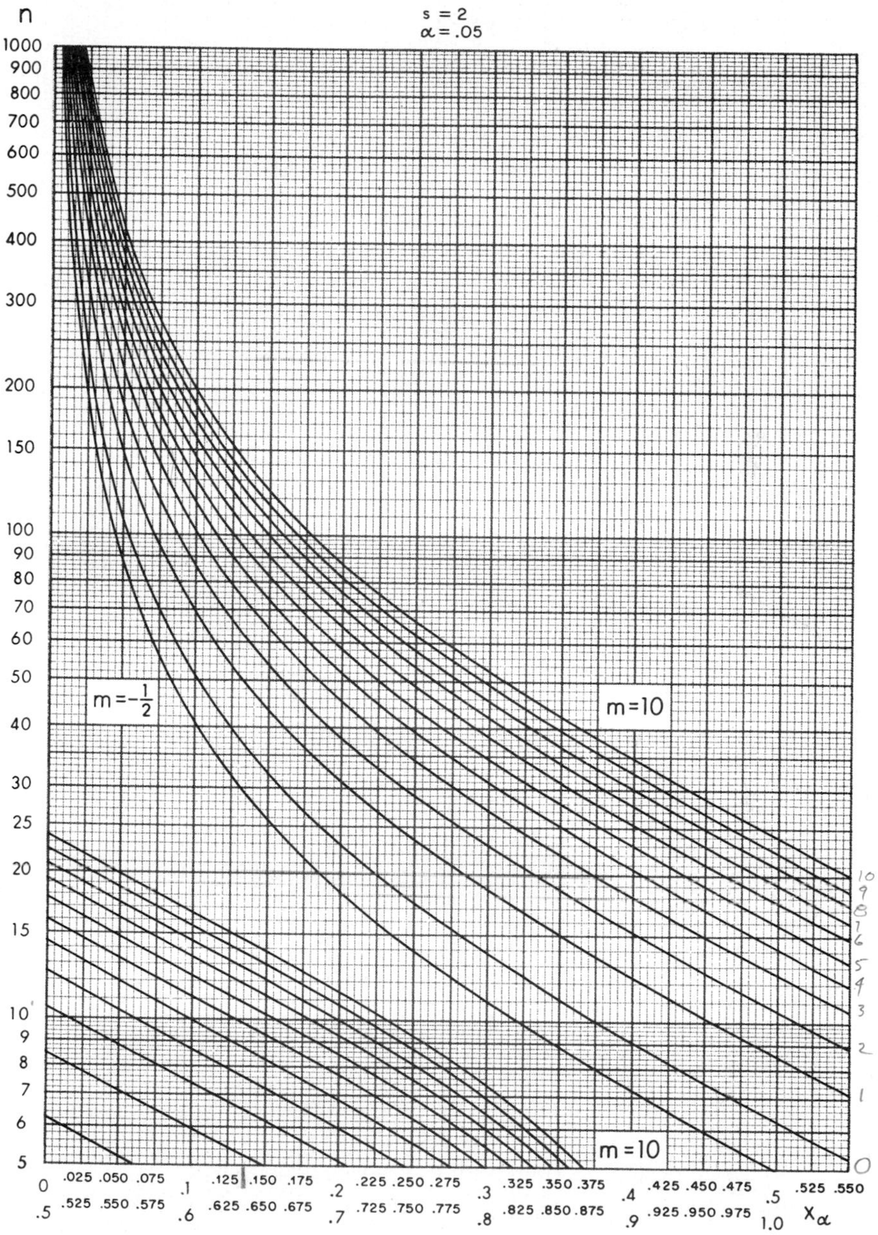

Reprinted from D. L. Heck: Charts of some upper percentage points of the distribution of the largest characteristic root, *The Annals of Mathematical Statistics*, vol. 31 (1960), pp. 625–642, with the permission of the author and the publisher.

Chart 10

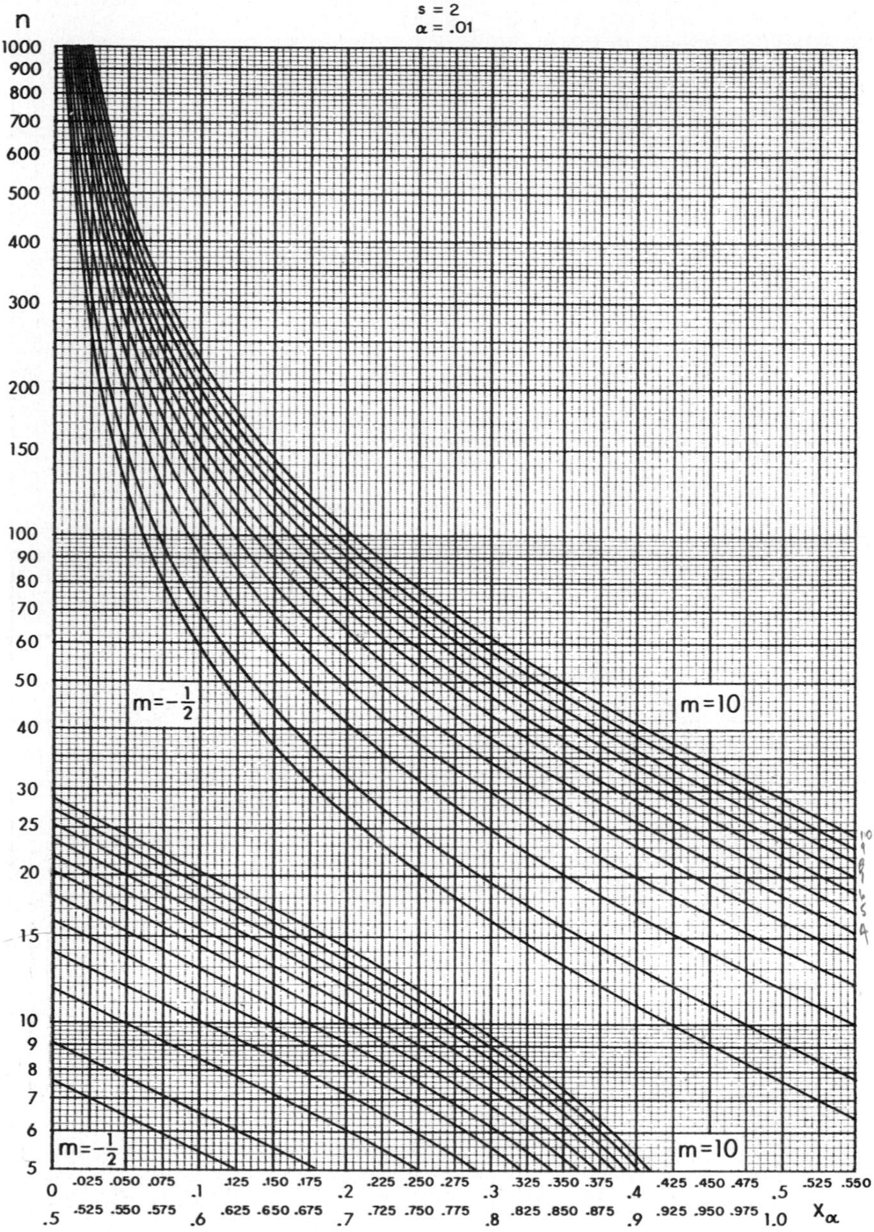
s = 2
α = .01
n
1000
900
800
700
600
500
400
300
200
150
100
90
80
70
60
50
40
30
25
20
15
10
9
8
7
6
5
m=−1/2
m=10
m=−1/2
m=10
0 .025 .050 .075 .1 .125 .150 .175 .2 .225 .250 .275 .3 .325 .350 .375 .4 .425 .450 .475 .5 .525 .550
.5 .525 .550 .575 .6 .625 .650 .675 .7 .725 .750 .775 .8 .825 .850 .875 .9 .925 .950 .975 1.0
$x_α$

Chart 11

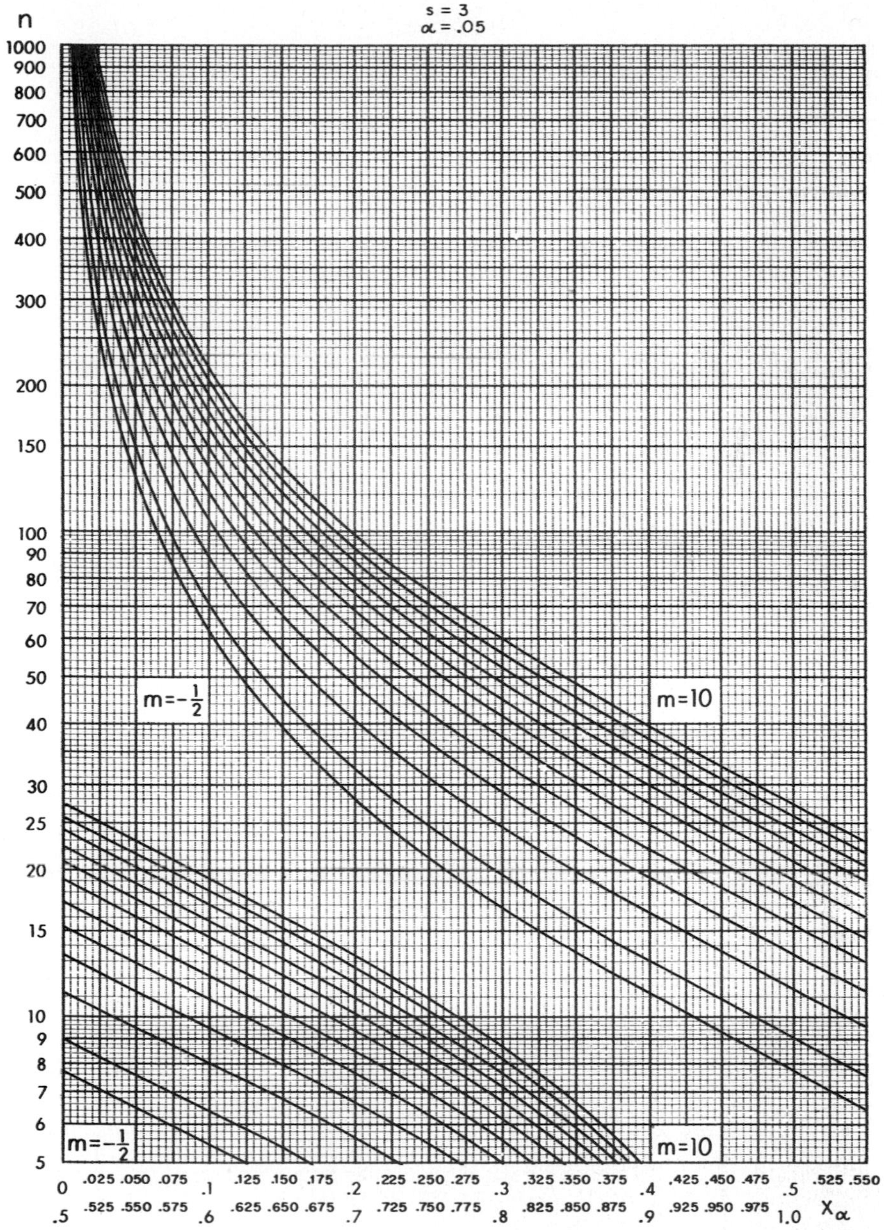
n
s = 3
α = .05
1000
900
800
700
600
500
400
300
200
150
100
90
80
70
60
50
40
30
25
20
15
10
9
8
7
6
5
m=-1/2
m=10
m=-1/2
m=10
0 .025 .050 .075 .1 .125 .150 .175 .2 .225 .250 .275 .3 .325 .350 .375 .4 .425 .450 .475 .5 .525 .550
.5 .525 .550 .575 .6 .625 .650 .675 .7 .725 .750 .775 .8 .825 .850 .875 .9 .925 .950 .975 1.0
Xα

Chart 12

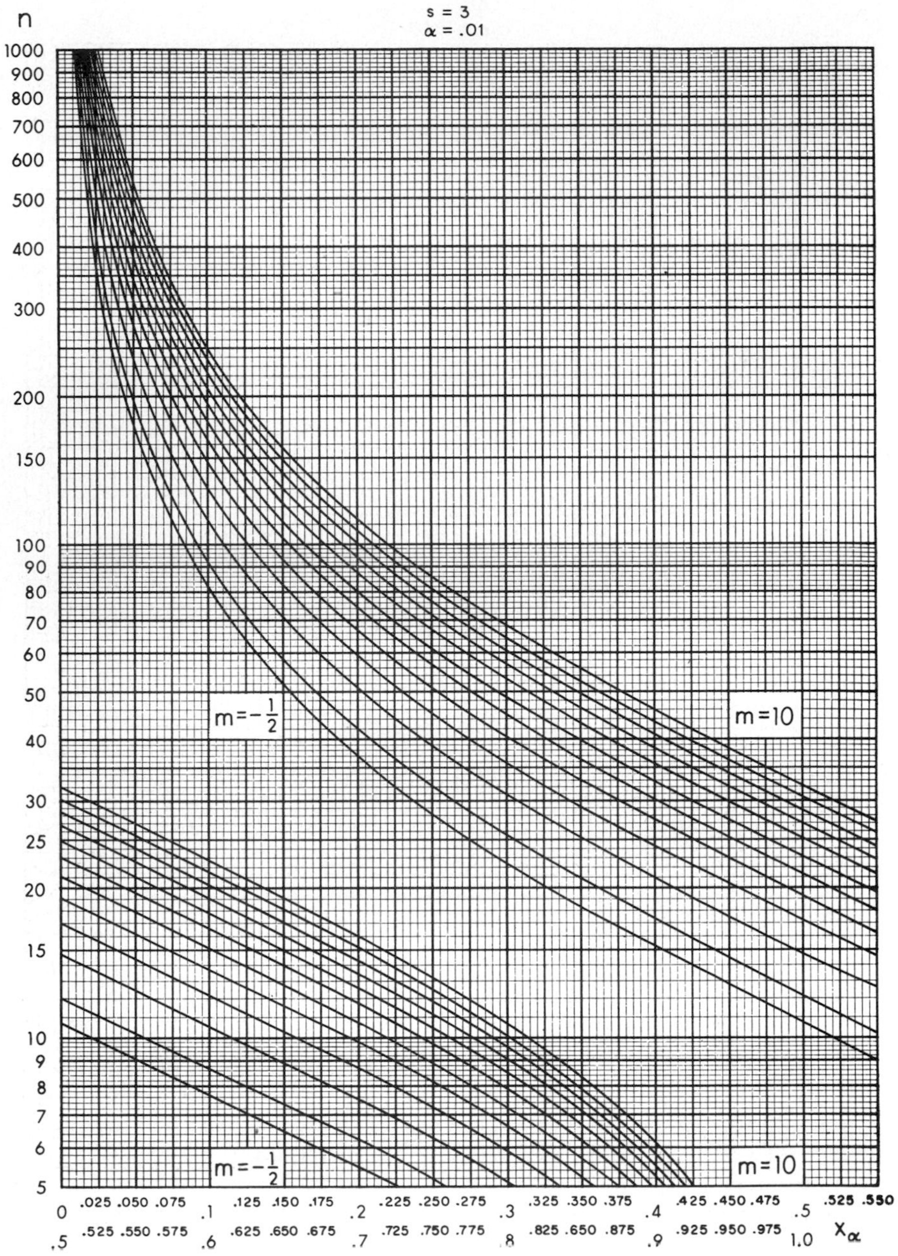
s = 3
α = .01
n
1000
900
800
700
600
500
400
300
200
150
100
90
80
70
60
50
40
30
25
20
15
10
9
8
7
6
5
m = −1/2
m = 10
m = −1/2
m = 10
0 .025 .050 .075 .1 .125 .150 .175 .2 .225 .250 .275 .3 .325 .350 .375 .4 .425 .450 .475 .5 .525 .550
.5 .525 .550 .575 .6 .625 .650 .675 .7 .725 .750 .775 .8 .825 .650 .875 .9 .925 .950 .975 1.0
Xα

Chart 13

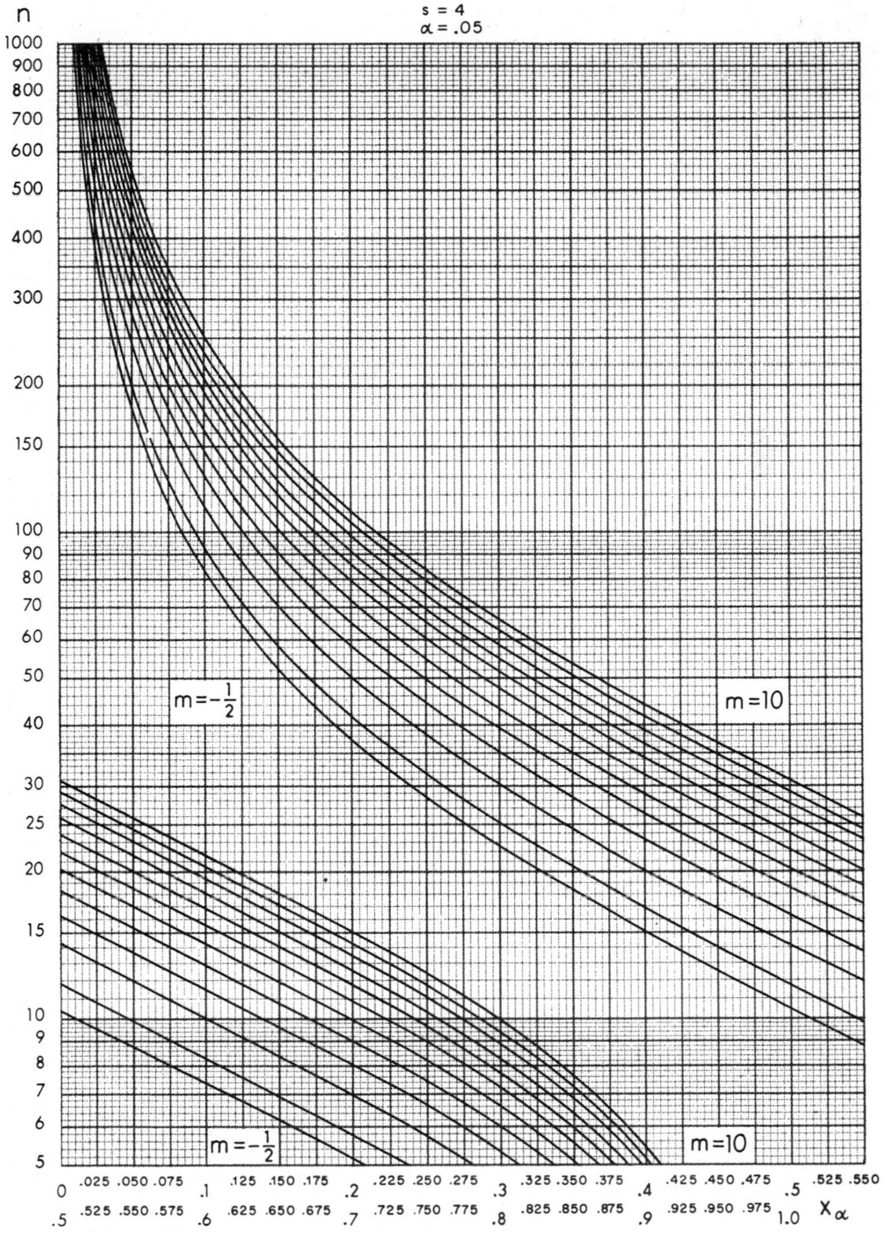
n
s = 4
α = .05
1000
900
800
700
600
500
400
300
200
150
100
90
80
70
60
50
40
30
25
20
15
10
9
8
7
6
5
m = -1/2
m = 10
m = -1/2
m = 10
0 .025 .050 .075 .1 .125 .150 .175 .2 .225 .250 .275 .3 .325 .350 .375 .4 .425 .450 .475 .5 .525 .550
.5 .525 .550 .575 .6 .625 .650 .675 .7 .725 .750 .775 .8 .825 .850 .875 .9 .925 .950 .975 1.0
x_α

Chart 14

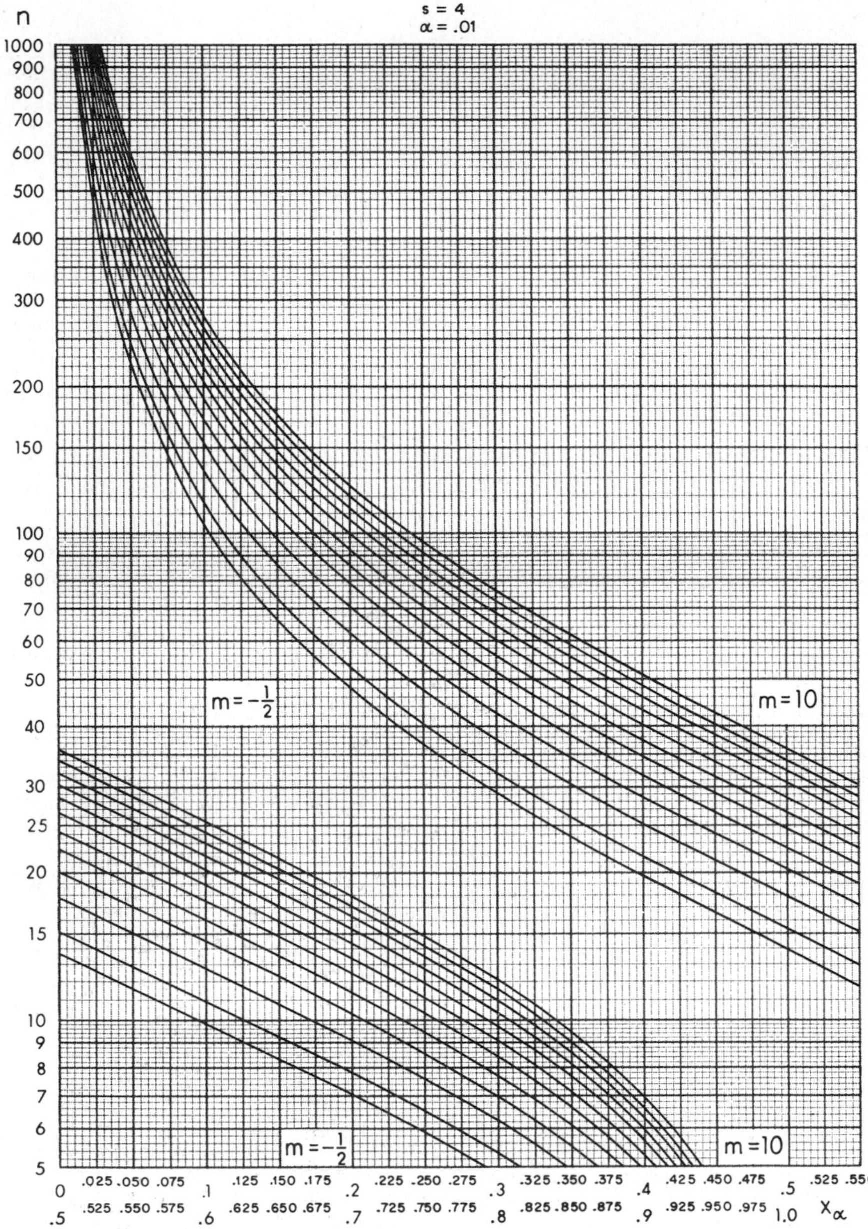
s = 4
α = .01
n
1000
900
800
700
600
500
400
300
200
150
100
90
80
70
60
50
40
30
25
20
15
10
9
8
7
6
5
m = −1/2
m = 10
m = −1/2
m = 10
0 .025 .050 .075 .1 .125 .150 .175 .2 .225 .250 .275 .3 .325 .350 .375 .4 .425 .450 .475 .5 .525 .550
.5 .525 .550 .575 .6 .625 .650 .675 .7 .725 .750 .775 .8 .825 .850 .875 .9 .925 .950 .975 1.0
x_α

Chart 15

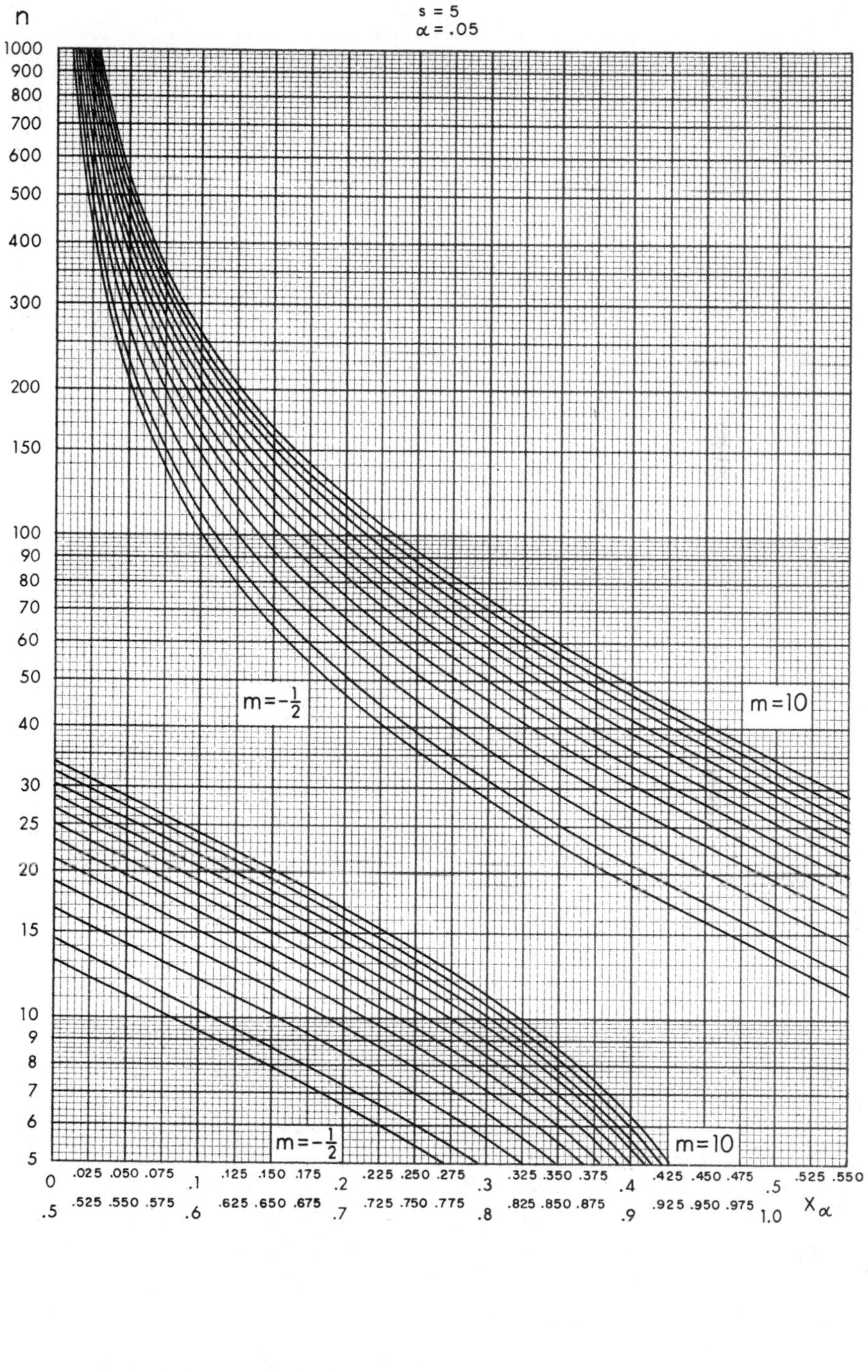
n
s = 5
α = .05
1000
900
800
700
600
500
400
300
200
150
100
90
80
70
60
50
40
30
25
20
15
10
9
8
7
6
5
m=-½
m=10
m=-½
m=10
0 .025 .050 .075 .1 .125 .150 .175 .2 .225 .250 .275 .3 .325 .350 .375 .4 .425 .450 .475 .5 .525 .550
.5 .525 .550 .575 .6 .625 .650 .675 .7 .725 .750 .775 .8 .825 .850 .875 .9 .925 .950 .975 1.0
X_α

Chart 16

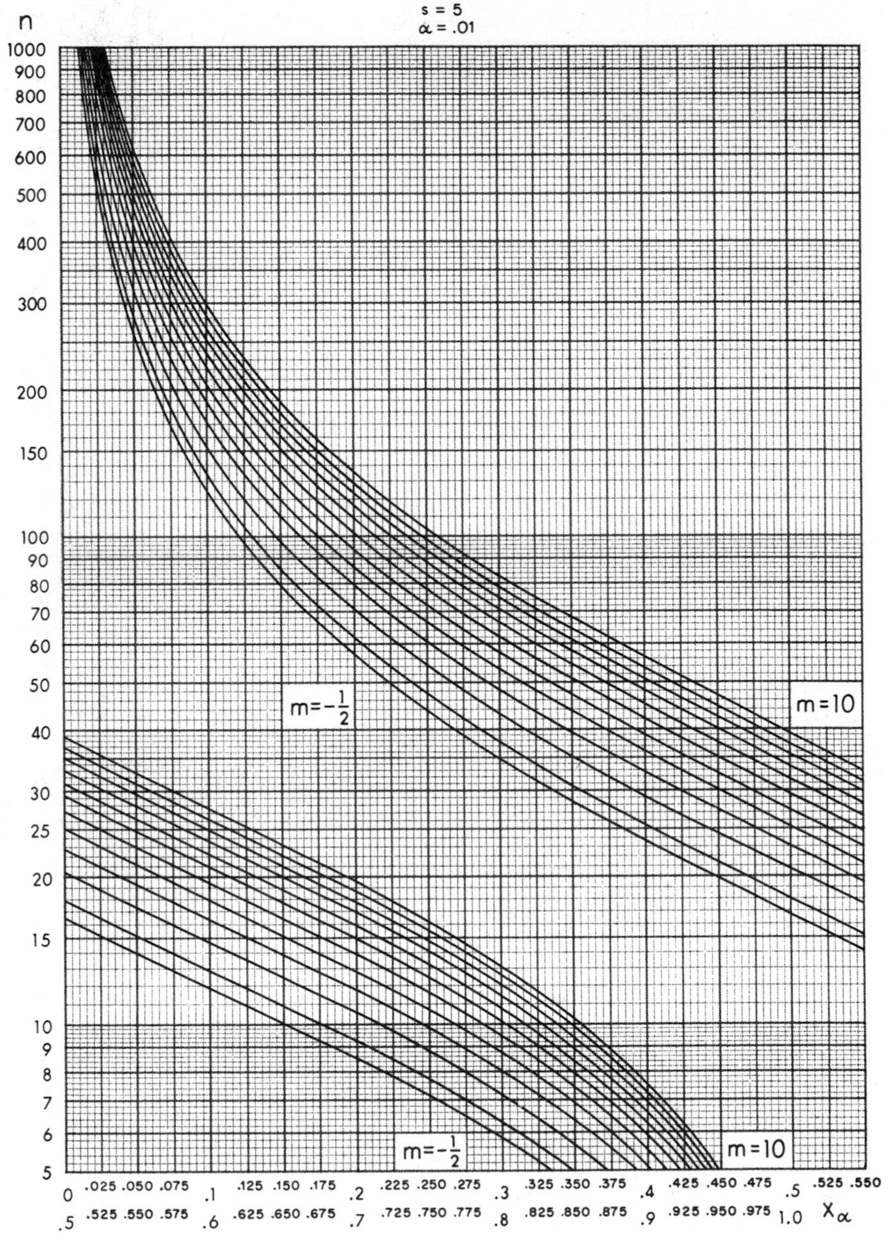
s = 5
α = .01
n
1000
900
800
700
600
500
400
300
200
150
100
90
80
70
60
50
40
30
25
20
15
10
9
8
7
6
5
m=−1/2
m=10
m=−1/2
m=10
0 .025 .050 .075 .1 .125 .150 .175 .2 .225 .250 .275 .3 .325 .350 .375 .4 .425 .450 .475 .5 .525 .550
.5 .525 .550 .575 .6 .625 .650 .675 .7 .725 .750 .775 .8 .825 .850 .875 .9 .925 .950 .975 1.0
Xα

Table 6. **Upper Percentage Points of the Largest Characteristic Root: $s = 6$***

$\alpha = 0.05$

n \ m	0	1	2	3	4
5	0.8246	0.8499	0.8685	0.8830	0.8945
10	0.6552	0.6917	0.7206	0.7442	0.7639
15	0.5371	0.5758	0.6077	0.6346	0.6577
20	0.4535	0.4912	0.5231	0.5505	0.5746
25	0.3918	0.4276	0.4583	0.4852	0.5091
30	0.3447	0.3782	0.4074	0.4332	0.4564
40	0.2775	0.3069	0.3329	0.3563	0.3776
60	0.1995	0.2225	0.2433	0.2624	0.2801
80	0.1556	0.1745	0.1916	0.2075	0.2224
100	0.1275	0.1434	0.1580	0.1716	0.1843
130	0.10036	0.11319	0.12504	0.13615	0.14666
160	0.08272	0.09348	0.10388	0.11284	0.12175
200	0.06702	0.07586	0.08409	0.09186	0.09926
300	0.04545	0.05156	0.05728	0.06281	0.06790
500	0.02765	0.03143	0.03498	0.03835	0.04160
1,000	0.01397	0.01590	0.01772	0.01946	0.02113

$\alpha = 0.01$

n \ m	0	1	2	3	4
5	0.8745	0.8929	0.9065	0.9169	0.9255
10	0.7173	0.7482	0.7724	0.7922	0.8086
15	0.5986	0.6334	0.6619	0.6858	0.7063
20	0.5111	0.5462	0.5757	0.6010	0.6231
25	0.4450	0.4790	0.5081	0.5335	0.5559
30	0.3936	0.4261	0.4542	0.4789	0.5011
40	0.3194	0.3484	0.3739	0.3969	0.4177
60	0.2315	0.2548	0.2757	0.2948	0.3125
80	0.1814	0.2006	0.2181	0.2342	0.2493
100	0.1491	0.1654	0.1803	0.1942	0.2072
130	0.11762	0.13091	0.14314	0.15457	0.16536
160	0.09713	0.10830	0.11901	0.12834	0.13754
200	0.07880	0.08803	0.09659	0.10466	0.11232
300	0.05355	0.05996	0.06594	0.07160	0.07701
500	0.03270	0.03661	0.04034	0.04388	0.04727
1,000	0.01651	0.01855	0.02046	0.02229	0.02405

* Reproduced from K. C. S. Pillai and C. G. Bantegui: On the distribution of the largest of six roots of a matrix in multivariate analysis, *Biometrika*, vol. 46 (1959), pp. 237–240, with the permission of the authors and the editor of *Biometrika*.

Table 7. **Upper Percentage Points of the Largest Characteristic Root: $s = 7$***

$\alpha = 0.05$

n \ m	0	1	2	3	4	5	7	10
5	0.85229	0.87214	0.88715	0.89893	0.90846	0.91632	0.9288	0.9445
10	0.69490	0.72561	0.75028	0.77064	0.78778	0.80243	0.8264	0.8540
15	0.57912	0.61295	0.64111	0.66505	0.68575	0.70387	0.7342	0.7695
20	0.49436	0.52818	0.55698	0.58197	0.60396	0.62353	0.6570	0.6970
25	0.43049	0.46310	0.49132	0.51617	0.53832	0.55827	0.5929	0.6352
30	0.38090	0.41189	0.43903	0.46319	0.48495	0.50472	0.5395	0.5825
40	0.30923	0.33684	0.36144	0.38367	0.40398	0.42267	0.4561	0.4987
60	0.22433	0.24644	0.26650	0.28496	0.30209	0.31811	0.3474	0.3858
80	0.17590	0.19414	0.21088	0.22642	0.24098	0.25471	0.2801	0.3141
100	0.14463	0.16011	0.17441	0.18776	0.20036	0.21230	0.2345	0.2647
130	0.11417	0.12676	0.13845	0.14945	0.15987	0.16981	0.1885	0.2141
160	0.094297	0.104892	0.114774	0.124101	0.132975	0.14147	0.15750	0.17965
200	0.076532	0.085273	0.093455	0.101205	0.108603	0.11570	0.12917	0.14792
300	0.052023	0.058098	0.063813	0.069251	0.074465	0.079493	0.08909	0.10258
500	0.031710	0.035480	0.039040	0.042442	0.045714	0.048883	0.05496	0.06357
1,000	0.016046	0.017979	0.019811	0.021566	0.023261	0.024905	0.02807	0.03259

$\alpha = 0.01$

n \ m	0	1	2	3	4	5	7	10
5	0.89470	0.90908	0.91991	0.92839	0.93522	0.9408	0.9498	0.9614
10	0.75082	0.77656	0.79714	0.81405	0.82824	0.8403	0.8600	0.8818
15	0.63628	0.66646	0.69144	0.71260	0.73083	0.7467	0.7732	0.8038
20	0.54905	0.58029	0.60677	0.62966	0.64973	0.6675	0.6978	0.7336
25	0.48171	0.51253	0.53909	0.56238	0.58308	0.6016	0.6338	0.6727
30	0.42859	0.45835	0.48430	0.50732	0.52798	0.5467	0.5795	0.6198
40	0.35059	0.37767	0.40170	0.42335	0.44306	0.4612	0.4934	0.5342
60	0.25649	0.27866	0.29872	0.31712	0.33415	0.3500	0.3789	0.4168
80	0.20204	0.22055	0.23748	0.25317	0.26783	0.2816	0.3071	0.3409
100	0.16660	0.18243	0.19700	0.21058	0.22336	0.2354	0.2579	0.2882
130	0.13187	0.14483	0.15684	0.16810	0.17875	0.1889	0.2079	0.2338
160	0.109113	0.120064	0.130255	0.139852	0.148966	0.15767	0.17406	0.19664
200	0.088695	0.097764	0.106235	0.114242	0.121870	0.12918	0.14300	0.16220
300	0.060419	0.066754	0.072702	0.078351	0.083759	0.08896	0.09888	0.11277
500	0.036892	0.040839	0.044561	0.048111	0.051521	0.05482	0.06113	0.07005
1,000	0.018692	0.020724	0.022645	0.024483	0.026254	0.02798	0.03127	0.03597

* Reproduced from K. C. S. Pillai: On the distribution of the largest of seven roots of a matrix in multivariate analysis, *Biometrika*, vol. 51 (1964), pp. 270–275, with the permission of the author and the editor of *Biometrika*.

Table 8. **Upper Percentage Points of the Largest Characteristic Root: $s = 8$***

$\alpha = 0.05$

n \ m	0	1	2	3	4	5	7	10	15
5	0.87386	0.88974	0.90198	0.91173	0.91968	0.92630	0.93670	0.94773	0.95948
10	0.72804	0.75412	0.77534	0.79300	0.80798	0.82085	0.84191	0.86546	0.89206
15	0.61550	0.64525	0.67024	0.69164	0.71022	0.72656	0.75402	0.78589	0.82359
20	0.53065	0.56108	0.58718	0.60997	0.63010	0.64806	0.67886	0.71565	0.76068
25	0.46544	0.49524	0.52122	0.54420	0.56475	0.58331	0.61561	0.65503	0.70464
30	0.41408	0.44274	0.46800	0.49059	0.51099	0.52957	0.56228	0.60290	0.65518
40	0.33876	0.36473	0.38799	0.40909	0.42841	0.44623	0.47816	0.51881	0.57296
60	0.24794	0.26912	0.28844	0.30627	0.32286	0.33838	0.36679	0.40411	0.45600
80	0.19537	0.21303	0.22931	0.24447	0.25871	0.27215	0.29703	0.33034	0.37787
100	0.16115	0.17623	0.19022	0.20334	0.21572	0.22748	0.24942	0.27915	0.32232
130	0.12759	0.13993	0.15145	0.16231	0.17263	0.18247	0.20099	0.22638	0.26392
160	0.10559	0.11602	0.12579	0.13504	0.14386	0.15230	0.16827	0.19034	0.22335
200	0.085849	0.094482	0.10260	0.11032	0.11769	0.12478	0.13824	0.15698	0.18531
300	0.058496	0.064526	0.070225	0.075663	0.080887	0.085929	0.095564	0.109119	0.12993
500	0.035726	0.039482	0.043047	0.046463	0.049755	0.052946	0.059074	0.067768	0.08119†
1,000	0.018105	0.020038	0.021878	0.023645	0.025355	0.027016	0.030219	0.03480†	0.04201†

† Value extrapolated.

$\alpha = 0.01$

n \ m	0	1	2	3	4	5	7	10	15
5	0.91031	0.92176	0.93055	0.93754	0.94323	0.94796	0.95537	0.96320	0.97152
10	0.77855	0.80027	0.81787	0.83248	0.84482	0.85541	0.87267	0.89189	0.91349
15	0.66867	0.69502	0.71708	0.73589	0.75218	0.76647	0.79039	0.81803	0.85053
20	0.58250	0.61043	0.63429	0.65505	0.67334	0.68961	0.71741	0.75045	0.79066
25	0.51467	0.54266	0.56696	0.58839	0.60751	0.62472	0.65456	0.69080	0.73614
30	0.46038	0.48773	0.51176	0.53317	0.55245	0.56997	0.60071	0.63869	0.68727
40	0.37948	0.40480	0.42741	0.44786	0.46653	0.48371	0.51438	0.55327	0.60475
60	0.28011	0.30123	0.32046	0.33814	0.35456	0.36990	0.39787	0.43447	0.48507
80	0.22175	0.23958	0.25597	0.27120	0.28547	0.29892	0.32375	0.35686	0.40386
100	0.18344	0.19878	0.21298	0.22626	0.23878	0.25064	0.27271	0.30251	0.34559
130	0.14565	0.15829	0.17007	0.18115	0.19165	0.20166	0.22043	0.24609	0.28386
160	0.12076	0.13149	0.14152	0.15100	0.16002	0.16865	0.18492	0.20734	0.24074
200	0.098336	0.10725	0.11562	0.12356	0.13114	0.13841	0.15219	0.17132	0.20013
300	0.067151	0.073412	0.079320	0.084948	0.090347	0.095551	0.105476	0.119400	0.14070
500	0.041086	0.045003	0.048715	0.052266	0.055685	0.058994	0.065337	0.074316	0.08822†
1,000	0.020850	0.022872	0.024794	0.026638	0.028420	0.030148	0.033477	0.03824†	0.04573†

† Value extrapolated.

* Tables 8 to 10 have been reproduced from K. C. S. Pillai: On the distribution of the largest characteristic root of a matrix in multivariate analysis, *Biometrika*, vol. 52 (1965), pp. 405–414, with the permission of the author and the editor of *Biometrika*.

Table 9. **Upper Percentage Points of the Largest Characteristic Root: $s = 9$**

$\alpha = 0.05$

n \ m	0	1	2	3	4	5	7	10	15
5	0.89098	0.90390	0.91402	0.92217	0.92889	0.93453	0.94348	0.95305	0.96347
10	0.75598	0.77833	0.79670	0.81213	0.82529	0.83666	0.85538	0.87647	0.90052
15	0.64727	0.67357	0.69584	0.71503	0.73178	0.74656	0.77150	0.80063	0.83527
20	0.56307	0.59053	0.61426	0.63508	0.65353	0.67006	0.69848	0.73257	0.77449
25	0.49716	0.52447	0.54841	0.56969	0.58878	0.60606	0.63622	0.67314	0.71978
30	0.44455	0.47111	0.49465	0.51577	0.53491	0.55237	0.58319	0.62156	0.67108
40	0.36633	0.39077	0.41278	0.43280	0.45118	0.46816	0.49864	0.53752	0.58940
60	0.27040	0.29069	0.30929	0.32650	0.34254	0.35757	0.38512	0.42136	0.47178
80	0.21408	0.23118	0.24699	0.26177	0.27567	0.28880	0.31315	0.34578	0.39237
100	0.17713	0.19182	0.20550	0.21836	0.23052	0.24208	0.26368	0.29297	0.33552
130	0.14066	0.15275	0.16408	0.17480	0.18499	0.19472	0.21305	0.23822	0.27542
160	0.11663	0.12689	0.13654	0.14570	0.15444	0.16283	0.17869	0.20065	0.23349
200	0.094986	0.10351	0.11156	0.11922	0.12656	0.13362	0.14704	0.16575	0.19404
300	0.064875	0.070856	0.076532	0.081961	0.087184	0.092232	0.10189	0.11548	0.13636
500	0.039699	0.043440	0.047005	0.050427	0.053733	0.056939	0.063104	0.071859	0.08542*
1,000	0.020149	0.022080	0.023925	0.025702	0.027423	0.029097	0.032330	0.036952	0.04416*

* Value extrapolated.

$\alpha = 0.01$

n \ m	0	1	2	3	4	5	7	10	15
5	0.92264	0.93192	0.93917	0.94499	0.94979	0.95381	0.96017	0.96696	0.97430
10	0.80179	0.82030	0.83548	0.84818	0.85899	0.86832	0.88362	0.90079	0.92029
15	0.69676	0.71994	0.73951	0.75631	0.77093	0.78381	0.80548	0.83067	0.86048
20	0.61221	0.63727	0.65886	0.67774	0.69444	0.70936	0.73493	0.76547	0.80282
25	0.54440	0.56992	0.59221	0.61197	0.62965	0.64561	0.67339	0.70724	0.74977
30	0.48940	0.51462	0.53691	0.55685	0.57486	0.59126	0.62013	0.65591	0.70182
40	0.40632	0.43003	0.45131	0.47064	0.48833	0.50464	0.53383	0.57092	0.62013
60	0.30247	0.32262	0.34104	0.35804	0.37386	0.38866	0.41570	0.45114	0.50020
80	0.24060	0.25778	0.27365	0.28844	0.30232	0.31542	0.33965	0.37199	0.41796
100	0.19966	0.21454	0.22837	0.24134	0.25359	0.26521	0.28688	0.31617	0.35853
130	0.15901	0.17134	0.18288	0.19377	0.20411	0.21398	0.23252	0.25789	0.29524
160	0.13209	0.14260	0.15248	0.16183	0.17075	0.17929	0.19541	0.21766	0.25082
200	0.10775	0.11652	0.12479	0.13265	0.14017	0.14739	0.16110	0.18015	0.20885
300	0.073760	0.079948	0.085812	0.091413	0.096795	0.10199	0.11191	0.12584	0.14716
500	0.045220	0.049108	0.052807	0.056355	0.059777	0.063093	0.069460	0.078480	0.09243*
1,000	0.022983	0.024997	0.026918	0.028767	0.030555	0.032293	0.035644	0.040426	0.04788*

* Value extrapolated.

Table 10. **Upper Percentage Points of the Largest Characteristic Root: $s = 10$**

$\alpha = 0.05$

n \ m	0	1	2	3	4	5	7	10	15
5	0.90483	0.91547	0.92393	0.93083	0.93656	0.94141	0.94916	0.95759	0.96720
10	0.77978	0.79907	0.81509	0.82864	0.84026	0.85037	0.86708	0.88605	0.90789
15	0.67519	0.69855	0.71848	0.73575	0.75090	0.76431	0.78704	0.81372	0.84561
20	0.59217	0.61704	0.63867	0.65772	0.67468	0.68990	0.71619	0.74784	0.78695
25	0.52606	0.55114	0.57325	0.59297	0.61073	0.62684	0.65503	0.68966	0.73355
30	0.47263	0.49728	0.51923	0.53901	0.55697	0.57339	0.60245	0.63872	0.68566
40	0.39214	0.41517	0.43599	0.45499	0.47248	0.48865	0.51774	0.55494	0.60465
60	0.29180	0.31125	0.32914	0.34574	0.36124	0.37580	0.40248	0.43765	0.48664
80	0.23209	0.24864	0.26400	0.27839	0.29194	0.30477	0.32857	0.36050	0.40613
100	0.19260	0.20690	0.22028	0.23287	0.24480	0.25616	0.27739	0.30622	0.34812
130	0.15339	0.16524	0.17638	0.18693	0.19699	0.20660	0.22473	0.24963	0.28647
160	0.12743	0.13752	0.14704	0.15610	0.16476	0.17307	0.18882	0.21063	0.24327
200	0.10396	0.11237	0.12034	0.12794	0.13523	0.14226	0.15562	0.17427	0.2025*
300	0.071167	0.077098	0.082744	0.088157	0.093372	0.098418	0.10808	0.12169	0.1426*
500	0.043634	0.047358	0.050918	0.054344	0.057657	0.060875	0.067067	0.07585*	0.08948*
1,000	0.022179	0.024107	0.025955	0.027739	0.029470	0.031155	0.03441*	0.03905*	0.04632*

* Value extrapolated.

$\alpha = 0.01$

n \ m	0	1	2	3	4	5	7	10	15
5	0.93258	0.94020	0.94624	0.95116	0.95524	0.95869	0.96419	0.97016	0.97680
10	0.82149	0.83740	0.85058	0.86170	0.87122	0.87948	0.89311	0.90854	0.92620
15	0.72134	0.74183	0.75927	0.77433	0.78751	0.79916	0.81886	0.84189	0.86931
20	0.63873	0.66133	0.68093	0.69814	0.71344	0.72714	0.75072	0.77900	0.81378
25	0.57137	0.59468	0.61519	0.63344	0.64982	0.66465	0.69054	0.72222	0.76217
30	0.51602	0.53932	0.56002	0.57862	0.59547	0.61085	0.63798	0.67172	0.71517
40	0.43131	0.45356	0.47362	0.49189	0.50867	0.52416	0.55195	0.58733	0.63440
60	0.32368	0.34292	0.36058	0.37692	0.39216	0.40644	0.43257	0.46688	0.51445
80	0.25867	0.27524	0.29059	0.30495	0.31844	0.33120	0.35482	0.38640	0.43132
100	0.21530	0.22974	0.24321	0.25588	0.26786	0.27924	0.30049	0.32925	0.37088
130	0.17197	0.18401	0.19532	0.20602	0.21619	0.22592	0.24420	0.26925	0.30617
160	0.14313	0.15344	0.16315	0.17238	0.18119	0.18963	0.20560	0.22765	0.26054
200	0.11696	0.12559	0.13375	0.14153	0.14898	0.15615	0.16977	0.18871	0.2173*
300	0.080258	0.086375	0.092191	0.097760	0.10312	0.10830	0.11820	0.13213	0.1534*
500	0.049301	0.053160	0.056844	0.060386	0.063808	0.067128	0.073509	0.08258*	0.09661*
1,000	0.025096	0.027100	0.029020	0.030871	0.032666	0.034411	0.03778*	0.04260*	0.05013*

* Value extrapolated.

Table 11. **Upper Percentage Points of the Largest Characteristic Root: $s = 14$***

$\alpha = 0.05$

n \ m	0	1	2	3	4	5	7	10	15
5	0.9403	0.9457	0.9503	0.9541	0.9574	0.9602	0.9649	0.9701	0.9761
10	0.8470	0.8584	0.8681	0.8766	0.8840	0.8906	0.9017	0.9146	0.9299
15	0.7590	0.7742	0.7875	0.7992	0.8096	0.8189	0.8350	0.8543	0.8779
20	0.6834	0.7006	0.7159	0.7296	0.7419	0.7531	0.7726	0.7966	0.8267
25	0.6196	0.6378	0.6542	0.6689	0.6824	0.6947	0.7165	0.7436	0.7784
30	0.5657	0.5843	0.6011	0.6165	0.6305	0.6435	0.6666	0.6958	0.7340
40	0.4807	0.4990	0.5158	0.5313	0.5457	0.5590	0.5832	0.6144	0.6565
60	0.3683	0.3848	0.4001	0.4145	0.4279	0.4407	0.4641	0.4952	0.5387
80	0.2980	0.3125	0.3261	0.3390	0.3512	0.3628	0.3844	0.4136	0.4554
100	0.2500	0.2629	0.2751	0.2866	0.2976	0.3081	0.3278	0.3546	0.3938
130	0.2013	0.2122	0.2226	0.2325	0.2419	0.2510	0.2682	0.2919	0.3271
160	0.1685	0.1779	0.1869	0.1955	0.2038	0.2117	0.2269	0.2479	0.2795
200	0.1383	0.1463	0.1539	0.1612	0.1683	0.1751	0.1881	0.2064	0.2340
300	0.09556	0.1013	0.1068	0.1121	0.1172	0.1222	0.1318	0.1454	0.1662
500	0.05904	0.06269	0.06621	0.06962	0.07293	0.07616	0.08240	0.09130	0.1052
1,000	0.03019	0.03210	0.03395	0.03575	0.03750	0.03921	0.04253	0.04730	0.05482

$\alpha = 0.01$

n \ m	0	1	2	3	4	5	7	10	15
5	0.95784	0.96171	0.96493	0.96764	0.96996	0.97197	0.97527	0.97897	0.98317
10	0.87660	0.88590	0.89386	0.90075	0.90678	0.91210	0.92107	0.93152	0.94385
15	0.79444	0.80757	0.81903	0.82914	0.83812	0.84617	0.86001	0.87655	0.89674
20	0.72109	0.73655	0.75023	0.76245	0.77346	0.78342	0.80082	0.82205	0.84866
25	0.65775	0.67449	0.68948	0.70300	0.71529	0.72651	0.74634	0.77095	0.80248
30	0.60342	0.62078	0.63645	0.65071	0.66376	0.67577	0.69718	0.72412	0.75929
40	0.51631	0.53380	0.54979	0.56452	0.57815	0.59082	0.61373	0.64320	0.68278
60	0.39883	0.41495	0.42993	0.44392	0.45705	0.46941	0.49218	0.52229	0.56430
80	0.32422	0.33862	0.35212	0.36484	0.37687	0.38830	0.40957	0.43817	0.47908
100	0.27291	0.28576	0.29788	0.30937	0.32030	0.33073	0.35028	0.37689	0.41558
130	0.22042	0.23140	0.24181	0.25173	0.26122	0.27033	0.28753	0.31121	0.34623
160	0.18481	0.19435	0.20343	0.21212	0.22045	0.22848	0.24372	0.26486	0.29651
200	0.15203	0.16012	0.16786	0.17528	0.18243	0.18933	0.20250	0.22090	0.24874
300	0.10529	0.11114	0.11675	0.12217	0.12741	0.13249	0.14224	0.15602	0.17718
500	0.065179	0.068928	0.072541	0.076040	0.079437	0.082746	0.089131	0.098227	0.11239
1,000	0.033379	0.035350	0.037256	0.039107	0.040910	0.042671	0.046084	0.050981	0.058685

* Tables 11 to 14 have been reproduced from K. C. S. Pillai: Upper percentage points of the largest root of a matrix in multivariate analysis, *Biometrika*, vol. 54 (1967), pp. 189–194, with the permission of the author and the editor of *Biometrika*.

Table 12. **Upper Percentage Points of the Largest Characteristic Root: $s = 16$**

$\alpha = 0.05$

n \ m	0	1	2	3	4	5	7	10	15
5	0.9510	0.9551	0.9586	0.9615	0.9641	0.9663	0.9701	0.9743	0.9793
10	0.8695	0.8786	0.8864	0.8933	0.8993	0.9047	0.9139	0.9248	0.9378
15	0.7892	0.8017	0.8127	0.8225	0.8313	0.8392	0.8529	0.8695	0.8901
20	0.7178	0.7324	0.7454	0.7571	0.7677	0.7774	0.7945	0.8154	0.8421
25	0.6561	0.6719	0.6861	0.6990	0.7108	0.7216	0.7409	0.7650	0.7963
30	0.6031	0.6194	0.6342	0.6478	0.6603	0.6719	0.6926	0.7189	0.7536
40	0.5177	0.5342	0.5493	0.5633	0.5764	0.5886	0.6107	0.6393	0.6781
60	0.4018	0.4170	0.4312	0.4446	0.4572	0.4691	0.4910	0.5202	0.5613
80	0.3276	0.3413	0.3541	0.3663	0.3778	0.3889	0.4094	0.4372	0.4771
100	0.2763	0.2886	0.3001	0.3111	0.3217	0.3317	0.3506	0.3765	0.4143
130	0.2237	0.2341	0.2441	0.2537	0.2628	0.2716	0.2883	0.3113	0.3456
160	0.1878	0.1969	0.2056	0.2140	0.2221	0.2298	0.2446	0.2652	0.2962
200	0.1547	0.1625	0.1699	0.1771	0.1840	0.1907	0.2035	0.2214	0.2487
300	0.1074	0.1130	0.1184	0.1236	0.1287	0.1337	0.1432	0.1566	0.1774
500	0.06656	0.07017	0.07366	0.07705	0.08035	0.08357	0.08981	0.09872	0.1127
1,000	0.03413	0.03603	0.03788	0.03968	0.04143	0.04315	0.04649	0.05129	0.05887

$\alpha = 0.01$

n \ m	0	1	2	3	4	5	7	10	15
5	0.96542	0.96833	0.97078	0.97288	0.97469	0.97628	0.97892	0.98193	0.98540
10	0.89495	0.90231	0.90868	0.91425	0.91916	0.92353	0.93097	0.93973	0.95022
15	0.82050	0.83127	0.84075	0.84918	0.85672	0.86352	0.87530	0.88951	0.90705
20	0.75190	0.76491	0.77652	0.78696	0.79641	0.80501	0.82011	0.83869	0.86220
25	0.69124	0.70562	0.71858	0.73034	0.74108	0.75094	0.76842	0.79028	0.81850
30	0.63825	0.65340	0.66716	0.67975	0.69132	0.70200	0.72113	0.74534	0.77716
40	0.55156	0.56719	0.58155	0.59484	0.60718	0.61868	0.63956	0.66653	0.70294
60	0.43151	0.44634	0.46017	0.47314	0.48534	0.49686	0.51813	0.54634	0.58585
80	0.35351	0.36698	0.37966	0.39164	0.40301	0.41382	0.43398	0.46118	0.50018
100	0.29912	0.31128	0.32278	0.33372	0.34415	0.35412	0.37285	0.39840	0.43564
130	0.24286	0.25335	0.26335	0.27289	0.28204	0.29084	0.30749	0.33045	0.36449
160	0.20433	0.21352	0.22230	0.23071	0.23881	0.24662	0.26147	0.28212	0.31308
200	0.16862	0.17647	0.18399	0.19123	0.19822	0.20498	0.21789	0.23597	0.26338
300	0.11730	0.12302	0.12854	0.13387	0.13903	0.14405	0.15371	0.16736	0.18838
500	0.072890	0.076586	0.080162	0.083631	0.087008	0.090300	0.096667	0.10575	0.11992
1,000	0.037439	0.039394	0.041291	0.043138	0.044941	0.046704	0.050127	0.055047	0.062803

Table 13. **Upper Percentage Points of the Largest Characteristic Root: $s = 18$**

$\alpha = 0.05$

n \ m	0	1	2	3	4	5	7	10	15
5	0.9590	0.9622	0.9649	0.9673	0.9693	0.9711	0.9741	0.9777	0.9819
10	0.8874	0.8947	0.9011	0.9067	0.9118	0.9163	0.9240	0.9332	0.9444
15	0.8140	0.8244	0.8336	0.8419	0.8494	0.8561	0.8679	0.8824	0.9004
20	0.7469	0.7593	0.7705	0.7806	0.7898	0.7982	0.8132	0.8317	0.8554
25	0.6876	0.7013	0.7137	0.7250	0.7354	0.7450	0.7622	0.7837	0.8119
30	0.6358	0.6502	0.6633	0.6754	0.6866	0.6969	0.7156	0.7394	0.7709
40	0.5509	0.5657	0.5794	0.5921	0.6040	0.6151	0.6354	0.6617	0.6976
60	0.4326	0.4467	0.4599	0.4724	0.4841	0.4952	0.5158	0.5433	0.5820
80	0.3553	0.3682	0.3803	0.3918	0.4028	0.4132	0.4328	0.4592	0.4974
100	0.3012	0.3128	0.3239	0.3344	0.3444	0.3541	0.3723	0.3972	0.4336
130	0.2450	0.2551	0.2647	0.2739	0.2827	0.2912	0.3074	0.3298	0.3631
160	0.2064	0.2152	0.2237	0.2318	0.2397	0.2472	0.2617	0.2818	0.3121
200	0.1706	0.1781	0.1854	0.1924	0.1991	0.2057	0.2183	0.2359	0.2628
300	0.1189	0.1244	0.1297	0.1349	0.1399	0.1448	0.1542	0.1675	0.1882
500	0.07397	0.07753	0.08100	0.08437	0.08765	0.09086	0.09708	0.1060	0.1199
1,000	0.03804	0.03993	0.04177	0.04357	0.04533	0.04705	0.05040	0.05522	0.06285

$\alpha = 0.01$

n \ m	0	1	2	3	4	5	7	10	15
5	0.97112	0.97336	0.97528	0.97693	0.97838	0.97965	0.98180	0.98428	0.98723
10	0.90947	0.91539	0.92057	0.92513	0.92919	0.93283	0.93906	0.94649	0.95550
15	0.84185	0.85079	0.85873	0.86584	0.87223	0.87803	0.88814	0.90045	0.91580
20	0.77780	0.78886	0.79880	0.80778	0.81596	0.82344	0.83663	0.85299	0.87387
25	0.71996	0.73240	0.74369	0.75399	0.76343	0.77212	0.78762	0.80712	0.83249
30	0.66859	0.68189	0.69405	0.70521	0.71551	0.72505	0.74221	0.76405	0.79293
40	0.58298	0.59701	0.60996	0.62198	0.63318	0.64366	0.66272	0.68746	0.72103
60	0.46148	0.47515	0.48796	0.49999	0.51135	0.52209	0.54196	0.56842	0.60560
80	0.38083	0.39345	0.40537	0.41666	0.42740	0.43763	0.45675	0.48261	0.51979
100	0.32383	0.33534	0.34628	0.35669	0.36664	0.37617	0.39411	0.41863	0.45446
130	0.26424	0.27429	0.28388	0.29306	0.30188	0.31038	0.32647	0.34873	0.38179
160	0.22307	0.23193	0.24041	0.24857	0.25643	0.26401	0.27847	0.29862	0.32887
200	0.18465	0.19226	0.19959	0.20664	0.21347	0.22008	0.23273	0.25047	0.27741
300	0.12901	0.13462	0.14003	0.14527	0.15036	0.15531	0.16485	0.17837	0.19921
500	0.080467	0.084114	0.087650	0.091089	0.094440	0.097713	0.10405	0.11311	0.12727
1,000	0.041452	0.043392	0.045280	0.047121	0.048921	0.050684	0.054112	0.059048	0.066841

Table 14. **Upper Percentage Points of the Largest Characteristic Root: $s = 20$**

$\alpha = 0.05$

n \ m	0	1	2	3	4	5	7	10	15
5	0.9653	0.9677	0.9699	0.9718	0.9735	0.9749	0.9774	0.9804	0.9840
10	0.9018	0.9078	0.9131	0.9178	0.9220	0.9258	0.9323	0.9402	0.9499
15	0.8346	0.8434	0.8512	0.8583	0.8647	0.8705	0.8807	0.8933	0.9092
20	0.7716	0.7823	0.7920	0.8008	0.8088	0.8162	0.8294	0.8458	0.8670
25	0.7149	0.7269	0.7378	0.7478	0.7570	0.7655	0.7808	0.8002	0.8256
30	0.6646	0.6773	0.6890	0.6998	0.7098	0.7192	0.7360	0.7575	0.7863
40	0.5806	0.5940	0.6065	0.6181	0.6289	0.6391	0.6577	0.6819	0.7151
60	0.4611	0.4742	0.4864	0.4980	0.5090	0.5194	0.5387	0.5646	0.6011
80	0.3814	0.3934	0.4049	0.4158	0.4261	0.4360	0.4546	0.4799	0.5163
100	0.3248	0.3358	0.3464	0.3564	0.3660	0.3753	0.3927	0.4166	0.4517
130	0.2655	0.2751	0.2844	0.2933	0.3018	0.3100	0.3257	0.3474	0.3798
160	0.2244	0.2329	0.2411	0.2490	0.2566	0.2640	0.2781	0.2977	0.3274
200	0.1859	0.1933	0.2004	0.2072	0.2138	0.2202	0.2326	0.2499	0.2763
300	0.1301	0.1355	0.1408	0.1459	0.1509	0.1557	0.1650	0.1782	0.1987
500	0.08127	0.08480	0.08823	0.09158	0.09484	0.09804	0.1042	0.1131	0.1271
1,000	0.04192	0.04380	0.04563	0.04742	0.04918	0.05090	0.05426	0.05910	0.06678

$\alpha = 0.01$

n \ m	0	1	2	3	4	5	7	10	15
5	0.97552	0.97728	0.97880	0.98013	0.98130	0.98234	0.98411	0.98618	0.98870
10	0.92115	0.92599	0.93026	0.93405	0.93744	0.94050	0.94578	0.95214	0.95995
15	0.85957	0.86708	0.87379	0.87984	0.88531	0.89030	0.89904	0.90977	0.92329
20	0.79980	0.80928	0.81785	0.82565	0.83277	0.83931	0.85090	0.86538	0.88402
25	0.74479	0.75564	0.76553	0.77459	0.78293	0.79064	0.80444	0.82191	0.84480
30	0.69519	0.70694	0.71773	0.72768	0.73689	0.74544	0.76089	0.78065	0.80695
40	0.61113	0.62377	0.63549	0.64640	0.65660	0.66616	0.68361	0.70635	0.73736
60	0.48906	0.50169	0.51356	0.52475	0.53533	0.54536	0.56396	0.58878	0.62380
80	0.40637	0.41821	0.42943	0.44008	0.45023	0.45991	0.47805	0.50264	0.53810
100	0.34717	0.35809	0.36848	0.37840	0.38790	0.39701	0.41418	0.43771	0.47217
130	0.28465	0.29428	0.30349	0.31233	0.32083	0.32902	0.34459	0.36614	0.39822
160	0.24109	0.24963	0.25783	0.26573	0.27336	0.28073	0.29480	0.31443	0.34398
200	0.20016	0.20755	0.21468	0.22156	0.22822	0.23468	0.24706	0.26446	0.29092
300	0.14045	0.14594	0.15125	0.15640	0.16142	0.16630	0.17571	0.18908	0.20972
500	0.087921	0.091519	0.095016	0.098422	0.10175	0.10500	0.11130	0.12033	0.13446
1,000	0.045423	0.047349	0.049227	0.051061	0.052857	0.054618	0.058048	0.062993	0.070818

NAME INDEX

SUBJECT INDEX